CAMBRIDGE PHYSICAL SERIES

A TREATISE

ON

THE THEORY OF ALTERNATING CURRENTS

CAMBRIDGE UNIVERSITY PRESS

C. F. CLAY, MANAGER

London: FETTER LANE, E.C.

Edinburgh: 100 PRINCES STREET

London: H. K. LEWIS AND CO., LTD., 136 GOWER STREET, W.C.

London: WILLIAM WESLEY AND SON, 28 ESSEX STREET, STRAND

New York: G. P. PUTNAM'S SONS

Bombay, Calcutta and Madras: MACMILLAN AND CO., LTD.

Toronto: J. M. DENT AND SONS, LTD.

Tokyo: THE MARUZEN-KABUSHIKI-KAISHA

A TREATISE

ON

THE THEORY OF ALTERNATING CURRENTS

BY

ALEXANDER RUSSELL, M.A., D.Sc., M.I.E.E.

Principal of Faraday House, London

Vice-President of the Institution of Electrical Engineers

Volume II

SECOND EDITION

Cambridge:
at the University Press
1916

First Edition 1906
Second Edition 1916

PREFACE TO THE FIRST EDITION

IN this volume I have endeavoured to give a sketch of the theory of the working of alternating apparatus in the hope that it will prove helpful to engineers, teachers and advanced students. In addition to the more elementary parts of the theory, an introduction is given to several of the more difficult problems which arise in practical work.

The questions of armature reaction, of phase swinging and of free and forced oscillations, of the magnetic effects produced by various types of windings, etc., have often been discussed at the meetings of technical societies in this and other countries. In some of the papers which are published in the proceedings of these societies, theorems are quoted from books or journals which are not readily accessible, and in others an advanced theoretical knowledge is assumed. It was thought, therefore, that an introduction to the theory would prove useful to many.

Formulae obtained from admittedly imperfect theory are often used in the practical design of electrical machinery, and it is of great importance to know their limitations. The utility of many of the theorems given below has been amply proved by modifications of the design of several well-known types of apparatus. I have to thank many engineers for their kind permission to make use of their papers or for furnishing me with experimental data. In particular I wish to thank the Maschinenfabrik Oerlikon.

In the first two chapters the theory of single and polyphase alternators is set forth. Great credit is due to Mr J. Swinburne for his early recognition of the importance of armature reaction in the working of these machines. Many of the phenomena which puzzled the early electricians are easily explained when this is taken into account. For the proofs of the formulae for armature reaction given in Chapters I and XIII I am indebted to Professor C. F. Guilbert. I am also deeply indebted to Professor André Blondel for the instructive oscillograms illustrating the working of two and three phase machines given in Chapter II.

The experimental methods of analysing E.M.F. waves given in Chapter III, particularly that due to Mr H. Armagnat, are useful in practice. The theory of synchronous motors developed in Chapters IV and V is an easy application of the methods used by J. Hopkinson. It is shown how the V-curves, first described by Mr W. M. Mordey, could have been predicted easily by elementary theory. The development of his father's theory by Professor B. Hopkinson given in Chapter VI is particularly interesting, and the theoretical method used will be found helpful in many allied problems.

The question of the cause of the fracture of shafts, coupling engines and alternators has been briefly discussed and a simple explanation, due to Dr C. Chree, of the whirling of shafts is also given.

The theory of the alternating current transformer is set forth at length, as it is in excellent accord with experiment. In this connection I have to acknowledge my indebtedness to Professor J. A. Fleming. The theory of the induction motor is developed on the lines laid down by A. Potier. In writing Chapter XIV, describing the effects of harmonics in the E.M.F. and flux waves on the working of induction motors, I have received great help from papers by Mr E. Noaillon and Mr M. B. Field. The theory of the commutator motor, enunciated in Chapter XV, is practically that used by many French engineers.

To Mr de Marchena, the engineer to the *Campagnie française Thomson-Houston,* I am particularly indebted for some of the theorems and experimental data given on rotary converters. In the slight sketch of the theory of the electric transmission of power given in Chapter XVII I have elaborated a theorem due to Professor J. Perry, and I am also under obligations to Mr Oliver Heaviside.

Considerations of space have compelled me to omit many problems of theoretical interest and practical importance. The student, however, by studying the analogous problems set forth in this volume will find that it is not difficult to make a practical working theory for himself. For instance a practical solution of the problem of the stability of the motion of three alternators coupled in series—a method of getting three phase currents which has been proposed by Mr C. P. Steinmetz—can easily be found by a slight extension of J. Hopkinson's method.

In conclusion I have to thank several friends who have assisted me in revising the proofs or by making suggestions. My best thanks are due to Dr C. Chree, F.R.S., for discussing with me several of the problems contained in this work and for revising many of the proof sheets. I am also deeply indebted to Mr F. J. Dykes, Fellow of Trinity College, Cambridge, and lately Professor of Electro-technics at the Royal Naval Schools, Portsmouth, for reading all the slip proofs, and to Mr Clifford Paterson, A.M.I.C.E., late of the Oerlikon Works and now of the National Physical Laboratory, for reading several of the earlier chapters. I have again the pleasure of thanking Mr W. C. D. Whetham, F.R.S., for the care with which he has edited this work.

A. R.

2, Bellevue Place,
Richmond, Surrey.
October, 1906.

PREFACE TO THE SECOND EDITION

THE progress made during the last ten years in perfecting the theory of alternating current machinery has made several additions and alterations necessary. The theory of armature reaction has been rewritten, the method adopted being founded on that first suggested by Professor Lyle. Approximate solutions only are given, as until we can modify our mathematical equations so as to take hysteresis accurately into account, the additional labour involved in using more accurate solutions is not warranted.

In Chapter III the Author has developed a novel method of graphical harmonic analysis which is useful in practice and has the advantage of enabling the value of each harmonic to be computed independently of the others.

The theory of the induction motor and the theory of power transmission have been rewritten. In order to assist the reader to master the latter, Chapter XIX, on hyperbolic trigonometry, has been added. The Author desires to acknowledge his indebtedness to Professor A. E. Kennelly of Harvard University for helping him to appreciate the practical value of complex hyperbolic functions in this connection.

For the purpose of illustrating theory, brief descriptions of the La Cour motor converter, the synchronous booster, the split-pole converter, the frequency transformer and similar comparatively novel machines have been added, but a full discussion of their value would be out of place in this work.

In revising this edition the Author has had the help of his colleague Mr F. T. Chapman, B.Sc., Assoc.M.Inst.C.E., A.M.I.E.E., who has read the proofs and made many helpful suggestions. He has also to thank the officials of the University Press for their kindly and valuable assistance.

In conclusion he desires to place on record his gratitude to the late Colonel G. P. Seligmann-Lui, Inspector-General of Telegraphs in France, for the honour he did him by translating his work into French.

A. R.

FARADAY HOUSE,
LONDON, W.C.
April, 1916.

CONTENTS

SYMBOLS

A, effective value of an alternating current.
A', area of E.M.F. wave.
B, magnetic induction; a constant.
$B_{\text{max.}}$, maximum value of the magnetic induction.
C, direct exciting current; direct current; effective load current on the bus bars.
D, a constant.
E, maximum value of the alternating voltage.
F, force; symbol for 'function of.'
G, average torque.
H, magnetic force; heat in calories.
I, maximum value of an alternating current when it follows the harmonic law.
K, capacity between the mains.
L, self-inductance; leakage inductance of armature.
L, L', self coefficients of stator and rotor of induction motors.
$L_{p.q}$, coefficient of magnetic induction.
M, mutual inductance; mutual coefficient between stator and rotor; mass.
Mk^2, moment of inertia.
N, number of turns of wire in series on the whole armature.
N_1, number of armature turns per field magnet pole.
N', number of conductors joined in series on the armature.
P, power, $\pm(\alpha + \iota\beta)$.
Q, quantity of electricity.
R, resistance.
R_1, resistance of the primary coil of a transformer; resistance of a single main.
R_1, R_1', R_1'', resistances of the primary coils of a three phase transformer.
S, area of cross section.
T, periodic time.
$V, V_1, \ldots$, effective voltages.
$V_{1.2}, V_{2.3}, \ldots$, effective mesh voltages.
W, energy; number of pounds of steam.
X, excitation losses.
Z, impedance.

a, pitch of poles; length.
b, breadth of the polar flux entering the armature; breadth of a coil.
b', breadth of the armature coil.
c, breadth of side of coil.
e, instantaneous value of E.M.F.
e_1, e_1', e_1'', primary voltages.
e_2, e_2', e_2'', secondary voltages.
f, frequency; symbol for 'function of.'
g, instantaneous torque.
i, instantaneous current.
i_1, i_1', i_1'', primary star currents.
i_2, i_2', i_2'', secondary star currents.
k, form factor; capacity between the mains per unit length.
k_m, form factor for mesh voltage.
k_s, form factor for star voltage.
l, self-inductance per unit length.
m, mass; a constant.
n, number of turns; a constant.
p, half the number of poles.
q, number of phases.
r, resistance; resistance per unit length.
r_2, resistance of the secondary coil of a transformer.
r_2, r_2', r_2'', resistances of the secondary coils of a three phase transformer.
s, leakage conductivity per unit length; slip.
t, time in seconds.
v, potential difference; velocity; $1/\sqrt{lk}$; 3×10^{10} cms. per sec.
v_1, v_1', v_1'', primary mesh voltages.
v_2, v_2', v_2'', secondary mesh voltages.
y, $(r^2 + l^2\omega^2)^{\frac{1}{2}}$.
z, $(s^2 + k^2\omega^2)^{\frac{1}{2}}$.

α, attenuation factor.
β, wave length factor.
α, β, numbers.
$\alpha, \beta, \gamma, \delta$, angles.
$\gamma, \theta, \xi, \psi$, phase differences.
ϵ, base of Neperian logarithms.
η, Steinmetz's coefficient; efficiency.
ι, $\sqrt{-1}$.
λ, dielectric coefficient.
μ, magnetic permeability; rigidity.
π, 3·14159....
ρ, resistivity; density.
σ, resistivity of insulating material; leakage factor $= 1 - M^2/L_1L_2$.
τ, time constant.
ϕ, instantaneous value of flux.
ω, angular velocity; $2\pi \times$ frequency of supply.

$\Gamma(n)$, the gamma function of n.
Σ, the symbol for summation.
Φ, maximum value of the flux when it follows the sine law.
$\Phi_{\text{max.}}$, maximum value of the flux.
Φ_A, flux of induction from a pole entering the armature.
Φ_a, leakage flux.
Ω, $2\pi \times$ frequency of supply.

$\mathscr{I}$, mean magnetising force in ampere-turns.
$\mathscr{I}_t$, mean transverse magnetising force.
$4\pi\mathfrak{R}/10$, reluctance.
$4\pi\mathfrak{R}_a/10$, leakage reluctance.
$4\pi\mathfrak{R}_f/10$, reluctance of field magnets.
$4\pi\mathfrak{R}_g/10$, air-gap reluctance.

$aN_1A \sin\psi$, the demagnetising turns per pole due to the armature current.
$\beta N_1 A \cos\psi$, the transverse magnetising turns per pole due to the armature current.

CHAPTER I

Dynamo electric machines. Stator and rotor. Various types of single phase alternators. Frequency. Armature with bar winding. Single coil winding. Disk armatures. Inductor machines. Distribution of magnetic flux. Effect of the armature currents on the field. Open circuit electromotive force formulae. Effect of the breadth of the armature coils. Equivalent windings. Irregular magnetic field. Open circuit characteristic. Flux curves. Short circuit characteristic. Wave windings. Lap windings. The armature reactions. Formula for the demagnetising effect of the lagging component of the current. Formula for the compensating ampere-turns required for the field magnets. The compensating ampere-turns required to keep the flux in the field magnets constant. Transverse magnetisation of the field. Numerical example. Load characteristics. Regulation. Methods of measuring the regulation. Theory of the alternator. The potential difference at the terminals. Theoretical characteristics. Temperature curves of alternators. References.

Dynamo electric machines.

WHEN a moving wire cuts lines of magnetic induction, an electromotive force is generated in it. If the wire form part of a closed circuit, a current will flow in the circuit, and, as Lenz pointed out, the current will produce electromagnetic forces tending to stop the motion. Hence, to overcome this resistance to the motion, mechanical work must be expended on the wire, and this work, by the Conservation of Energy, will be the equivalent of the electrical work generated. This method of converting mechanical energy into electrical energy is the method utilised in dynamo electric machines. In a direct current dynamo, the current always flows in the same direction round the external circuit, but, in an alternating current dynamo, the direction of the flow of the current in the external circuit is continually reversed. In a direct current machine, however, the current induced in an armature coil is flowing in one direction when it is moving past a north pole and in the other direction when it is moving past a south pole. Hence the current in the coil must be reversed in some intermediate position. In the process of reversal the coil is first short circuited by one of the brushes which press on the commutator. The currents flowing in the armature coils of a direct current machine are thus really alternating currents, the frequency of which equals the product of half the number of poles multiplied by the number of revolutions of the armature per second.

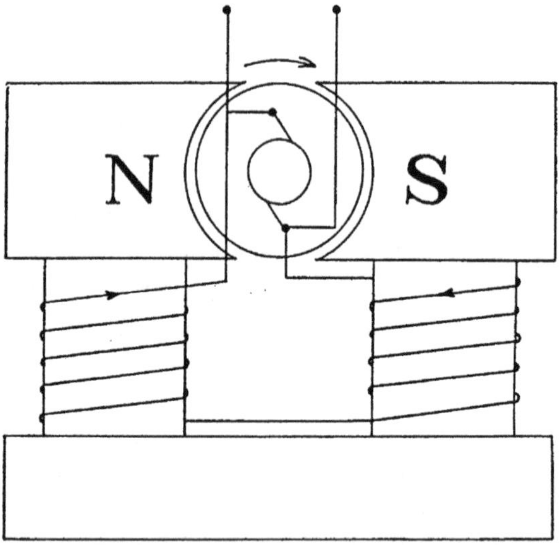

Fig. 1 (*a*). Shunt Wound Exciter.

In an alternating current dynamo, or as it is generally called, an alternator, the coils of the armature are connected in such a way that the electromotive forces generated in them are all acting in the same direction at any instant, the direction of the resultant electromotive force altering every time a coil passes a pole. If the electromagnets which produce the field rotate, the ends of the armature windings are connected directly with the terminals of the machine, the rotation of the exciting magnetic field maintaining an alternating potential difference between these terminals. If the armature rotates and the field magnets are stationary, then the ends of the armature windings are connected with metal rings fixed on the shaft, but insulated from it, on which press copper or carbon brushes connected with the terminals of the machine. These rings are called slip rings or collector rings.

One advantage that direct current machines have over alternators is that they are self-exciting. After the magnets have once been excited, their residual magnetism is sufficient to produce a weak field in the air-gap. If the dynamo is shunt wound, the field magnet windings are in parallel with the external load but in series with the armature winding. When the armature rotates, either on open or closed circuit, the low E.M.F. generated in it by the residual field will send a small current round the field magnet windings. This current excites the field magnets and increases the induction in the air-gap. Both the E.M.F. and the current, therefore, will go on increasing until the E.M.F. generated in the armature conductors only suffices to produce the magnetising current required to maintain the magnetic field giving that E.M.F. In a series dynamo, the field magnet windings and the armature windings are connected in series between the terminals of the machine, and thus, on open circuit, no current will flow in the field magnet windings and the potential difference between the terminals will be due merely to the residual field. When, however, the terminals are connected through an outside load, a current will flow, and the magnetic field and the electromotive force generated will both increase until equilibrium is attained in the same way as in a shunt machine. In a compound wound dynamo we have both shunt and series windings and by designing them so that the ampere-turns of each have the right value we can ensure either a constant voltage at the terminals or a terminal voltage that increases as the external current increases.

In almost every type of alternator, on the other hand, we require a small direct current dynamo to provide the current required to excite the field magnets. This dynamo, which is called the exciter, is sometimes mounted on the shaft of the alternator. Its capacity is usually about two per cent. of that of the alternator. The exciters of modern alternators are shunt or compound wound. The voltage of the exciter, and therefore the strength of the alternator's field, can be regulated by varying the resistance of a rheostat in the shunt circuit of the exciter.

It has to be remembered however that the voltage does not

respond instantaneously to an alteration in the value of the resistance of the rheostat. For instance, when a resistance equal to three times that of the external circuit was inserted in the shunt circuit of a certain exciter the terminal voltage took about five seconds to fall from 125 to 25. The sluggishness of this action is of importance in the theory of automatic regulators.

Except in the case when the speed of the prime mover keeps almost perfectly constant it is inadvisable to mount the exciter

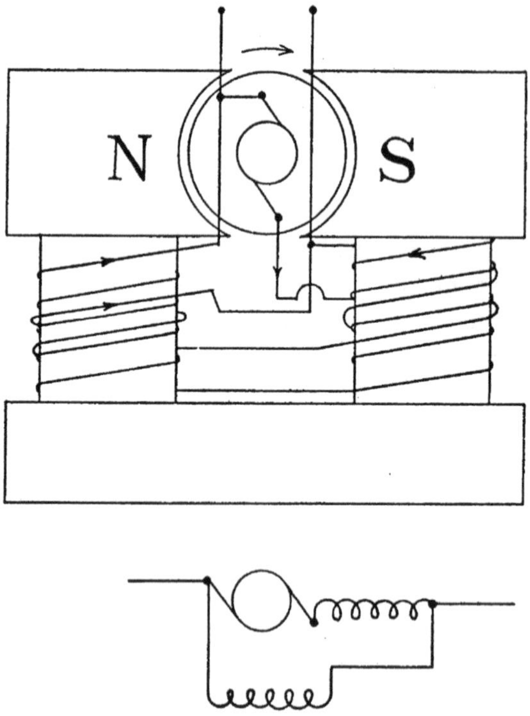

Fig. 1 (*b*). Compound Wound Exciter.

on the shaft of the alternator, as any alteration in the speed alters the exciting current, and this, owing to a variation in the permeability of the iron of the field magnets, may cause a much larger percentage variation in the electromotive force generated.

In large central stations the exciters are sometimes wound for 500 volts as the machines and switches required are smaller than those for 100 volts. For example the commutators of the 500 volt machines would only be about half as large as those of the 100 volt machines. Central Stations usually have three exciters, two of them being kept running in parallel and one being kept in reserve. Recent practice seems to favour compound wound exciters, as they work better in parallel and divide the load between them more evenly. A battery of storage cells is often used in addition, so as to reduce the risk of a break-down in the exciting circuit to a minimum.

Stator and rotor.

In an alternator either the field magnets or the armature may rotate. It is convenient therefore to refer to the rotating part of a machine as the rotor, and to the stationary part as the stator.

If the armature coils are connected in series, and if $\phi_1, \phi_2, \ldots$ be the instantaneous values of the fluxes linked with them and the coils have $N_1, N_2, \ldots$ turns of wire respectively, the electromotive force e generated at any instant is given by

$$e = N_1 \frac{\partial \phi_1}{\partial t} + N_2 \frac{\partial \phi_2}{\partial t} + \ldots.$$

Various types of single phase alternators.

The magnetic flux through a coil can be altered mechanically in several ways, and we can classify alternating current generators according to the method utilised for varying the flux. The first class of alternator comprises those which have rotating armatures and fixed field magnets. In the second class, the armatures are fixed and the field magnets rotate; and in the third, both the field magnets and the armature are fixed. The types of alternator belonging to the second class are those most commonly employed in practice. Since the armatures are stationary, they can easily be wound for high pressures. The large moment of inertia of the revolving field magnets promotes steady running by diminishing the effect on the speed of any irregularities in the driving torque. In this respect its action is similar to that of a flywheel. In the first two types of alternator the poles of the field magnets are evenly distributed round the circumference of the stator or rotor, and adjacent poles are of opposite polarity. The field magnet coils are often formed by a single layer of copper strip wound edgewise round the field magnet, and insulated by a fibrous material between the turns. The exterior surface of the windings is merely protected by an insulating varnish which allows the heat generated in the field coils to be radiated away rapidly. In order to avoid appreciable losses due to eddy currents, the armature is built up of thoroughly annealed soft iron or steel plates, which are generally insulated from each other either by means of thin paper pasted on one side of each plate or by a suitable varnish. The polar 'pieces' or 'shoes' which form the poles of the field magnets are also built up of thin plates of iron or steel.

Frequency.

If the rotor of an alternator be made to revolve, the value of the magnetic flux embraced by an armature coil continually alters. When the armature rotates, the magnetic flux embraced by a coil on it goes through all its cyclical values in the time the coil takes to pass two adjacent poles, and when the poles rotate, the period of the varying flux is the time taken by two adjacent poles to pass the coil. Hence, the frequency is independent of the armature windings and depends only on the number of field poles and the number of revolutions per minute of the rotor. If $2p$ be the number of poles, so that p is the number of pairs of poles, and if N be the number of revolutions of the rotor per minute, then the frequency f is given by

$$f = \frac{pN}{60}.$$

In both the first and the second type of alternator, the magnetic flux embraced by an armature coil alternates between equal positive and negative values. The field magnet poles are also invariably similar and the armature is situated symmetrically with regard to them. Hence the positive and negative halves of the E.M.F. waves produced are of exactly the same shape.

Armature with bar winding.

A simple form of armature winding for a twenty pole alternator is shown diagrammatically in Fig. 2. The field magnets point inwards and neighbouring poles are of opposite polarity. The thick lines represent copper bars placed in slots on a cylinder built up of iron stampings. This forms the armature. For clearness of illustration the bars are drawn radially, but in reality they are perpendicular to the plane of the paper, that is, parallel to the shaft of the machine. Since the electromotive forces generated in neighbouring bars, as the armature rotates, are of opposite sign, we shall have all the E.M.F.s generated acting in the same direction if we connect the ends of the bars alternately as in the figure. The ends of the circuit are connected to two slip rings S_1 and S_2, and so an alternating potential difference is maintained at the terminals of

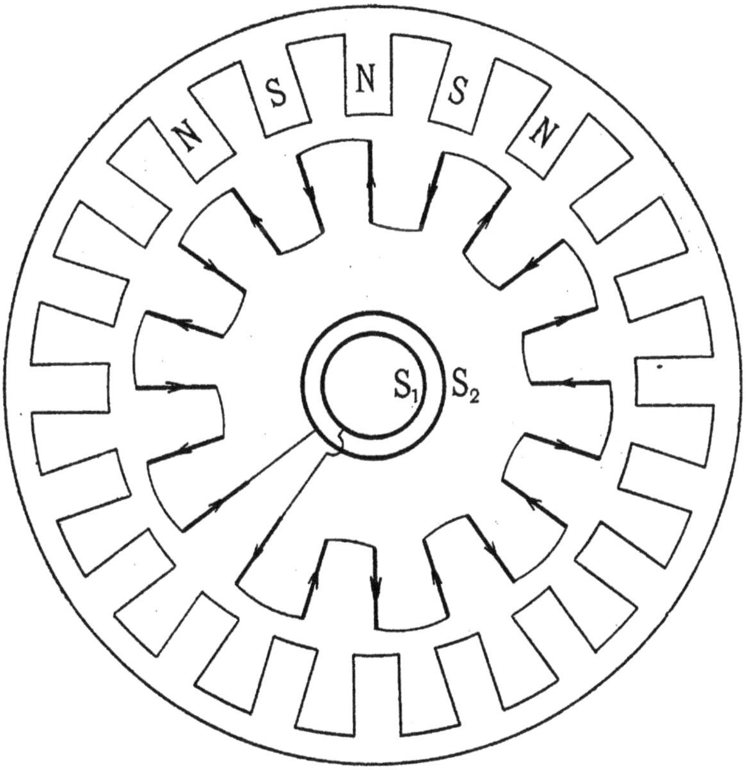

Fig. 2. Armature winding of twenty pole single phase alternator. The current is collected from the slip rings S_1 and S_2 by contact brushes.

the machine, which are in electrical connection with brushes pressing on the rings. It is to be noticed that the two coils connected with the slip rings are adjacent and the voltage difference between them is practically that of the machine. Special care, therefore, has to be taken in insulating them so as to prevent a disruptive discharge taking place between them.

In Fig. 2 we may suppose that the field magnets revolve. In this case the direct current required for the excitation of the field magnets would be collected by slip rings, and the windings of the armature would be connected to fixed terminals.

Single coil winding.

In the single coil winding shown in Fig. 3, the coils are connected in series, and as their E.M.F.s are all in phase with one another, the terminal voltage of the machine is the sum of all the E.M.F.s generated in the coils. Since

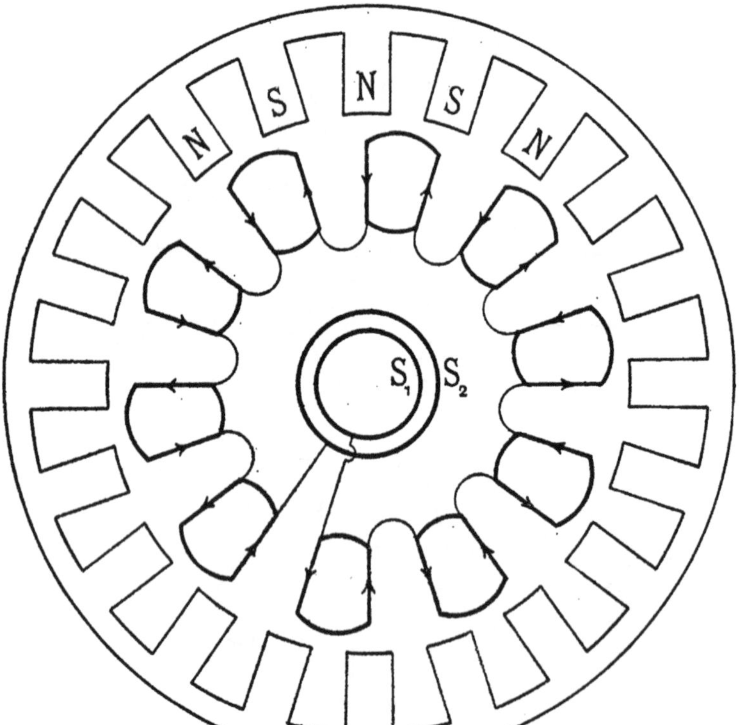

Fig. 3. Twenty pole alternator with single coil winding.

the voltage is proportional to the number of turns in each coil, the machine illustrated in Fig. 3 can easily be constructed to give a much higher voltage than that shown in Fig. 2.

In both the above machines we have iron in the armature. Most of the magnetic lines common to two adjacent poles complete their paths through the iron cylinder on which the copper conductors are placed. There will be two air-gaps in the path of the magnetic flux, one immediately under each pole. The magnetic lines where they leave the polar faces are pointing approximately in the radial direction, that is, towards the axis of the shaft, and the conductors cutting them are parallel to this axis.

Disk armatures.

An alternative form of construction is to have the armature conductors pointing radially and the lines of force parallel to the axis of the shaft. This kind of armature is called a disk armature, and iron need not be used in its construction. The armature windings, shown in Figs. 2 and 3, illustrate also the 'wave' and 'coil' windings respectively for disk armatures. The conductors are of copper strip and are wound on non-magnetic frames, generally of laminated brass. Consecutive turns of the copper strip in a coil winding are insulated from one another by some suitable material. All the coils are bolted together and are mounted on the circumference of the armature wheel so that the axes of the coils are at right angles to the plane of the wheel. The field frame supports two rings of magnetic poles facing one another. The axis of each pole is parallel to the shaft, and the poles facing one another are of opposite polarities, and so also are the poles adjacent to one another on the same ring. If N_1 and S_1 be two adjacent poles on the first ring, and S_1' and N_1' be the two opposite poles on the second ring, then, neglecting leakage, half the flux leaving N_1 crosses the air-gap and goes through S_1'. From S_1' it goes through part of the second ring to N_1' and crosses the air-gap to S_1, and finally it returns, through part of the first ring, to N_1. There will thus be two air-gaps in the path of the flux. The other half of the flux leaving N_1 has a similar path on the side of N_1 remote from S_1. The size of the air-gaps is only sufficient to allow the armature coils to rotate safely. As adjacent fields are of opposite polarities we get alternating electromotive forces set up in the armature coils, which may be connected with one another as in Fig. 3. It is customary, in practice, to place another set of coils between those indicated in Fig. 3 and exactly similar to them. The two sets of coils are generally connected in parallel.

In constructing this type of alternator it is difficult to make the armature of sufficient mechanical strength to withstand the appreciable mechanical stresses to which it is subjected when running. None of the insulating materials employed in practice, such as micanite, fibre, slate, ebonite, stabilit, presspahn, etc., have any great mechanical strength. This type is now rarely used.

Inductor machines.

Machines belonging to the third class are called inductor machines. In the commonest type of this class the rotor consists of a wheel carrying on its rim blocks of laminated iron which, in certain positions, make the reluctance of the magnetic circuit, common to the field and the armature, exceedingly small. If Φ be the total flux, and nC the exciting ampere-turns round a magnetic circuit, then (Vol. I, p. 72) we have

$$\Phi = \frac{4\pi nC/10}{\text{Reluctance}}.$$

Hence, if we vary the reluctance, C remaining constant, Φ will vary, and therefore an E.M.F. will be set up in any coil embracing this magnetic circuit. In some inductor machines, the armature coils and the exciting coils are wound on the same polar projections. In this case the flux merely undulates between a maximum and a minimum value. In actual machines of the undulating type the ratio of the maximum to the minimum flux varies between three and ten.

In other inductor machines the flux periodically reverses in direction. To see how this is done consider the diagrams 4, 5, and 6.

The polar projections N and S (Fig. 4) are excited by direct currents flowing in coils wound round them. A represents a polar projection which is generally made of laminated steel with an armature winding coiled round it. The blocks of laminated iron on the circumference of the rotor are marked M. In Fig. 4, the flux is leaving M and entering A, whilst in Fig. 6 the flux has been completely reversed. In some intermediate position (Fig. 5) the algebraical sum of the fluxes entering A must be zero. We see that, when M advances over the step between the centres of two polar projections, the alternating current has gone through half of its values. Hence, the frequency of the alternating current is

$pN/60$ where $2p$ is the number of polar projections on the circumference of the stator, and N is the number of revolutions of the

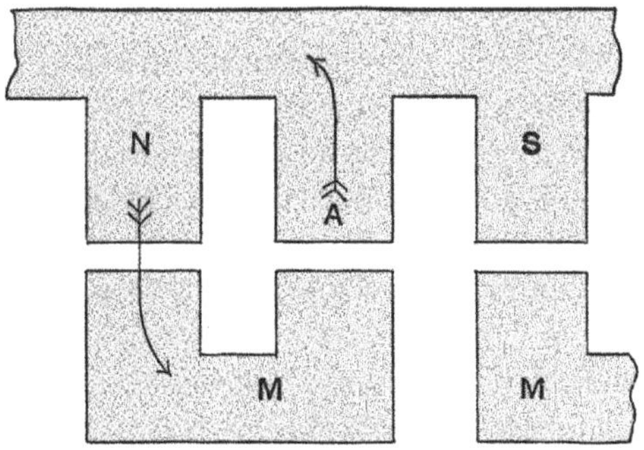

Fig. 4. Principle of inductor machine. Initial position; flux entering *A*.

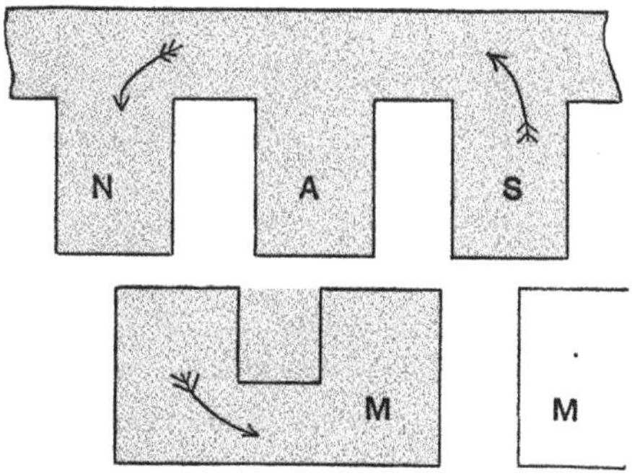

Fig. 5. Principle of inductor machine. Intermediate position.

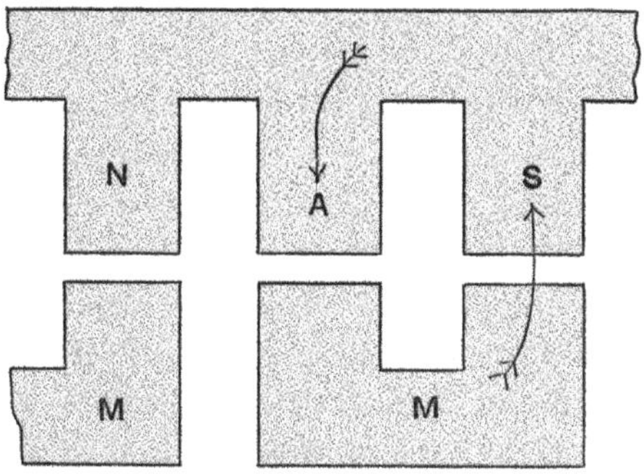

Fig. 6. Principle of inductor machine. Half a period later than in Fig. 4; flux leaving *A*.

rotor per minute. As the flux in the field magnets of inductor machines is continually varying, an alternating current will be superposed on the direct current exciting the magnets.

Distribution of magnetic flux.

Before we can find a formula for the electromotive force generated by an alternator we must make some supposition as to the distribution of the magnetic flux in the air-gap. Unfortunately, this distribution varies in a complicated manner in practice owing to the slots in the armature in which the conductors are placed, the different

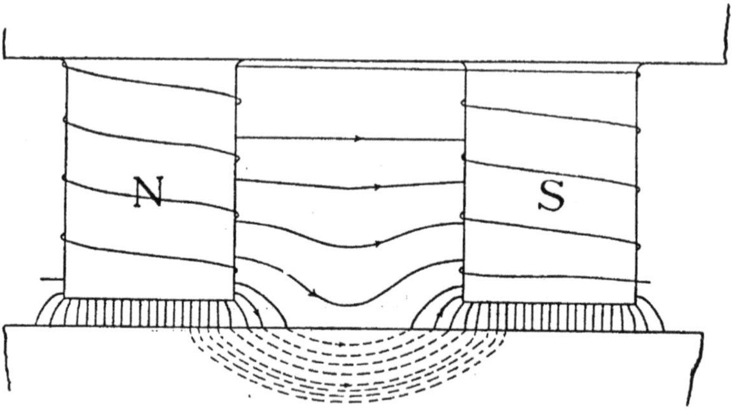

Fig. 7. Lines of force in the air-gap of an alternator.

ratios of the distance between the poles to the polar breadth, etc. If we suppose that the poles are rectangular and that the distance between them is approximately ten times the air-gap, then the distribution of the magnetic flux would be approximately as shown in Fig. 7. The lines of magnetic induction in more complicated cases can be found by drawing the lines of flow between copper electrodes of suitable shape placed on a sheet of tinfoil and maintained at a constant potential. We could also find in this manner the lines of magnetic induction in the neighbourhood of a slot (Fig. 8). It is important to notice that very few lines penetrate far into the slot, hence, unless it be very shallow, its depth has very little effect on the distribution of the magnetic lines in the air-gap.

When currents flow in the armature of an alternator, they may distort the magnetic field very considerably. Later on in this chapter we shall find formulae for the demagnetising and cross magnetising forces produced by these currents. At present we shall consider the problem from an elementary point of view. Suppose that a wire carrying a current is placed parallel to an infinite plate of iron (Fig. 9). The magnetic field produced is similar to that shown in

Effect of the armature currents on the field.

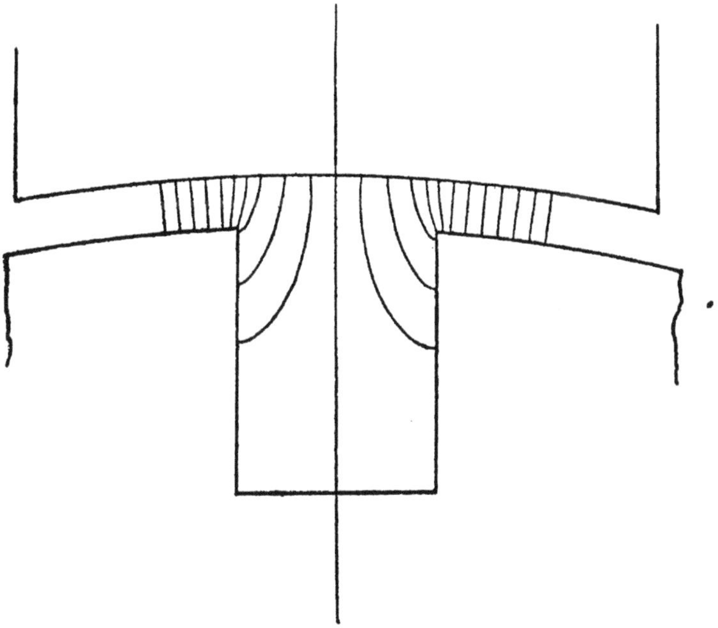

Fig. 8. Lines of force in the neighbourhood of a slot.

the figure. It will be seen that many of the lines of force in the air meet the iron and complete the rest of their circuit round the wire as lines of induction in the iron. The shape of the lines of force shows that the wire will be attracted towards the iron.

In Fig. 10 are shown the lines of induction round a wire embedded in an iron plate and parallel to its surface. The plate is supposed to be very thick compared with the depth of the embedded wire. It is to be noticed that the lines of force are nearly perpendicular to the surface of the iron. Both the Figures 9 and 10, which are due to G. F. C. Searle, illustrate what is called the refraction of lines of force on entering iron. Searle has pointed out one most important advantage gained by leading the wire through a tunnel in the armature instead of placing it on the surface, namely, that the mechanical force experienced by the wire in this case is much less than it would be if it were on the surface of the armature, although the torque required to drive the armature and the electromotive force developed in its windings

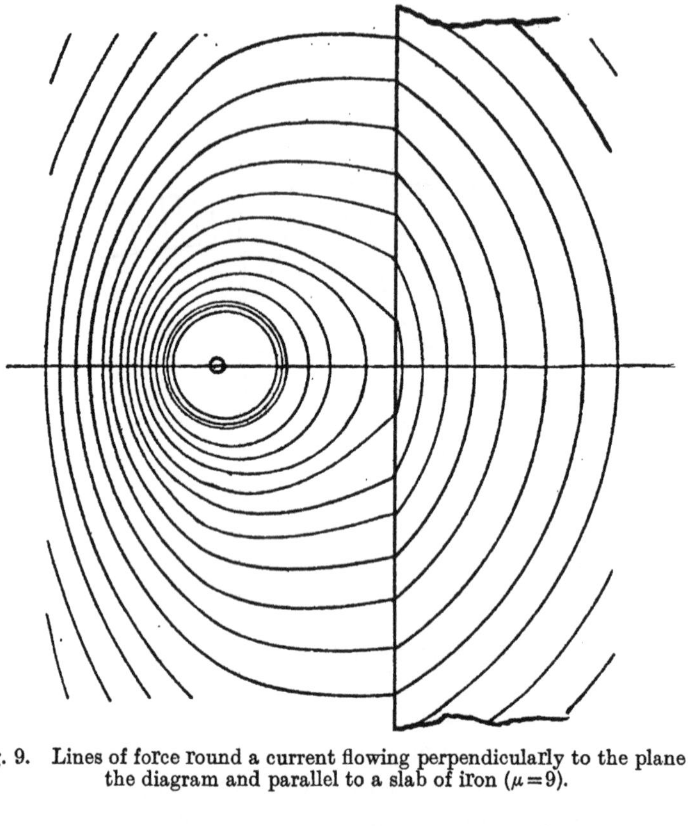

Fig. 9. Lines of force round a current flowing perpendicularly to the plane of the diagram and parallel to a slab of iron ($\mu = 9$).

are practically the same in the two cases. The iron experiences all the force that would otherwise come on the conductors. The same advantage applies in a slightly diminished degree when we have slots on the surface of the armature instead of tunnels through its substance. In this case, any tendency to slip is at once checked mechanically by the sides of the slots.

Open circuit electromotive force formulae.

Suppose that the breadth of the polar pitch, that is, the distance from the centre of one pole to the centre of the next measured along a circle which has its centre in the axis of rotation, is a, and suppose that b is the breadth of the pole. If we suppose, in addition, that the

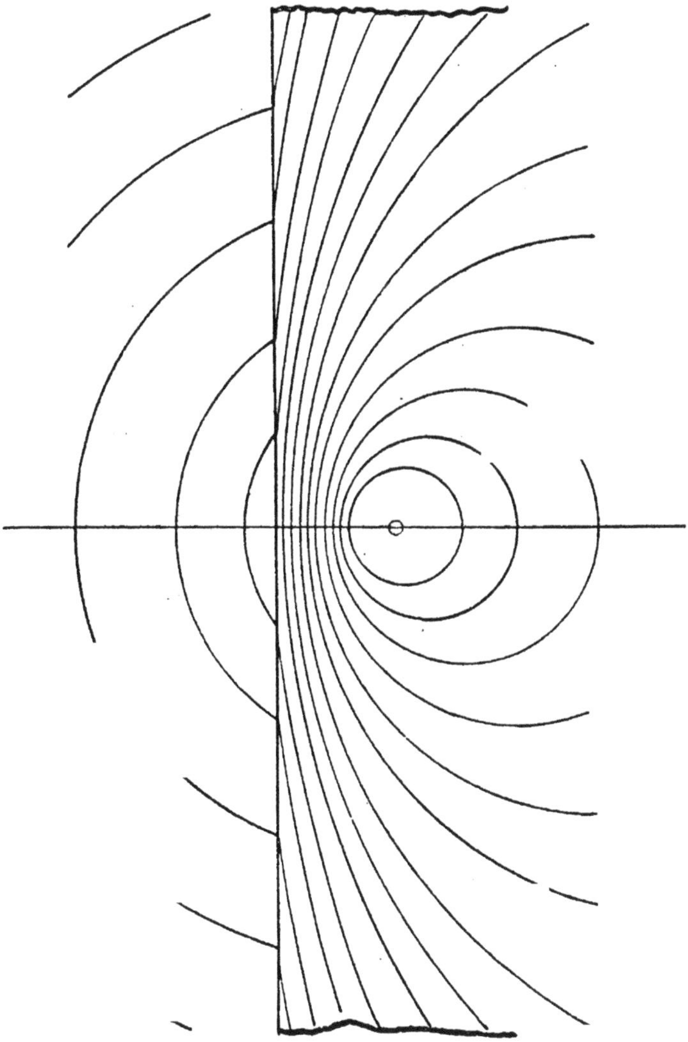

Fig. 10. Lines of magnetic induction round a wire carrying a current and embedded in an iron plate. The wire is supposed to be perpendicular to the plane of the paper and parallel to the surface of the iron ($\mu = 9$).

induction density is constant over the polar face, the density of the magnetic flux in the air-gap will be given approximately by the ordinates of the curve shown in Fig. 11.

The equations to give the magnetic flux at any point in the air-gap are

$$\left.\begin{aligned} y &= \left(\frac{2x}{a-b}\right)^n h \quad \text{from } x = 0 \text{ to } x = \tfrac{1}{2}(a-b) \\ y &= h \quad \text{from } x = \tfrac{1}{2}(a-b) \text{ to } x = \tfrac{1}{2}(a+b) \\ y &= \left\{\frac{2(a-x)}{a-b}\right\}^n h \quad \text{from } x = \tfrac{1}{2}(a+b) \text{ to } x = a \end{aligned}\right\} \ldots(a).$$

If n be unity, the curves in Fig. 11 become straight lines.

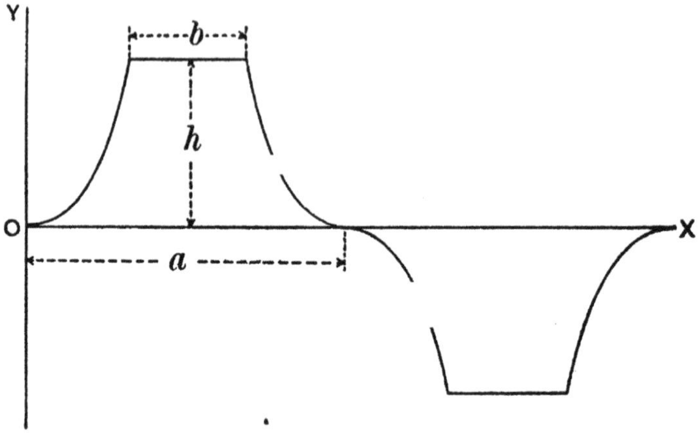

Fig. 11. Induction wave in the air-gap.

If n be less than unity, the curves are concave to the axis of x, and if n be greater than unity, they are similar to the curves shown in the figure. In the particular case, when n is zero, we get a rectangle for the curve of flux. When n is infinite, the density is constant directly under the poles and zero elsewhere. We can thus get important practical cases by giving various values to n.

Let x be the distance of a conductor, measured along the circumference of the rotating armature, from a fixed point O in the air-gap. We suppose that O is infinitely near to the surface of the armature, and that at all points on a line through O parallel to the shaft and parallel, therefore, to the conductors on the armature surface, the induction density is zero.

Let the density y of the field at x be as shown in Fig. 11, then, the electromotive force e generated in the conductor is given by (see Vol. I, p. 41)

$$e = ly\frac{\partial x}{\partial t} \times 10^{-8} \text{ volts} \quad \ldots\ldots\ldots\ldots\ldots\ldots (\beta),$$

where l is the active length of the conductor in centimetres and $\frac{\partial x}{\partial t}$ is its velocity in centimetres per second. If the armature is rotating with constant angular velocity, $\partial x/\partial t$ is constant, and hence the shape of the E.M.F. wave in a simple bar winding will be the same as the shape of the wave of flux in the air-gap.

If T be the period of the electromotive force generated, we get, from (β), by integration

$$\int_0^{\frac{T}{2}} e\partial t = l\int_0^a y\partial x \times 10^{-8}$$
$$= \Phi_A \times 10^{-8},$$

where Φ_A is the flux of induction which enters the armature from one pole.

Now
$$\int_0^{\frac{T}{2}} e\partial t = \frac{T}{2} e_m$$
$$= \frac{V}{2f} \cdot \frac{e_m}{V},$$

where V is the effective voltage, e_m the mean value of e, and f the frequency of the alternating E.M.F. Hence, if there are N' bars joined in series on the armature, as in a simple bar winding (Fig. 2), we have

$$V = 2fN'\frac{V}{e_m}\Phi_A \times 10^{-8}.$$

It is convenient to call the ratio of the effective value V to the mean value e_m the form factor of the wave. We shall denote this ratio V/e_m by k, so that

$$V = 2fN'k\Phi_A \times 10^{-8}.$$

If we have N turns of wire in series, as in Fig. 3, then

$$V = 4fNk\Phi_A \times 10^{-8},$$

since each turn of wire has two active bars in series.

The above formulae show that it is not sufficient to know the total flux entering the armature from a pole and the number of windings on the armature in order to determine the effective electromotive force. We must know, in addition, how the flux is distributed.

Let us suppose that the distribution of the flux is represented by the curve shown in Fig. 11; then by the equation (β) given above, we have

$$\frac{V}{e_m} = \frac{Y}{y_m} = k,$$

where Y is the effective value of y.

Now, from the equations (α)

$$a y_m = \int_0^{\frac{a-b}{2}} \left(\frac{2x}{a-b}\right)^n h\,\partial x + bh + \int_{\frac{a+b}{2}}^{a} \left\{\frac{2(a-x)}{a-b}\right\}^n h\,\partial x,$$

and therefore $$y_m = \frac{a+nb}{a+na} h.$$

Again $$aY^2 = 2\left(\frac{a-b}{2}\right)\frac{h^2}{2n+1} + bh^2.$$

Therefore $$Y^2 = \frac{a+2nb}{a(2n+1)} h^2,$$

and $$Y = \sqrt{\frac{a+2nb}{a(2n+1)}}\, h.$$

Hence $$k = \frac{Y}{y_m} = \frac{a(n+1)}{a+nb}\sqrt{\frac{a+2nb}{a(2n+1)}}.$$

The values of k in a few special cases are given in the following table.

n	k	k when a is $2b$	k when a is $1{\cdot}5\,b$
0	1	1	1
$\frac{1}{2}$	$\frac{3a}{2a+b}\sqrt{\frac{a+b}{2a}}$	1·04	1·03
1	$\frac{2a}{a+b}\sqrt{\frac{a+2b}{3a}}$	1·09	1·06
2	$\frac{3a}{a+2b}\sqrt{\frac{a+4b}{5a}}$	1·16	1·10
∞	$\sqrt{\frac{a}{b}}$	1·41	1·22

If we assume that the shape of the curve giving the distribution of the flux in the air-gap is rounded, a useful equation to take for it is

$$y = B\{x(a-x)\}^n.$$

When n is zero we get a rectangle, when n is $\frac{1}{2}$ we get an ellipse and when n is 1 we get a parabola. When n is greater than 1 we get curves similar to the tallest curve in Fig. 12.

Now
$$ay_m = B\int_0^a x^n (a-x)^n \partial x$$
$$= Ba^{2n+1}\frac{\{\Gamma(n+1)\}^2}{\Gamma(2n+2)},$$

where $\Gamma(n)$ is the Gamma function (see Vol. I, p. 131).

We have also
$$aY^2 = B^2\int_0^a x^{2n}(a-x)^{2n}\partial x$$
$$= B^2 a^{4n+1}\frac{\{\Gamma(2n+1)\}^2}{\Gamma(4n+2)}.$$

Therefore
$$\frac{Y}{y_m} = \frac{\Gamma(2n+1)\,\Gamma(2n+2)}{\{\Gamma(n+1)\}^2\sqrt{\Gamma(4n+2)}}.$$

Notice that $\Gamma(n+1) = n\Gamma(n)$, $\Gamma(1) = 1$ and $\Gamma(\frac{1}{2}) = \sqrt{\pi}$. Other values of $\Gamma(n)$ can be found by means of the following table.

n	1·1	1·2	1·3	1·4	1·5	1·6	1·7	1·8	1·9
$\Gamma(n)$	0·951	0·918	0·897	0·887	0·886	0·894	0·909	0·931	0·962

In Fig. 12 the areas of the curves are all the same, so that Φ_A is constant. The value of k for the ellipse (a) is 1·04, for the parabola (b) k is 1·10, and for the biquadratic curve (c) k is 1·20.

For a sine shaped distribution of the flux, k would be $\pi/(2\sqrt{2})$, that is, 1·111.

In this case, for a bar winding, the formula for the effective voltage would be

$$V = 2{\cdot}222 f N' \Phi_A \times 10^{-8},$$

and so also

$$V = 4{\cdot}443 f N \bar{\Phi}_A \times 10^{-8}.$$

For example, if V were 1000 volts, the frequency 50 and the number of bars joined in series round the armature 200, then Φ_A would be $4 \cdot 5 \times 10^6$ C.G.S. units nearly.

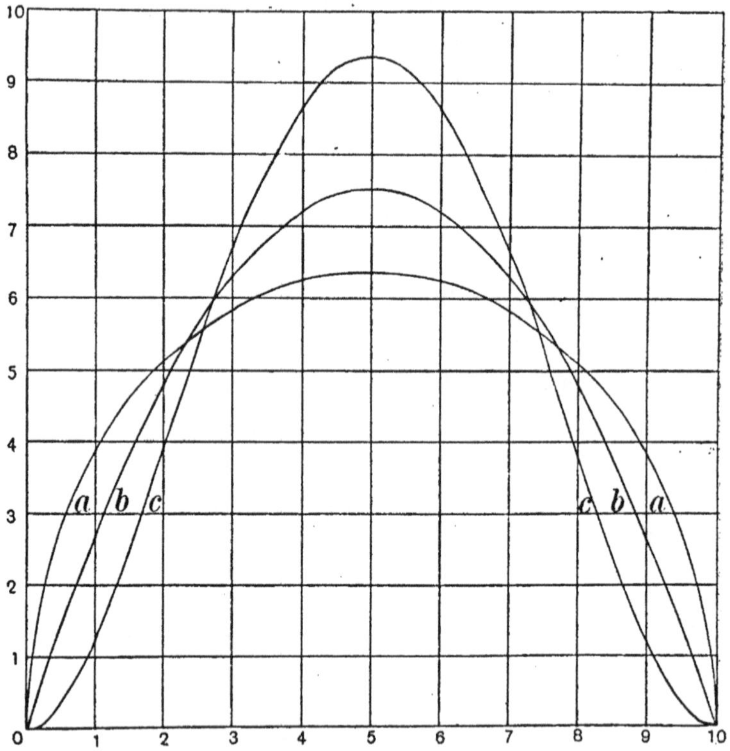

Fig. 12. Curves giving the flux density in the air-gap for a given total flux. The areas of the curves are equal. (*a*) is an ellipse, (*b*) a parabola, and (*c*) a biquadratic.

Effect of the breadth of the armature coils.

When an armature coil consists of many turns of wire, it is obvious that some of the windings will have a greater breadth than others, and hence, all the electromotive forces generated in the various turns of the coil windings will not be in the same phase.

If $e_1, e_2, \ldots e_n$ be the electromotive forces generated in each turn of the coil, and e be the resultant electromotive force at its terminals, we have

$$e = e_1 + e_2 + e_3 + \ldots + e_n.$$

By squaring and taking the mean of the values for a whole

period, we get

$$V^2 = \Sigma V_1^2 + 2\Sigma V_1 V_2 \cos \alpha_{1.2},$$

where $\alpha_{1.2}$ is the phase difference between e_1 and e_2. Since, by hypothesis, all these phase differences are not zero,

$$\Sigma V_1^2 + 2\Sigma V_1 V_2 \cos \alpha_{1.2} \text{ is less than } (V_1 + V_2 + \ldots + V_n)^2.$$

We see, therefore, that V is less than $V_1 + V_2 + \ldots + V_n$, and hence, the effect of the electromotive forces in the various turns not being in phase with one another is to diminish the effective value of the resultant electromotive force generated.

It has to be remembered that the quantities V_1, V_2, ... V_n only compound together according to the polygon law in certain very special cases (see Vol. I, Chap. XII), and hence it is not correct to say that the above theorem follows geometrically from the polygon construction.

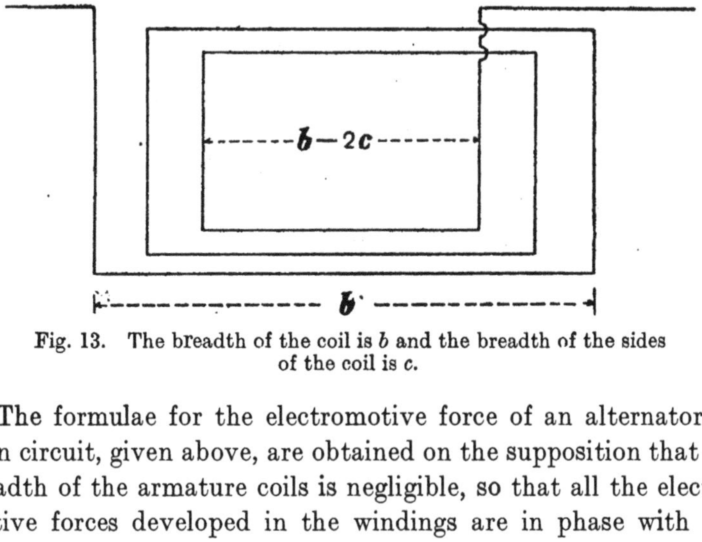

Fig. 13. The breadth of the coil is b and the breadth of the sides of the coil is c.

The formulae for the electromotive force of an alternator on open circuit, given above, are obtained on the supposition that the breadth of the armature coils is negligible, so that all the electromotive forces developed in the windings are in phase with one another. These formulae, therefore, fix the maximum possible values of the open circuit electromotive force. In order to find a formula which will take into account the breadth of the coil, we must, as formerly, make some assumption as to the shape of the flux. If we assume that the distribution of the flux is represented by the curve shown in Fig. 11, then, it is easy to show that the shape of the resultant electromotive force wave would be different from this curve. This makes the calculation of the formula for the electromotive force more laborious. We shall assume, therefore, for the present, a sine distribution of the flux, since, in this case, the resultant electromotive force wave is of the same shape as its components.

Let us suppose that the armature is cylindrical in shape and that it is the rotor. We shall suppose that the flux density at right angles to the surface of the rotor on a line, parallel to the shaft, at a distance x, measured along the circumference of the rotor, from a parallel fixed line, tangential to the surface of the rotor, is given by $B \sin \pi x/a$. The fixed line, therefore, is midway between two consecutive poles as the radial magnetic force is zero at all points along it. Let us now suppose that the armature coils are similar to the coil represented in Fig. 13. The breadth of this coil is b, and the breadth of the sides of the coil is c, so that $b - 2c$ is the breadth of the narrowest winding of the coil. We suppose that these breadths are all measured along the circumference of the rotor. If there are n layers in the side of a coil (there are 3 in Fig. 13), and if k be the distance between consecutive layers we shall have $(n-1)\,h$ equal to c. If m be the number of wires in a layer, mn will be the total number of windings in the coil. If we now make the assumption that the E.M.F. developed in a conductor is independent of its radial depth, we get for the instantaneous value e of the E.M.F., in volts, generated in a coil

$$e\,.\,10^8 = mlB \sin\frac{\pi x}{a}\frac{\partial x}{\partial t} + mlB \sin\frac{\pi(x-h)}{a}\frac{\partial x}{\partial t} + \ldots$$

$$+ mlB \sin\frac{\pi\{x-(n-1)\,h\}}{a}\frac{\partial x}{\partial t}$$

$$- mlB \sin\frac{\pi(x-b)}{a}\frac{\partial x}{\partial t} - mlB \sin\frac{\pi\{x-(b-h)\}}{a}\frac{\partial x}{\partial t} - \ldots.$$

where x is the distance of the end layer of the coil from the fixed line, and a length l of each of the conductors is supposed to cut the flux. Summing this series we get

$$e\partial t\,.\,10^8 = mlB\frac{\sin\frac{n\pi h}{2a}}{\sin\frac{\pi h}{2a}}\left[\sin\left\{\frac{\pi x}{a} - \frac{\pi(n-1)\,h}{2a}\right\}\right.$$

$$\left. - \sin\left\{\frac{\pi(x-b)}{a} + \frac{\pi(n-1)\,h}{2a}\right\}\right]\partial x$$

$$= 2mlB \frac{\sin \dfrac{n\pi h}{2a}}{\sin \dfrac{\pi h}{2a}} \sin \left\{\frac{\pi b}{2a} - \frac{\pi (n-1) h}{2a}\right\} \cos \frac{\pi (2x - b)}{2a} \partial x.$$

If e_m denote the mean value of e we get, on integrating over the half of a period, and noticing that the limits on the right-hand side are from $(-a+b)/2$ to $(a+b)/2$,

$$\frac{T}{2} e_m \,.\, 10^8 = 2m \,.\, \frac{2lBa}{\pi} \,.\, \frac{\sin \dfrac{n\pi h}{2a}}{\sin \dfrac{\pi h}{2a}} \sin \left\{\frac{\pi b}{2a} - \frac{\pi (n-1) k}{2a}\right\}.$$

Again, since we suppose that the velocity of the rotor is uniform, e is sine shaped, and thus

$$e_m = \frac{2\sqrt{2}}{\pi} V_1,$$

where V_1 is the effective value of e. Also if Φ_A be the value of the flux entering the armature from one pole, we have

$$\Phi_A = \frac{2}{\pi} lBa.$$

Substituting these values of e_m and Φ_A in the above equation, and noticing that T equals $1/f$, we find that

$$V_1 \,.\, 10^8 = \pi \sqrt{2} fm \Phi_A \frac{\sin (n\pi h/2a)}{\sin (\pi h/2a)} \sin \left\{\frac{\pi b}{2a} - \frac{\pi (n-1) h}{2a}\right\}$$

and thus

$$V \,.\, 10^8 = 4{\cdot}443 f N \Phi_A \frac{\sin (n\pi h/2a)}{n \sin (\pi h/2a)} \sin \left\{\frac{\pi b}{2a} - \frac{\pi (n-1) h}{2a}\right\},$$

where N denotes the total number of turns in series on the armature and V is the value of the resultant E.M.F. When n equals unity and b equals a, we see that this agrees with the formula for a simple bar winding when the flux is sine shaped.

If n' be the total number of slots per pole, then $n'h = a$, and since in practice $b - c = a$, the formula becomes

$$V \,.\, 10^8 = 4{\cdot}443 f N \Phi_A \frac{\sin (n\pi/2n')}{n \sin (\pi/2n')}.$$

We see that $\dfrac{\sin (n\pi/2n')}{n \sin (\pi/2n')}$ is the factor by which $4{\cdot}443 f N \Phi_A \times 10^{-8}$ must be multiplied in order to get the true value of V. This factor is called the 'breadth factor' of the winding. The values of this quantity for various values of n/n' are given in the following table.

n/n'	0·05	0·10	0·15	0·20	0·25	0·30	0·35	0·40	0·45	0·50
$\frac{2n'}{\pi n}\sin\frac{\pi n}{2n'}$	0·999	0·996	0·991	0·984	0·975	0·963	0·950	0·935	0·919	0·900
n/n'	0·55	0·60	0·65	0·70	0·75	0·80	0·85	0·90	0·95	1·00
$\frac{2n'}{\pi n}\sin\frac{\pi n}{2n'}$	0·880	0·859	0·835	0·811	0·784	0·757	0·729	0·699	0·668	0·637

In practice it is customary to use about two-thirds of the slots for the windings, the remainder of them being left empty.

Equivalent windings.

Since there are no electromotive forces developed in the end connections an inspection of the coil winding indicated in Fig. 14 and the lap winding indicated in Fig. 15 will show that they are electrically equivalent. It will

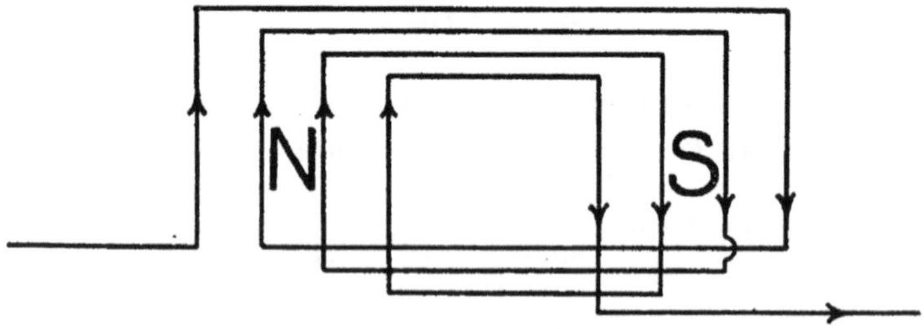

Fig. 14. Actual winding, one coil per pair of poles.

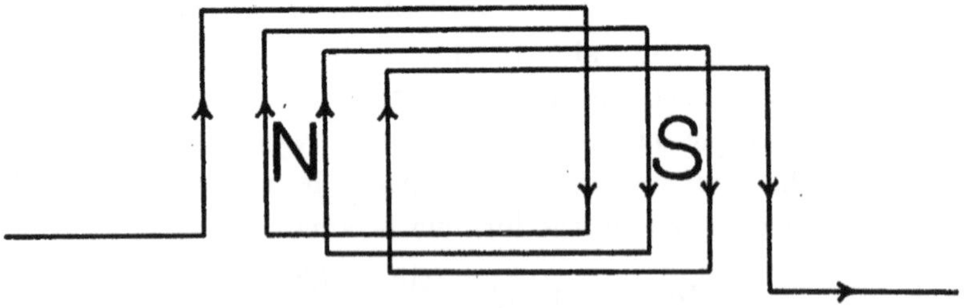

Fig. 15. Equivalent full pitch coil per pair of poles.

be seen that the breadth of the coils in the lap winding equals the polar step, they are thus 'full pitch' windings. We shall now show that replacing the coil winding by the lap winding enables us to give a simpler proof of the formula for the E.M.F. developed.

Let us suppose that m of the n' slots are occupied by windings and that $E \sin \omega t$ would be the E.M.F. developed if the winding had no breadth and were full pitch. The actual E.M.F. e developed is given by

$$e = (E/m) \sin \omega t + (E/m) \sin (\omega t - \pi/n') + \dots + (E/m) \sin \{\omega t - (m-1)\pi/n'\}$$
$$= \frac{E}{m} \frac{\sin (m\pi/2n')}{\sin (\pi/2n')} \sin \{\omega t - (m-1)\pi/(2n')\}.$$

Thus the value of the maximum E.M.F. can be found by multiplying E by the breadth factor.

In practice it is usually more convenient to have one coil per pole (Fig. 16). In this case adjacent coils are wound in opposite

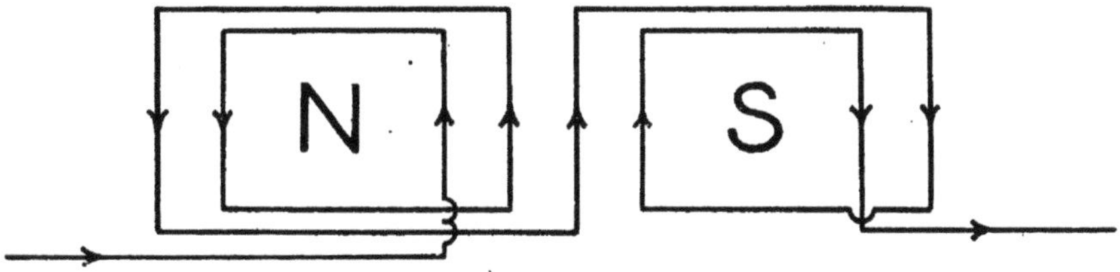

Fig. 16. One coil per pole.

directions. The E.M.F. developed in this winding is the same as that developed in the windings shown in Figs. 14 and 15, the breadth factor is thus the same.

Irregular magnetic field.

When the magnetic field in the air-gap does not follow the harmonic law the problem becomes more complicated. When the breadth of the coils is zero we have seen that the E.M.F. wave on open circuit is the same as that of the graph of the magnetic flux in the air-gap. Hence in this case resolving the E.M.F. wave into its harmonics we get by Fourier's theorem (see Chapter III)

$$e = E_1 \sin (\omega t - \epsilon_1) + E_3 \sin (3\omega t - \epsilon_3) + \dots.$$

If now we have a distributed winding, m slots per pole being used and $n' - m$ being left empty, we get

$$e = (E_1/m)\sin(\omega t - \epsilon_1) + (E_3/m)\sin(3\omega t - \epsilon_3) + \dots$$
$$+ (E_1/m)\sin(\omega t - \epsilon_1 - \pi/n') + (E_3/m)\sin(3\omega t - \epsilon_3 - 3\pi/n') + \dots$$
$$+ \dots \qquad + \dots \qquad + \dots$$
$$+ (E_1/m)\sin\{\omega t - \epsilon_1 - (m-1)\pi/n'\} + (E_3/m)\sin\{3\omega t - \epsilon_3 - 3(m-1)\pi/n'\} + \dots,$$

and thus

$$e = E_1 \frac{\sin(m\pi/2n')}{m\sin(\pi/2n')}\sin\left\{\omega t - \epsilon_1 - \frac{(m-1)\pi}{2n'}\right\}$$
$$+ E_3 \frac{\sin(3m\pi/2n')}{m\sin(3\pi/2n')}\sin\left\{3\omega t - \epsilon_3 - \frac{(m-1)3\pi}{2n'}\right\}$$
$$+ E_5 \frac{\sin(5m\pi/2n')}{m\sin(5\pi/2n')}\sin\left\{5\omega t - \epsilon_5 - \frac{(m-1)5\pi}{2n'}\right\} + \dots \qquad \dots\dots(A).$$

If in this formula $m/n' = 2/3$, which is a usual value, we see that all terms of order $3q$ vanish, where q is any integer.

As an example, let us consider the case when the flux follows the rectangular law. We shall see in Chapter III that the E.M.F. e is given by

$$e = \frac{4}{\pi}E\left(\sin\omega t + \frac{1}{3}\sin 3\omega t + \frac{1}{5}\sin 5\omega t + \dots\right)$$

when the breadth of the windings can be neglected. Let us now suppose that $m = 4$ and $n' = 6$. Substituting in (A) we get

$$e = \frac{4E}{\pi}\left[\frac{\sin(\pi/3)}{4\sin(\pi/12)}\sin(\omega t - \epsilon_1 - \pi/4) + \frac{1}{5}\left\{\frac{\sin(5\pi/3)}{4\sin(5\pi/12)}\sin(5\omega t - \epsilon_5 - 5\pi/4)\right\} + \dots\right]$$
$$= \frac{\sqrt{3}E}{2\pi\sin(\pi/12)}\{\sin(\omega t - \epsilon_1 - \pi/4) - 0{\cdot}054\sin(5\omega t - \epsilon_5 - 5\pi/4)$$
$$+ 0{\cdot}038\sin(7\omega t - \epsilon_7 - 7\pi/4)$$
$$- 0{\cdot}091\sin(11\omega t - \epsilon_{11} - 11\pi/4) + \dots\}$$

and $\sqrt{3}/\{2\pi\sin(\pi/12)\} = 1{\cdot}065$.

If we had a rectangular shaped magnetic field, therefore, the amplitudes of the fifth, seventh, ... harmonics in the generated E.M.F. wave would be appreciable although the harmonics of orders 3, 9, 15, ... would be absent.

In proving formula (A) we have made the assumption that the magnetic induction in the air-gap remains constant as the rotor revolves. Owing, however, to the slots in the armature the reluctance of the magnetic circuits is appreciably different in different positions of the rotor. If there are n' slots per pole the flux density at any given point in the air-gap will fluctuate over a small range with frequency $2n'f$. In the preceding paragraph when the coils had no breadth we supposed that the flux linkages ϕ with the armature windings were given by

$$\phi = (E_1/\omega)\cos(\omega t - \epsilon_1) + (E_3/3\omega)\cos(3\omega t - \epsilon_3) + \dots .$$

If we suppose that the flux density B at any given point in the air-gap varies between $1 + \lambda$ and $1 - \lambda$, where λ, in practice, is small compared with unity, we may write the flux linkages in the form

$$\phi\,\{1 + \lambda \sin 2n'\omega t\}, \text{ that is, } \phi + \lambda\phi \sin 2n'\omega t.$$

The differential coefficient of this expression with regard to t gives us the electromotive force. The terms multiplied by λ give the variation of the E.M.F. due to the pulsation of the reluctance.

Since

$$2 \sin 2n'\omega t \cos\{(2q+1)\,\omega t - \epsilon_{2q+1}\}$$
$$= \sin\{(2n' + 2q + 1)\,\omega t - \epsilon_{2q+1}\} + \sin\{(2n' - 2q - 1)\,\omega t + \epsilon_{2q+1}\}$$

we see that the term in ϕ of order $2q + 1$ gives rise to two components in the E.M.F. wave the frequencies of which are

$$(2n' + 2q + 1)\,\omega/2\pi \quad\text{and}\quad (2n' - 2q - 1)\,\omega/2\pi$$

respectively.

Similarly if the axis of the stator be not exactly coincident with the axis of the rotor, other terms the order of which is even will be introduced into the expression for the E.M.F. wave.

Open circuit characteristic.

When an alternator is running on open circuit, the potential difference between the terminals varies with the excitation of the field magnets. The variation of the voltage with the excitation is different in different machines, and it is necessary that the connection between the two should be known. This can be found easily by experiment and is generally shown by a curve which has the voltage between the terminals on open circuit for ordinates and the ampere-turns per field magnet spool for abscissae. To find this curve we proceed as follows. An ammeter is placed in the exciting circuit, and an electrostatic voltmeter is placed across the terminals of the machine. The alternator is then run at its normal speed, and, as the excitation is increased from zero to its maximum value, simultaneous readings of the ammeter and voltmeter are taken. These values, when plotted as described above, give the open circuit characteristic. The curve is sometimes also called the open circuit saturation curve. In Fig. 17 the curve OA is typical of an open circuit characteristic curve. It will be noticed that almost up to the full working pressure it is a straight line. It then bends downwards.

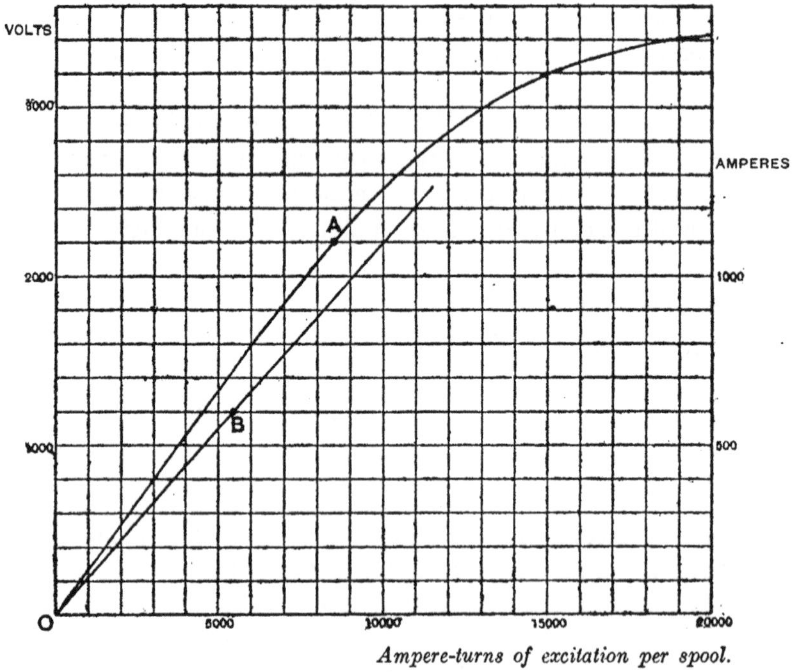

Fig. 17. OA is the open circuit characteristic of a 1250 kilo-volt ampere alternator. OB is its short circuit characteristic. The points A and B give the full load volts and amperes.

If we assume that an alteration of the ampere-turns in the field magnet windings does not alter appreciably the shape of the curve representing the flux in the air-gap, but only alters the scale of the ordinates of this curve, then, when the machine is run at constant speed, the voltage on open circuit will be proportional to the flux per pole linked with the armature. To find a formula connecting the voltage and the exciting ampere-turns nC of the field magnet we find, first of all, a formula connecting nC and the flux Φ_A entering the armature from a pole. In Vol. I, p. 72, we obtained the equation

$$\text{Flux} = \frac{4\pi nC/10}{\text{Reluctance}},$$

where the reluctance is calculated by the formula $l/\mu S$, l denoting the length of the path of the flux, S its cross sectional area and μ the permeability at the given flux density. Now, in practice, we are given the permeability curve of the iron, and so, if we know the flux, and therefore the flux density, we can calculate the reluctance. Similarly, in this case, when we know the magnetic force we can find μ, and thus we can find the reluctance and the magnetic flux. In proving the above formula we considered the case of an infinite solenoid so that the magnetic force is assumed constant at every point on the cross section. We saw, however, that in the case of a finite circuit, like an anchor ring uniformly wound with insulating wire carrying a current, the magnetic forces to which the iron is subjected are not constant but are greater at points near the inner circumference of the ring than they are at points near the outer circumference. If the permeability corresponding to the given magnetising forces be represented by a point on the steep part of the permeability curve, so that a small variation in the value of the magnetic forces makes a large variation in the value of the permeability, then the variation of the flux density over the cross section of the ring may be large. It follows that $l/\mu S$, where μ is the permeability corresponding to the density Φ/S, may not give the true value of the reluctance. The formula, therefore, which is used in practice for the magnetic circuit is only approximately correct.

In a dynamo, the path of the flux in a field magnet is partly in the iron and partly in the air. It is customary (see Vol. I, p. 73) to extend the magnetic analogy of Ohm's law to this case, the reluctances of the paths in air and the paths in iron of the flux being calculated separately by the formula $l/\mu S$, and the sum of these quantities being given as the total reluctance of the magnetic circuit. In practice, it is only possible to find these reluctances approximately, and in the case of armatures provided with slots to hold the armature windings the calculation is a difficult one.

The methods of approximating to the values of these reluctances are explained in treatises on the design of direct current dynamos. Formulae containing reluctances, which can only be calculated roughly, have a limited use. They are, however, a help to the designer as they show him the relative effects produced by alterations in the various parts of the magnetic circuit. In what follows we shall assume that the armature has a smooth surface.

In practice, it is customary to consider ampere-turns nC instead of magnetomotive force $4\pi nC/10$. To simplify the formulae, therefore, we shall denote the reluctance of a magnetic circuit by $4\pi\mathcal{R}/10$ so that the fundamental equation becomes

$$\text{Flux} = \frac{nC}{\mathcal{R}}.$$

Let us now consider the magnetic flux linked with two adjacent poles N_2 and S_2 in a multipolar field $N_1, S_1, N_2, S_2, N_3, \ldots$. The flux proceeding from N_2 is linked with both S_1 and S_2, half of it coming back by S_1 and half by S_2. We shall consider the flux linked with N_2 and S_2. This flux will be half the total flux leaving the pole N_2. On leaving N_2 an amount $\Phi_a/2$ of this portion of the flux will leak directly through the air to the pole S_2. Let the reluctance of the path in the air of this leakage flux be $4\pi\mathcal{R}_a/10$. The remainder $\Phi_A/2$ of the flux leaving N_2 and linked with S_2 will after passing across the air-gap, the reluctance of which we will denote by $4\pi\mathcal{R}_g/10$, enter the armature. Let the reluctance of the path of the flux $\Phi_A/2$ in the armature between N_2 and S_2 be $4\pi\mathcal{R}_A/10$. After crossing a second air-gap ($4\pi\mathcal{R}_g/10$) this flux will enter S_2. The flux $(\Phi_A + \Phi_a)/2$ will complete its path through the pole S_2, then through that part of the iron ring (the yoke) to which N_2 and S_2 are both fixed, and finally through N_2 to the polar surface from which it started. We shall denote the reluctance of this part of the magnetic circuit traversed by $(\Phi_A + \Phi_a)/2$ by $4\pi\mathcal{R}_f/10$. It is to be noted that the reluctance of the air-gap from the whole of the pole of N_2 to the armature will be $4\pi\mathcal{R}_g/20$, and in calculating the reluctance of the path of $(\Phi_A + \Phi_a)/2$ in the field magnets we assume that this path occupies half of the iron of the field magnets.

Making use of the electrical analogy, shown in Fig. 18, we get the following equations:

$$\tfrac{1}{2}(\Phi_A + \Phi_a) = \frac{2nC}{\mathcal{R}_f + \mathcal{R}'},$$

where
$$\mathcal{R}' = \frac{\mathcal{R}_a(\mathcal{R}_A + 2\mathcal{R}_g)}{\mathcal{R}_a + \mathcal{R}_A + 2\mathcal{R}_g}.$$

We also have
$$(\mathcal{R}_A + 2\mathcal{R}_g)\,\Phi_A = \mathcal{R}_a\Phi_a,$$

and thus
$$\Phi_A = \frac{4nC}{\mathcal{R}_f + \mathcal{R}_A + 2\mathcal{R}_g + \mathcal{R}_f(\mathcal{R}_A + 2\mathcal{R}_g)/\mathcal{R}_a}.$$

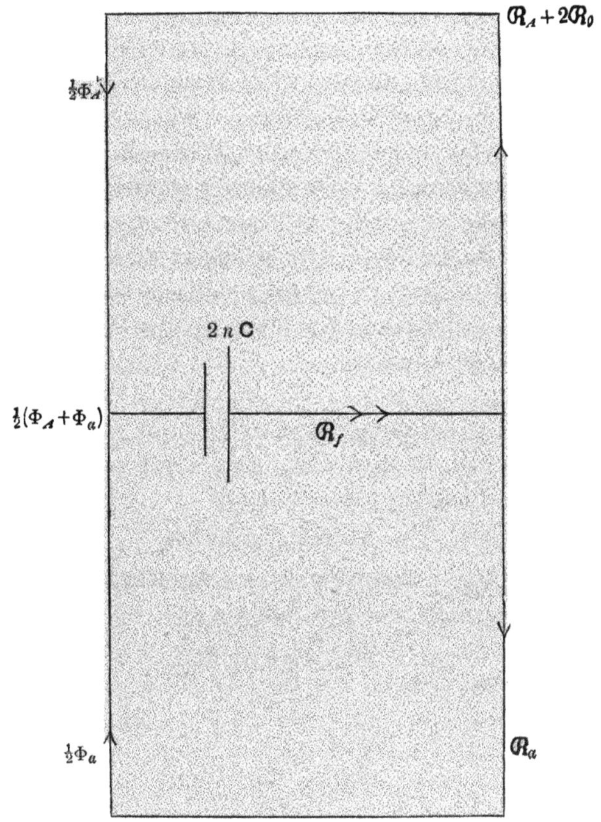

Fig. 18. Electrical analogy with the magnetic circuits linked with a field magnet and the two field magnets adjacent to it.

In practice $\mathcal{R}_a$ is large compared with $\mathcal{R}_f(\mathcal{R}_A + 2\mathcal{R}_g)$, and therefore, we have approximately

$$\frac{4nC}{\Phi_A} = \mathcal{R}_f + \mathcal{R}_A + 2\mathcal{R}_g.$$

Now, when the magnetising force is not large, the permeability of the iron or steel is very high, and thus, since reluctance is inversely proportional to permeability, $\mathcal{R}_f + \mathcal{R}_A$ is small in comparison with $2\mathcal{R}_g$. But the value of $2\mathcal{R}_g$ is independent of the magnetising force, and therefore, the ratio of nC to Φ_A is very approximately constant if the magnetising forces are not large. Thus the curve having Φ_A for ordinates and nC for abscissae will be very approximately a straight line until nC becomes large. When nC is large the iron in the path of the flux becomes saturated, and so $\mathcal{R}_f + \mathcal{R}_A$ becomes appreciable and the curve giving the armature flux Φ_A in terms of nC begins to bend downwards. If the relative distribution of the flux in the air-gap does not alter as nC is increased, V will be proportional to Φ_A, and thus we should expect the open circuit characteristic to be similar to the curve A shown in Fig. 17. In machines with large air-gaps, the open circuit characteristic is almost an exact straight line. In machines with insufficient iron, or iron of inferior quality, in the field magnets the characteristic curve bends down rapidly and the loss of power due to the large excitation required is excessive.

Flux curves.

If we neglect the breadth of the coils in the armature circuit of an alternator, the formula for the open circuit electromotive force, namely

$$V = 4kfN\Phi_A \times 10^{-8},$$

enables us to find Φ_A. From Fig. 18 we see that

$$\mathcal{R}_a\Phi_a = (\mathcal{R}_A + 2\mathcal{R}_g)\,\Phi_A,$$

and therefore
$$\Phi_A + \Phi_a = \frac{\mathcal{R}_A + 2\mathcal{R}_g + \mathcal{R}_a}{\mathcal{R}_a}\,\Phi_A$$

$$= \nu\Phi_A,$$

where
$$\nu = \frac{\mathcal{R}_A + 2\mathcal{R}_g + \mathcal{R}_a}{\mathcal{R}_a}.$$

The coefficient ν is sometimes called the 'leakage coefficient,' and sometimes the 'dispersion coefficient.' Hence, if we multiply the ordinates of the open circuit characteristic by $10^8\nu/4kNf$, we get the curve showing the total flux in the field magnets for various excitations.

In practice, the calculation of ν is difficult. It is to be noticed that whether the armature or the field magnets rotate, the path of the flux in the field magnets is continually rotating relatively to the iron sheets of which the armature is built up. It follows that the polarity of the molecules of the iron in the armature alternates with the frequency of the alternating E.M.F., and if the flux density in it be high, the loss in the iron of the armature due to hysteresis and eddy currents will be considerable. This will affect the accuracy of the fundamental magnetic equation. Assuming, however, that this introduces no serious error, we must calculate the values of $\mathcal{R}_a$, $\mathcal{R}_A$ and $\mathcal{R}_g$ in order to find ν. This calculation is very difficult as the paths of the flux are not simple geometrical curves and the permeability of the iron in the various parts of the magnetic circuit is not accurately known. We can, therefore, as a rule, only make a rough approximation to the value of ν by calculation. An average value for ν in good modern machines would be about 1·1, but occasionally it is 1·2 or even higher. For a particular type of machine, however, designers can estimate its value with fair accuracy, and thus, by the aid of the formulae given above, it would be possible to predetermine the open circuit voltage of the machine for any excitation and at any speed. We could therefore predetermine the open circuit characteristic curves of the machine for various speeds.

Short circuit characteristic.

If we short circuit the terminals of an alternator through an ammeter when the field magnets are only feebly excited, the current will not be large. This is due to the small value of the electromotive force generated and the appreciable impedance of the armature itself. If we now gradually increase the excitation, the machine running at its normal speed, we can get a series of simultaneous readings of an ammeter in the exciting circuit and of the ammeter short circuiting the alternator. Plotting out a curve (*OB*, Fig. 17), having ampere-turns of excitation per field magnet spool for abscissae, and short circuit amperes for ordinates, we get the short circuit characteristic. The curve is practically a straight line. The excitation required for the electromotive force to produce the full load current in the short circuited armature is much less than that required to produce the full load current in the armature together with the voltage required for an external non-inductive load. The phase difference, however, between the current and the E.M.F. generated is greater in the case of the short circuited armature and one effect of a lagging current is to demagnetise the field. In some machines this effect is very marked, and appreciable magnetising currents are required in order to get the short circuit characteristic. In order to understand why a lagging current tends to demagnetise the field magnets, we shall consider in detail some simple armature windings.

Wave windings. A simple method of studying cylindrical (drum) armature windings is to imagine that the winding is cut across parallel to the axis of the drum and developed out into a plane. In Fig. 19 a diagram of a four pole alternator is shown expanded in this fashion. If the field magnets rotate, we

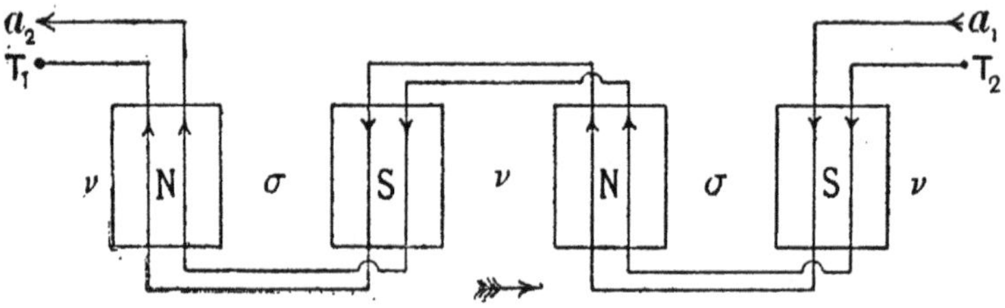

Fig. 19. Expanded diagram of a four pole alternator. Simple wave winding. The arrow indicates the direction of rotation of the poles. The Greek letters indicate the magnetic effects of the currents in the armature.

can suppose that the poles are moving underneath the windings in the direction of the arrow. The electromotive forces developed in the wires will act in the directions of the arrow heads. These directions are the same as if the field magnets were stationary and the armature rotated in the opposite direction. Hence by applying Fleming's rule we find them at once. T_1 and T_2 are the terminals of the machine and the points a_1 and a_2 coincide, when the winding is on the armature.

If the adjacent active conductors of the winding $T_1 a_1$ (Fig. 19) be at a distance from one another equal to the polar step, the electromotive forces developed in adjacent conductors will be in exact opposition in phase, and thus, since the conductors are connected so that the E.M.F.s act in the same direction round the winding, the effective value of the resultant E.M.F. between T_1 and a_1 will be equal to the sum of the effective values of the E.M.F.s developed in the four active conductors between T_1 and a_1. Similarly, the effective value of the resultant E.M.F. between T_2 and a_2 will be the sum of the effective values of the E.M.F.s developed in the four active conductors between these points. The electromotive forces, however, developed in the two windings $T_1 a_1$ and $a_2 T_2$ will only be in phase when the windings are superposed. Thus a differential action between the various E.M.F.s developed can only be avoided by using a simple bar winding.

If the distance between any two conductors which pass across the face of a pole in Fig. 19 be greater than the minimum distance between the poles, then, when one conductor is leaving one pole the other will be over the next and, at this instant, the arrow heads, indicating the direction of the E.M.F.s developed in the two conductors, will be pointing in opposite ways. The differential action, therefore, will be excessive. In practice, the displacement of the two windings relatively to one another is made less than the minimum distance between the poles. We can also have any number of windings similar to $T_1 a_1$ and $a_2 T_2$ in Fig. 19, but the displacement of the two which are farthest apart should be less than the minimum distance between the poles. This simple form of winding is called a 'distributive' wave winding.

When the terminals of the machine are connected through a large non-inductive resistance, the currents in the conductors will be flowing in the direction of the arrow heads (Fig. 19), and their values will be large at the instant pictured in the diagram. The armature current will produce a magnetic flux leaving the paper perpendicularly at ν and entering it perpendicularly at σ. It will thus produce a transverse magnetisation of the field in the same way that the corresponding effect is produced in direct current machines. The magnetic flux on the trailing side of the pole pieces will be strengthened and that on the leading side weakened. We should expect, therefore, that this transverse magnetisation would have an appreciable effect on the shape of the wave of the electromotive force generated, and on the distribution of the heat generated by eddy currents and hysteresis in the pole pieces. This is found to be the case in practice.

A quarter of a period after the armature current has its maximum value, the poles will lie between the windings, as in Fig. 21, and the current will be zero. The current now changes sign and at the end of the next quarter of a period it attains a maximum value. Hence it is easy to see that in this case, namely, when the load is non-inductive, the mean value of the magnetising force exerted by the armature currents on the field magnets is zero.

Lap windings.

In Fig. 20 a simple lap winding is shown for the alternator represented diagrammatically in Fig. 19. It will be seen that, so far as the electrical effects produced are concerned, the lap windings and wave windings are identical. It

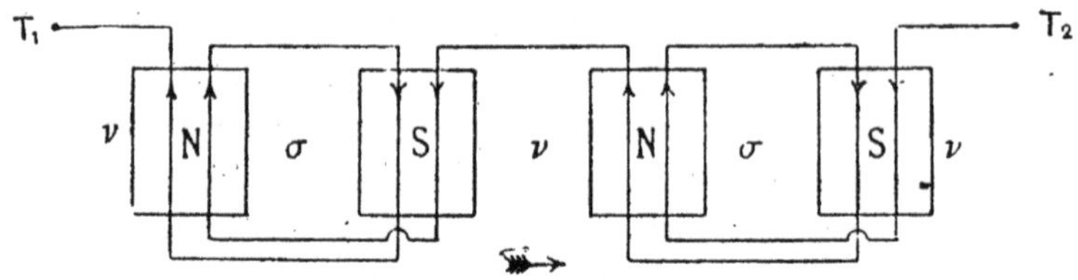

Fig. 20. Expanded diagram of four pole alternator. Simple lap winding. The Greek letters indicate the magnetising effect of armature currents which are in phase with the E.M.F.

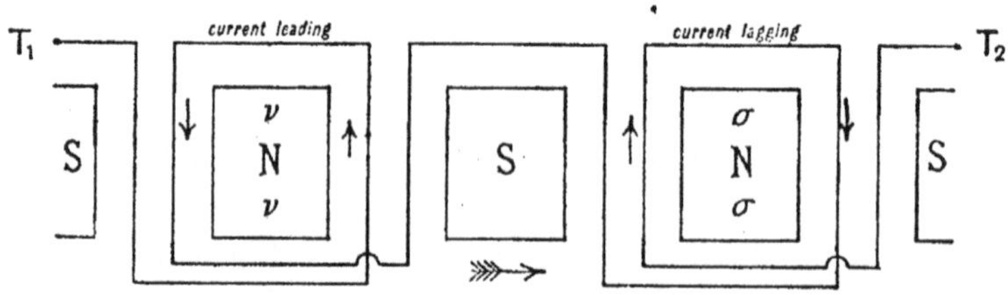

Fig. 21. The diagram of the four pole alternator (Fig. 20) as it would appear at a quarter of a period later. The magnetising effect on the field magnets of leading currents is shown at $\nu N \nu$. The demagnetising effect on the field magnets of lagging currents is shown at $\sigma N \sigma$.

is to be noticed that a very narrow or a very broad winding adds little to the total E.M.F. generated in the coil.

In Fig. 21 the developed diagram of this alternator is shown a quarter of a period later, when the electromotive force is zero. If the current is ninety degrees in advance of the electromotive force, it will be seen from Fig. 21 that the magnetomotive force increases the magnetisation of the field. If, on the other hand, the current is lagging by ninety degrees, the magnetising effect of the armature coils directly opposes that of the field coils and so the flux in the air-gap is weakened.

The armature reactions.

It will be seen from Fig. 20 that the armature currents distort the magnetic field in the air-gap and sometimes also they magnetise or demagnetise the field magnets (Fig. 21). In computing the electromotive force generated in the armature therefore when the machine is working under load, it is necessary to take these effects into account. In practice the problem is a difficult one as armatures have slots on their surface in which the conductors are embedded and the outlines of the pole shoes are not simple geometrical figures. Blondel and others have shown that by making suitable assumptions so as to simplify the problem useful formulae can be obtained.

Let us first consider the case of an armature coil of N turns having resistance R and inductance L. We shall suppose that it rotates with uniform speed in a magnetic field produced by a direct current C flowing in a suitably shaped coil. We shall also suppose that the mutual inductance between the armature coil and the field coil is $M \cos \omega t$. If the impedance of the field coil is so great that C is practically unaffected by the electromotive force induced in it by the moving magnetic field due to the armature current, then the E.M.F. induced in the armature circuit is $-(\partial/\partial t)(MC \cos \omega t)$, that is $M\omega C \sin \omega t$, and so we may denote the armature current by $I \sin(\omega t - \psi)$, where I is a constant and ψ is only zero when $\omega L/R$ is negligibly small.

The flux linked with the armature at the time t is given by

$$MC \cos \omega t + LI \sin(\omega t - \psi),$$

which equals

$$(MC - LI \sin \psi) \cos \omega t + LI \cos \psi \sin \omega t.$$

By Faraday's law the rate at which this flux varies gives the E.M.F. which drives the armature current i through the resistance R.

It will be seen that the flux linked with the armature coil consists of two components which are in quadrature. The amplitude of the first component is $MC - LI \sin \psi$, and it has this value when t is zero, that is, when the mutual inductance between the armature and field coils is a maximum. The amplitude of the second component is $LI \cos \psi$ and it has this value when

$\omega t = \pi/2$, that is when the mutual inductance is zero. The first component is called the direct flux and the second is called the transverse or cross flux. The E.M.F. due to the first component is zero when the mutual inductance is a maximum and has its maximum value when the mutual inductance is zero. If we can compute the values of these two fluxes we can readily find the E.M.F. in the armature. This method of finding the E.M.F. depending on what Blondel calls the principle of two reactions is much used in practice.

When the armature current I is zero MC is the value of the amplitude of the direct flux. This can be found by evaluating $nC/\mathcal{R}$, where n is the number of turns of the field coil and $\mathcal{R}$ multiplied by $4\pi/10$ is the reluctance of the magnetic circuit (see p. 30) linking the armature and field coils at the instant when the mutual inductance is a maximum. When the armature current i lags behind the mutual inductance by an angle $\pi/2 + \psi$, the amplitude of the direct flux is $(nC - aNA \sin \psi)/\mathcal{R}$, where a is a constant which varies with the nature of the winding and A is the effective value $(I/\sqrt{2})$ of the armature current. As we have supposed that the self-inductance L is constant in all positions of the armature, the amplitude of the transverse flux in this case will be $aNA \cos \psi/\mathcal{R}$.

When the current i in the armature follows the harmonic law it may be written in the form $I \sin (\omega t - \psi)$ which equals

$$I \cos \psi \sin \omega t + I \sin \psi \sin (\omega t - \pi/2).$$

Hence we may regard i as the sum of the two currents having maximum values $I \cos \psi$ and $I \sin \psi$ respectively and differing in phase by ninety degrees. The current the amplitude of which is $I \cos \psi$ has its maximum value when the mutual inductance is zero. It produces only a transverse magnetising effect on the field at this instant, and it is easy to see that the average direct magnetising or demagnetising effect on the field magnets from the instant the current is zero until it is again zero vanishes. The average transverse magnetising effect of this component however is not zero. Similarly the component of the current which has $I \sin \psi$ for its maximum value produces its maximum demagnetising effect when the mutual inductance is a maximum and the average transverse magnetising effect produced is zero. In practice in getting approximate formulae for the E.M.F. in the armature we may consider that it is due to the conductors cutting two fluxes, the direct and the transverse, which are at right angles to one another. The average mean values of the ampere-turns producing these two fluxes can be computed approximately by the method shown below and the reluctances of the paths can also be approximately found by the ordinary methods. Hence approximate values of the magnitude of these fluxes can be found. We shall now show how to find an approximate value of the demagnetising ampere-turns. If we had a condenser in the main circuit so that the angle ψ was negative, then noticing that

$$\sin(-\psi) = -\sin\psi,$$

we see that the armature magnetising turns help the field magnetising turns and so the magnetic flux in the field magnets is increased.

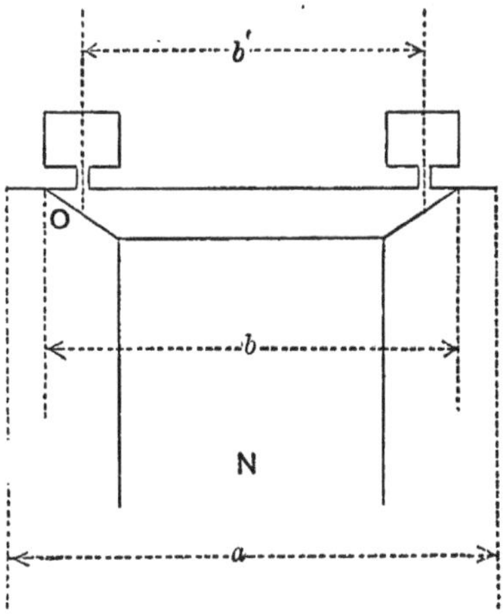

Fig. 22. The polar step equals a. The breadth of the polar flux entering the armature equals b. The distance between the axes of the slots equals b'.

Formula for the demagnetising effect of the lagging component of the current.

We shall consider the case of a machine with a simple coil winding on the armature, as in Fig. 3, and we shall calculate the mean value of the demagnetising ampere-turns acting on the field due to the current in the armature. Let there be $2N_1$ conductors in a slot, and therefore N_1 turns per pole. We shall suppose that the magnetomotive force, due to the current in an armature coil, acting on a given tube of magnetic flux in a field magnet, changes from $(4\pi/10)\, N_1 I \sin\psi \sin(\omega t - \pi/2)$ to zero, or *vice versâ*, when the tube passes through the axis of a slot. We shall assume that the breadth of the arc intercepted on the cylindrical armature by the flux leaving a pole is b, and, since the flux leaving a pole spreads out in the air-gap, b will be greater than the breadth of the pole. Let us assume also that the density of the flux entering the armature is uniform. Let b' be the breadth of an armature coil, measured along the circumference of the armature, between the axes of the two slots (Fig. 22) and let a be the polar step, that is, the distance between the middle points of consecutive poles measured along the arc of the circle on which these middle points lie. As the air-gap is very narrow compared with the radius of the rotor we may, without sensible error, assume

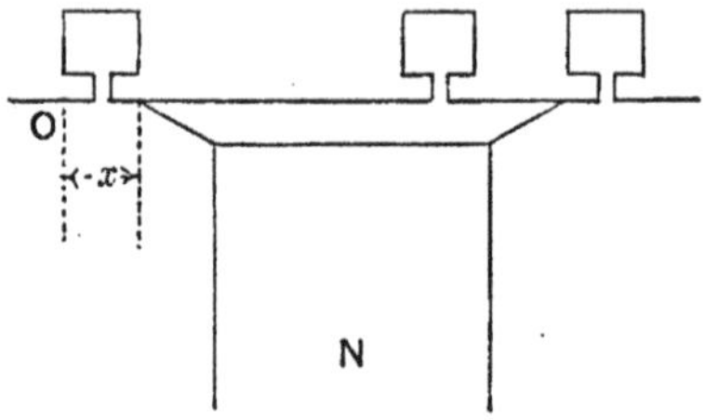

Fig. 23. In this diagram x is greater than $\frac{b-b'}{2}$ but is less than $a - \frac{b+b'}{2}$.

that the circumference of the armature is $2pa$, where $2p$ is the number of the field poles. For convenience of drawing, we have shown the sections of the polar and armature faces as if they were straight. Suppose now that the faces of the poles of the field magnets are moving with a linear velocity $2a/T$, and let O (Fig. 22) be taken as the origin from which the distance x (Fig. 23) of the end of the trailing flux is measured. If the space x be described in time t, then, since the air-gap is very narrow, we may write

$$x = 2at/T = a\omega t/\pi, \text{ since } \omega = 2\pi/T,$$

and therefore $\omega t = \pi x/a$.

Hence, the lagging component of the current may be written in the form $I \sin\psi \sin(\pi x/a - \pi/2)$ or $-I \sin\psi \cos \pi x/a$.

We only need to calculate the mean value of the magnetising turns produced by this current during a quarter of a period, as this will be the same as over a whole period, for the frequency of this magnetising force acting on a pole is double that of the alternating current. We shall first find expressions for the magnetising force during various intervals of the quarter period, and then calculate its mean value.

The first interval of time is the time taken by x (Fig. 22) to increase from zero to $\frac{1}{2}(b - b')$. In this case, the demagnetising turns of the armature coils act only on the fraction b'/b of the total flux entering the armature. The magnetising ampere-turns, therefore, from x equal to zero to x equal to $\frac{1}{2}(b - b')$, are

$$-(b'/b)\, N_1 I \sin\psi \cos \pi x/a.$$

When x (Fig. 23) is greater than $\frac{1}{2}(b - b')$ but less than

$$\tfrac{1}{2}(a - b) + \tfrac{1}{2}(a - b'),$$

the coil embraces the fraction $[b' - \{x - \frac{1}{2}(b - b')\}]/b$ of the total flux. The value of the magnetising ampere-turns, from x equal to $\frac{1}{2}(b - b')$ to x equal to $a - \frac{1}{2}(b + b')$, is therefore equal to

$$-\frac{1}{b}\{\tfrac{1}{2}(b + b') - x\}\, N_1 I \sin\psi \cos\frac{\pi x}{a}.$$

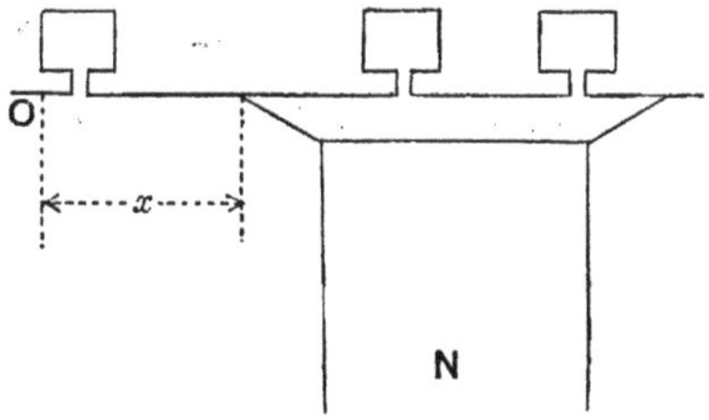

Fig. 24. In this diagram x is greater than $a - \dfrac{b + b'}{2}$ but is less than $\dfrac{a}{2}$.

When x (Fig. 24) is greater than $a - \frac{1}{2}(b + b')$, some of the flux from the pole is embraced by the adjacent coil which tends to magnetise it in the opposite direction. Hence the magnetising ampere-turns, from x equal to $a - \frac{1}{2}(b + b')$ to x equal to $\frac{1}{2}a$, are

$$-\frac{1}{b}[b' - \{x - \tfrac{1}{2}(b - b')\} - \{x - (a - \tfrac{1}{2}\overline{b + b'})\}]\, N_1 I \sin\psi \cos\frac{\pi x}{a},$$

which is equal to

$$-\left(\frac{a - 2x}{b}\right) N_1 I \sin\psi \cos\frac{\pi x}{a}.$$

If A, therefore, denote the effective value $I/\sqrt{2}$ of the armature current during the time the field magnet takes to move from 0 to $a/2$, and if $aN_1A\sin\psi$ denote the mean value of the demagnetising ampere-turns, we have

$$aN_1A\sin\psi\,.\,\frac{a}{2}=\int_0^{\frac{b-b'}{2}}\frac{b'}{b}N_1I\sin\psi\cos\frac{\pi x}{a}\partial x$$

$$+\int_{\frac{b-b'}{2}}^{a-\frac{b+b'}{2}}\frac{1}{b}\left(\frac{b+b'}{2}-x\right)N_1I\sin\psi\cos\frac{\pi x}{a}\partial x$$

$$+\int_{a-\frac{b+b'}{2}}^{\frac{a}{2}}\left(\frac{a-2x}{b}\right)N_1I\sin\psi\cos\frac{\pi x}{a}\partial x.$$

Putting θ equal to $\pi x/a$ and simplifying, we have

$$\frac{\pi b}{2\sqrt{2}}a=b'\int_0^{\frac{\pi}{2a}(b-b')}\cos\theta\partial\theta+\int_{\frac{\pi}{2a}(b-b')}^{\pi-\frac{\pi}{2a}(b+b')}\left(\frac{b+b'}{2}-\frac{a\theta}{\pi}\right)\cos\theta\partial\theta$$

$$+\int_{\pi-\frac{\pi}{2a}(b+b')}^{\frac{\pi}{2}}\left(a-\frac{2a\theta}{\pi}\right)\cos\theta\partial\theta.$$

Noticing that $\int\theta\cos\theta\partial\theta=\theta\sin\theta+\cos\theta,$

we easily find that

$$\frac{\pi b}{2\sqrt{2}}a=b'\Big[\sin\theta\Big]_0^{\frac{\pi}{2a}(b-b')}$$

$$+\left[\frac{b+b'}{2}\sin\theta-\frac{a\theta}{\pi}\sin\theta-\frac{a}{\pi}\cos\theta\right]_{\frac{\pi}{2a}(b-b')}^{\pi-\frac{\pi}{2a}(b+b')}$$

$$+\left[a\sin\theta-\frac{2a\theta}{\pi}\sin\theta-\frac{2a}{\pi}\cos\theta\right]_{\pi-\frac{\pi}{2a}(b+b')}^{\frac{\pi}{2}}$$

$$=b'\sin\frac{\pi}{2a}(b-b')$$

$$+\frac{b+b'}{2}\left\{\sin\frac{\pi}{2a}(b+b')-\sin\frac{\pi}{2a}(b-b')\right\}$$

$$-\frac{a}{\pi}\left\{\left(\pi-\frac{\pi}{2a}\overline{b+b'}\right)\sin\frac{\pi}{2a}(b+b')-\frac{\pi}{2a}(b-b')\sin\frac{\pi}{2a}(b-b')\right\}$$

$$+\frac{a}{\pi}\left\{\cos\frac{\pi}{2a}(b+b')+\cos\frac{\pi}{2a}(b-b')\right\}$$
$$+a\left\{1-\sin\frac{\pi}{2a}(b+b')\right\}$$
$$-\frac{2a}{\pi}\left\{\frac{\pi}{2}-\left(\pi-\frac{\pi}{2a}\overline{b+b'}\right)\sin\frac{\pi}{2a}(b+b')+\cos\frac{\pi}{2a}(b+b')\right\}$$
$$=\frac{a}{\pi}\left\{\cos\frac{\pi}{2a}(b-b')-\cos\frac{\pi}{2a}(b+b')\right\}$$
$$=\frac{2a}{\pi}\sin\frac{\pi b'}{2a}\sin\frac{\pi b}{2a}.$$

Therefore
$$\alpha=\frac{4\sqrt{2}a}{\pi^2 b}\sin\frac{\pi b'}{2a}\sin\frac{\pi b}{2a}.$$

This formula enables us to find the values of α, and thus, when the current A in the armature and the angle ψ of lag of this current behind the electromotive force are known, we can readily find the mean value $\alpha N_1 A\sin\psi$ of the demagnetising ampere-turns per pole. If we suppose that b is equal to a, then, in the case of a simple bar winding α is 0·57 nearly.

When ψ is negative, that is, when the current is leading, $\alpha N_1 A\sin\psi$ is also negative, and thus, in this case, the armature reaction strengthens the field.

If the breadth of the coil be c and if it contain many conductors, we can get a more accurate formula as follows. Let h be the distance between the axes of two neighbouring wires which are equidistant from the axis of the rotor, and let nh be equal to c. Calculating the demagnetising force for each turn separately, and adding the results, we find that

$$\alpha N_1 A\sin\psi=\frac{4a}{\pi^2 b}N_1 I\sin\psi\sin\frac{\pi b}{2a}$$
$$\times\frac{\sin\frac{\pi(b''-h)}{2a}+\sin\frac{\pi(b''-3h)}{2a}+\ldots+\sin\frac{\pi(b''-\overline{2n-1}h)}{2a}}{n},$$

where b'' is the distance between the outside wires of the sides of the coil. Summing the series we get

$$\alpha=\frac{4a\sqrt{2}}{\pi^2 b}\sin\frac{\pi b}{2a}\,\frac{\sin\left\{\frac{\pi b''}{2a}-\frac{\pi c}{2a}\right\}\sin\frac{\pi c}{2a}}{n\sin\frac{\pi c}{2an}}.$$

Now b'' equals $c + b'$, where b' is the distance between the middle wires of the sides of the coil. Thus, when n is large, so that we can write $\pi c/2an$ for $\sin \pi c/2an$, we get

$$a = \frac{4a\sqrt{2}}{\pi^2 b} \sin \frac{\pi b}{2a} \sin \frac{\pi b'}{2a} \left(\sin \frac{\pi c}{2a}\right) \Big/ \frac{\pi c}{2a}.$$

It will be seen that the factor $(2a/\pi c) \sin (\pi c/2a)$ corrects for the breadth of the coil.

Formula for the compensating ampere-turns required for the field magnets.

An approximate value for the ampere-turns round the field magnets required to compensate for the demagnetising effects due to the current in the armature, when $\sin \psi$ is not zero, can be found as follows. We have seen that the mean value of the demagnetising ampere-turns per field magnet pole is $aN_1A \sin \psi$. The compensating ampere-turns per pole must be greater than this since all the flux generated in the field magnets does not pass through the armature coils. If we amplify the electrical analogy shown in Fig. 18 we get Fig. 25. In this diagram $n'C'$ represents the ampere-turns, of the compensating coil on every field magnet, required to keep the flux in the armature constant, and $aN_1A \sin \psi$ denotes the demagnetising ampere-turns per pole, due to the current A in the armature windings, when the power factor is $\cos \psi$. Let the reluctances, divided by $4\pi/10$, of the path in the armature, of the leakage paths in air, of the two air-gaps, and of the path in the field magnets, traversed by the flux linked with two adjacent field magnets, be denoted by $\mathcal{R}_A$, $\mathcal{R}_a$, $2\mathcal{R}_g$ and $\mathcal{R}_f$ respectively (p. 30). Let also Φ_A be the flux from a pole, linked with the armature, and let Φ_a be the leakage flux from a pole, on open circuit. The magnetic equations on open circuit are

$$4nC - \mathcal{R}_f(\Phi_A + \Phi_a) = (\mathcal{R}_A + 2\mathcal{R}_g)\,\Phi_A = \mathcal{R}_a\Phi_a \;\ldots\ldots(1),$$

where n is the number of turns on one pole of a field magnet winding, and C is the exciting current. Let us suppose that the number of coils in the armature equals the number of field magnet poles, and that N_1 is the number of windings in a coil, so that $aN_1A \sin \psi$ is the measure of the mean demagnetising effect of an armature coil on a pole, when the current in it is A, and $\cos \psi$ is the power factor. Let $n'C'$ be the compensating ampere-turns on each field magnet and let Φ_a' be the leakage flux in the air. Our equations in this case are

$$4(nC + n'C') - \mathcal{R}_f(\Phi_A + \Phi_a') = (\mathcal{R}_A + 2\mathcal{R}_g)\Phi_A + 4aN_1A\sin\psi = \mathcal{R}_a\Phi_a' \quad \dots(2),$$

since the flux in the armature is made the same in the two cases and we suppose that $\mathcal{R}_f$ remains constant.

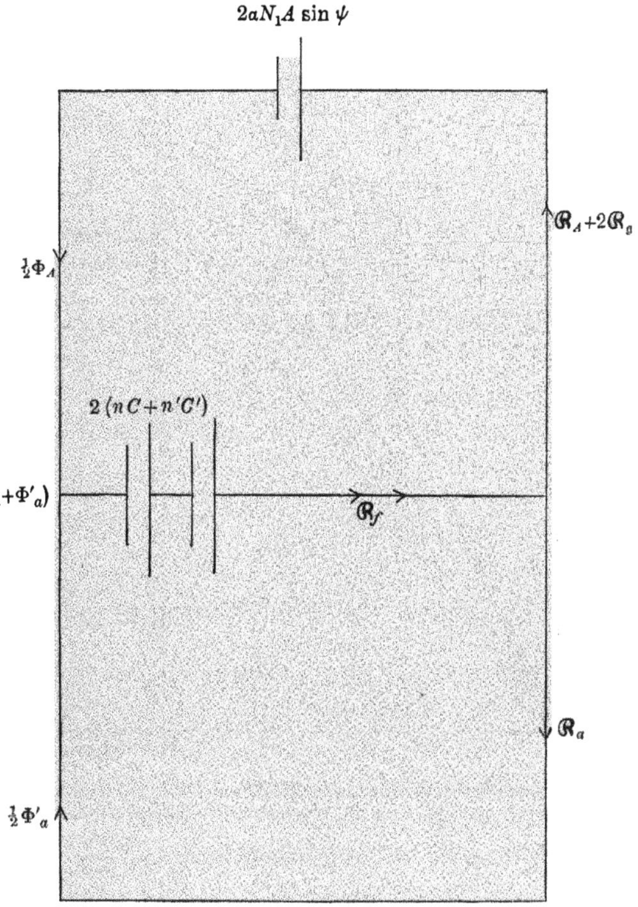

Fig. 25. $n'C'$ represents the ampere-turns per spool required to keep the flux in the armature constant. $aN_1A\sin\psi$ represents the demagnetising ampere-turns due to lagging currents in an armature coil.

Subtracting (1) from (2) we get

$$4n'C' - \mathcal{R}_f(\Phi_a' - \Phi_a) = 4aN_1A\sin\psi = \mathcal{R}_a(\Phi_a' - \Phi_a),$$

and therefore $$n'C' = \tfrac{1}{4}(\mathcal{R}_f + \mathcal{R}_a)(\Phi_a' - \Phi_a)$$
$$= \left(1 + \frac{\mathcal{R}_f}{\mathcal{R}_a}\right) a N_1 A \sin\psi.$$

Let us now suppose that $\mathcal{R}_f$ does not remain constant, and that its value is $\mathcal{R}_f + \Delta\mathcal{R}_f$ when the leakage flux is Φ_a'. In this case equations (2) must be written in the form

$$4(nC + n'C') - (\mathcal{R}_f + \Delta\mathcal{R}_f)(\Phi_A + \Phi_a') = (\mathcal{R}_A + 2\mathcal{R}_g)\Phi_A + 4aN_1A\sin\psi$$
$$= \mathcal{R}_a\Phi_a'.$$

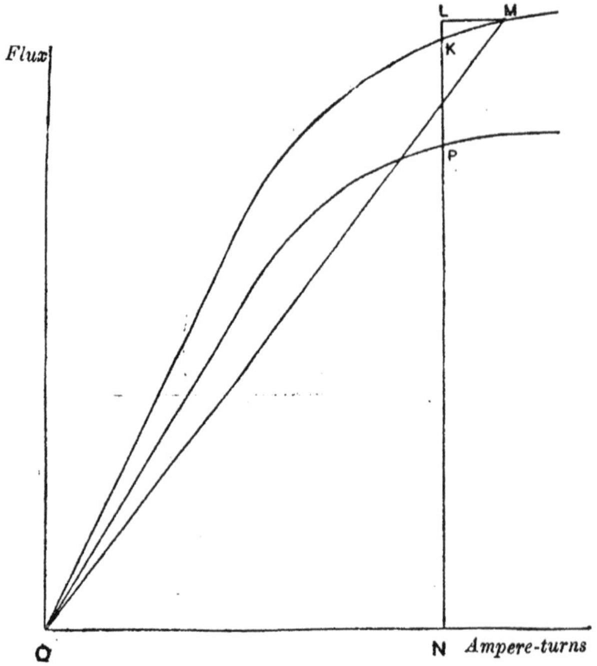

Fig. 26. The ordinates of the curves OKM and OP give the total field flux per pole and the flux per pole passing into the armature respectively.
$$KL = 4aN_1A\sin\psi/\mathcal{R}_a \text{ and } \cot MON = \mathcal{R}_f + \Delta\mathcal{R}_f + \frac{\mathcal{R}_a(\mathcal{R}_A + 2\mathcal{R}_g)}{\mathcal{R}_a + \mathcal{R}_A + 2\mathcal{R}_g}.$$

Hence, by means of (1), we get

$$4n'C' - (\mathcal{R}_f + \Delta\mathcal{R}_f)(\Phi_a' - \Phi_a) - \Delta\mathcal{R}_f(\Phi_A + \Phi_a) = 4aN_1A\sin\psi.$$

From the equations given above we easily find that

$$4aN_1A\sin\psi = \mathcal{R}_a(\Phi_a' - \Phi_a)$$

and $$4nC = (\Phi_A + \Phi_a)\left\{\mathcal{R}_f + \frac{\mathcal{R}_a(\mathcal{R}_A + 2\mathcal{R}_g)}{\mathcal{R}_a + \mathcal{R}_A + 2\mathcal{R}_g}\right\}.$$

Thus, by substituting for $(\Phi_a' - \Phi_a)$ and $(\Phi_A + \Phi_a)$ their values and simplifying, we get

$$n'C' = \left(1 + \frac{\mathcal{R}_f + \Delta\mathcal{R}_f}{\mathcal{R}_a}\right) a N_1 A \sin\psi$$
$$+ nC \,.\, \Delta\mathcal{R}_f \Big/ \left\{\mathcal{R}_f + \frac{\mathcal{R}_a(\mathcal{R}_A + 2\mathcal{R}_g)}{\mathcal{R}_a + \mathcal{R}_A + 2\mathcal{R}_g}\right\}.$$

We can find $\Delta\mathcal{R}_f$ from the open circuit characteristic by the following construction. The flux curve *OP* (Fig. 26) can be constructed from the open circuit characteristic when we know the form factor k of the wave of the electromotive force. Also if we can calculate $\mathcal{R}_a$, $\mathcal{R}_g$ and $\mathcal{R}_A$ we know ν, and thus by p. 32 we can construct the curve of total flux *OK* (Fig. 26). When the exciting ampere-turns are represented by *ON*, *NP* will be Φ_A and *NK* will be $\Phi_A + \Phi_a$. Therefore *PK* is equal to Φ_a. Make *KL* equal to $4aN_1A \sin\psi/\mathcal{R}_a$, which is equal to $\Phi_a' - \Phi_a$, then *PL* will be Φ_a'. Hence *N*L will represent the total flux in the field magnet, and if we draw *LM* parallel to *ON* to meet the curve of the total flux in *M*, the abscissa of the point *M* will give the ampere-turns required to produce the flux $\Phi_A + \Phi_a'$ in the field magnet. But

$$\cot MON = \frac{\text{ampere-turns}}{\text{flux}}$$
$$= \mathcal{R}_f + \Delta\mathcal{R}_f + \frac{\mathcal{R}_a(\mathcal{R}_A + 2\mathcal{R}_g)}{\mathcal{R}_a + \mathcal{R}_A + 2\mathcal{R}_g},$$

and hence, when $\mathcal{R}_f$ is known, $\Delta\mathcal{R}_f$ can be found.

The compensating ampere-turns required to keep the flux in the field magnets constant.

Let us suppose that, instead of keeping the magnetic flux through the armature constant, we maintain the flux in the field magnets constant. In this case, we can find a simple formula for the compensating ampere-turns per pole. Using the same notation as before, our equations are

$$4nC - \mathcal{R}_f(\Phi_A + \Phi_a) = (\mathcal{R}_A + 2\mathcal{R}_g)\,\Phi_A = \mathcal{R}_a\Phi_a,$$

and

$$4nC + 4n'C' - \mathcal{R}_f(\Phi_A' + \Phi_a')$$
$$= (\mathcal{R}_A + 2\mathcal{R}_g)\,\Phi_A' + 4aN_1A\sin\psi$$
$$= \mathcal{R}_a\Phi_a',$$

and, by hypothesis,

$$\Phi_A + \Phi_a = \Phi_A' + \Phi_a'.$$

Therefore
$$\begin{aligned} 4n'C' &= (\mathcal{R}_A + 2\mathcal{R}_g)(\Phi_A' - \Phi_A) + 4aN_1A\sin\psi \\ &= -(\mathcal{R}_A + 2\mathcal{R}_g)(\Phi_a' - \Phi_a) + 4aN_1A\sin\psi \\ &= \mathcal{R}_a(\Phi_a' - \Phi_a) \\ &= \frac{\mathcal{R}_a}{\mathcal{R}_A + \mathcal{R}_a + 2\mathcal{R}_g} 4aN_1A\sin\psi \end{aligned}$$

and thus
$$n'C' = \frac{a}{\nu} N_1A\sin\psi,$$

where ν is the leakage coefficient.

We also have
$$\Phi_A' - \Phi_A = -\frac{4aN_1A\sin\psi}{\mathcal{R}_A + \mathcal{R}_a + 2\mathcal{R}_g}.$$

Thus, in order to prevent the flux in the field magnets falling below its no-load value, when the current flowing in the armature windings is A and the power factor is $\cos\psi$, the ampere-turns acting on a field magnet must be increased by $aN_1A\sin\psi/\nu$. The flux, however, entering the armature from a pole will be diminished by $4aN_1A\sin\psi/(\mathcal{R}_A + \mathcal{R}_a + 2\mathcal{R}_g)$.

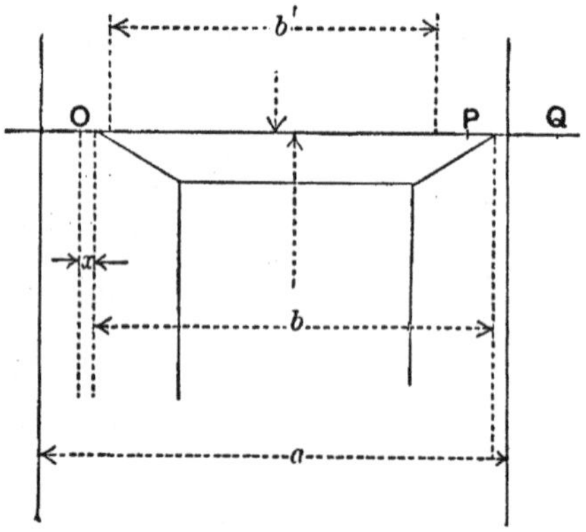

Fig. 27. Relative positions of the pole and the armature coil when x lies between 0 and $\frac{1}{2}(b - b')$.

Transverse magnetisation of the field.

We shall now consider the effect of the component $I\cos\psi\sin\omega t$ of the armature current which is in phase with the electromotive force. This component produces a transverse magnetisation of the field magnets, so that, in rotating field machines, the field in the air-gap under the leading polar horn, that is the leading end of the polar piece, is weakened. and that under the trailing horn is strengthened by this component of the current. When the armature rotates, a similar distortion of the field is produced; in this case, however, the other end of the polar piece is the more strongly magnetised, as the effect is the same as if the armature were at rest and the poles rotated in the opposite direction. In order to get a measure of this distorting effect, we will find a formula for the difference between the magnetising ampere-turns due to this current acting on the fluxes in the two sides of a field magnet pole.

Let the arc intercepted by the polar flux on the circumference of the armature be b, and suppose that this arc is greater than b', the distance, measured along the circumference, between the axes of two slots in the armature in each of which there are N_1 conductors. If a be the polar step, we can express the component of the current which is in phase with the E.M.F. in the form $I \cos\psi \sin \pi x/a$, where x (Fig. 27) is the distance of the end of the trailing flux from a fixed point O on the armature. We shall consider the difference of the effective magnetising ampere-turns of the coil, acting on the fluxes coming from the leading and the lagging half of the polar face of a field magnet, as this difference will be a measure of the distorting forces acting on the field. We shall find expressions for this difference during the quarter of a period, starting from the instant when it is zero. It is to be noticed that the pulsations of the cross magnetising force on the poles go through all their values in the half of a period.

Let us first suppose that x (Fig. 27) is less than $(b - b')/2$. In this case, the armature coil b' remains inside the polar flux. The effective number of ampere-turns acting on the flux traversing the trailing half of the polar face will be

$$(b'/2 + x)/b \,.\, N_1 I \cos\psi \sin(\pi x/a),$$

and the effective number of ampere-turns acting on the flux traversing the leading half of the polar face will be

$$(b'/2 - x)/b \,.\, N_1 I \cos\psi \sin(\pi x/a).$$

The difference, therefore, between the magnetising turns on each half of the polar flux will be

$$\frac{1}{b}\left\{\frac{b'}{2} + x - \left(\frac{b'}{2} - x\right)\right\} N_1 I \cos\psi \sin\frac{\pi x}{a} = \frac{1}{b}(2x)\, N_1 I \cos\psi \sin\frac{\pi x}{a}.$$

Now, in the figure, we have made b' greater than $a - \frac{1}{2}b$, and thus $b'/2$ is greater than $a - \frac{1}{2}(b + b')$, hence, when x (Fig. 28) is less

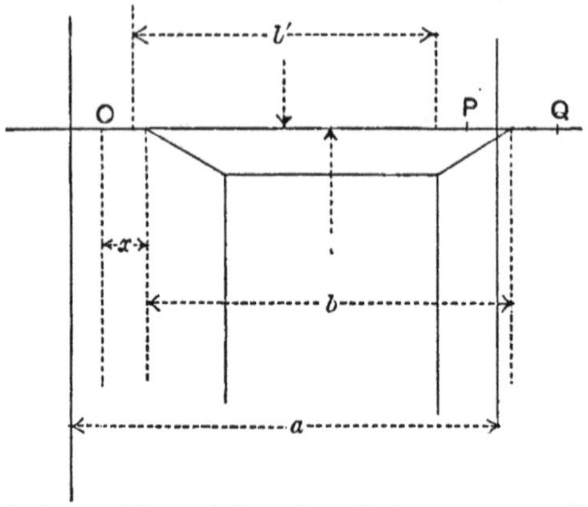

Fig. 28. Relative positions of the pole and the armature coil when x lies between $\frac{1}{2}(b - b')$ and $a - \frac{1}{2}(b + b')$.

than $a - \frac{1}{2}(b + b')$ but is greater than $(b - b')/2$, the difference of the number of ampere-turns

$$= \frac{1}{b}\left\{\frac{b}{2} - \left(\frac{b'}{2} - x\right)\right\} N_1 I \cos\psi \sin\frac{\pi x}{a}$$

$$= \frac{1}{b}\{x + \tfrac{1}{2}(b - b')\} N_1 I \cos\psi \sin\frac{\pi x}{a}.$$

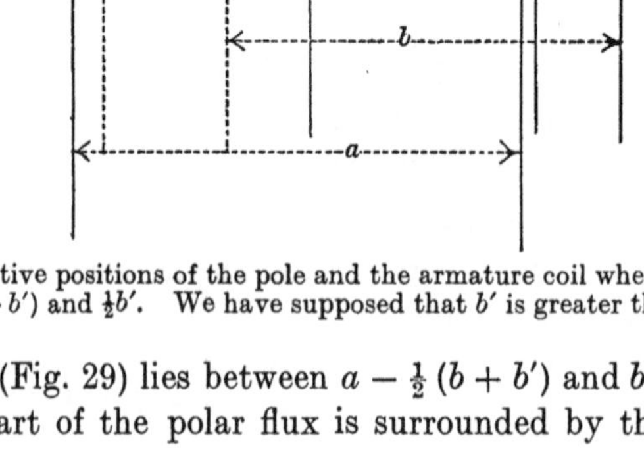

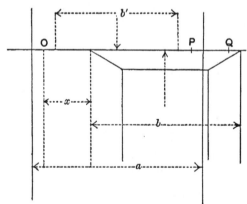

Fig. 29. Relative positions of the pole and the armature coil when x lies between $a - \frac{1}{2}(b + b')$ and $\frac{1}{2}b'$. We have supposed that b' is greater than $a - \frac{1}{2}b$.

When x (Fig. 29) lies between $a - \frac{1}{2}(b + b')$ and $b'/2$, it will be seen that part of the polar flux is surrounded by the current in the next coil, a side of which passes down the slot at Q, and the magnetising ampere-turns due to this coil act in the opposite direction to those of the first coil. Hence the difference of the ampere-turns

$$=\frac{1}{b}\left\{\frac{b}{2}-\left(\frac{b'}{2}-x\right)+x-\left(a-\frac{b+b'}{2}\right)\right\} N_1 I \cos\psi \sin\frac{\pi x}{a}$$

$$=\frac{1}{b}(2x-a+b) N_1 I \cos\psi \sin\frac{\pi x}{a}.$$

When x lies between $b'/2$ and $a/2$ (Fig. 30), the difference of the ampere-turns is

$$\frac{1}{b}\left\{\frac{b}{2}-\left(x-\frac{b'}{2}\right)+x-a+\frac{b+b'}{2}\right\} N_1 I \cos\psi \sin\frac{\pi x}{a}$$

or

$$\frac{1}{b}(b+b'-a) N_1 I \cos\psi \sin\frac{\pi x}{a}.$$

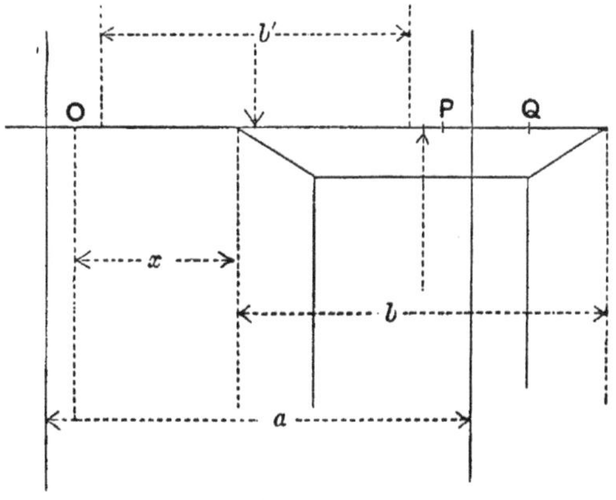

Fig. 30. Relative positions of the pole and the armature coil when x lies between $\frac{1}{2}b'$ and $\frac{1}{2}a$.

Hence, if $\beta N_1 A \cos\psi$, where A is the effective current, denote the mean value of the difference of the ampere-turns acting on each half of the polar flux over the quarter of a period, we have

$$\frac{ab}{2\sqrt{2}}\beta=\int_0^{\frac{b-b'}{2}} 2x\sin\frac{\pi x}{a}\,\partial x+\int_{\frac{b-b'}{2}}^{a-\frac{b+b'}{2}}\left(x+\frac{b-b'}{2}\right)\sin\frac{\pi x}{a}\,\partial x$$

$$+\int_{a-\frac{b+b'}{2}}^{\frac{b'}{2}}(2x-a+b)\sin\frac{\pi x}{a}\,\partial x+(b+b'-a)\int_{\frac{b'}{2}}^{\frac{a}{2}}\sin\frac{\pi x}{a}\,\partial x.$$

4—2

Therefore

$$\beta = \frac{4\sqrt{2}a}{\pi^2 b} \sin \frac{\pi b'}{2a}\left(1 - \cos\frac{\pi b}{2a}\right).$$

It is easy to see that the mean value of $\beta N_1 A \cos\psi$ over the whole period is the same as over the quarter of the period. This expression, therefore, gives us the mean value of the magnetising turns of the armature current which act so as to distort the magnetic field in the air-gap. It is due to C. F. Guilbert.

We can also show that the above formula for $\beta N_1 A \cos\psi$ is true when the distance b' between the axes of the slots is greater than the breadth b of the polar flux entering the armature, and also when b' is less than $a - \frac{1}{2}b$. It is therefore always true.

Let us now consider how α and β vary with the breadth of the polar flux and with the breadth of the coils. We have shown that

$$\alpha = \frac{4\sqrt{2}a}{\pi^2 b} \sin \frac{\pi b'}{2a} \sin \frac{\pi b}{2a},$$

and

$$\beta = \frac{4\sqrt{2}a}{\pi^2 b} \sin \frac{\pi b'}{2a}\left(1 - \cos\frac{\pi b}{2a}\right),$$

where b is the breadth of the arc intercepted on the armature by the flux leaving a pole, and b' is the breadth, measured along the circumference, of an armature coil. The greater the value of b', as long as it does not exceed the polar pitch a, the greater will be the values of α and β. We can see also that the greater the value of b, that is, the broader the poles, the greater will be the values of both α and β. Again, we have

$$\frac{\beta}{\alpha} = \tan\frac{\pi b}{4a},$$

and thus, the broader the poles the greater will be the ratio of the transverse magnetising coefficient β to the direct magnetising coefficient α.

In the following table the values of α and β for various values of b/a are given for the case when the breadth of the coil equals the polar pitch, as, for example, in a simple wave winding.

In this case, we have

$$\alpha = \frac{4\sqrt{2}}{\pi^2}\frac{a}{b}\sin\frac{\pi b}{2a} = 0{\cdot}573\frac{a}{b}\sin\frac{\pi b}{2a},$$

and
$$\beta = \alpha\tan\frac{\pi b}{4a}.$$

$\frac{b}{a}$	1	0·9	0·8	0·7	0·65	0·6	0·55	0·5
α	0·573	0·629	0·681	0·730	0·752	0·773	0·792	0·811
β	0·573	0·537	0·495	0·447	0·421	0·394	0·365	0·336

When the breadth b' of the coil is not equal to the polar pitch a, we have to multiply the values of α and β given in the preceding table by $\sin(\pi b'/2a)$. The following table shows how $\sin(\pi b'/2a)$ varies with the ratio of b' to a.

$\frac{b'}{a}$	1	0·9	0·8	0·7	0·65	0·6	0·55	0·5
$\sin\frac{\pi b'}{2a}$	1	0·988	0·951	0·891	0·853	0·809	0·760	0·707

Numerical example.

Let us suppose that the distance between the axes of the slots is 16 inches and that the breadth of the polar flux entering the armature is 18 inches. Let the polar step be 20 inches, the number of conductors in a slot 48 and the effective value of the current, which we assume follows the harmonic law, 100 amperes. Then, in our notation,

$$a = 20,\quad b = 18,\quad b' = 16,\quad N_1 = 24 \text{ and } A = 100.$$

The ampere-turns $\alpha N_1 A \sin\psi$ acting on the direct flux

$$= \frac{4\sqrt{2}a}{\pi^2 b} N_1 A \sin\psi \sin\frac{\pi b'}{2a}\sin\frac{\pi b}{2a}$$

$$= 1435 \sin\psi.$$

The ampere-turns $\beta N_1 A\cos\psi$ acting on the transverse flux

$$= \alpha N_1 A\cos\psi\tan\frac{\pi b}{4a}$$

$$= 1225\cos\psi.$$

The following table gives the values of $aN_1A \sin \psi$ and $\beta N_1 A \cos \psi$ for various power factors:

0	0·1	0·2	0·3	0·4	0·5	0·6	0·7	0·8	0·9
90	84·3	78·5	72·5	66·4	60	53·1	45·6	36·9	25·8
1435	1428	1406	1369	1314	1243	1149	1024	860·6	625·6
0	122·5	245	367·5	490	612·5	735	857·5	980	1102·5

Load characteristics.

If we vary the load connected with an alternator which is running at constant speed, and if the exciting current in the windings of the field magnets be constant, we find that the potential difference between the terminals of the machine varies with the load. Let the load in the external circuit be varied, always keeping the power factor constant and equal to $\cos \psi'$, and let simultaneous readings of the voltage at the terminals of the machine and of the current in the external circuit be taken. If we plot these readings in a curve, having the terminal voltages for ordinates and the currents in amperes for abscissae, we get the load characteristic for a power factor of $\cos \psi'$.

Regulation.

Let V' be the terminal voltage of an alternator at a given load and power factor and let V be the terminal voltage when the load is switched off, the speed and excitation of the machine being adjusted so that it is the same in the two cases. The regulation of the machine at the given load and power factor is defined to be the ratio of the voltage rise to the initial voltage. In symbols, the percentage regulation is, therefore, $100 (V - V')/V'$. For example, if the regulation for a given load is 20 per cent., the voltage rise when this load is switched off, the excitation and speed being made the same, will be 20 per cent.

Methods of measuring the regulation.

There are three methods in everyday use for finding the regulation of an alternator at a given load and power factor. In one method the voltages V' and V are found directly by experiment, care being taken when reading V to adjust both the speed and the excitation to the values they had when V' was read. The regulation $(V - V')/V'$ can thus be computed. Owing to the large amount of power required to run a large generator at full load, it is often impossible to measure the regulation directly in the factory. In this case it can be found approximately by either of the following methods.

In the first of these methods we obtain experimentally the open circuit characteristic and the full load characteristic at zero power factor (Fig. 31). For this purpose very little power is required. The zero power factor characteristic can be found with

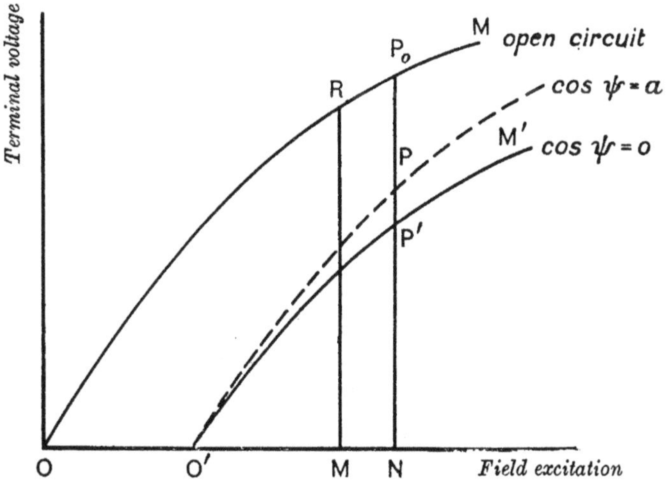

Fig. 31. Characteristic curves.

sufficient accuracy by running the alternator on a load of under excited synchronous motors (Chapter VI) connected in parallel and unloaded. The curve $O'M'$ obtained under these conditions is practically indistinguishable from the zero power factor characteristic. In the figure when the excitation is ON the terminal voltage on open circuit is NP_0 and on full load current at zero power factor it is NP'.

We have now to find what the terminal voltage PN is when the power factor is a, where $a = \cos\psi$. Let us first suppose that when the armature is short circuited the losses in it are so small that we can assume that the E.M.F. in it is in quadrature with the current passing through it. In this case the vector diagram (Fig. 32) of the E.M.F.s in the armature circuit can be easily constructed as follows.

In Fig. 32 draw the line EBN and mark off $EB = P'P_0$. Make the angle $NBO = \pi/2 - \psi$ and with centre E and radius equal to NP_0 draw a circle cutting BO in O. Then ON drawn perpendicular to EN gives the phase of the current and OB gives the voltage at the given current and power factor.

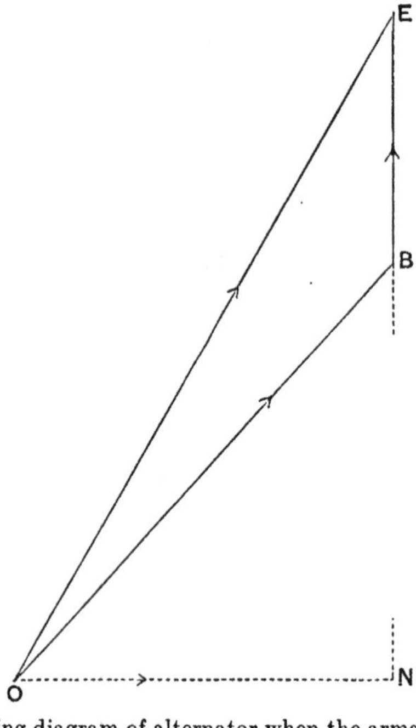

Fig. 32. Working diagram of alternator when the armature resistance can be neglected.

When the losses in the armature are appreciable we can take them into account when the angle of lag ψ_a between the armature current and E.M.F. is known. The construction is shown in Fig. 33.

As formerly, $EB = P'P_0$, the angle $EBM = \psi_a$ and so the right-angled triangle EBM can be constructed. Then drawing BO making an angle $\pi/2 - \psi$ with EM and making $EO = P_0N$, we find the terminal voltage OB. Finding in this way the terminal voltage for various excitations we can construct the characteristic for the given current and power factor $\cos\psi$.

The second method is only used when it is impossible to obtain the zero power factor characteristic. In this case the point O'

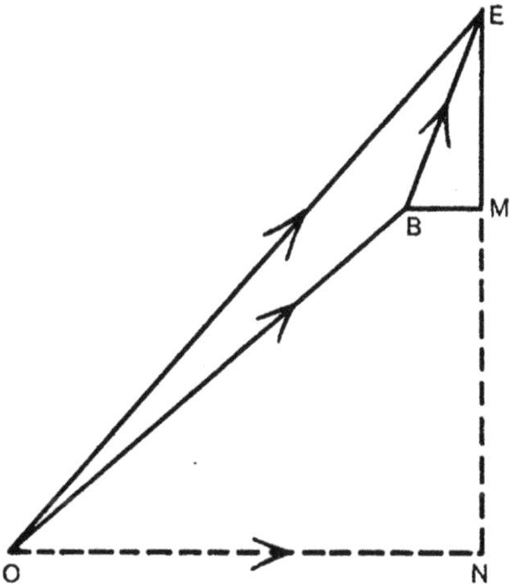

Fig. 33. Working diagram of alternator.

(Fig. 34) is obtained from the short circuit characteristic and the curve $O'M'$ is obtained by moving the curve OM parallel to itself

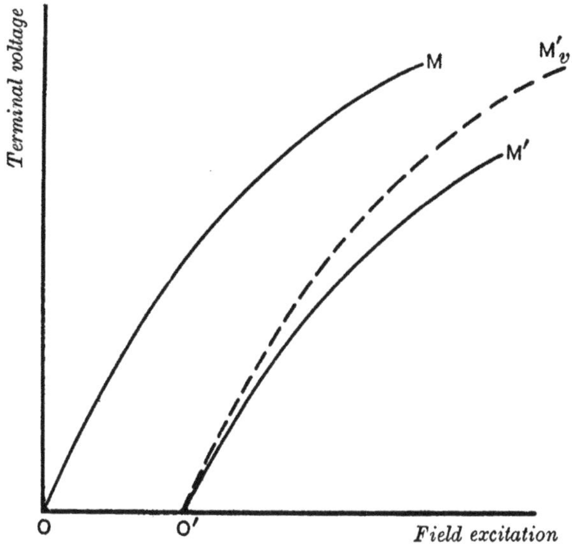

Fig. 34.

through a distance OO'. The rest of the construction is the same as for the method described above. If the machine have not low saturation and low reactance this method is inaccurate. In high speed machines as turbo-alternators or high speed water wheel generators it is found that the zero power factor characteristic lies quite close to $O'M'$. In many cases however the true zero power factor characteristic $O'M_v'$ deviates from $O'M'$ and the higher the saturation, the reactance or the leakage, the greater will be the deviation.

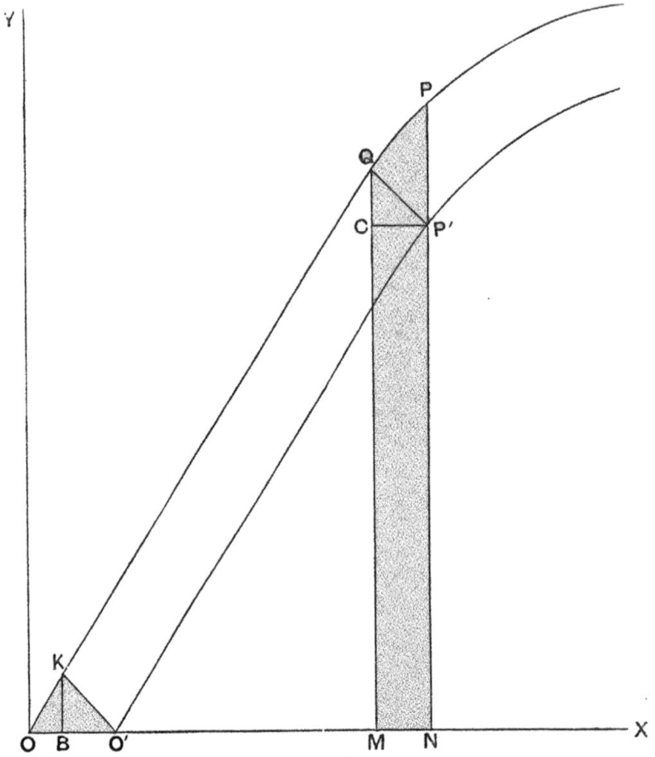

Fig. 35. *OP* is the open circuit characteristic. *PN* gives the terminal volts when the magnetising turns of the field magnets are represented by *ON*. *O'P'* is the characteristic on a wattless inductive load when the current in the armature is maintained constant. $KB = QC$ = the armature leakage volts, $BO' = CP'$ = the demagnetising turns of the armature current.

The methods described above which depend on the open circuit and zero power factor characteristics only give approximate results. One reason of this is that the armature reaction the value of which depends on $\cos\psi$ is only partially taken into account. We shall now give a more rigorous graphical method of constructing the zero power factor characteristic for a current A taking armature reaction into account, but neglecting changes in the value of ν.

Let OO' (Fig. 35) be equal to the abscissa corresponding to the ordinate A on the short circuit characteristic. Calculate the demagnetising ampere-turns aN_1A by Guilbert's formula and make $O'B$ equal to aN_1A. Erect the ordinate BK and join $O'K$, then BK will be very approximately equal to the E.M.F. required to drive the current through the armature. Now draw any ordinate QM and make QC equal to BK. Draw CP' parallel to OX and equal to BO'. Then P' will be a point on the wattless characteristic through O'. To prove this we notice that when the field magnet excitation is ON, the real excitation is

$$ON - aN_1A = ON - MN = OM,$$

and MQ is the E.M.F. generated at this excitation. The external potential difference is $QM - BK$ and this equals $P'N$. Similarly we can find other points on the wattless characteristic and so construct the curve. Since QP' is equal and parallel to KO', it will be seen that if we move the open circuit characteristic through a distance equal and parallel to KO' it will coincide with the wattless characteristic.

Theory of the alternator.

In deducing the method of armature reactions one of the assumptions made was that the exciting current remained practically constant. In other words we neglected the current produced in the exciting circuit by the varying and moving flux due to the armature current which is linked with the field circuit. It is known however that when the armature reaction is great an appreciable alternating current component flows in the exciting circuit. We shall therefore now discuss the armature reaction more rigorously.

In order to simplify the theory as much as possible we shall assume that the magnetic circuit of the alternator has constant permeability. We shall also assume that the armature rotates with constant angular velocity ω and that its inductance L_a and resistance R_a are constant. The ends of the armature coil are connected with slip rings mounted on the shaft and we suppose that a fixed inductive coil (R_e, L_e) representing the load is placed between the connecting brushes pressing on these slip rings.

We shall denote the total resistance and inductance of the armature circuit by R and L respectively, so that

$$R = R_a + R_e \text{ and } L = L_a + L_e.$$

Let us suppose also that a fixed coil (R_2, L_2) carries the exciting current for the magnetic field and that it is connected across a battery of accumulators of negligible resistance and constant electromotive force E.

When the armature rotates, the self-inductance L of the armature circuit remains constant, but the value of the mutual inductance m between it and the field circuit continually alters.

The electromotive force induced in the armature is due to the periodic variation in the value of mi' where i' is the current in the field circuit. There is also an electromotive force induced in the field circuit due to the variation in the value of mi where i is the current in the armature circuit.

By the definitions of self and mutual inductance it follows that the linkages of magnetic flux and current in the armature and field circuit are $Li + mi'$ and $L_2i' + mi$ respectively. Hence by Ohm's and Faraday's laws we get

$$Ri + \frac{\partial}{\partial t}(Li + mi') = 0 \qquad \text{(3)},$$

and

$$R_2 i' + \frac{\partial}{\partial t}(L_2 i' + mi) = E \qquad \text{(4)}.$$

It readily follows from equations (3) and (4) that

$$Ei' + ii' \frac{\partial m}{\partial t} = Ri^2 + R_2 i'^2 + \frac{\partial}{\partial t}(\tfrac{1}{2}Li^2 + mii' + \tfrac{1}{2}L_2 i'^2).$$

By Joule's law the first two terms on the right-hand side of this equation give the power expended in heating the two coils and the last term gives the rate at which energy is being stored in the electromagnetic field. The term Ei' gives the output of the battery at the time t and thus, by the principle of the conservation of energy, $ii'\,(\partial m/\partial t)$ must be the value of the mechanical power given by the prime mover to the armature at the same instant.

As our object is to simplify the theory as much as possible we shall suppose that $m = M \cos \omega t$, the time being reckoned from the instant when m has its maximum value. In actual alternators the higher harmonics in m are always present owing to the hysteresis in the iron and the slots in the armature. We can imagine however an ideal machine for which this assumption would be permissible. The armature, for instance, may consist of a circle of wire rotating about a diameter which is perpendicular to the magnetic field produced inside a long cylindrical coil the terminals of which are connected with the battery. Equations (3) and (4) may now be written

$$Ri + L\frac{\partial i}{\partial t} + M\frac{\partial}{\partial t}(i' \cos \omega t) = 0 \quad \ldots\ldots\ldots\ldots\ldots(5),$$

and

$$R_2 i' + L_2\frac{\partial i'}{\partial t} + M\frac{\partial}{\partial t}(i \cos \omega t) = E \quad \ldots\ldots\ldots\ldots\ldots(6).$$

When t becomes $t + T/2$, $\cos \omega t$ becomes $-\cos \omega t$. In order therefore that these equations may still retain the same form, it is necessary that i becomes $-i$ and i' remains i'. Hence i is a function of t which changes sign when t becomes $t + T/2$, but i' remains unaltered when this happens. Hence the expression for the armature current can contain only odd harmonics and the expression for the field current only even harmonics.

The solution of the equations.

If we write $i' = C + i_2$, where $C = E/R_2$, equations (5) and (6) become

$$Ri + L\frac{\partial i}{\partial t} + M\frac{\partial}{\partial t}(i_2 \cos \omega t) = M\omega C \sin \omega t \quad \ldots\ldots(7),$$

and

$$R_2 i_2 + L_2\frac{\partial i_2}{\partial t} + M\frac{\partial}{\partial t}(i \cos \omega t) = 0 \quad \ldots\ldots\ldots\ldots\ldots\ldots(8).$$

Since i_2 is generally small, equation (7) suggests writing

$$i = (M\omega C/Z_1)\sin(\omega t - a_1) + i_3,$$

where

$$Z_1{}^2 = R^2 + \omega^2 L^2 \text{ and } \tan a_1 = \omega L/R.$$

Substituting this value for i in (7) and (8) we get

$$Ri_3 + L\frac{\partial i_3}{\partial t} + M\frac{\partial}{\partial t}(i_2 \cos \omega t) = 0 \quad \ldots\ldots\ldots\ldots(9)$$

and

$$R_2 i_2 + L_2\frac{\partial i_2}{\partial t} + M\frac{\partial}{\partial t}(i_3 \cos \omega t) = -\frac{(M\omega)^2 C}{Z_1}\cos(2\omega t - a_1) \ldots(10).$$

We next assume that

$$i_2 = -\frac{(M\omega)^2 C}{Z_1 Z_2}\cos(2\omega t - a_1 - a_2) + i_4,$$

where $\quad Z_2^2 = R_2^2 + (2\omega)^2 L_2^2$ and $\tan a_2 = 2\omega L_2/R_2$.

Hence substituting in (9) we get

$$Ri_3 + L\frac{\partial i_3}{\partial t} + M\frac{\partial}{\partial t}(i_4 \cos \omega t) = -\frac{(M\omega)^3 C}{2Z_1 Z_2}\{3 \sin (3\omega t - a_1 - a_2) + \sin (\omega t - a_1 - a_2)\} \quad \text{...(11)}.$$

The next assumption we make is

$$i_3 = -\frac{\frac{3}{2}(M\omega)^3 C}{Z_1 Z_2 Z_3} \sin (3\omega t - a_1 - a_2 - a_3) - \frac{(M\omega)^3 C}{2Z_1^2 Z_2} \sin (\omega t - 2a_1 - a_2) + i_5 \quad \text{.........(12)}.$$

Proceeding in this way we see how i and i' can be obtained in terms of infinite series. It will be seen that the series we have deduced for these currents verify that i contains only odd harmonics and that i' contains only even harmonics.

When $(M\omega)^2$ is small compared with $Z_1 Z_2$ the series converge rapidly. We shall now assume that i_4, i_6, ... can be neglected compared with C and that i_5, i_7, ... can be neglected compared with i_1. The approximate values of i and i' may be written as follows:

$$i = \frac{M\omega C}{Z_1} B \sin (\omega t - a_1 + \beta) - \frac{\frac{3}{2}(M\omega)^3 C}{Z_1 Z_2 Z_3} \sin (3\omega t - a_1 - a_2 - a_3) \quad \text{...(13)}$$

and

$$i' = C - \frac{(M\omega)^2 C}{Z_1 Z_2} \cos (2\omega t - a_1 - a_2) \quad \text{......................(14)},$$

where

$$B^2 = 1 - \frac{(M\omega)^2}{Z_1 Z_2} \cos (a_1 + a_2) + \frac{(M\omega)^4}{4Z_1^2 Z_2^2} \quad \text{....................(15)},$$

and

$$\tan \beta = \frac{(M\omega)^2 \sin (a_1 + a_2)}{2Z_1 Z_2 - (M\omega)^2 \cos (a_1 + a_2)} \quad \text{............................(16)}.$$

If ϕ denote the flux linked with the armature circuit we have

$$\begin{aligned}\phi &= Li + M \cos \omega t \,.\, i' \\ &= LIB \sin (\omega t - a_1 + \beta) + MC \cos \omega t \\ &\quad - MI\,(M\omega/2Z_2)\,\{\cos (3\omega t - a_1 - a_2) + \cos (\omega t - a_1 - a_2)\} \\ &\quad - 3LI\,(M^2\omega^2/2Z_2 Z_3) \sin (3\omega t - a_1 - a_2 - a_3),\end{aligned}$$

where $I = M\omega C/Z_1$, so that IB equals the amplitude of the fundamental harmonic of the armature current.

If $M^2\omega^2/Z_1Z_2$ can be neglected compared with unity, $B = 1$, $\beta = 0$, and we get as formerly (p. 37)

$$\phi = (MC - LI \sin \alpha_1) \cos \omega t + LI \cos \alpha_1 \sin \omega t.$$

If we take the third harmonic into account, we have

$$\begin{aligned}\phi = &\{MC - LBI \sin(\alpha_1 - \beta) - MI\,(M\omega/2Z_2) \cos(\alpha_1 + \alpha_2)\} \cos \omega t \\ &+ \{LBI \cos(\alpha_1 - \beta) + MI\,(M\omega/2Z_2) \sin(\alpha_1 + \alpha_2)\} \sin \omega t \\ &- \{MI\,(M\omega/2Z_2) - 3LI\,(M^2\omega^2/2Z_2Z_3) \sin \alpha_3\} \cos(3\omega t - \alpha_1 - \alpha_2) \\ &- 3LI\,(M^2\omega^2/2Z_2Z_3) \cos \alpha_3 \sin(3\omega t - \alpha_1 - \alpha_2) \quad \text{............(17)}.\end{aligned}$$

From the definitions of α_1 and α_2 we see that neither of them can be greater than $\pi/2$. In practice we can assume without appreciable error that α_2 is $\pi/2$. Let us first consider the case when the armature circuit is purely inductive so that α_1 is also $\pi/2$. In general this is approximately true when the armature is short circuited. We see that in this case $I = (M/L)\,C$,

$$(M\omega)^2/(Z_1Z_2) = M^2/(2LL_2), \quad (M\omega)^2/Z_2Z_3 = M^2/(6LL_2),$$

and by (15) and (16), $B = 1 + M^2/(4LL_2)$ and $\beta = 0$. Hence we find that

$$i = -I\,\{1 + M^2/(4LL_2)\} \cos \omega t - I\,\{M^2/(4LL_2)\} \cos 3\omega t \text{... (18)},$$

$$i' = C\,[1 + \{M^2/(2LL_2)\} \cos 2\omega t] \quad \text{..............................(19)},$$

and

$$\phi = 0 \quad \text{..............................(20)}.$$

We see that the flux due to self-induction is exactly neutralised by the flux due to mutual induction. If A be the effective value of the short circuit current, we have

$$A = (I/\sqrt{2})\,\{1 + M^2/(2LL_2) + M^4/(8L^2L_2^2)\}^{\frac{1}{2}}.$$

The current on short circuit is therefore greater than $I/\sqrt{2}$, the current obtained on the assumption that i' is constant and equal to C. We also see that the disturbance of the exciting current i' is greater, the greater the value of $M^2/(2LL_2)$.

When a condenser of capacity K is placed across the terminals of the machine, we have

$$\tan \alpha_1 = \omega L/(R_a + R_e), \text{ where } L = L_a - 1/(K\omega^2),$$

and $$\tan \alpha_3 = 3\omega L'/(R_a + R_e), \text{ where } L' = L_a - 1/\{K\,(3\omega)^2\}.$$

Hence if $R_a + R_e$ be zero and $L_aK\,(3\omega)^2$ be less than unity, both

a_1 and a_3 will equal $-\pi/2$. Assuming as before that a_2 equals $\pi/2$, we get

$$B = 1 + M^2/(4LL_2), \quad \beta = 0, \quad I = -(M/L)\,C$$

and $$(M\omega)^2/(Z_2Z_3) = -M^2/(6L'L_2).$$

Hence

$$i = I\,\{1 + M^2/(4LL_2)\}\cos\omega t + I\,\{M^2/(4L'L_2)\}\cos 3\omega t \quad \text{...(21)},$$

$$i' = C\,[1 + \{M^2/(2LL_2)\}\cos 2\omega t] \quad \text{...(22)},$$

and

$$\phi = -\{M^2I/(4L_2)\}\,(1 - L/L')\cos 3\omega t \quad \text{...(23)}.$$

In this case the first harmonic component of the flux due to self-induction is exactly neutralised by the flux due to mutual induction. In the formulae given above, both L and L' are negative quantities. We see that the amplitude of the first harmonic component of the current is less than it would be if i' were constant and equal to C.

The potential difference at the terminals.

If v denote the instantaneous value of the potential difference at the terminals, we have

$$v = \left(R_e + \mathrm{L}_e\frac{\partial}{\partial t}\right) i,$$

where i is given by (13). Hence denoting $R_e^2 + \omega^2L_e^2$ by Z_e^2, and $R_e^2 + (3\omega)^2L_e^2$ by $Z_e'^2$, and writing $\tan\psi = \omega L_e/R_e$ and $\tan\psi' = (3\omega)\,L_e/R_e$, we get

$$v = M\omega C\,(Z_e/Z_1)\,B\sin(\omega t - a_1 + \psi + \beta)$$
$$- \frac{3\,(M\omega)^3\,C}{2Z_1Z_2}\cdot\frac{Z_e'}{Z_3}\sin(3\omega t - a_1 - a_2 - a_3 + \psi') \quad \text{...(24)}.$$

If therefore V be the effective value of the potential difference between the terminals of the alternator, and we write $I = M\omega C/Z_1$, we get

$$V^2 = \left(\frac{I}{\sqrt{2}}Z_e\right)^2\left\{1 - \frac{(M\omega)^2}{Z_1Z_2}\cos(a_1 + a_2) + \frac{(M\omega)^4}{4Z_1^2Z_2^2}\right\}$$
$$+ \left(\frac{I}{\sqrt{2}}Z_e'\right)^2\frac{9M^4\omega^4}{4Z_2^2Z_3^2}.$$

It is to be noticed that $I/\sqrt{2}$ is the effective value of the current in the armature obtained on the assumption that i' remained constant and equal to C as the armature rotated. In this case the value of V would be $(I/\sqrt{2})\,Z_e$.

Hence since in practice $\alpha_1 + \alpha_2$ is usually greater than 90° so that $\cos(\alpha_1 + \alpha_2)$ is a negative quantity we see that the true value of the current for a given external voltage is less than that obtained on the assumption that there is no armature reaction.

Theoretical characteristics.

It will have been seen that the cross magnetising and demagnetising effects of the armature currents considerably complicate the problem of predetermining the pressure drop at the terminals of an alternator for a given load. In some machines, however, these effects are not large, and so it is instructive to find the relations between the various voltages and the current, on the assumption that these effects are negligible. The shapes of the curves obtained, on this assumption, are similar to those obtained by experiment on most forms of alternator.

Let us suppose that v is the instantaneous value of the E.M.F. generated in the armature, v' the external potential difference and v_a the E.M.F. required to drive the current i through the armature. We have $v = v_a + v'$ at every instant, and thus

$$v^2 = v_a^2 + 2v_a v' + v'^2.$$

Hence by taking mean values we get

$$V^2 = V_a^2 + 2V_a V' \cos(\psi_a - \psi_e) + V'^2 \quad \text{.........(i)},$$

where ψ_a is the phase difference between v_a and i, ψ_e is the phase difference between v' and i, and we are making sine wave assumptions.

If Z_a denote the impedance of the armature and A be the effective value of the armature current we have $V_a = Z_a A$. Hence from (i)

$$V^2 = \{V' + Z_a A \cos(\psi_a - \psi_e)\}^2 + Z_a^2 A^2 \sin^2(\psi_a - \psi_e)$$

and hence

$$V - V' = Z_a A \cos(\psi_a - \psi_e) + \{Z_a^2 A^2 / 2V\} \sin^2(\psi_a - \psi_e) \quad \text{approximately ...(ii)},$$

when V is great compared with $Z_a A$.

In many cases we can write $\psi_a = 90°$, and so

$$V - V' = Z_a A \sin\psi_e + \{Z_a^2 A^2 / 2V\} \cos^2\psi_e, \text{ approximately ...(iii)}.$$

On a non-inductive load $\psi_e = 0$ and thus

$$V - V' = (Z_a A)^2 / (2V).$$

When $\psi_e = 45°$, we have

$$V - V' = Z_a A/\sqrt{2} + (Z_a A)^2/(4V),$$

and when $\psi_e = 90°$, we have

$$V - V' = Z_a A.$$

On a non-inductive load, therefore, the voltage drop increases as the square of the current, and on an inductive load it increases as the current.

If we denote the angle of phase difference between v and i by ψ, then since $v^2 = vv' + vv_a$ we get by taking mean values

$$V = V' \cos(\psi - \psi_e) + Z_a A \cos(\psi_a - \psi).$$

Similarly we have

$$V' = V \cos(\psi - \psi_e) - Z_a A \cos(\psi_a - \psi_e).$$

Hence by subtracting the two equations we get

$$V - V' = Z_a A \frac{\cos(\psi_a - \psi) + \cos(\psi_a - \psi_e)}{1 + \cos(\psi - \psi_e)}$$

$$= Z_a A \frac{\cos\{\psi_a - (\psi + \psi_e)/2\}}{\cos\{(\psi - \psi_e)/2\}} \quad \text{..............(iv).}$$

In this equation $\tan \psi_a = \omega L_a/R_a$, $\tan \psi_e = \omega L_e/R_e$ and

$$\tan \psi = \omega(L_a + L_e)/(R_a + R_e).$$

Hence ψ lies in value between ψ_a and ψ_e. Equation (iv) can also be proved directly from Fig. 33. In comparing equations (iv) and (ii) it has to be remembered that (iv) is an exact equation. When $\psi_a = 90°$, we get

$$V - V' = Z_a A \frac{\sin\{(\psi + \psi_e)/2\}}{\cos\{(\psi - \psi_e)/2\}} \quad \text{..............(v).}$$

If in equation (i) we write $V' = y$, $V_a = Z_a A = ax$, where $x = A$, and $h = Z_a \cos(\psi_a - \psi_e)$, we get

$$a^2x^2 + 2hxy + y^2 = V^2,$$

which is the equation to an ellipse. Hence the angles θ_1 and θ_2 that the axes of this ellipse make with OX are determined from the equation

$$\tan 2\theta = \frac{2h}{a^2 - 1}.$$

If, therefore, the impedance is unity, θ is 45 degrees. In order to simplify the equation as much as possible, let us suppose that R_a is zero and $L_a\omega$ is 1. Then Z_a is 1 and k is $\sin\psi_e$. Hence the equation becomes

$$x^2 + 2xy\sin\psi_e + y^2 = V^2.$$

Solving this equation for y (the voltage) we get

$$y = -x\sin\psi_e \pm \sqrt{V^2 - x^2\cos^2\psi_e}.$$

As both x and y must be positive, we need only consider the part

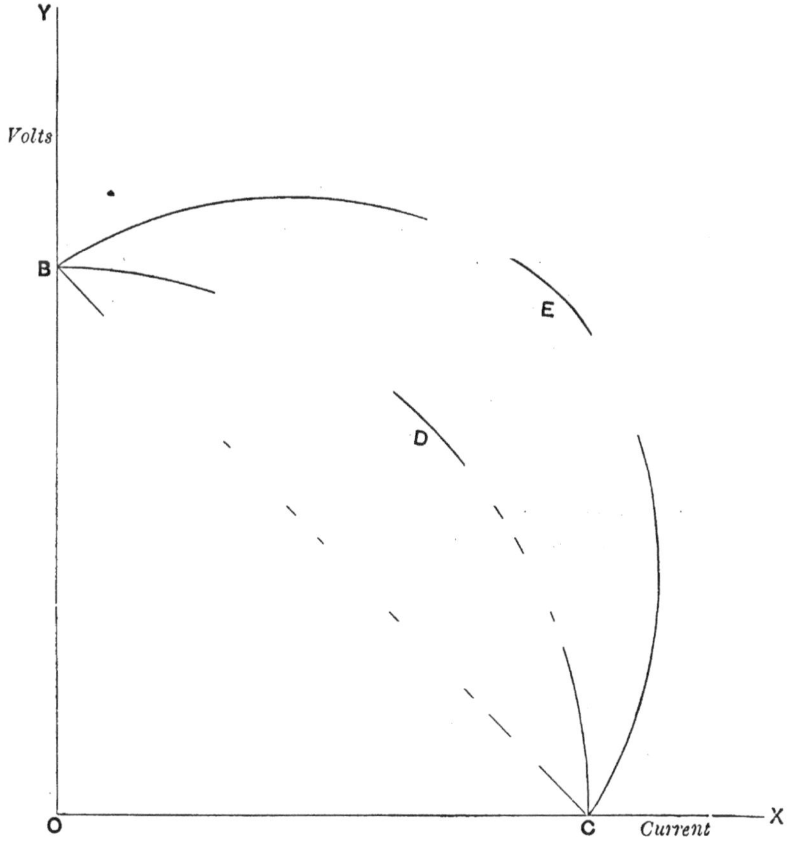

Fig. 36. Theoretical characteristic curves of armature electromotive force and current.
BC is the curve on an inductive load.
BDC is the curve on a non-inductive load.
BEC is the curve on a condenser load.

of the ellipse lying in the first quadrant. When ψ_e is negative, that is, on a condenser load, y attains a maximum value $V/\cos\psi_e$ when x is $-V\tan\psi_e$. If ψ_e be a large angle, we see that the potential difference between the terminals may be very high on a condenser load.

On a very inductive load ψ_e is ninety degrees, and the equation becomes

$$x + y = V.$$

This is the equation of the line BC shown in Fig. 36. On a condenser load the equation is

$$x - y = V,$$

which represents a line through B at right angles to BC. On a non-inductive load we should get the circle BDC, and on a load which gave a leading current, the ellipse BEC. In many cases the curves obtained by experiment are very like those shown in the figure.

Temperature curves of alternators.

The output of alternators is usually expressed as the number of kilovolt-amperes (kVA) available at the terminals. The reason why kilovolt-amperes and not kilowatts are chosen to give the output is because the current in the armature and the consequent temperature rise of the armature conductors etc. fixes the rating of the machine. The temperatures of the various parts of an alternator must not exceed certain specified values otherwise permanent damage may be done to the machine. For instance if the conductors be cotton covered their temperature should never be allowed to exceed 80° C. The temperature of the bearings also should not exceed 70° C. On a very inductive load the excitation has to be largely increased in order to balance the armature reaction and hence the heating of the field magnet windings may be excessive. This may limit the permissible output on an inductive load.

It takes in most cases many hours for a machine to attain its steady thermal state. When it has attained this state the heat generated in it is exactly equal to the heat carried off by convection currents and lost by radiation. The rating of a machine for an intermittent load therefore is higher than for a continuous load. Owing to the long time taken before the temperatures of the various parts of the machine attain their steady values many attempts have been made to devise methods of predicting the final values by taking observations over a brief period of time. The following is the theory of the usual method of procedure.

Let us suppose that a mass m of metal in air is being heated by electric currents. If W be the electric power in watts given to it the heat generated in it will be $W/4{\cdot}2$ calories per second. Part of this heat is lost by convection currents and by radiation. The remainder of the heat raises the temperature of the metal. Owing to the high thermal conductivity of metal no appreciable error is made by the assumption that the temperature θ has practically the same value throughout its mass. If the temperature of the surrounding air be θ_0 then by Newton's law of cooling the rate at which heat is lost by convection will equal $c(\theta - \theta_0)$ very approximately where c depends on the shape of the surface, the velocity of the cooling draught, the thermal conductivity and specific heat of the air, etc. If $\theta - \theta_0$ be small the rate at which heat is lost by radiation will be $k_r(\theta - \theta_0)$ approximately, where k_r is a constant. If the temperatures be expressed in degrees Centigrade, by Stefan's law the radiation loss will be $k_r' \{(273 + \theta)^4 - (273 + \theta_0)^4\}$, where k_r' is a constant. Experiment proves however that the heat lost by radiation is often only one or two per cent. of that lost by convection. We shall therefore neglect it. Hence if s be the specific heat of the metal, the rate at which heat is being stored up in it equals the difference between the rate at which heat is given to it and the rate it is being carried away by convection currents. In symbols, we have

$$\frac{\partial}{\partial t}\{ms(\theta - \theta_0)\} = \frac{W}{4{\cdot}2} - c(\theta - \theta_0).$$

Assuming that c is a constant, it is easy to verify that

$$\theta - \theta_0 = \frac{W}{4{\cdot}2c}\{1 - \epsilon^{-(c/ms)t}\}$$

is a solution of this differential equation.

If Θ denote the maximum temperature rise, that is, the value of $\theta - \theta_0$ when t is infinite, we have

$$\theta - \theta_0 = \Theta\{1 - \epsilon^{-t/\tau}\}$$

where $\tau = (ms)/c =$ the thermal time constant.

If we measure temperatures θ_1, θ_2 and θ_3 at equal intervals of time, it is easy to show from the equation given above, that

$$\frac{1}{\theta_3 - \theta_2} - \frac{1}{\theta_2 - \theta_1} = \frac{1}{\Theta - \theta_2 + \theta_0},$$

and thus
$$\Theta = \theta_2 - \theta_0 + \frac{(\theta_3 - \theta_2)(\theta_2 - \theta_1)}{2\theta_2 - \theta_1 - \theta_3}.$$

Experiment proves that this formula is only of limited use when applied to alternators. The insulating materials used for the windings are bad conductors of heat and so the assumption of uniform temperature throughout is not always permissible. The cooling draught also is not constant for stationary windings but increases as $\theta - \theta_0$ increases. As the value of c is directly proportional to the square root of V, where V is the velocity of the air flowing past it, we see that to assume that c remains constant is admissible only in special cases.

The temperature of the field magnet windings is usually found from the increase of their resistance due to their rise of temperature. For example, if R_t denote the resistance of the windings at a temperature of $t°$ C. and $R_{t'}$ denote the resistance at $t'°$ C., we have for standard annealed copper,

$$R_t = R_0(1 + a_0 t) \text{ and } R_{t'} = R_0(1 + a_0 t')$$

where a_0 is the temperature coefficient 'at constant mass' at zero degrees Centigrade. Hence

$$R_t = R_{t'}\{1 + a_t(t - t')\}$$

where
$$a_t = a_0/(1 + a_0 t').$$

Therefore
$$t - t' = \text{the temperature rise}$$
$$= (R_t - R_{t'})/R_{t'}a_t.$$

The value of a_0 adopted internationally is $0{\cdot}00426_5$. The following table will be found useful.

t	5	10	15	20	25	30	35	40
a_t	0·00417	0·00409	0·00401	0·00393	0·00385	0·00378	0·00371	0·00364

In the case of field magnet windings a thermometer is only used to measure the temperature when the resistance method cannot be used.

It has to be remembered that it is the maximum temperature

attained at a certain part or parts of a machine that affects its life. The resistance method merely gives us the mean temperature of the copper in the windings and if the coils have several layers this mean temperature may differ appreciably from the 'hottest spot' temperature. Hence a low value must be specified for this temperature so as to allow of a sufficient 'factor of safety.'

To measure the temperature of permanently short circuited windings, bearings and local temperatures generally, a thermometer is employed. In practice the permissible temperature rise is affected by the temperature of the surrounding air and the height of the barometer, and suitable corrections to take this into account must be made.

The problem of the prevention of excessive rise of temperature of machines by means of suitable ventilation ducts is one of great importance to the manufacturer. In an alternator driven by a reciprocating engine the surrounding air has easy access to the ventilating ducts and hence the ventilation can easily be made effective. In a turbo-alternator, however, that is, in an alternator directly coupled to a steam turbine, the cooling must be done by forced ventilation. In large machines there are usually both radial and axial air ducts in the stator and rotor. The air is drawn into the enclosed machine by 'blowers' attached to the ends of the rotor and blows directly on the end connections of the stator. Air filters are used to prevent dirt accumulating in the air passages. In practice the alternators are of the rotating field type and owing to the high speed at which they run they have to be made very strong mechanically and the amount of copper that can be used in their construction is therefore limited. One effect is that the armature reaction is extremely high and to counteract it the exciting current has to be high also. Hence it is usually the heating of the rotor that limits the output of the machine.

REFERENCES.

SILVANUS THOMPSON, *Dynamo Electric Machinery.*

H. F. PARSHALL and H. M. HOBART, *Armature Windings.*

C. P. STEINMETZ, *Alternating Current Phenomena.*

G. F. C. SEARLE, 'On the Magnetic Field due to a Current in a Wire placed parallel to the Axis of a Cylinder of Iron,' *The Electrician*, Vol. 40, p. 453, 1898.

F. W. CARTER, 'Air-Gap Induction,' *The Electrical World and Engineer*, Vol. 38, p. 884, 1901.

C. F. GUILBERT, 'The Armature Reaction of Alternators,' *The Electrical World and Engineer*, Vol. 40, p. 658, 1902.

R. GOLDSCHMIDT, 'Temperature Curves and the Rating of Electrical Machinery,' *Journ. of the Inst. of Elect. Engin.*, Vol. 34, p. 660, 1905.

E. H. RAYNER, 'Report on Temperature Experiments, carried out at the National Physical Laboratory,' *Journ. of the Inst. of Elect. Engin.*, Vol. 34, p. 613, 1905.

T. R. LYLE, 'Theory of the Alternate-current Generator,' *Phil. Mag.*, Vol. 18, p. 45, 1909.

R. POHL, 'Notes on National and International Standards for Electrical Machinery,' *Journ. of the Inst. of Elect. Engin.*, Vol. 48, p. 174, 1911.

W. R. COOPER, 'Short Heat Tests on Electrical Machines,' *The Electrician*, Vol. 68, p. 638, 1913.

S. P. SMITH and R. S. H. BOULDING, 'The Shape of the Pressure Wave in Electrical Machinery,' *Journ. of the Inst. of Elect. Engin.*, Vol. 53, p. 205, 1915.

Standardization Rules of the American Institute of Electrical Engineers, 1915.

British Standardization Rules for Electrical Machinery, 1915.

CHAPTER II

Three phase machines. Effect of star or mesh connection of the armature on the output of the machine. Current in a mesh connected armature on no load. Connection of the armature of a three phase machine so as to give single phase currents. Diagram of a three phase winding. Armature reaction. Illustrations. Examples. The terminal potential difference. The shapes of the star and mesh voltage waves. The P.D. wave on closed circuit. Inductive characteristics. Tests of a three phase machine. Characteristic curves. Oscillograph records. Two phase machines. Armature current on no load. Armature reaction in a two phase machine. Tests of a two phase machine. Characteristic curves. Oscillograph records. Tests of a large three phase generator. Load losses. The efficiency of the exciter. References.

Three phase machines.

WE saw in Volume I that the armature of a three phase machine has three windings, which may be connected either in star or in mesh fashion. In a two phase machine we can have two windings which are quite separate from one another, or we can have four windings which may be connected in star or in mesh. We shall only consider three phase and two phase alternators, as these are the only practical forms of polyphase machines. In a three phase machine there are, when the armature is the stator, three terminals, and, when the armature is the rotor, three slip rings from which the alternating current is collected; just as in a single phase machine we have two terminals or two slip rings. In a two phase machine there are generally only three terminals or slip rings, when the armature has two separate windings, and a three wire system of distribution is used (Vol. I, Chap. XVI); in all other cases there must be four terminals or slip rings. We shall first consider three phase machines. In Figs. 37 and 38 are shown the simplest forms of mesh and star windings for three phase armatures. The three circles in the centre of Fig. 37 represent the slip rings. The slip rings are mounted on the shaft and insulated from it. The current is collected from them by means of copper or carbon brushes. The arrow heads show the directions of the currents in the various conductors. The conductors are drawn radially so as to make the diagram clearer, but they are really parallel to the shaft, and are placed in slots in the circumference of the armature. The armature

Fig. 37. Three phase armature with bar winding mesh connected.

is built up of thin circular iron sheets placed at right angles to its axis. These sheets are insulated from one another, and are pressed together between end plates, the whole being firmly keyed to the shaft. In some machines the armature rotates, but more commonly the field magnets rotate. In the latter case no slip rings are required for the alternating current, but slip rings are required to bring the direct current to the exciting coils of the rotating field magnets.

In Fig. 38 the windings are indicated for a machine which has a star connected armature. It will be seen that the winding is practically identical with that shown in Fig. 37.

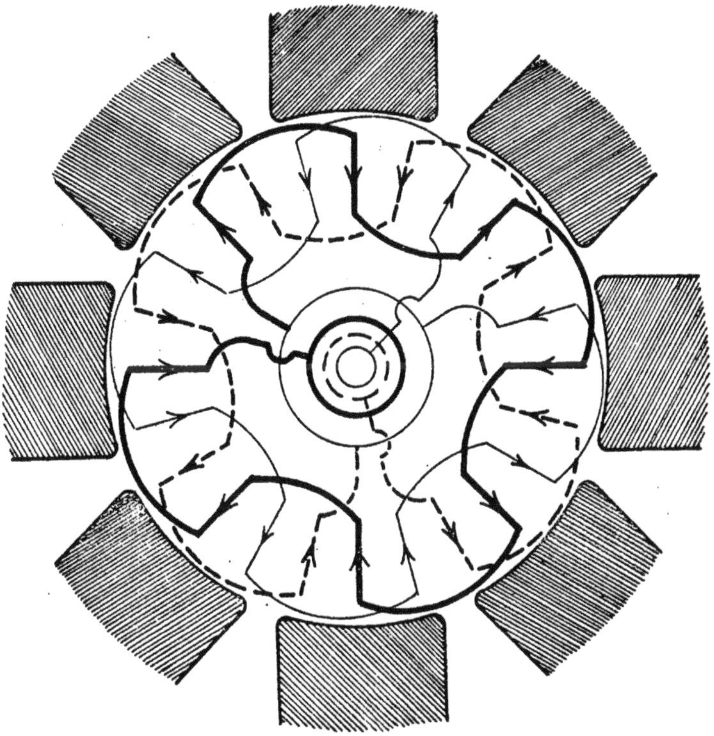

Fig. 38. Three phase armature with bar winding star connected.

Let us first consider a machine with a mesh connected armature.

Effect of star or mesh connection of an armature on the output of a machine.

When the load is balanced, the currents in the external mains will each be equal to $A\sqrt{3}$ (Fig. 39), where A is the effective current in a phase winding of the armature. This follows because we can regard the current in the main *Bb*, for example, as the resultant of the currents flowing in *AB* and *CB* respectively. Now we know (see Vol. I, p. 366) that the currents in *CB* and *BA* differ in phase by 120 degrees, and therefore the currents in the directions *CB* and *AB* differ in phase by 60 degrees. It follows that the current in the main is the resultant of two currents each having an effective value A, the phase difference between them being 60 degrees. Hence the current in each main is $A\sqrt{3}$. If V be the effective voltage between any two of the slip rings on this machine, the effective voltage between the mains will also be V. We can show in a similar manner that the currents in each arm of the balanced load *abc* (Fig. 39) are each equal to A.

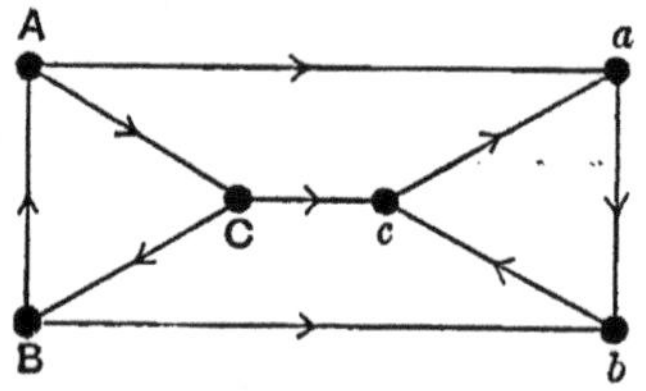

Fig. 39. Mesh connected armature ABC. When the load is balanced the current in each main is $\sqrt{3}$ times the current in an armature winding.

If the mains be very short so that the 'voltage drop' along them is negligible, the voltage across the arms of the load will be V. The power given to the load therefore is $3VA \cos\psi$, where $\cos\psi$ is the power factor of each arm of the load. When the load is non-inductive the power given to it is $3VA$.

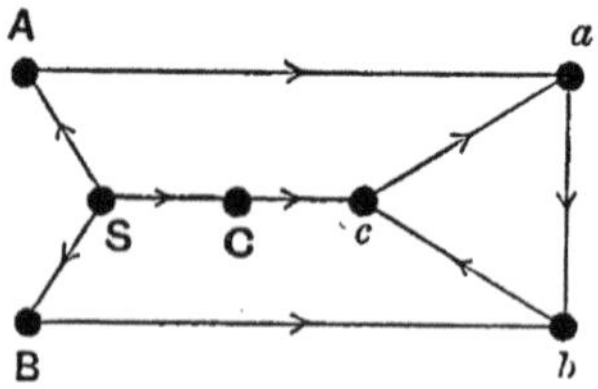

Fig. 40. Star connected armature ABC. When the load is balanced the voltage between any two of the terminals A, B, and C, equals $\sqrt{3}$ times the voltage between A and S, where S is the centre of the star.

A diagram of the armature when it is star connected is shown in Fig. 40. In this case the current in the main is the same as the current in a winding, but the effective voltage between the slip rings on a balanced load will now be $V\sqrt{3}$ since V is the potential difference between A and S (Fig. 40). The output of the machine is $3 \times V\sqrt{3} \times (A \cos\psi)/\sqrt{3}$, that is, $3VA \cos\psi$. It is therefore the same as when the armature is mesh connected.

The maximum output of a machine is limited either by the

rise of temperature in the armature or by the voltage drop at the terminals. In the first case, whether the armature be connected in star or in mesh, the output is $3VA \cos\psi$, and is limited by the maximum permissible value of $3RA^2$. Thus, if V and $\cos\psi$ be constant, the maximum output in the two cases is the same. We must note, however, that the voltage between the mains with the star connected armature is $\sqrt{3}$ times the voltage with the mesh connected armature.

We shall see in the next paragraph that local currents, which will lower the efficiency of the machine, may be generated in a mesh connected armature. In this respect only is the mesh connection inferior to the star connection. For equal power and voltage we require $\sqrt{3}$ times the number of windings when the armature is mesh connected compared with what is necessary when it is star connected. In the latter case, however, the cross section of the wire needs to be $\sqrt{3}$ times as great, and thus the copper required for the armature winding is much the same in the two cases.

Current in a mesh connected armature on no load.

It is to be noticed that, with the mesh winding, if we start from any slip ring 1, we get metallic connection with the slip ring 2 through the winding (1, 2), then with 3 through the winding (2, 3), and finally back again to the first slip ring through the winding (3, 1). The three windings thus form a closed metallic circuit, and, if the three E.M.F.s are not balanced at every instant, we get a local current circulating in the windings.

If the slots are arranged symmetrically and if the E.M.F. in one winding be $f(t)$, then, if the resultant E.M.F. round the circuit of the armature coils always vanishes, we must have

$$f(t) + f(t + T/3) + f(t + 2T/3) = 0.$$

Solving this equation (see Vol. I, p. 367) we find

$$f(t) = X \sin(2\pi t/T + Y) \quad \text{.................(i)},$$

where X and Y are functions of t that do not alter when

$$t + T/3,\ t + T/2 \text{ or } t + 2T/3$$

is written for t. An example of such a function would be $\sin 6(2\pi t/T)$. If $f(t)$ be a sine curve or any other of the curves

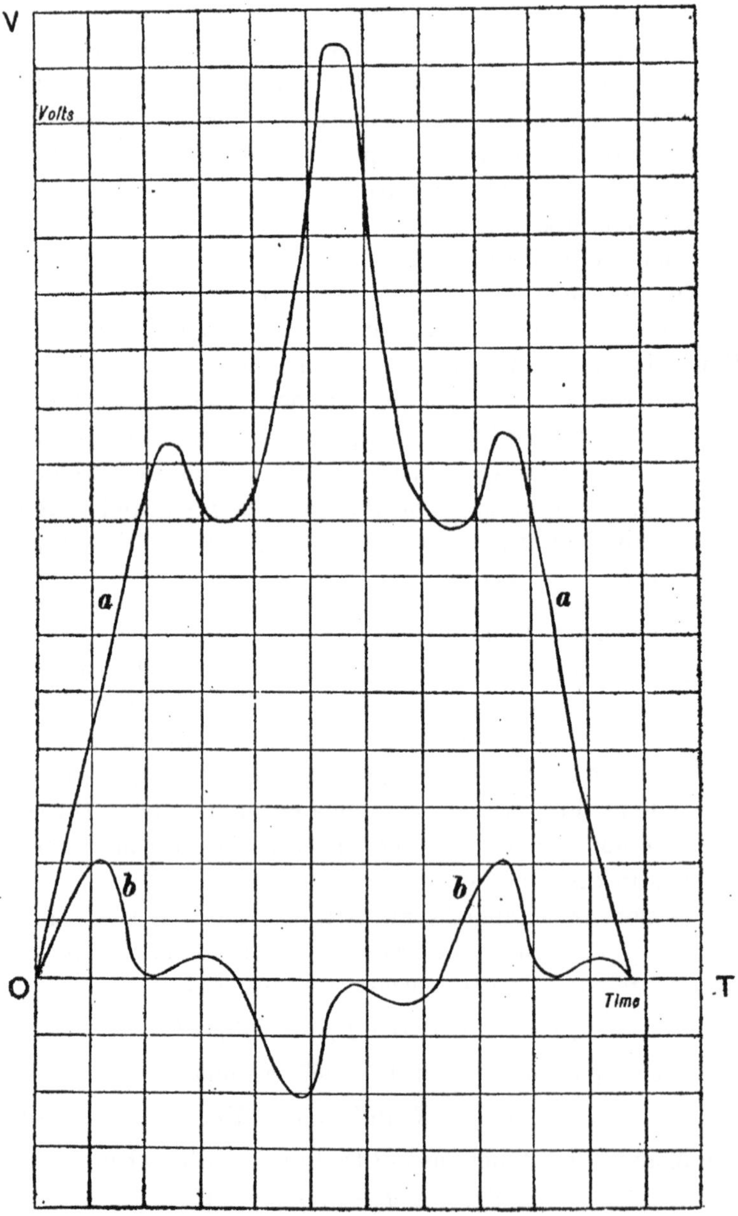

Fig. 41. '*a*' is the mesh E.M.F. of a General Electric 'A.T.' machine at no load. '*b*' is the resultant E.M.F. wave round the circuit formed by the mesh windings.

given by the equation (i), there will be no currents in the armature at no load.

If the E.M.F. wave $f(t)$ generated do not satisfy the equation (i), the resultant E.M.F. $\phi(t)$ will be given by

$$\phi(t) = f(t) + f(t + T/3) + f(t + 2T/3),$$

and hence

$$\begin{aligned}\phi(t + T/6) &= f(t + T/6) + f(t + T/2) + f(t + 5T/6)\\ &= -f(t + 2T/3) - f(t) - f(t + T/3)\\ &= -\phi(t).\end{aligned}$$

Therefore

$$\phi(t + T/3) = -\phi(t + T/6) = \phi(t).$$

Hence the frequency of the circulating current is at least three times as rapid as the frequency of the alternating current given out by the machine. Since the reactance of the three windings of the armature, in series, is always high compared with the circuital E.M.F., the current and therefore the loss due to this cause is small. In Fig. 41 'a' is the mesh voltage of a General Electric 'A.T.' machine at no load. 'b' is a curve obtained graphically by finding the resultant of three curves similar to 'a' and having time lags of 0, $T/3$ and $2T/3$ respectively. A very small change in the shape of 'a' may produce a considerable change in the shape and magnitude of 'b.'

Connection of the armature of a three phase machine so as to give single phase currents.

We shall now consider how to connect the coils of a star wound armature so as to get single phase currents. If the windings X, Y and Z (Fig. 42) be separated from the common junction and connected as in Fig. 42, the phases and the magnitudes of the component effective voltages x, y and z may be represented by lines as in the figure. When x, y and z are each equal to V, the resultant voltage will be $2V$, provided that the sum of the instantaneous values of the E.M.F.s generated in the armature coils is zero when they are connected in mesh. For let the instantaneous values of the E.M.F.s be represented by $f(t)$, $f(t + T/3)$ and $f(t + 2T/3)$, then the resultant E.M.F. when they are connected as in the figure is

$$f(t) - f(t + T/3) + f(t + 2T/3),$$

and this is equal to $-2f(t+T/3)$,

if $f(t)+f(t+T/3)+f(t+2T/3)=0$.

Thus the effective value of the resultant voltage is $2V$.

Now, in most practical cases, the maximum output of an alternator is governed either by the maximum permissible heating of the armature or by the maximum current the armature windings can carry. In the first case, let us suppose that the armature, when it has attained its highest permissible

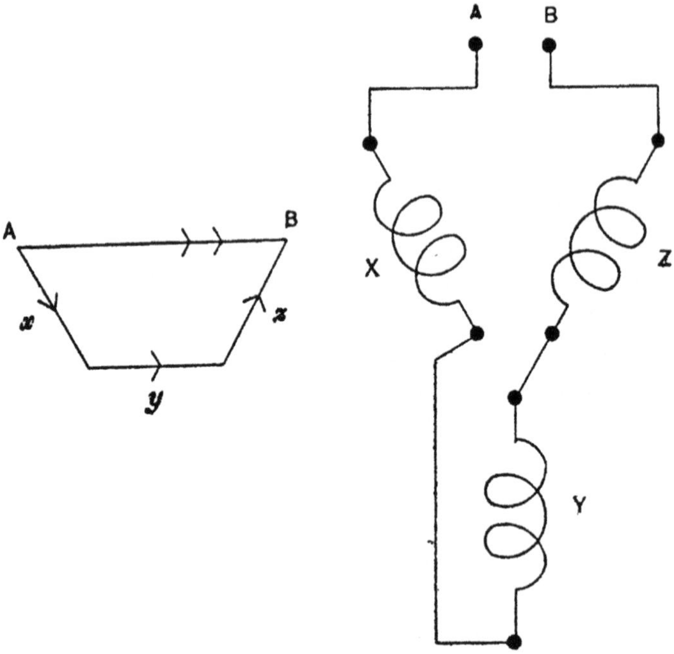

Fig. 42. Three phase armature connected so as to give single phase currents. Resultant voltage between A and B is $2V$, where V is the effective voltage generated in each winding.

temperature, can radiate an amount of heat which is equivalent to H joules per second. Let A_3 be the largest effective current which it is safe to take from each winding. Then, neglecting the iron losses, we have

$$H=3RA_3^2,$$

where R is the resistance of one winding. If V be the potential difference between the terminals, the output is

$$3VA_3\cos\psi=\sqrt{3}V\sqrt{(H/R)}\cos\psi.$$

When it works as a single phase machine (Fig. 42), the output is $2VA_1 \cos\psi$, and

$$H = 3RA_1^2,$$

and therefore the output equals

$$\frac{2}{\sqrt{3}} V\sqrt{(H/R)}\cos\psi = \frac{2}{3}\cdot\sqrt{3}V\sqrt{(H/R)}\cos\psi$$
$$= 1{\cdot}15V\sqrt{(H/R)}\cos\psi \text{ nearly.}$$

When the armature is connected in the manner shown in Fig. 40, the output is, therefore, only two-thirds the output of the machine when giving three phase currents.

Sometimes single phase currents are obtained by merely

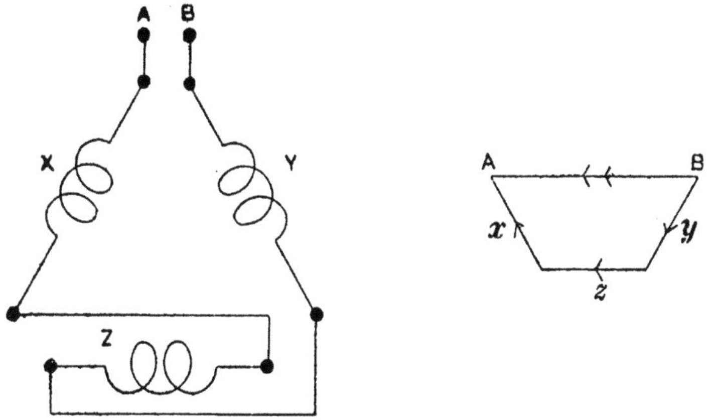

Fig. 43. Armature with windings in 'mesh,' connected so as to give single phase currents. Resultant voltage between A and B is $2V$, where V is the effective voltage generated in each winding.

loading one phase of the three phase generator. If we assume that the maximum current the armature winding can carry is A, then the output is $\sqrt{3}VA\cos\psi$, as compared with $2VA\cos\psi$ when the machine is connected as in Fig. 40.

If, however, we assume that the output is governed by the heating of the armature, we have

$$H = 2RA^2,$$

and the output $= \sqrt{3}VA\cos\psi$

$$= \sqrt{(3/2)}\,V\sqrt{(H/R)}\cos\psi$$
$$= 1{\cdot}22V\sqrt{(H/R)}\cos\psi \text{ nearly.}$$

In this case the current in each of the active windings is about twenty per cent. greater than when connected as in Fig. 42, and the output is about six per cent. greater.

A method of connecting a three phase mesh connected armature so as to get single phase currents is shown in Fig. 43. It will be seen from the diagram that the problem is practically identical with the preceding one. Thus, if we take the heating of the armature as the governing factor, the output as a single phase machine is only two-thirds of the output as a polyphase machine.

Diagram of a three phase winding.

A conventional method of representing the windings in a three phase armature is shown in Fig. 44. It will be seen that there are three slots on the armature per pole, or in other words one slot per pole and per phase. Only the end connections are shown, the armature conductors being perpendicular to the plane of the paper.

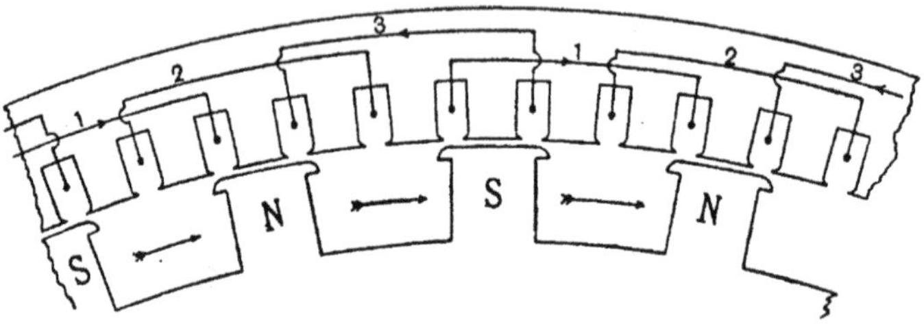

Fig. 44. Three phase alternator with rotating field. Currents in phase with the armature electromotive force.

If the current be in phase with the armature electromotive force, then, with the field magnets in the position shown in the diagram, the currents in the wires marked 2 will be zero and the currents in the wires marked 1 and 3 respectively will be equal in magnitude but opposite in sign. The effect of these currents is to produce both a direct and a transverse magnetisation of the field.

The position of the field magnets, an eighth of a period later, is shown in Fig. 45. The arrow heads indicate the directions of the currents when they are in phase with the armature electromotive forces. It will be seen that transverse and direct magnetising effects on the field are still being produced. In Chapter I, p. 43, formulae were found for the mean demagnetising effect of the component of the current which is ninety degrees different in phase from the armature electromotive force, and formulae were also found for the mean ampere-turns due to the component of the current in phase with the armature electromotive force tending to magnetise the field transversely.

We shall now consider how these formulae have to be modified for the case of three phase machines.

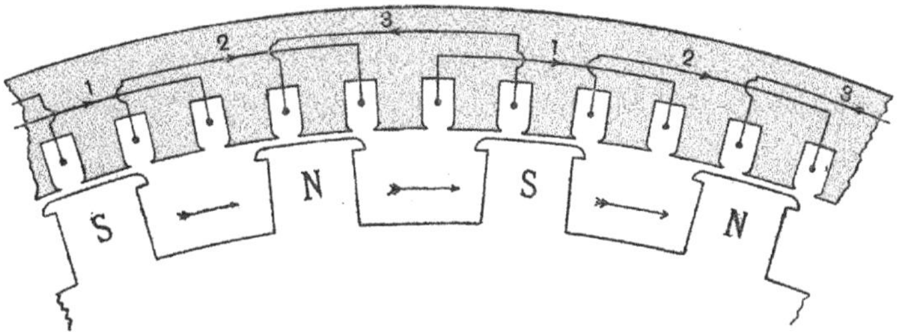

Fig. 45. Three phase alternator with rotating field. The directions of the currents an eighth of a period later than in Fig. 44.

Armature reaction.

Let us first consider the effect produced on the magnetic field by the current in a coil forming one phase of a three phase winding. Let AB (Fig. 46) be the plane of the coil and let us assume that t is zero when OK

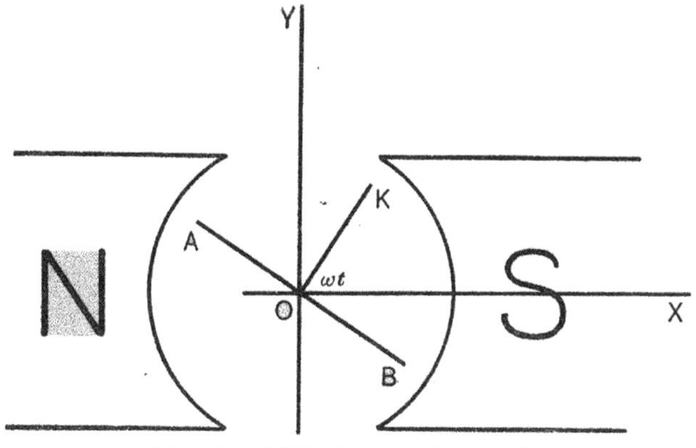

Fig. 46. OK is the axis of the coil.

the axis of the coil coincides with OX and that the current in the coil is $I \sin(\omega t - \psi)$. If N_1 be the number of turns in the coil, then the component ampere-turns in the direction OX are equal to

$$N_1 I \sin(\omega t - \psi) \cos \omega t = (N_1 I/2)\{\sin(2\omega t - \psi) - \sin\psi\}.$$

Similarly we find that the component ampere-turns in the direction OY are equal to

$$(N_1 I/2)\{\cos\psi - \cos(2\omega t - \psi)\}.$$

We see therefore that the direct magnetising ampere-turns fluctuate between

$$-(N_1 I/2)(1+\sin\psi) \text{ and } (N_1 I/2)(1-\sin\psi).$$

Similarly the transverse ampere-turns fluctuate between

$$(N_1 I/2)(1+\cos\psi) \text{ and } -(N_1 I/2)(1-\cos\psi).$$

The mean value of the direct ampere-turns is $-(N_1 I/2)\sin\psi$ and of the transverse ampere-turns $(N_1 I/2)\cos\psi$.

In this proof we have assumed that the current in the winding follows the harmonic law. We saw in the last chapter that this was only strictly permissible in cases where the armature reaction was weak. If we make this assumption we see that for a three phase machine on a balanced load the component ampere-turns in the direction OX will be

$$(N_1 I/2)\{\sin(2\omega t-\psi)+\sin(2\omega t-\psi-2\pi/3)$$
$$+\sin(2\omega t-\psi-4\pi/3)-3\sin\psi\} = -3(N_1 I/2)\sin\psi.$$

In a similar way we find that the component ampere-turns in the direction OY will be $3(N_1 I/2)\cos\psi$. These components, therefore, are absolutely constant, and hence on our assumptions there would be no ripple induced in the current in the field magnet windings. If nC be the ampere-turns on open circuit of the field magnet windings, the exciting current being equal to C, the resultant ampere-turns on the magnetic field will be

$$nC - 3(N_1 I/2)\sin\psi,$$

when the wattless currents in the windings are each equal to $I\sin\psi$. This result can also be deduced readily from the theory of rotating fields given in Vol. I, Chapter XVIII. To enable us to compute the E.M.F. developed in a three phase winding a more rigorous investigation is necessary.

Let us take the case of an ideal three phase machine connected in mesh and working on a balanced load. Our equations are

Analytical method.

$$0 = Ri_1 + L\frac{\partial i_1}{\partial t} - M'\frac{\partial i_2}{\partial t} - M'\frac{\partial i_3}{\partial t} + M\frac{\partial}{\partial t}(i'\cos\omega t),$$

$$0 = Ri_2 + L\frac{\partial i_2}{\partial t} - M'\frac{\partial i_3}{\partial t} - M'\frac{\partial i_1}{\partial t} + M\frac{\partial}{\partial t}\{i'\cos(\omega t - 2\pi/3)\},$$

$$0 = Ri_3 + L\frac{\partial i_3}{\partial t} - M'\frac{\partial i_1}{\partial t} - M'\frac{\partial i_2}{\partial t} + M\frac{\partial}{\partial t}\{i'\cos(\omega t - 4\pi/3)\},$$

where R and L are the resistance and inductance respectively of a phase winding and its external circuit, $-M'$ is the mutual inductance between the first phase winding and either of the others, and $M \cos \omega t$, $M \cos (\omega t - 2\pi/3)$, and $M \cos (\omega t - 4\pi/3)$ are the mutual inductances between the phase windings and the exciting circuit.

By addition we have $Ri + (L - 2M')(\partial i/\partial t) = 0$, where $i = i_1 + i_2 + i_3$. Therefore $i = A\epsilon^{-\{R/(L-2M')\}t}$, where A is a constant, and hence when the steady working state is reached i is zero.

If E be the E.M.F. in the exciting circuit the constants of which are R_2 and L_2, we have

$$E = R_2 i' + L_2 \frac{\partial i'}{\partial t} + M \frac{\partial}{\partial t} \{i_1 \cos \omega t + i_2 \cos (\omega t - 2\pi/3) + i_3 \cos (\omega t - 4\pi/3)\}.$$

Let us suppose that $i_1 = I \sin (\omega t - \psi)$ is a solution and, since everything is symmetrical, that consequently

$$i_2 = I \sin (\omega t - \psi - 2\pi/3),$$

and
$$i_3 = I \sin (\omega t - \psi - 4\pi/3).$$

Substituting these values in the equation for the exciting current we get $E = R_2 i' + L_2 (\partial i'/\partial t)$, and thus when the steady state is reached $i' = E/R_2 = C$. Substituting these values for the currents in the first of the equations given above and rearranging the terms we get

$$\{RI \cos \psi + (L + M')\omega I \sin \psi - M\omega C\} \sin \omega t + \{(L + M')\omega I \cos \psi - RI \sin \psi\} \cos \omega t = 0.$$

As this must be true for all values of t, the coefficients of $\sin \omega t$ and $\cos \omega t$ must vanish, and thus

$$I = M\omega C/Z \text{ and } \tan \psi = (L + M')\omega/R,$$

where
$$Z^2 = R^2 + (L + M')^2 \omega^2.$$

The assumed values, therefore, for the currents give the solution of the problem when the steady state is attained. Unlike the corresponding single phase problem there are no higher harmonics in the current waves.

If ϕ_1 be the flux linked with the first phase winding we have

$$\phi_1 = Li_1 - M'i_2 - M'i_3 + MC\cos\omega t$$
$$= (L + M')\,i_1 + MC\cos\omega t$$
$$= \{MC - (L + M')\,I\sin\psi\}\cos\omega t + (L + M')\,I\cos\psi\sin\omega t.$$

Hence we may regard the flux as made up of two components, the direct flux, the amplitude of which is $\{MC - (L + M')\,I\sin\psi\}$, and the transverse flux, the amplitude of which is $(L + M')\,I\cos\psi$. These fluxes are in quadrature with one another, and the direct flux has its maximum value when the mutual inductance between it and the field magnet windings has its maximum value.

Similarly we see that the flux ϕ linked with the field circuit is given by

$$\phi = L_2C + M\,\{i_1\cos\omega t + i_2\cos(\omega t - 2\pi/3) + i_3\cos(\omega t - 4\pi/3)\}$$
$$= L_2C - (3/2)\,MI\sin\psi.$$

We see that for a given value of $I\sin\psi$, ϕ remains absolutely constant. It is to be noticed that ϕ is independent of M', and hence a knowledge of its value does not enable us to calculate the potential difference at the terminals of a phase winding.

If we suppose that $M' = L\cos 60° = L/2$, we have

$$\phi_1 = \{MC - (3/2)\,LI\sin\psi\}\cos\omega t + (3/2)\,LI\cos\psi\sin\omega t.$$

Comparing this with the corresponding formula for a single phase machine (p. 37) we see that if the demagnetising ampere-turns are $\alpha N_1 A\sin\psi$ and the transverse ampere-turns are $\beta N_1 A\cos\psi$, then α and β will be 1·5 times greater than the corresponding values given on p. 52. Hence

$$\alpha N_1 A\sin\psi = (1{\cdot}5)\,\frac{4\sqrt{2}}{\pi^2}\frac{a}{b}\,N_1 A\sin\psi\sin\frac{\pi b'}{2a}\sin\frac{\pi b}{2a}$$
$$= 0{\cdot}860\,\frac{a}{b}\,N_1 A\sin\psi\sin\frac{\pi b'}{2a}\sin\frac{\pi b}{2a},$$

and $$\beta N_1 A\cos\psi = 0{\cdot}860\,\frac{a}{b}\,N_1 A\cos\psi\sin\frac{\pi b'}{2a}\left(1 - \cos\frac{\pi b}{2a}\right),$$

where $2N_1$ is the number of armature conductors per pole and per phase, or in other words N_1 is the number of armature turns per pole and per phase, and the other symbols are defined as on pp. 39 and 40.

Illustrations.

To illustrate the method of applying these formulae, let us first consider the winding illustrated in Fig. 44. In this case the breadth of a coil b' equals the pitch of the poles a, and hence the formulae become

$$\alpha N_1 A \sin\psi = 0{\cdot}860\,\frac{a}{b}\,N_1 A \sin\psi \sin\frac{\pi b}{2a},$$

and $$\beta N_1 A \cos\psi = 0{\cdot}860\,\frac{a}{b}\,N_1 A \cos\psi\left(1 - \cos\frac{\pi b}{2a}\right).$$

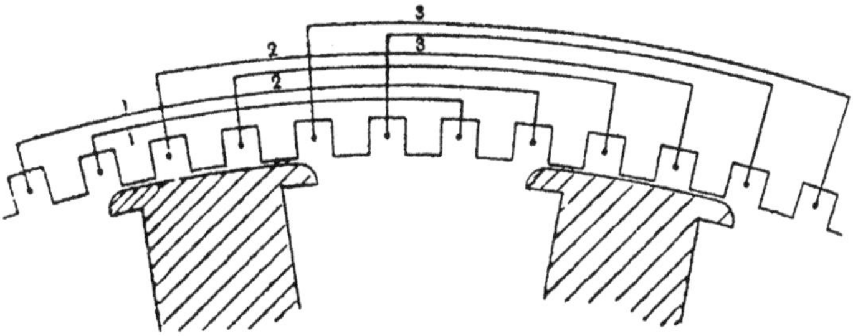

Fig. 47. Three phase alternator with rotating field having two slots per pole and per phase in the armature.

Let us next consider the winding illustrated in Fig. 47. In this case we have two slots per pole and per phase, and the breadth of the inner winding of a coil is $5a/6$ and of the outer $7a/6$. Now, for the inner coil, $\sin(\pi b'/2a)$ equals $\sin(5\pi/12)$, that is, $\sin 75°$, and for the outer coil, $\sin(\pi b'/2a)$ equals $\sin(7\pi/12)$ which is also equal to $\sin 75°$. Hence we get

$$\alpha N_1 A \sin\psi = (1{\cdot}5)\,\frac{4\sqrt{2}}{\pi^2}\,\frac{a}{b}\,N_1 A \sin\psi \sin\frac{\pi b}{2a}\sin 75°$$

$$= 0{\cdot}830\,\frac{a}{b}\,N_1 A \sin\psi \sin\frac{\pi b}{2a},$$

and $$\beta N_1 A \cos\psi = 0{\cdot}830\,\frac{a}{b}\,N_1 A \cos\psi\left(1 - \cos\frac{\pi b}{2a}\right).$$

If we have three slots per pole and per phase, the breadth of the middle winding of a phase is a and the breadths of the inner and outer windings $7a/9$ and $11a/9$ respectively. We have, therefore,

$$\alpha N_1 A \sin\psi = (1{\cdot}5)\,\frac{4\sqrt{2}}{\pi^2}\,\frac{a}{b}\,N_1 A \sin\psi \sin\frac{\pi b}{2a}\,\frac{\sin\frac{7\pi}{18} + \sin\frac{9\pi}{18} + \sin\frac{11\pi}{18}}{3}$$

$$= 0{\cdot}826\,\frac{a}{b}\,N_1 A \sin\psi \sin\frac{\pi b}{2a},$$

and $$\beta N_1 A \cos\psi = 0{\cdot}826\,\frac{a}{b}\,N_1 A \cos\psi\left(1 - \cos\frac{\pi b}{2a}\right).$$

In a 1400 kilovolt-ampere Creüsot alternator the armature has two slots per pole and per phase and has six conductors in each slot. Hence the number N_1 of turns per pole

Examples.

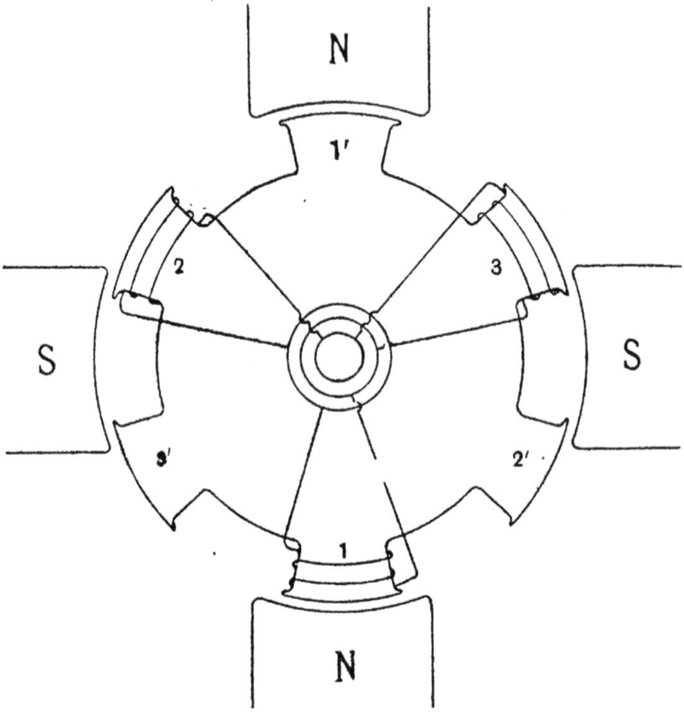

Fig. 48. Number of armature coils 6. Number of poles 4. The coils round 1', 2' and 3' (not shown in the diagram) may be connected either in parallel or series with the coils round 1, 2 and 3. (Obsolete type.)

and per phase will be 6. When a equals b and A is 156 amperes, we find that

$$\alpha N_1 A \sin\psi = 0{\cdot}830 \frac{a}{b} N_1 A \sin\psi \sin\frac{\pi b}{2a}$$
$$= 777 \sin\psi,$$

and $$\beta N_1 A \cos\psi = 777 \cos\psi.$$

In a 760 kilovolt-ampere Heyland alternator the armature has two tunnels per pole and per phase with three conductors per tunnel. In this case, N_1 is 3, and a equals b. When A is 200 amperes we have

$$\alpha N_1 A \sin\psi = 498 \sin\psi,$$

and $$\beta N_1 A \cos\psi = 498 \cos\psi.$$

A 2600 kilovolt-ampere Siemens and Halske alternator has three slots per pole and per phase, with one conductor in each slot, and a is equal to b. In this case N_1 is 1·5, and when A is 520 amperes

$$\alpha N_1 A \sin \psi = 644 \sin \psi,$$

and

$$\beta N_1 A \cos \psi = 644 \cos \psi.$$

It is to be noticed that in an alternator the number of armature coils is not necessarily a multiple of the number of poles. For instance, in the four pole machine illustrated in Fig. 48 we have six armature coils and four poles. In this case, since the number of turns per armature coil is three, N_1 would be 1·5.

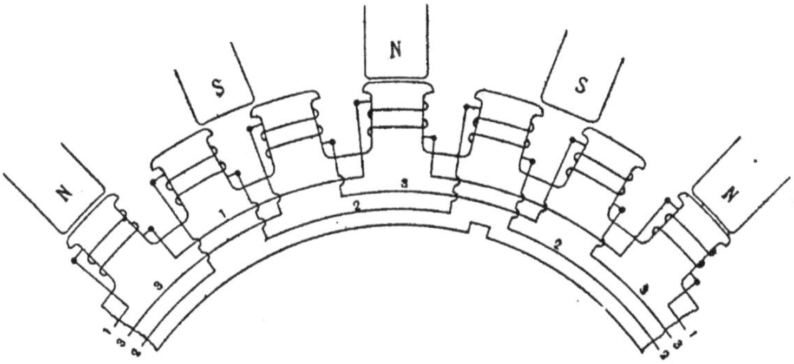

Fig. 49. Part of the armature of a three phase machine having sixteen poles and twenty-four armature coils. (Obsolete type.)

In Fig. 49 is shown part of the armature of a three phase machine which has sixteen poles and twenty-four armature coils. If there are n turns on each armature coil, then, N_1 will equal $8n/16$, that is, $n/2$. These types of winding are not now used.

The formulae can be applied even when the armature winding is complicated. The effect of the various windings of the armature shown in Fig. 50 in producing the potential differences between the slip rings will be understood from the diagram in Fig. 51, where the manner in which the potential differences combine vectorially is indicated. Since there are ten poles and twelve armature coils the number of armature coils per pole and per phase will be 4/10, and thus, if there are n_1 turns per armature coil, N_1 will equal $2n_1/5$.

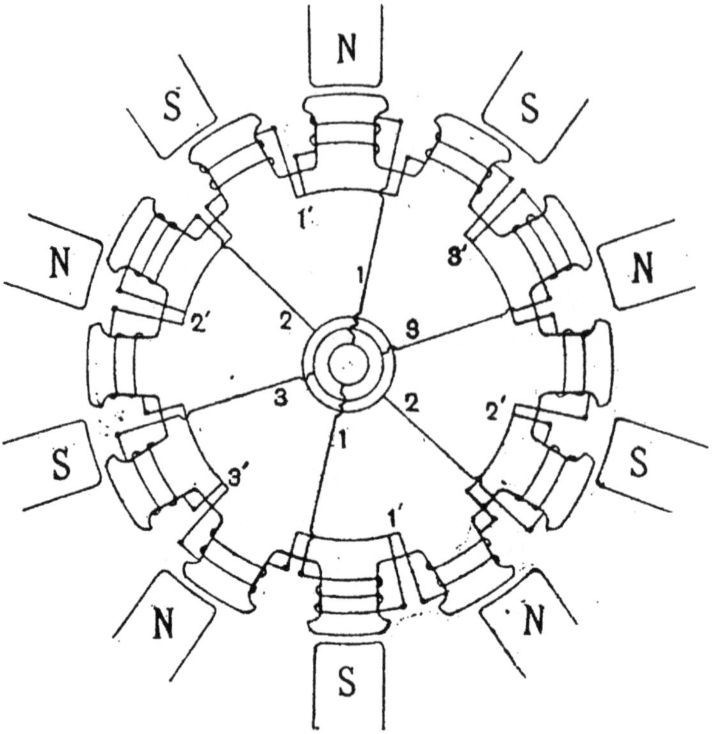

Fig. 50. Three phase alternator with ten poles and twelve armature coils. (Obsolete type of winding.)

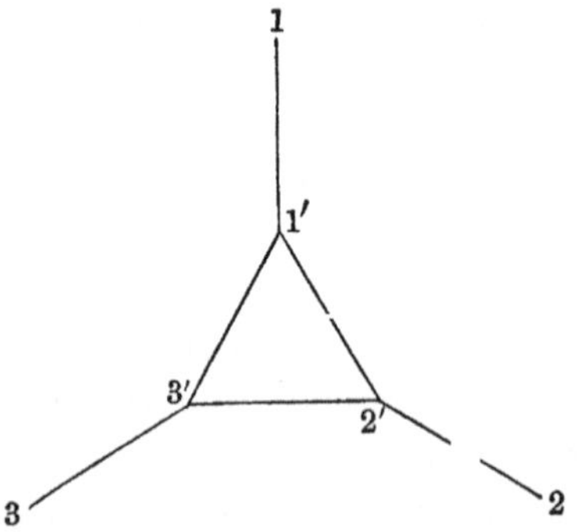

Fig. 51. Diagram of the E.M.F.s in the three phase alternator shown in Fig. 50.

The terminal potential difference.

The electromotive force, on open circuit, at the terminals of an armature coil of a three phase machine, can be calculated in the same way as the E.M.F. for a single phase machine. With our usual notation, if there are N' bars joined in series in one winding, then, the voltage V_0 at the terminals of the coil on open circuit is given by (p. 17)

$$V_0 \doteq 2fN' (V_0/e_m) \Phi_A \times 10^{-8}.$$

In order to find V_0, therefore, we must know the shape of the wave of electromotive force, and this can be predetermined when we know how the flux in the air-gap is distributed.

If the flux from the pole piece does not vary appreciably with the relative position of the armature and the pole, the shape of the E.M.F. wave can be found by plotting a graph of the flux density in the air-gap. The effective value H_e of the height of this curve can be found (Vol. I, p. 126), and the mean height H_m can be found from the area and breadth of the curve. Hence $V_0/e_m = H_e/H_m$ can be found, and V_0 follows from the formula.

In the case of the ideal machine considered on p. 85, if the constants of the load across the terminals of the phase winding be (R_e, L_e) we have

$$e_1 = \{R_e + L_e (\partial/\partial t)\} i$$
$$= (Z_e/Z) M\omega C \sin (\omega t - \psi + \psi'),$$

where $\tan \psi' = \omega L_e/R_e$ and $Z_e^2 = R_e^2 + L_e^2\omega^2$.

Hence we have $V_1 = (Z_e/Z) V_0$.

The shapes of the star and mesh voltage waves.

When the armature of the machine is star wound, the shape of the wave of P.D. between a terminal and the centre of a star winding is, in general, different from the shape of the wave between two terminals. If e_1, e_2 and e_3 be the instantaneous values of the P.D.s between the terminals 1, 2 and 3 and the centre of the winding, and if $v_{1.2}$, $v_{2.3}$ and $v_{3.1}$ be the P.D.s between the terminals, we have

$$v_{1\cdot 2} = e_1 - e_2, \quad v_{2.3} = e_2 - e_3, \quad v_{3.1} = e_3 - e_1.$$

On open circuit e_1, e_2 and e_3 can be calculated when the distribution of the flux in the air-gap is known. Let us suppose that the machine is symmetrical, so that we may write

$$e_1 = f(t), \quad e_2 = f(t + T/3) \text{ and } e_3 = f(t + 2T/3).$$

Then we have

$$v_{1.2} = f(t) - f(t + T/3) = f(t) + f(t - T/6).$$

We can, therefore, easily find $v_{1.2}$ graphically by adding together the ordinates of two periodic curves each equal to the curve $f(t)$ representing the star voltage and one having a time lag relative to the other of one-sixth of a period. An illustration of this theorem is shown in Fig. 52. The lower curve gives the star voltage wave and the upper curve the mesh voltage wave of a three phase generator.

Let us now suppose that

$$e_1 + e_2 + e_3 = 0,$$

and that V is the effective value of each of the star voltages. In this case their vectors will be inclined to one another at angles of 120 degrees, and so the effective value of $v_{1.2}$ will be $V\sqrt{3}$. Let us also suppose that $f(t)$ represents a symmetrical alternating wave (see Vol. I, p. 280), so that we have

$$f(t) = f(T/2 - t) = -f(-t).$$

Then since

$$v_{1.2} = f(t) + f(t - T/6),$$

we see that $v_{1.2}$ vanishes when t is $T/12$. Thus if A' be the area of the wave $v_{1.2}$, we have

$$\begin{aligned} A' &= \int_{T/12}^{7T/12} \{f(t) + f(t - T/6)\}\, \partial t \\ &= A - 2A_1 + A - 2A_1 \\ &= 2A - 4A_1, \end{aligned}$$

where A is the area of the positive half of the wave $f(t)$,

and $$A_1 = \int_0^{T/12} f(t)\, \partial t.$$

If we divide the base of the part of the curve $f(t)$, between 0 and $T/2$, into six equal parts, and erect ordinates at the five points of division, A_1 will be the area of either the first or the last of the segments into which the area has been divided.

Let k_s and k_m be the form factors of the star and the mesh wave respectively. Then we have

$$k_s = \frac{V}{A} \cdot \frac{T}{2}; \quad k_m = \frac{V\sqrt{3}}{2A - 4A_1} \cdot \frac{T}{2},$$

and therefore
$$\frac{k_m}{k_s} = \frac{\sqrt{3}}{2} \cdot \frac{A}{A - 2A_1} \quad \text{.....................}(a).$$

If A_1 be zero, then, whatever the shape of the rest of the wave,
$$\frac{k_m}{k_s} = \frac{\sqrt{3}}{2} = 0{\cdot}866.$$

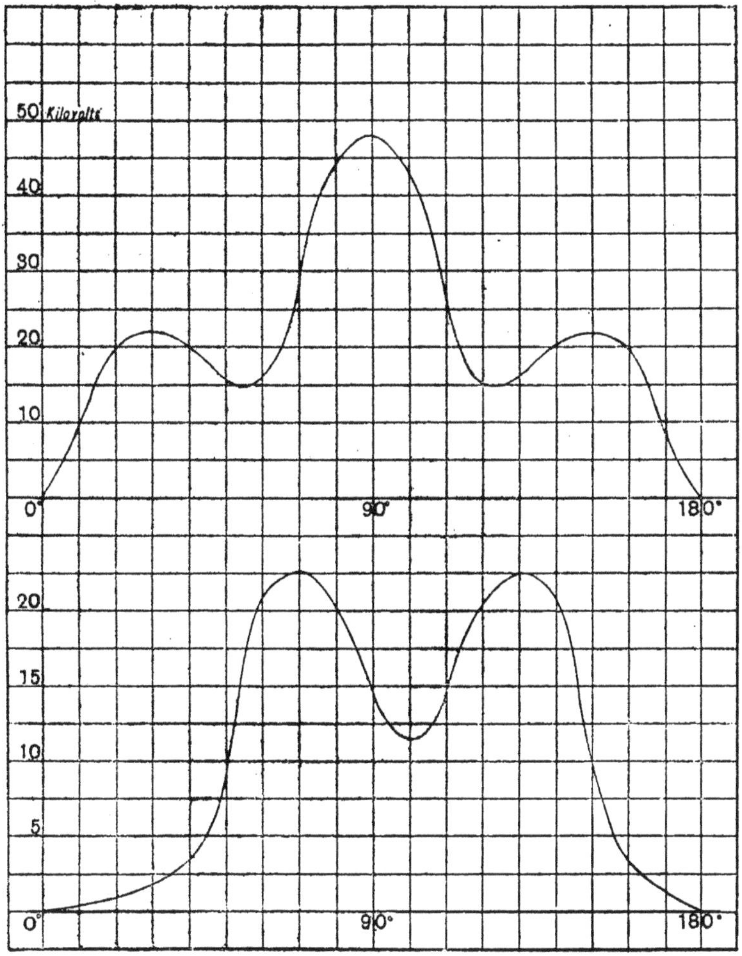

Fig. 52. The upper curve is the mesh voltage wave of an Oerlikon three phase generator (type 6065) and the lower curve is the star voltage wave of the same machine. The upper curve can be obtained by adding together two curves similar to the lower one and having a time lag of 60 degrees.

It is easy to see that the assumptions made in proving (a) are equivalent to assuming that $f(t)$ is given by a Fourier series of the form $\Sigma b_{6n\pm1} \sin(6n \pm 1)\,\omega t$. In the general case we must write (see Chap. III)

$$e_1 = \Sigma a_{2n+1} \cos(2n+1)\,\omega t + \Sigma b_{2n+1} \sin(2n+1)\,\omega t.$$

It readily follows that

$$2V^2 = \Sigma(a^2_{2n+1} + b^2_{2n+1}) \text{ and } 2V^2_{1.2} = 3\Sigma(a^2_{6n\pm1} + b^2_{6n\pm1}).$$

Hence $V_{1.2}$ is less than $\sqrt{3}\,.\,V$ and the formula giving the value of k_m/k_s is complicated.

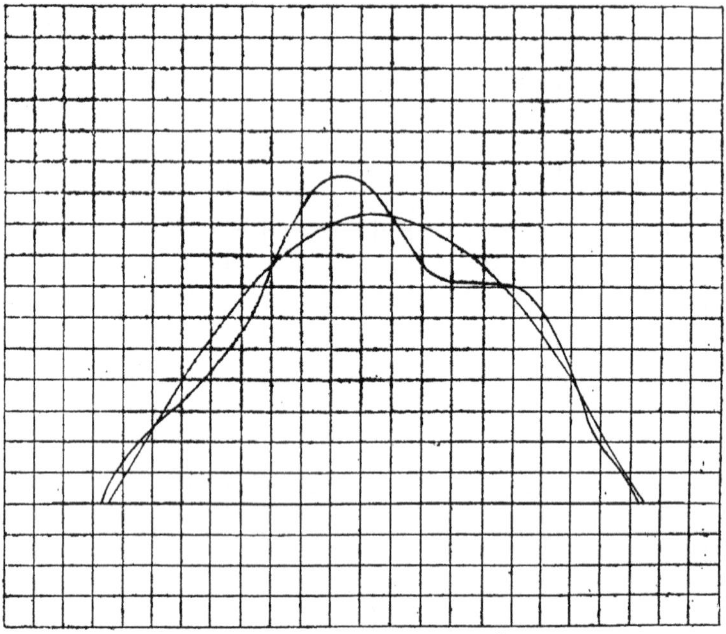

Fig. 53. The shape of the wave of the mesh potential difference across the terminals of a three phase generator (Oerlikon, type 6065) on a non-inductive load. Note the distortion due to the cross magnetisation of the field.

If the wave is symmetrical and the amplitude of $e_1 + e_2 + e_3$ is small compared with that of e_1, the formula (*a*) gives useful approximate values. For example for a triangular wave we find $k_m/k_s = 0{\cdot}974$. But from the table on p. 18 we get $k_s = 2/\sqrt{3}$, $(n = 1, b = 0)$ and similarly we find that $k_m = \frac{1}{2}\sqrt{5}$, $(n = 1, b = a/3)$ and thus $k_m/k_s = \sqrt{15}/4 = 0{\cdot}968$. The error due to using the approximate formula is therefore only about 0·6 of one per cent.

For a sine wave $k_m/k_s = 1$, and for a rectangular wave $k_s = 1$ and $k_m = \sqrt{(1\cdot5)} = 1\cdot225$. Even in this latter case, where the amplitude of $e_1 + e_2 + e_3$ is equal to that of e_1, the value given by (a) is only about 5 per cent. high.

The potential difference wave on closed circuit.

The effects of the transverse magnetisation of the field by the armature currents on the shape of the potential difference wave across the terminals of the machine are conspicuous in the P.D. wave shown in Fig. 53. This is the shape of the wave of the mesh potential

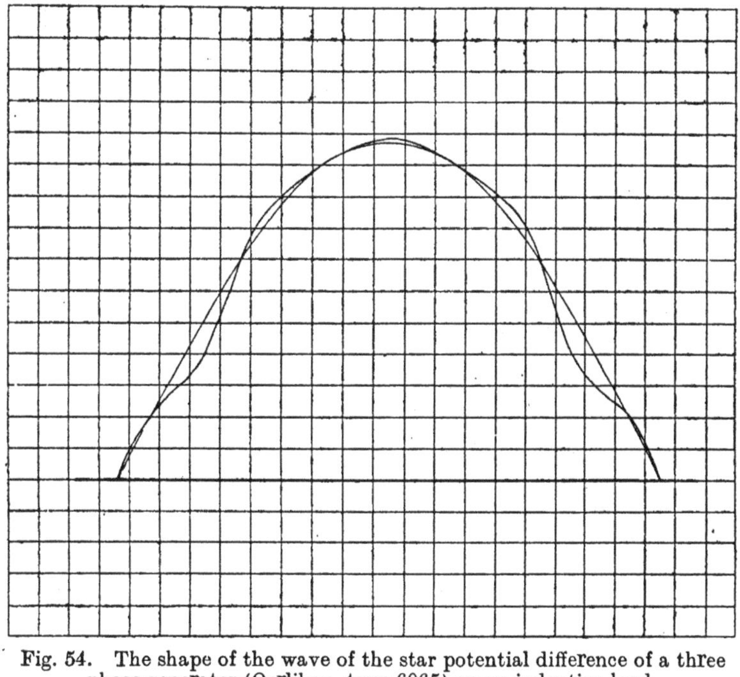

Fig. 54. The shape of the wave of the star potential difference of a three phase generator (Oerlikon, type 6065) on an inductive load.

difference of an Oerlikon three phase generator when working on a non-inductive load. In this machine the cross magnetising effect of the current in the phase winding in which the maximum electromotive force is being developed is large. In Fig. 54 the P.D. wave of the same machine when working on an inductive load is shown. It will be seen that the wave is nearer to a sine wave than when the machine is working on a non-inductive load.

In Fig. 55 the curve A is the open circuit characteristic of a 400 kilovolt-ampere generator with a mesh connected armature. It is evident from the figure that the iron in the field magnet windings is saturated when the exciting current is large. In small machines, owing to the large air-gap, this characteristic is often very nearly a straight line. The curve B gives the characteristic when the machine is

Inductive characteristics.

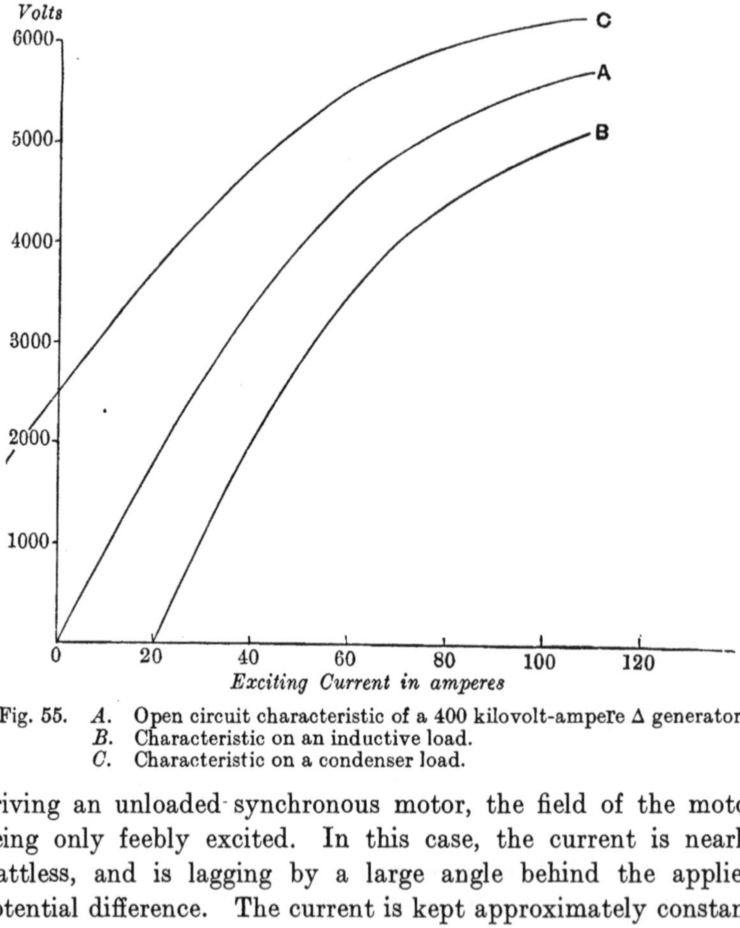

Fig. 55. *A*. Open circuit characteristic of a 400 kilovolt-ampere Δ generator.
B. Characteristic on an inductive load.
C. Characteristic on a condenser load.

driving an unloaded synchronous motor, the field of the motor being only feebly excited. In this case, the current is nearly wattless, and is lagging by a large angle behind the applied potential difference. The current is kept approximately constant and the field excitation of the alternator is varied. The curve obtained is similar to the corresponding curve for a single phase machine, and it can be utilised in a similar manner to find the leakage electromotive force of the armature. The curve C is obtained by over exciting the field of the synchronous motor, so that it acts like a condenser, and we have a wattless leading current. It will be noticed that in this case we can even reverse the direction of the current in the field magnet windings of the alternator, without the motor falling out of step as the field of the alternator is excited by the armature currents. These curves are useful, as by their aid we can determine approximately what the potential drop at the terminals of the machine will be with a given power factor (see p. 55), and as only a small amount of power is required to drive a machine on a wattless load, the makers are able to test it economically.

The characteristic curves of three phase machines are similar to those of single phase machines provided that the three phases are equally loaded. A difference arises, however, when the phases are unequally loaded, and it will be interesting to consider the curves obtained in this case. The following diagrams and data for a small three phase machine were obtained by André Blondel and will well repay study.

Tests of a three phase machine.

The armature of the three phase machine on which the experiments were carried out is star connected, and the full load current in the windings is 9 amperes, the pressure between the slip rings being 110 volts. The output of the machine on a balanced non-inductive load is therefore $\sqrt{3} \times 9 \times 110$ watts, that is 1·7 kilowatts. The following are the principal mechanical data.

Number of field magnet poles ...	4.
Area of polar face	100 square centimetres.
Diameter of armature	310 millimetres.
Number of slots	54.
Length of slots	110 millimetres.
Depth of slots	20 millimetres.
Greatest breadth of teeth	11 millimetres.
Number of conductors per slot ...	6.
Air-gap	3 millimetres.
Revolutions per minute	1350.
Frequency	45.

Characteristic curves.

The characteristic curves given in Fig. 56 were obtained in the usual manner, and their general shapes are in agreement with the curves obtained from first principles in Chapter I. The curve 11 is the open circuit characteristic, and, as it is a straight line, it proves that the iron of the field magnets is not magnetised strongly. The short circuit characteristic 22 is also a straight line. The

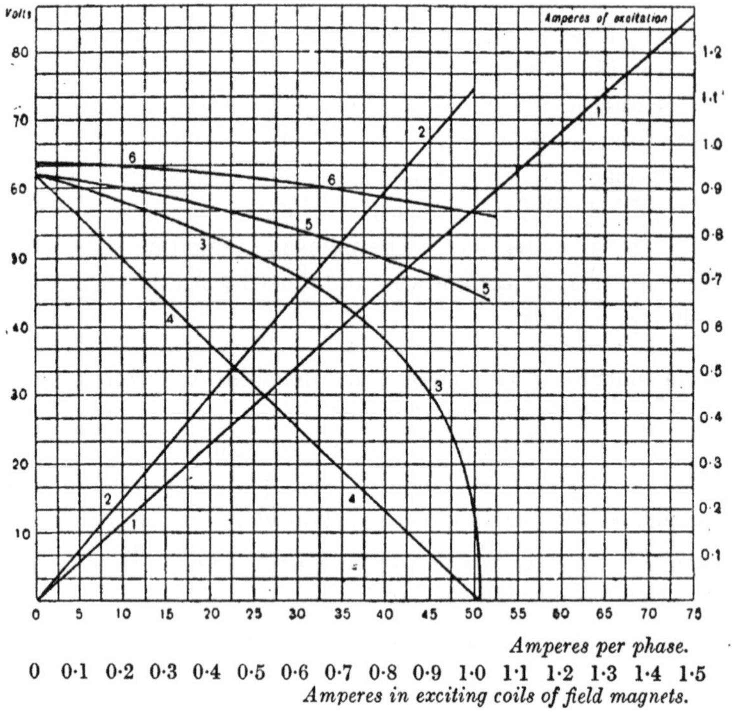

Fig. 56. Characteristic curves of a three phase alternator with star connected armature.

1. Open circuit characteristic.
2. Short circuit characteristic.
3. Characteristic on a non-inductive load when the three phases are loaded symmetrically.
4. Characteristic on an inductive load when the three phases are loaded symmetrically.
5. Voltage of the phase (1, 2) when it alone has a non-inductive load placed across it.
6. Voltage of the phases (2, 3) and (3, 1) in this case.

curve 33 gives the characteristic on a non-inductive load symmetrically balanced, when the excitation is 1·08 amperes. It is approximately an ellipse. The curve 44 is the characteristic on a purely inductive load; it is indistinguishable from a straight line. The curves 55 and 66 give the voltages between the slip rings 1 and 2, and between the slip rings 2 and 3, or 3 and 1, for various values of a non-inductive load connected between 1 and 2. The regular shapes of these curves might have led us to think that the shapes of the electromotive force and current waves do not vary much with the character of the load. The following oscillograph records, however, prove that they vary in an extraordinary manner.

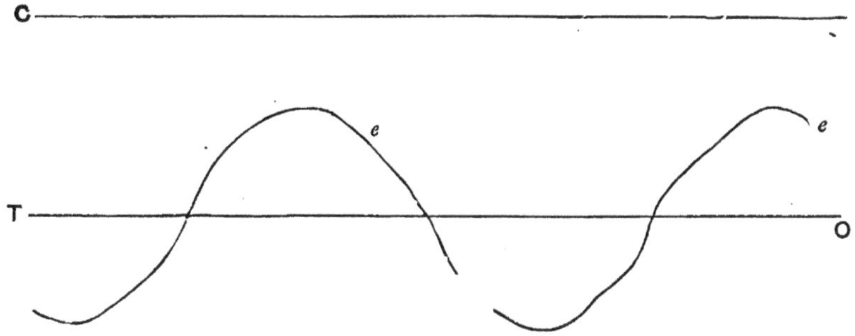

Fig. 57. *e*. Voltage wave across one phase of a star wound three phase machine on open circuit.
C. Oscillograph record of exciting current.

Oscillograph records.

In Fig. 57 the shape of the electromotive force wave *e* at the terminals of one phase on open circuit is given. It differs very little from that of a sine wave and is not rippled by the variations of the reluctance caused by the slots in the armature. This is due to the large air-gap making these variations in the reluctance small compared with the total reluctance of the gap. The effective value of *e* is 63 volts. The curve *C* (Fig. 57) is the record of the exciting current. The effective value of the exciting current in all the experiments was kept constant and equal to 1·08 amperes. On open circuit it is practically a straight line. Notice that in Fig. 57 and in the succeeding oscillograph records, the time is measured from right to left.

In Fig. 58 the curves of the exciting current *C* and the load current *i* are given when the three phases are equally loaded. The load in this case consisted of glow lamps, so that the shape of the electromotive force waves is the same as that of the current waves. The effective voltage between any of the slip rings and the neutral point common to the three windings is 57, and the current in each phase is 11·2 amperes. It will be noticed that C

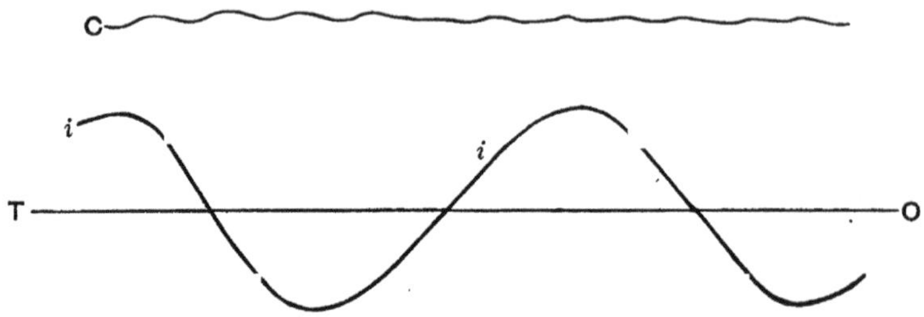

Fig. 58. i. Current wave in a phase winding of a star connected three phase machine when symmetrically loaded.
C. Oscillograph record of exciting current.

is a pulsatory current, the frequency of the pulsations being six times that of the alternating currents. This is due to the armature reaction. The magnetomotive force of the current in the armature

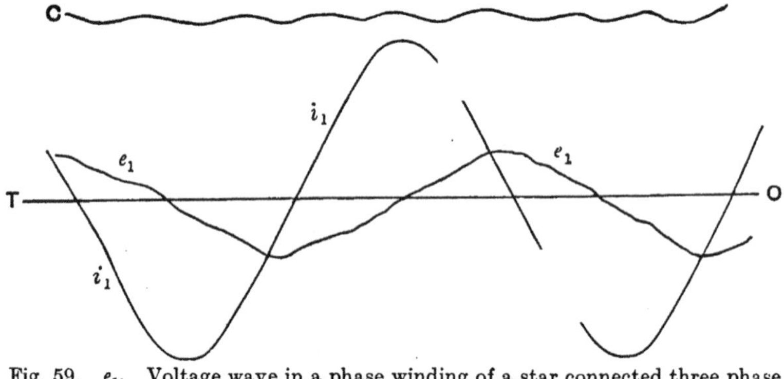

Fig. 59. e_1. Voltage wave in a phase winding of a star connected three phase machine when working on a symmetrical inductive load.
i_1. Current wave.
C. Exciting current.

windings which acts on the field magnets goes through all its values in one-sixth of a period.

The curves e_1 and i_1 in Fig. 59 show the shape of the voltage and current waves when the machine is working on a symmetrical inductive load. The effective value of i_1 is 32 amperes, and of e_1 24·1 volts.

In Fig. 60 the curve *i* practically gives the shape of the current wave on short circuit. The effective value of the potential difference between a slip ring and the neutral point is only one volt. The effective value of the current is 47·5 amperes. The effective value of *C* is 1·08 amperes, and the pulsations seem to be due to two disturbing causes, the frequencies of which are $2f$ and $6f$ respectively. This may be owing to the slightly greater demagnetising effect on the field magnets of one of the windings.

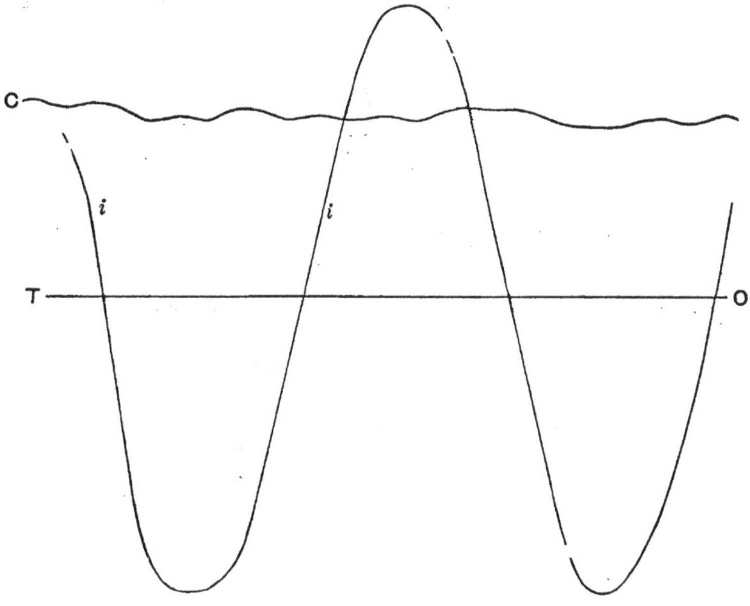

Fig. 60. *i*. Current wave when the slip rings are short circuited.
C. Exciting current.

The curves shown in Fig. 61 are very instructive, as they show the effect of loading one phase of a three phase machine. The load consisted of glow lamps in parallel with an electrolytic rheostat. The electrolyte was a solution of sulphate of zinc and the electrodes were zinc sheets. It was found experimentally that this rheostat acted to a certain extent like a condenser, the current wave leading the electromotive force wave. The load is connected across the slip ring joined to phase 1 and the neutral point. The effective value of the current in phase 1 is 50 amperes, and the effective potential difference across this phase is 44 volts.

In this case, the voltage across the second phase is 63·5, and the voltage across the third phase is 55·6. It will be seen that the shapes of the electromotive force waves in the three cases are quite different, the curve e_3 is more rounded than the curve e_1, and the curve e_2 is much more pointed. Since the resistance of the external circuit of phase 1 is less than one ohm, the current will lag by an appreciable angle ψ behind the phase of the armature electromotive force. If we denote the current by $I \sin(\omega t - \psi)$, then, by the principle of two reactions, the trans-

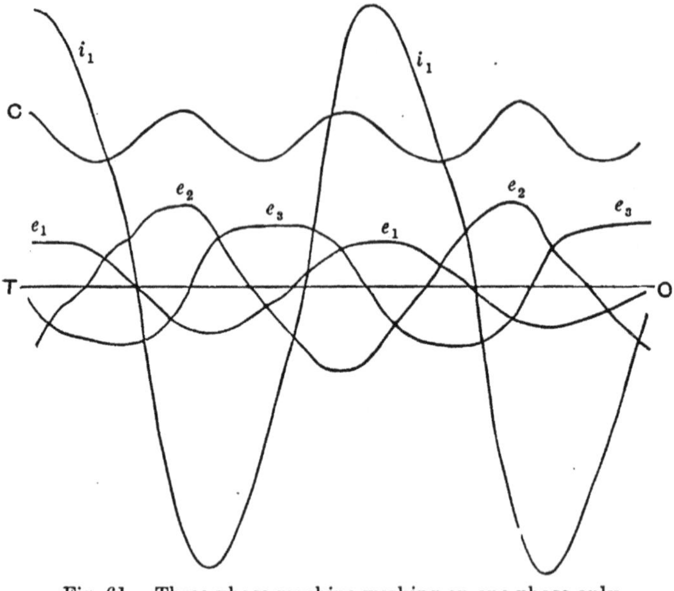

Fig. 61. Three phase machine working on one phase only.
e_1. Voltage wave of the loaded phase.
i_1. Current wave of this phase.
C. Exciting current.

verse magnetisation of the field will be proportional to $I \cos \psi$ and the demagnetising force acting on the field magnets will be proportional to $I \sin \psi$. The voltage of the second phase in this case is actually greater than the voltage on open circuit. This is due to the component $I \cos \psi \sin \omega t$ of the current in the phase 1 increasing the magnetisation of the sides of the poles nearly opposite the windings 2. Similarly this component weakens the flux density of the field on the other sides of the poles which are adjacent to the windings of phase 3. In addition, the component

$-I \sin\psi \cos\omega t$ of the armature current tends to demagnetise the field magnets. The voltage across the phase 2 being greater than on open circuit proves that the increased flux due to the transverse magnetisation more than compensates for the demagnetising effect of the lagging component of the current.

The pulsations of the exciting current C (Fig. 61) in this case are large, and, just as in single phase machines, their frequency is twice that of the frequency of the alternating currents. Although the effective value of the exciting current is 1·08 amperes, the same as on open circuit, yet the effective potential difference across the field magnet windings is now 95 volts, whilst on open circuit it is only 90 volts. This is due to the alternating electromotive force induced in the exciting circuit by the armature reaction.

Two phase machines.

The theory of two phase machines is practically the same as that of three phase machines. The four armature windings may be connected either in star or in mesh, and either the field magnets or the armature may rotate. There is, however, one case in which there is an important difference, namely when the armature has two separate windings as in Figs. 62 and 63. The armature of the machine represented diagrammatically in Fig. 62 rotates, and the effective value of the potential difference between the two outer slip rings equals that between the two inner rings, but differs from it in phase by 90 degrees. Hence the currents supplied respectively by the two pairs of slip rings to two symmetrical loads will differ in phase by 90 degrees. We may replace any two slip rings not attached to the same winding by a single slip ring without affecting the working of the machine. Suppose, for example, the slip rings 2 and 3 are replaced by a single slip ring x, and let $V_{1.x}$ denote the effective value of the volts between 1 and x. Then $V_{1.x}$ and $V_{4.x}$ and $V_{1.4}$ form an isosceles right-angled triangle; we have, therefore,

$$V_{1.x} = V_{4.x} = V_{1.4}/\sqrt{2} = 0{\cdot}7071\ V_{1.4}.$$

The phase difference between $V_{1.x}$ and $V_{4.x}$ is 90 degrees, and between $V_{1.4}$ and $V_{1.x}$ or $V_{4.x}$ is 135 degrees.

The electromotive forces can be calculated approximately by

the same formulae as for single phase machines, and so also can the transverse and demagnetising forces of the armature currents.

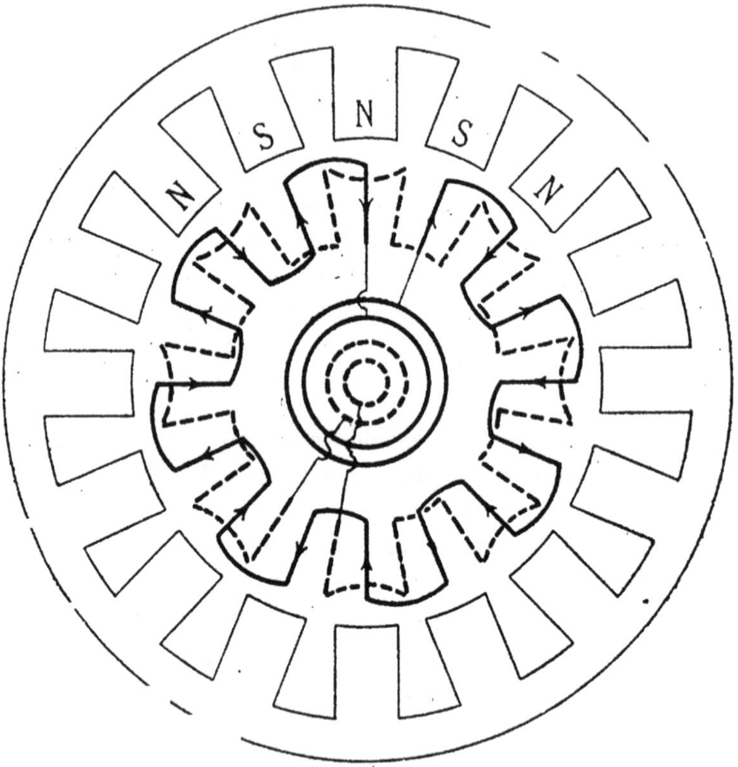

Fig. 62. Two phase armature with bar winding for a sixteen pole machine.

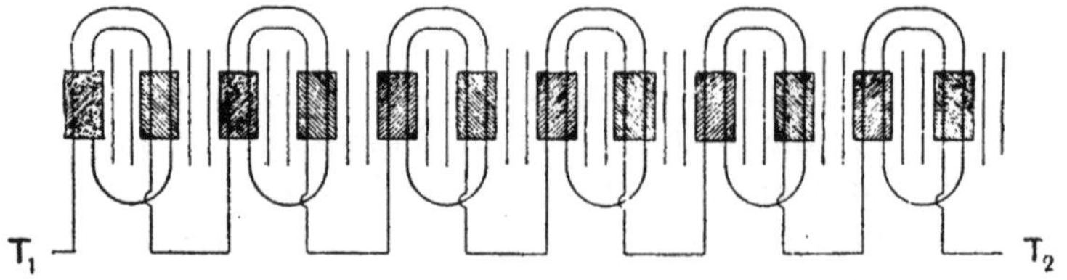

Fig. 63. Developed diagram of the winding of the armature of a two phase alternator. The winding of one phase only is shown.

In Fig. 63 a developed diagram of the winding of one phase of a two phase machine is shown. The field magnets rotate and T_1 and T_2 are two of the four terminals of the machine.

In Fig. 64 a coil winding for an armature is shown, consecutive coils of one phase being wound in reverse directions. When the coils are wound in the same directions, the connections of adjacent coils must be reversed.

If the armature be wound with four coils, the electromotive force generated in each of which differs in phase by ninety degrees from the E.M.F. generated in the two coils adjacent to it, then these windings may be connected in star or in mesh. If V be the effective voltage generated, and A the current flowing in each

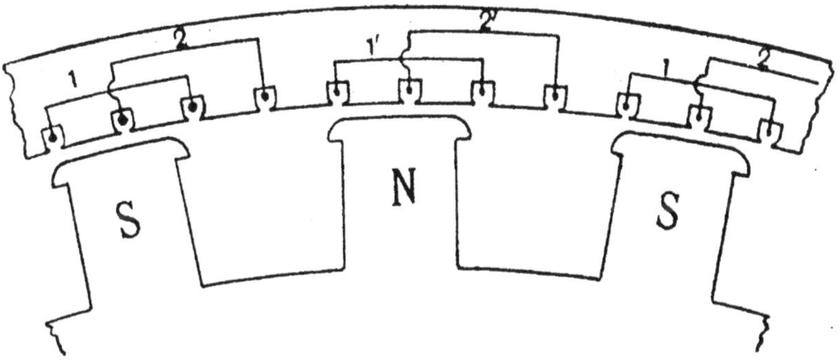

Fig. 64. Coil winding for the armature of a two phase generator. Coils 1 and 2 are wound in the reverse direction to coils 1′ and 2′.

coil when the machine is symmetrically loaded, the maximum output is $4VA$, whether the coils be connected in star or mesh fashion. When the coils are star connected, the effective voltage $V_{1.2}$ between adjacent mains is $\sqrt{2}V$, and the currents in the mains are each equal to A. When the coils are mesh connected, the effective voltage $V_{1.2}$ between adjacent mains is V, but the currents in the mains are now $\sqrt{2}A$.

If we connected the coils 1 and 3 in series and also the coils 2 and 4, we should have a two phase machine with two separate windings. In this case the voltage $V_{1.3}$ would be $2V$, and the maximum output would be $4VA$, the same as before.

Armature current on no load.

If $f(t)$ be the electromotive force generated in one phase of a mesh connected armature, the resultant E.M.F. round the armature windings will be

$$f(t) + f(t + T/4) + f(t + T/2) + f(t + 3T/4).$$

Now, whatever the shape of the wave, we have, if the field magnets and the armature windings are symmetrically situated,

$$f(t) = -f(t + T/2),$$

and $$f(t + T/4) = -f(t + 3T/4).$$
Hence the resultant E.M.F. is always zero.

A slight lack of symmetry, however, in the four windings might introduce a small local armature current at all loads.

Armature reaction in a two phase machine.

If we make our usual assumptions, the problem of the armature reaction in a two phase machine on a balanced load becomes very simple. Let us suppose that the machine has two separate windings and that the mutual inductance between the windings is zero. With our usual notation (p. 60) we have

$$0 = Ri_1 + L\frac{\partial i_1}{\partial t} + M\frac{\partial}{\partial t}(i' \cos \omega t),$$

$$0 = Ri_2 + L\frac{\partial i_2}{\partial t} + M\frac{\partial}{\partial t}(i' \sin \omega t),$$

and $$E = R_2 i' + L_2\frac{\partial i'}{\partial t} + M\frac{\partial}{\partial t}\{i_1 \cos \omega t + i_2 \sin \omega t\}.$$

It is easy to verify that

$$i_1 = I\sin(\omega t - \psi), \quad i_2 = -I\cos(\omega t - \psi), \text{ and } i' = E/R_2 = C,$$

where

$$I = M\omega C/Z, \quad \tan\psi = L\omega/R \text{ and } Z^2 = R^2 + L^2\omega^2$$

satisfy the above equations. They therefore represent the particular integral, and give the working of the machine when the steady state is attained.

The flux ϕ_1 linked with the first phase winding is given by

$$\phi_1 = Li_1 + MC\cos\omega t$$
$$= \{MC - LI\sin\psi\}\cos\omega t + LI\cos\psi\sin\omega t.$$

Hence the flux can be resolved into two components which are mutually at right angles and can be computed approximately in exactly the same way as for a single phase machine. The terminal potential difference can thus be found.

It is to be noticed that the E.M.F. induced in the first phase winding is $-M\omega C\sin\omega t$ whatever the load or power factor may be. The effective value of the terminal P.D. V_1 is given by

$$V_1 = \{(R_e^2 + L_e^2\omega^2)/(R^2 + L^2\omega^2)\}^{\frac{1}{2}}(M\omega C/\sqrt{2}).$$

The flux linked with the magnetising circuit at any instant is

$$L_2 C - MI\sin\psi.$$

It therefore depends only on the value of the magnetising current and the value of the wattless component of the armature current in either phase.

Tests of a two phase machine.

The following tests on a two phase alternator were made by Blondel. The two phase machine experimented on had two separate windings, and the normal current in each was 14 amperes at 110 volts, so that the full load output was 3·08 kilowatts. The principal data of this machine are given below.

Number of field magnet poles ...	4.
Area of polar face	100 square centimetres.
Diameter of the armature	310 millimetres.
Number of slots	52.
Length of slots	110 millimetres.
Depth of slots	24 millimetres.
Greatest breadth of teeth	11 millimetres.
Number of conductors per slot ...	7.
Air-gap	3 millimetres.
Revolutions per minute	1350.
Frequency	45.

Characteristic curves.

In Fig. 65 the characteristic curves of this machine are given, and it will be seen that their general appearance is similar to that of the three phase curves shown in Fig. 56.

The curve 11 is the open circuit characteristic, and is practically a straight line, showing that the iron is far from being saturated. The characteristic 22 on a balanced non-inductive load is approximately an ellipse, and on a purely inductive load it is a straight line 33. The characteristics when one phase only is loaded are shown in A_4, B_4, and A_5, B_5 respectively. A_4 is the characteristic of the loaded phase when working on a non-inductive resistance and A_5 its characteristic on a purely inductive load.

Oscillograph records.

In Fig. 66 the shape of the electromotive force wave of this machine on open circuit is shown. The ripples in the wave are due to the slots, and the equation to the wave, making sine curve assumptions, would be of the form

$$e = E \sin \omega t\,(1 + \lambda \sin 2n\omega t),$$

where λ is a small fraction and $2n$ is the number of armature teeth in the polar step. The effective value of the electromotive force in this case is 110 volts, and the exciting current is 1·02 amperes, the potential difference across the field magnet terminals being 89 volts.

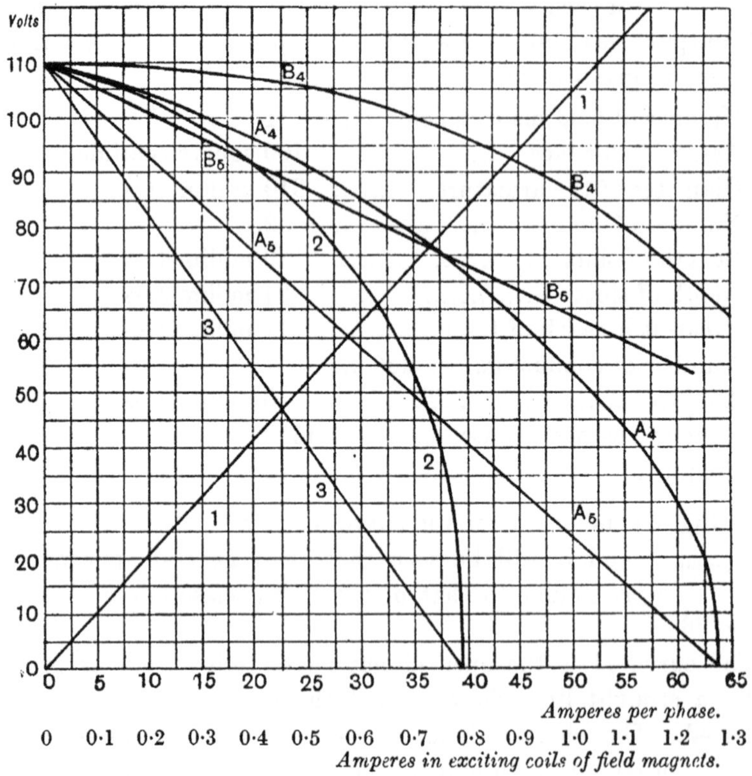

Fig. 65. Characteristic curves of a two phase alternator.

1. Open circuit characteristic.
2. Characteristic on a non-inductive load when the two circuits are equally loaded.
3. Characteristic on an inductive load when both circuits are equally loaded.

A_4. Characteristic of the loaded circuit when the other circuit is open. The volts of the open circuit are shown by B_4. The load is non-inductive.

A_5 and B_5. The same characteristics when the oaded phase is working on a purely inductive load.

It will be seen from the following diagrams that the ripples still remain in the potential difference waves when the machines are loaded, but they disappear from the current wave when the current becomes large (see Vol. I, p. 135).

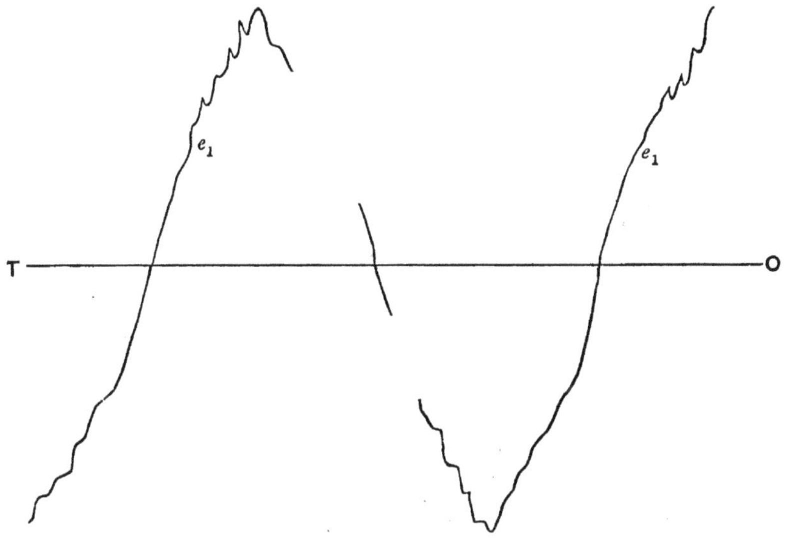

Fig. 66. Shape of the electromotive force wave e_1 of a two phase machine on open circuit.

In Fig. 67 the shapes of the potential difference wave e_1 and of the current wave i_1 are shown when the two circuits are working

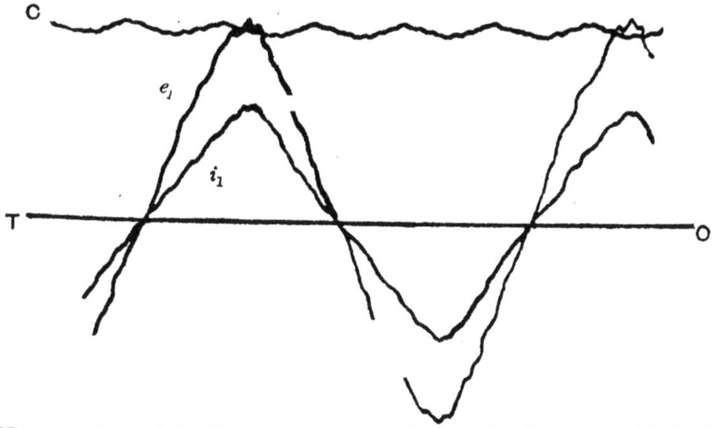

Fig. 67. e_1. Potential difference wave when the two circuits are equally loaded.
i_1. Current wave in the same case.
C. Exciting current.
The load consisted of glow lamps.

on equal non-inductive loads. The effective potential difference in each circuit is 99 volts and the effective current is 14·5 amperes. The ripples in the exciting current show that there is some armature reaction, and, although the effective value is 1·02 amperes, as before, yet the effective voltage at the terminals of the field magnet windings is 89·5 instead of 89, the value it has on open circuit. The frequency of the alternating component

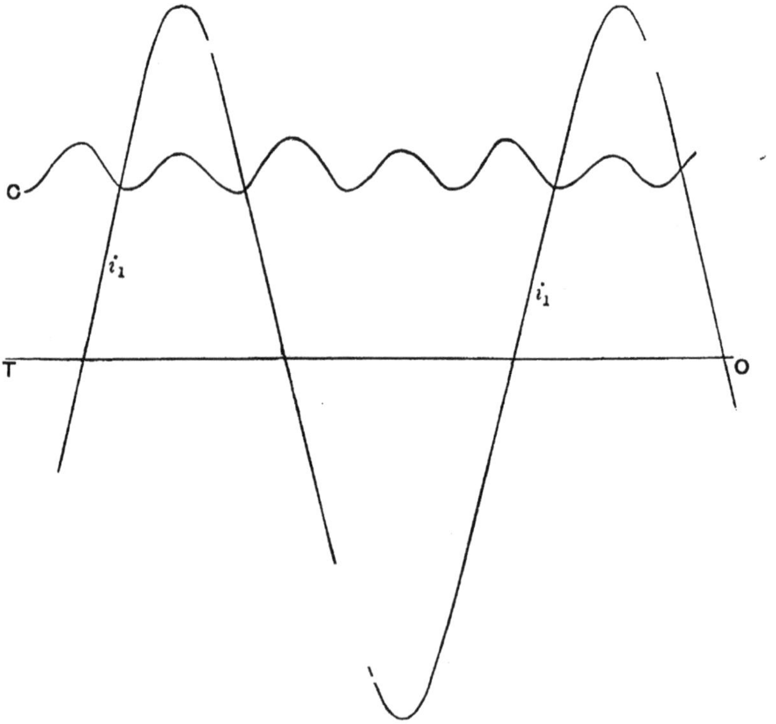

Fig. 68. i_1. Shape of the current wave when both circuits of a two phase machine are short circuited.
C. Exciting current.

in the exciting current is four times the frequency of the alternating current.

The shape of the current wave i_1 (Fig. 68) when the armature windings are short circuited is approximately a sine curve. This is due to the large inductance and small resistance of the circuits in this case. For this reason the high frequency components of the electromotive force produce only very minute currents, which are not apparent on the resultant current wave i_1. The effective value of i_1 is 39·5 amperes.

When one circuit is closed through a choking coil and the other is open, the difference between the shapes of the potential difference waves (Fig. 69) is very marked. The wave across the working circuit is very flat, and its effective value is 50 volts, whilst the wave across the unloaded circuit is peaky and has an effective value of 76 volts. The effective value of the current in the choking coil is 34·2 amperes, and the frequency of the

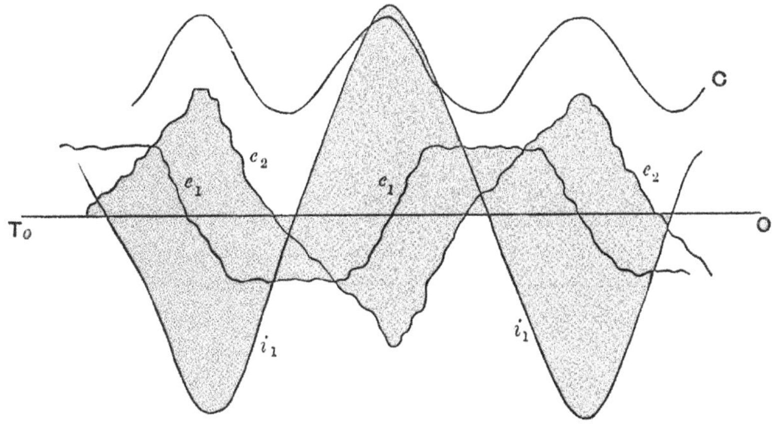

Fig. 69. e_1. Potential difference wave when the first circuit is working on an inductive load, the second circuit being open.
i_1. The current wave.
e_2. Potential difference wave across the unloaded circuit.
C. Exciting current.

alternating component of the exciting current, as in single phase machines, is twice that of the alternating current.

In Fig. 70 the current and potential difference waves are shown when one winding is short circuited. In this case the current wave is curiously distorted, and the exciting current has a very large alternating current component. The effective value of e_2, which is 110 when the other phase is open circuited, is now only 60 volts. The effective value of the current in the first winding is 62·5 amperes. If the current wave had been a sine curve, we should have expected that the maximum demagnetising forces would have occurred at the instants when the armature current had its maximum values. Since the frequency of the induced electromotive force in the field magnet windings is double that of the alternating current, this induced E.M.F. will have its maximum value one-eighth of a period after the armature current

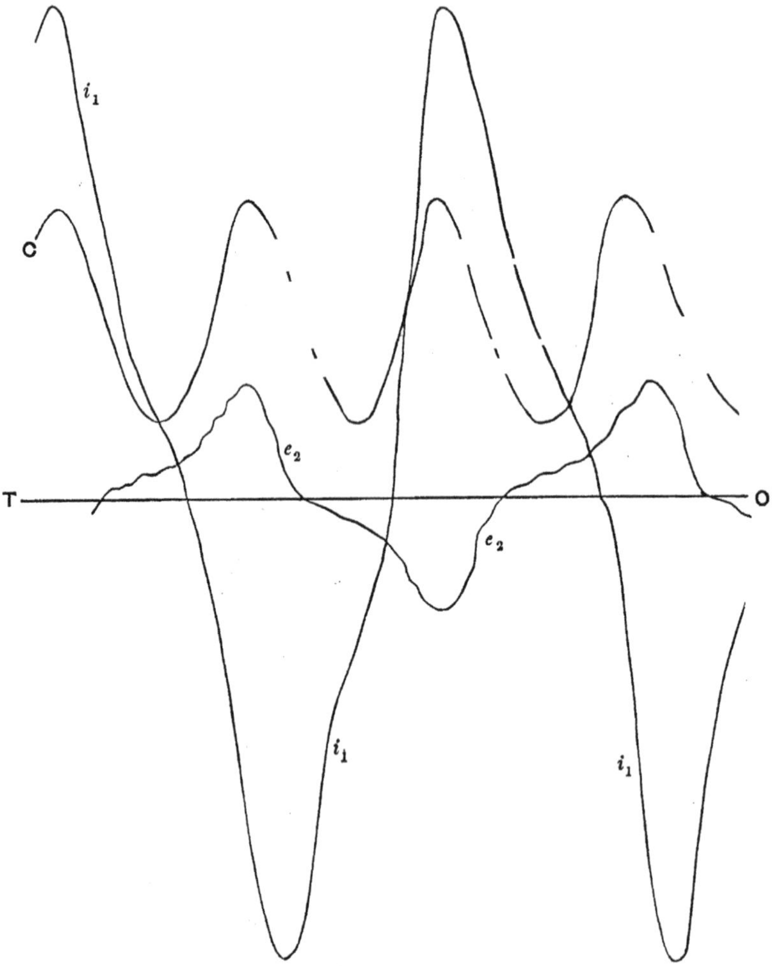

Fig. 70. Two phase machine with two separate windings, the first of which is short circuited and the second is on open circuit.

i_1. Current in short circuited windings.
e_2. Potential difference acıoss unloaded winding.
C. Exciting current.

has its maximum value. The phase of the alternating current component of C, however, will depend on the magnetic leakage and hence, even when the current i_1 is sine shaped, it would be difficult to determine the time lags between the maximum values of i_1 and C. We see from the figure that, for the given machine, one set of maximum values of C occur nearly at the same instants as the maximum positive values of i_1, and the other maximum values occur about one-eighth of a period later than the maximum negative values of i_1.

Fig. 71. 2330 K.V.A. Caffaro generator.

Tests of a large three phase generator.

The three phase generator shown in Fig. 71 is one of several which were supplied by the Oerlikon Company to the generating station of the power transmission line at Caffaro in the north of Italy. Each generator is used in conjunction with a transformer immersed in oil, which is kept cool by pipes through which water flows. One of these transformers is shown in Chapter XII, and its efficiency curve is given. The generator voltage is raised to 40,000 by means of the transformers, and this is the voltage between each of the transmission wires. The total weight of each generator is 82,000 pounds, and the weight of the revolving field system is 30,000 pounds. The output of each machine is 233 amperes at 10,000 volts, and the rotor makes 315 revolutions per minute.

In Fig. 72 the short circuit and open circuit characteristics of this machine are shown. The working pressures vary between 9000 and 10,500, so that points on the load characteristics corresponding to working values of the amperes and volts lie between the thick lines shown on the diagram. Parts of the

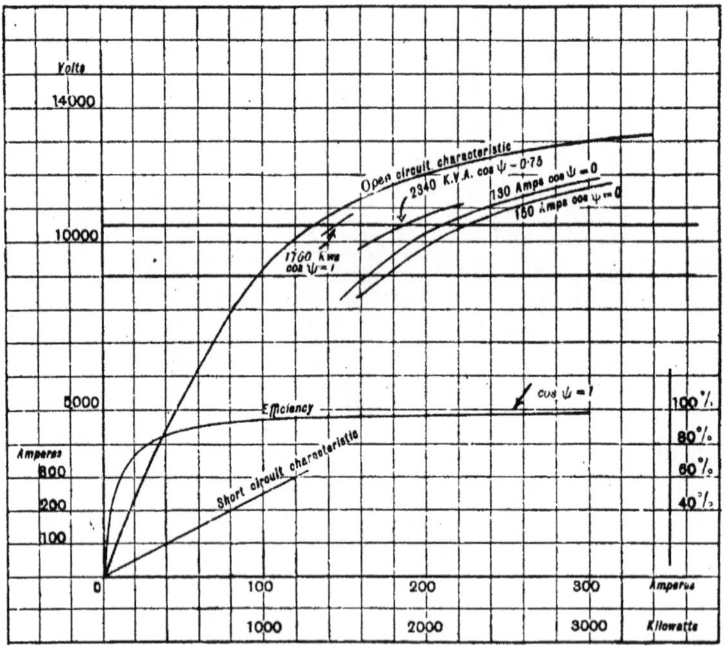

Fig. 72. Curves for 2330 kilovolt-ampere three phase generator (Oerlikon, type 6235). The machine runs at 315 revs. per minute, and its voltage is varied from 9000—10,500 as desired.

wattless characteristics at 130 and 150 amperes are shown, and parts also of the load characteristics when $\cos\psi$ is equal to 1 and when $\cos\psi$ is 0·75. The efficiency curve η, in terms of the load when the power factor is unity, is also given. When the power factor is 0·75, the efficiency curve is practically the same, at small loads, as the curve shown. At the maximum load on this power factor the ordinate of this curve is only diminished by about one per cent.

Load losses.

The losses of the Oerlikon generator for different loads are shown by the curves in Fig. 73. The resistance of one phase of the armature winding is 0·19 of an ohm, and the

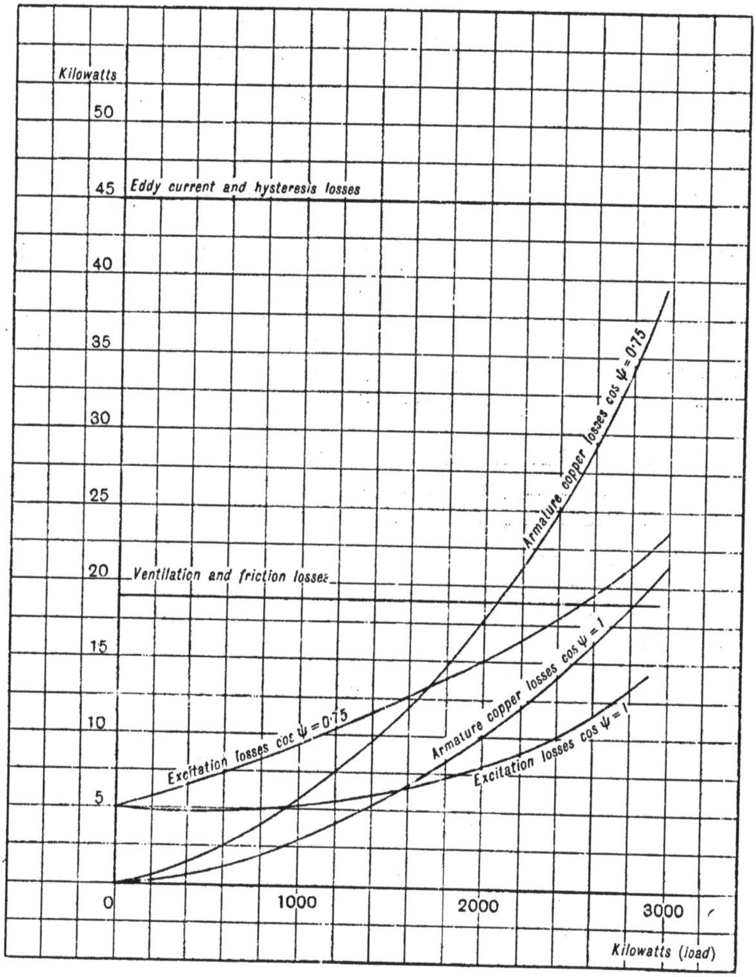

Fig. 73. Curves of the losses in a 2330 kilovolt-ampere three phase generator (Oerlikon, type 6235).

resistance of the windings of the field magnets is 0·52 of an ohm, both being measured when warm. It should be noticed how rapidly the copper losses in the armature increase on inductive loads.

The efficiency of the exciter.

The curves showing the performance of the hundred kilowatt separate exciter used in conjunction with the above machine are shown in Fig. 74. The machine is shunt wound; the armature resistance is 0·0015 of an ohm and the resistance of the shunt winding is 10·5 ohms, both being measured when warm. By varying the resistance of the rheostat

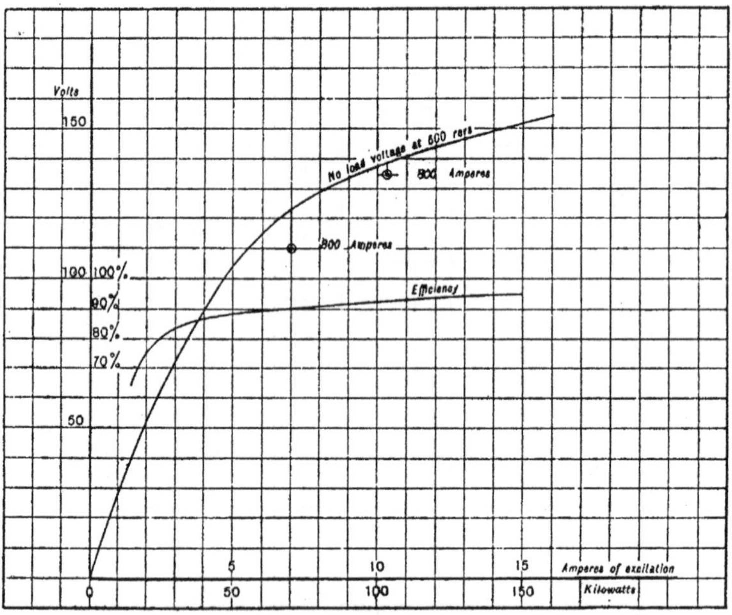

Fig. 74. Open circuit characteristic and efficiency curve of a 100 K.W. separate exciter. When running at 600 revs. per minute it has an output of 800 amperes at 125 volts.

Armature resistance		0·0015 ohm.
Shunt resistance	=	10·5 ohms.
Loss at no load	–	6·3 kilowatts.

in the shunt circuit, the armature being driven at a constant speed of 600 revolutions per minute, we get the no-load voltage curve by plotting out simultaneous readings of the volts and amperes for different positions of the contact maker of the rheostat. The difference of the P.D. drop at the terminals with a curıent of 800 amperes when the excitation is 7 and 10·4 amperes respectively should be noticed.

REFERENCES.

H. F. Parshall and H. M. Hobart, *Armature Windings of Electric Machines.*

C. C. Hawkins and F. Wallis, *The Dynamo.*

D. C. Jackson and J. P. Jackson, *Alternating Currents and Alternating Current Machinery.*

André Blondel, Dobkévitch, Duris, Farmer and Tcheruosvitoff, 'Application des Oscillographes à l'étude des Alternateurs,' *L'Éclairage Électrique*, Vol. 29, p. 391. Paper communicated to the International Congress of Electricians at Paris, August, 1900.

T. R. Lyle, 'Theory of the Alternate Current Generator,' *Proc. of the Phys. Soc.*, Vol. 21, p. 702, 1909.

Miles Walker, 'Short-circuiting of Large Electric Generators and the Resulting Forces on Armature Windings,' *Journ. of the Inst. of Elect. Engin.*, Vol. 45, p. 295, 1910.

—— 'The Design of Turbo Field Magnets for Alternate-current Generators, with Special Reference to Large Units at High Speeds,' *Journ. of the Inst. of Elect. Engin.*, Vol. 45, p. 319, 1910.

André Blondel, 'Analyse des réactions de l'induit dans les alternateurs,' *Comptes Rendus*, Vol. 158, p. 1961, 1914.

A. E. Clayton, 'The Wave Shapes obtaining with Alternating-Current Generators working under Steady Short-Circuit Conditions,' *Journ. of the Inst. of Elect. Engin.*, Vol. 54, p. 84, 1915.

CHAPTER III

Harmonic Analysis. Introductory. The *n*th harmonic. Proof of Fourier's theorem. Values of the constants. The practical problem. Practical methods. Weddle's rule. Series formulae involving areas. Series formulae involving ordinates. Comparison of the series methods. Approximate formulae. Semicircular wave. Special formulae. Formulae for alternating waves. Method of finding the form factor. Conclusion. References.

Introductory.

If the value of a function $f(x)$ recur over successive intervals λ of the variable x so that the equation $f(x) = f(x + n\lambda)$ is true for all integral values of n, then Fourier showed that we may write

$$f(x) = a_0 + a_1 \cos(2\pi/\lambda)x + a_2 \cos 2(2\pi/\lambda)x + \ldots$$
$$+ b_1 \sin(2\pi/\lambda)x + b_2 \sin 2(2\pi/\lambda)x + \ldots$$

where $a_0, a_1, a_2, \ldots b_1, b_2, \ldots$ are constants which he proved can be expressed in the form of definite integrals. He pointed out that these coefficients were as real constants of a periodic curve as, for instance, its area or its centroid.

The *n*th harmonic.

The sum of the two terms, $a_n \cos n(2\pi/\lambda)x + b_n \sin n(2\pi/\lambda)x$, gives the *n*th harmonic, or the harmonic of order n, of the periodic curve $y = f(x)$. The first harmonic is sometimes called the fundamental harmonic of the given curve. The *n*th harmonic can be written in the form $(a_n^2 + b_n^2)^{\frac{1}{2}} \sin\{n(2\pi/\lambda)x + \alpha_n\}$ where $\tan \alpha_n = a_n/b_n$. Fourier's theorem thus states that the curve $y = f(x)$ can be supposed to be obtained by adding together the ordinates of the graphs of an infinite number of curves the equations to which are

$$y = a_0, \quad y = (a_1^2 + b_1^2)^{\frac{1}{2}} \sin\{(2\pi/\lambda)x + \alpha_1\}, \ldots.$$

The amplitude of the *n*th harmonic is given by $(a_n^2 + b_n^2)^{\frac{1}{2}}$ and its phase by the angle α_n which equals $\tan^{-1}(a_n/b_n)$. It is therefore completely determined when we know a_n and b_n.

Fourier's theorem may be proved by the ordinary symbolical methods used in solving differential equations. We may write Taylor's theorem for the expansion of a function of two variables as follows,

Proof of Fourier's theorem.

$$f(x+\lambda) = \epsilon^{\lambda \frac{\partial}{\partial x}} f(x),$$

where ϵ is the base of Neperian logarithms. Hence we may write the equation $f(x+\lambda) = f(x)$ in the form

$$(\epsilon^{\lambda \frac{\partial}{\partial x}} - 1) f(x) = 0.$$

Regarding this as a differential equation, the roots of the auxiliary equation are given by

$$\epsilon^{\lambda p} = 1,$$

where p has been written for $\partial/\partial x$. Now by De Moivre's theorem

$$\epsilon^{\lambda(2m\pi/\lambda)\iota} = \cos 2m\pi + \iota \sin 2m\pi = 1,$$

and m can have any positive or negative integral value. Hence by the rules given for solving differential equations, the general solution is

$$\begin{aligned} f(x) &= a_0 + a_1'\epsilon^{(2\pi/\lambda)x\iota} + a_2'\epsilon^{2(2\pi/\lambda)x\iota} + \dots \\ &\quad + b_1'\epsilon^{-(2\pi/\lambda)x\iota} + b_2'\epsilon^{-2(2\pi/\lambda)x\iota} + \dots \\ &= a_0 + a_1\cos(2\pi/\lambda)x + a_2\cos 2(2\pi/\lambda)x + \dots \\ &\quad + b_1\sin(2\pi/\lambda)x + b_2\sin 2(2\pi/\lambda)x + \dots \end{aligned} \qquad \text{...............(1),}$$

and if $f(x)$ is real the constants must all be real. It will be seen that this is Fourier's theorem. Before, however, we are justified in using it in practice we should have to investigate whether such a series is uniformly and absolutely convergent. This is done in E. W. Hobson's *Theory of Functions of a Real Variable.*

Mathematical expressions for the values of the constants can be readily obtained. Multiply both sides of equation (1) by ∂x and integrate from 0 to λ; then, noticing that

Values of the constants.

$$\int_0^\lambda \cos n\frac{2\pi}{\lambda}x\,\partial x = 0 \text{ and } \int_0^\lambda \sin n\frac{2\pi}{\lambda}x\,\partial x = 0,$$

where n is an integer not less than unity, we find that

$$a_0 = \frac{1}{\lambda}\int_0^\lambda f(x)\,\partial x \qquad \text{.......................(2).}$$

Similarly, multiplying both sides of equation (1) by $\cos n\,(2\pi x/\lambda)\,.\,\partial x$ and integrating from 0 to λ, we get

$$a_n = \frac{2}{\lambda}\int_0^\lambda f(x)\cos n\left(\frac{2\pi x}{\lambda}\right)\partial x \quad \ldots\ldots\ldots\ldots(3),$$

where n equals 1, 2, ...

This readily follows since

$$2\int \cos m\left(\frac{2\pi x}{\lambda}\right)\cos n\left(\frac{2\pi x}{\lambda}\right)\partial x$$

$$= \frac{\sin (m+n)(2\pi x/\lambda)}{(m+n)(2\pi/\lambda)} + \frac{\sin (m-n)(2\pi x/\lambda)}{(m-n)(2\pi/\lambda)}$$

when m is not equal to n, and

$$= \frac{\sin 2n\,(2\pi x/\lambda)}{2n\,(2\pi/\lambda)} + x$$

when m equals n. Hence

$$\int_0^\lambda \cos m\left(\frac{2\pi x}{\lambda}\right)\sin n\left(\frac{2\pi x}{\lambda}\right)\partial x$$

equals zero when m and n are different and equals $\lambda/2$ when m equals n. In an exactly similar way we find that

$$b_n = \frac{2}{\lambda}\int_0^\lambda f(x)\sin n\left(\frac{2\pi x}{\lambda}\right)\partial x \quad \ldots\ldots\ldots\ldots(4).$$

When we know the mathematical expression for $f(x)$, approximate numerical values of the constants can generally be determined without much difficulty. In some cases the exact numerical values can be found.

The practical problem. In engineering practice we are as a rule given the graph of $f(x)$ and we have to determine the constants $a_0, a_1, b_1, \ldots$. The value of a_0 can be found at once by finding the area $\int_0^\lambda f(x)\,\partial x$ and dividing it by λ. The area is determined by one or other of the methods of mechanical quadrature in everyday use. It is to be noticed that this gives us an approximate value of the integral. In one method a certain number of equidistant ordinates are measured and the integral is expressed in terms of them. In exactly the same way we can find the values of a_n and b_n in terms of certain selected ordinates of the curves

$$y = f(x)\cos n\,(2\pi x/\lambda) \quad \text{and} \quad y = f(x)\sin n\,(2\pi x/\lambda)$$

respectively. It is unnecessary to draw the graphs of these curves as the values of the ordinates can be found at once by multiplying $f(x)$ by the required cosine or sine coefficient. Before showing the best way of doing this, we shall give a brief *résumé* of the methods ordinarily employed.

Practical methods.

There are three typical methods used in practice. In the first method all harmonics the order of which is higher than some value n are neglected. In this case the equation $y = f(x)$ contains $2n + 1$ constants. In order to determine these constants we require to know the values of at least $2n + 1$ ordinates. We get, therefore, $2n + 1$ equations of the form

$$y_1 = a_0 + a_1 \cos(2\pi x_1/\lambda) + a_2 \cos 2(2\pi x_1/\lambda) + \ldots$$
$$+ b_1 \sin(2\pi x_1/\lambda) + b_2 \sin 2(2\pi x_1/\lambda) + \ldots$$

and various ways are given for lightening the algebraical labour involved in finding a_0, a_1, b_1, ... from these equations. We thus find the equation to a curve having the same $2n + 1$ ordinates as the given curve.

Gauss showed that the required equation could be written down at once in the following form:

$$y = y_1 \frac{\sin\frac{1}{2}(x - x_2)\sin\frac{1}{2}(x - x_3)\ldots\sin\frac{1}{2}(x - x_{2n+1})}{\sin\frac{1}{2}(x_1 - x_2)\sin\frac{1}{2}(x_1 - x_3)\ldots\sin\frac{1}{2}(x_1 - x_{2n+1})}$$
$$+ y_2 \frac{\sin\frac{1}{2}(x - x_1)\sin\frac{1}{2}(x - x_3)\ldots\sin\frac{1}{2}(x - x_{2n+1})}{\sin\frac{1}{2}(x_2 - x_1)\sin\frac{1}{2}(x_2 - x_3)\ldots\sin\frac{1}{2}(x_2 - x_{2n+1})}$$
$$+ \ldots \quad \ldots\ldots\ldots\ldots\ldots\ldots\ldots\ldots\ldots\ldots\ldots\ldots\ldots\ldots(5),$$

where x_1, x_2, ... are the abscissae corresponding to the ordinates y_1, y_2, ...

To prove (5) it is sufficient to notice that when $x = x_1$, $y = y_1$; when $x = x_2$, $y = y_2$; etc., and that the coefficients of y_1, y_2, ... when expanded only contain sines and cosines of integral multiples of x which are not greater than n.

By supposing that n is infinitely great and noticing that when x increases by $2m\pi$ the value of y is unaltered we see that any periodic curve the wave length of which is 2π can be put in the form (5) and therefore in the form (1). This indicates another way of proving Fourier's theorem.

In practice it is laborious to find the coefficients of cos mx and sin mx in (5) even when n is small. The theorem, therefore, is not of much use for harmonic analysis although it is of value in interpolation.

Apart altogether from the labour involved in the methods described above, a serious drawback is the lack of any indication as to how far the calculated values of the harmonics differ from their true Fourier values. When n is large it is probable that the values of a_m and b_m found in this way are approximately correct when m is small. When however m is equal to or nearly equal to n it is highly probable that the computed values are quite different from the Fourier values.

In the second method an attempt is made to determine the values of the definite integrals given in (2), (3) and (4) by mechanical quadrature. For this purpose Weddle's rule for integration discussed below will be found very suitable. One of the advantages of this method is that the amplitude and phase of any given harmonic can be determined separately to any required degree of accuracy. The lower the order of the harmonic the less is the arithmetical labour involved in finding its value.

In the third method certain infinite series are obtained connecting the values of Fourier constants. When the harmonics diminish rapidly this method is a simple and easy one. In any case these formulae are useful as supplementary aids in checking the values of a_n and b_n found by other methods.

Weddle's rule.

An approximate value of the area A included between a curve, two ordinates and the axis of x ($BPQN_6O$ in Fig. 75) is

$$A = \int_0^x y\partial x = \frac{x}{20}[y_0 + y_2 + y_4 + y_6 + 5(y_1 + y_5) + 6y_3] \quad \ldots(6),$$

where $ON_1 = N_1N_2 = \ldots = N_5N_6 = x/6$, and $y_0, y_1, y_2, \ldots$ are the values of the ordinates at the points $O, N_1, N_2, \ldots$.

To prove this formula we first make the assumption that the equation to the curve shown in Fig. 75 can be put in the form

$$y = a_0 + a_1x + a_2x^2 + \ldots + a_nx^n \quad \ldots\ldots\ldots\ldots\ldots(7).$$

This curve can be made to pass through any $n + 1$ points on the given curve and it is the 'simplest' curve that can be drawn through these $n + 1$ points. Its equation may be written in the form

$$y = y_1 \frac{(x - x_2)(x - x_3) \ldots (x - x_{n+1})}{(x_1 - x_2)(x_1 - x_3) \ldots (x_1 - x_{n+1})}$$
$$+ y_2 \frac{(x - x_1)(x - x_3) \ldots (x - x_{n+1})}{(x_2 - x_1)(x_2 - x_3) \ldots (x_2 - x_{n+1})}$$
$$+ \ldots \qquad \ldots\ldots(8).$$

By expanding the coefficients of $y_1, y_2, \ldots$ in this equation and comparing with (7) the values of $a_0, a_1, a_2, \ldots$ are at once found.

By applying Weddle's rule to (7) we find that the term $a_m x^m$ adds to the value found for the area the expression

$$\frac{a_m x^{m+1}}{20 \cdot 6^m} [0^m + 2^m + 4^m + 6^m + 5(1 + 5^m) + 6.3^m] \ldots(9).$$

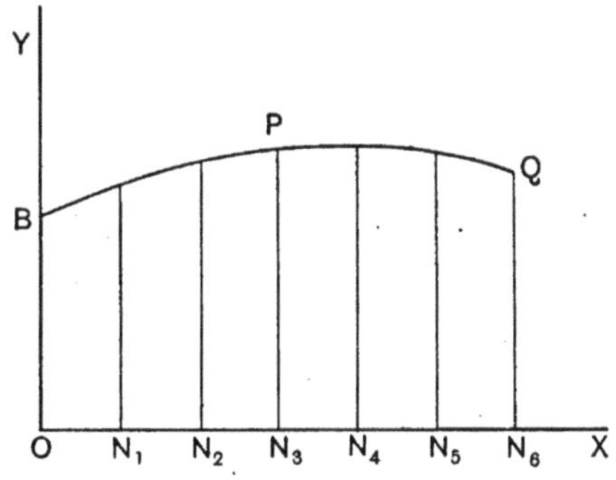

Fig. 75. Weddle's rule.

Hence, putting $m = 0, 1, 2, \ldots$ in (9) and simplifying, we find that Weddle's rule gives the following value for the area:

$$A = a_0 x + a_1 (x^2/2) + a_2 (x^3/3) + a_3 (x^4/4) + a_4 (x^5/5)$$
$$+ a_5 (x^6/6) + a_6 (x^7/7)(1 \cdot 00013)$$
$$+ a_7 (x^8/8)(1 \cdot 00051) + a_8 (x^9/9)(1 \cdot 00145)$$
$$+ a_9 (x^{10}/10)(1 \cdot 00343) + a_{10} (x^{11}/11)(1 \cdot 00691)$$
$$+ \ldots \qquad \ldots\ldots(10).$$

By the Integral Calculus the true value of the area is

$$A = \sum_{p=0}^{p=n} a_p x^{p+1}/(p + 1).$$

We see that, provided that n is not greater than 5, Weddle's rule is absolutely correct, and even for values of n as great as 10 the error *for individual terms* is well under one per cent.

For example, let us determine the value of $\int_0^{0\cdot 6} \sinh x \partial x$ by (6). We get

$$\int_0^{0\cdot 6} \sinh x \partial x = \frac{3}{100}[\sinh(0\cdot 2) + \sinh(0\cdot 4) + \sinh(0\cdot 6) + 5\{\sinh(0\cdot 1) + \sinh(0\cdot 5)\} + 6 \sinh(0\cdot 3)]$$
$$= 0\cdot 185\,465\,2 \ldots$$

which is correct to the last figure. On the other hand we see by (10) that the error in the value of $\int_0^x x^{10} \partial x$ found by quadrature is 0·69 of one per cent. It is advisable, therefore, to have means of checking the computed values.

In applying Weddle's rule to find the area of an isosceles triangle it will be found that the error is nearly seven per cent. The large error is due to the fact that the vertex is a point of discontinuity in the bounding line. In this case the rule should be applied to find the area of each half of the triangle in succession, and the sum will give the correct area. In general, if we have several points $P_1, P_2, P_3, \ldots$ of discontinuity in the bounding line, the rule should be applied to the areas whose upper boundaries are $P_1P_2, P_2P_3, \ldots$ successively. The sum of them will then give very approximately the value of the required area.

Series formulae involving areas.

By Euler's theorem we may write

$$\cos x = (\epsilon^{x\iota} + \epsilon^{-x\iota})/2,$$

and thus we see that

$$\cos x - \tfrac{1}{3}\cos 3x + \tfrac{1}{5}\cos 5x - \ldots$$
$$= \tfrac{1}{2}[\epsilon^{x\iota} - \tfrac{1}{3}\epsilon^{3x\iota} + \ldots + \epsilon^{-x\iota} - \tfrac{1}{3}\epsilon^{-3x\iota} + \ldots]$$
$$= \tfrac{1}{2}[\tan^{-1}\epsilon^{x\iota} + \tan^{-1}\epsilon^{-x\iota}]$$
$$= \tfrac{1}{2}\tan^{-1} \pm \infty.$$

Hence the series equals $\pi/4$, from $x = (2n - \frac{1}{2})\pi$ to $(2n + \frac{1}{2})\pi$, and equals $-\pi/4$ from $x = (2n + \frac{1}{2})\pi$ to $(2n + \frac{3}{2})\pi$. Similarly we can show that

$$\sin x + \frac{1}{3}\sin 3x + \frac{1}{5}\sin 5x + \ldots = \frac{\pi}{4}, \text{ from } x = 2n\pi \text{ to } (2n+1)\pi,$$

and $\qquad = -\frac{\pi}{4}$, from $x = (2n+1)\pi$ to $(2n+2)\pi$.

Hence by formula (3) we find that

$$a_1 - \frac{a_3}{3} + \frac{a_5}{5} - \dots = \frac{2}{\lambda}\int_0^\lambda f(x)\left\{\cos\frac{2\pi x}{\lambda} - \frac{1}{3}\cos 3\frac{2\pi x}{\lambda} + \dots\right\}\partial x$$

$$= \frac{\pi}{2\lambda}\left[\int_0^{\lambda/4} - \int_{\lambda/4}^{3\lambda/4} + \int_{3\lambda/4}^{\lambda}\right] y\,\partial x.$$

The series therefore on the left-hand side equals $(\pi/2\lambda)$ times the difference between the sum of the areas of the curve from 0 to $\lambda/4$ and from $3\lambda/4$ to λ, and the area of the curve from $\lambda/4$ to $3\lambda/4$.

Similarly we see that

$$a_m - \frac{a_{3m}}{3} + \frac{a_{5m}}{5} - \dots = \frac{\pi}{2\lambda}\left[\int_0^{\lambda/4m} - \int_{\lambda/4m}^{3\lambda/4m} + \int_{3\lambda/4m}^{5\lambda/4m} - \dots\right] y\,\partial x \qquad \dots\dots(11),$$

and

$$b_m + \frac{b_{3m}}{3} + \frac{b_{5m}}{5} + \dots = \frac{\pi}{2\lambda}\left[\int_0^{\lambda/2m} - \int_{\lambda/2m}^{2\lambda/2m} + \int_{2\lambda/2m}^{3\lambda/2m} - \dots\right] y\,\partial x \qquad \dots\dots(12).$$

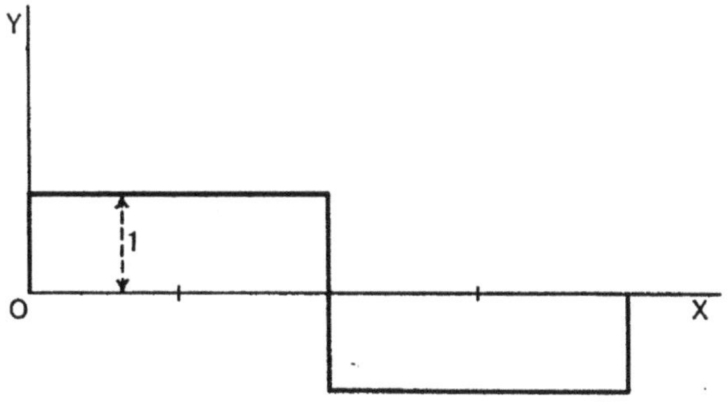

Fig. 76. Rectangular wave.

It will be noticed that the expressions on the right-hand side of (11) and (12) represent areas which can be found by any of the ordinary methods. Formulae (11) and (12), therefore, enable us to find approximate values for the Fourier coefficients when the amplitudes of the higher harmonics are small.

Let us first consider the case of the rectangular wave shown in Fig. 76. In this case y is $+1$ when x lies between 0 and $\lambda/2$ and -1 when x lies between $\lambda/2$ and λ. By (11) we find that

$$a_m - a_{3m}/3 + a_{5m}/5 - \dots = 0,$$

for all values of m, and thus the coefficients of all the cosine terms are zero. By (12) we find that

$$b_m + b_{3m}/3 + b_{5m}/5 + \ldots = \pi/(2m) \text{ or } 0,$$

according as m is odd or even.

If we neglect all harmonics whose orders are higher than 16, we have

$$b_1 + b_3/3 + b_5/5 + \ldots + b_{15}/15 = \pi/2,$$
$$b_3 + b_9/3 + b_{15}/5 = \pi/6,$$
$$b_5 + b_{15}/3 = \pi/10,$$
$$b_7 = \pi/14, \text{ etc.},$$

and hence we easily find the values of b_1, b_3, ... obtained on this assumption.

Harmonics	Computed values	True values	Harmonics	Computed values	True values
b_1	1·286	1·273	b_9	0·174	0·141
b_3	0·445	0·424	b_{11}	0·143	0·116
b_5	0·279	0·255	b_{13}	0·121	0·098
b_7	0·224	0·182	b_{15}	0·105	0·085

The true values have been calculated from the coefficients of the sine terms in the following well-known equation to the wave shown in Fig. 76,

$$y = \frac{4}{\pi}\left[\sin x + \tfrac{1}{3}\sin 3x + \tfrac{1}{5}\sin 5x + \ldots\right],$$

the wave length being 2π.

As the true values of the areas have been substituted in the formulae used, the errors are due to the neglect of the seventeenth and higher harmonics. It will be seen that the error in the value of b_1 is about one per cent. and the errors in the computed values of b_7 and higher harmonics are greater than 23 per cent.

As a further example let us consider the highly irregular wave shown in Fig. 77. The equation to this wave, when $\lambda = 2\pi$, is

$$y = \frac{2\sqrt{3}}{\pi}\left[\cos x - \tfrac{1}{5}\cos 5x + \tfrac{1}{7}\cos 7x - \tfrac{1}{11}\cos 11x + \ldots\right],$$

the harmonics the orders of which are multiples of 3 being absent. In this case, we have

$$a_1 - \frac{a_3}{3} + \frac{a_5}{5} - \dots = \frac{1}{4}\left[\int_0^{\pi/2} - \int_{\pi/2}^{3\pi/2} + \int_{3\pi/2}^{2\pi}\right] y\,\partial x$$

$$= \int_0^{\pi/2} y\,\partial x = \frac{\pi}{3},$$

$$a_3 - \frac{a_9}{3} + \frac{a_{15}}{5} - \dots = 0,$$

$$a_5 - \frac{a_{15}}{3} + \frac{a_{25}}{5} - \dots = -\frac{\pi}{15},$$

$$a_7 - \frac{a_{21}}{3} + \frac{a_{35}}{5} - \dots = \frac{\pi}{21}, \text{ etc.}$$

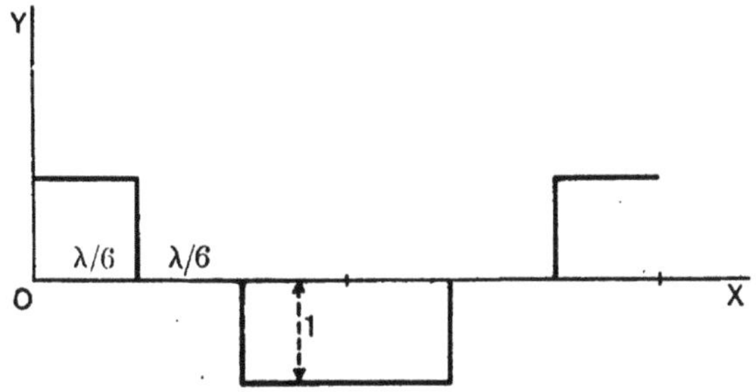

Fig. 77. Discontinuous rectangular wave.

We also have

$$b_m + b_{3m}/3 + b_{5m}/5 + \dots = 0,$$

for all values of m.

From the equations given above we see that $b_m = 0$. Neglecting a_m when m is greater than 24, we easily compute the numbers given in the following table.

Harmonics	Computed values	True values	Harmonics	Computed values	True values
a_1	1·100	1·103	a_{13}	0·081	0·085
$-a_5$	0·209	0·221	$-a_{17}$	0·062	0·065
a_7	0·150	0·158	a_{19}	0·055	0·058
$-a_{11}$	0·095	0·100	$-a_{23}$	0·046	0·048

The error in the value of a_1 is about 0·3 of one per cent. and in the value of the higher harmonics which do not vanish it is about 5 per cent.

We have from (3)

Series formulae involving ordinates.

$$a_1 + a_3 + a_5 + \dots$$
$$= \frac{2}{\lambda}\int_0^\lambda f(x)\left\{\cos\frac{2\pi x}{\lambda} + \cos 3\frac{2\pi x}{\lambda} + \dots\right\}\partial x$$

and hence, integrating by parts, we get

$$a_1 + a_3 + a_5 + \dots = \frac{1}{\pi}\left[f(x)\left\{\sin\frac{2\pi x}{\lambda} + \frac{1}{3}\sin 3\frac{2\pi x}{\lambda} + \dots\right\}\right]_0^\lambda$$
$$- \frac{1}{\pi}\int_0^\lambda f'(x)\left\{\sin\frac{2\pi x}{\lambda} + \frac{1}{3}\sin 3\frac{2\pi x}{\lambda} + \dots\right\}\partial x.$$

Noticing that the first expression on the right-hand side vanishes at both limits and that the series inside the bracket under the integral sign equals $\pi/4$ from 0 to $\lambda/2$ and $-\pi/4$ from $\lambda/2$ to λ, we get

$$a_1 + a_3 + a_5 + \dots = -(1/4)\,[f(\lambda/2) - f(0) - f(\lambda) + f(\lambda/2)]$$
$$= (1/2)\,[f(0) - f(\lambda/2)]$$
$$= (1/2)\,(y_0 - y_{\lambda/2}).$$

In general we have

$$a_m + a_{3m} + a_{5m} + \dots = (1/2m)\,\{y_0 - y_{\lambda/2m} + y_{2\lambda/2m} - \dots - y_{(2m-1)\lambda/2m}\}\dots(13),$$

and

$$b_m - b_{3m} + b_{5m} - \dots = (1/2m)\,\{y_{\lambda/4m} - y_{3\lambda/4m} + y_{5\lambda/4m} - \dots - y_{(4m-1)\lambda/4m}\}\dots(14).$$

To illustrate the use of formulae (13) and (14) let us consider the rectangular wave shown in Fig. 76. We see by (13) that all the coefficients of the cosine terms are zero. Neglecting as formerly the seventeenth and higher harmonics we get

$$b_{15} = 1/15,\quad b_{13} = 1/13,\quad b_{11} = 1/11,\quad b_9 = 1/9,\quad b_7 = 1/7,$$
$$b_5 - b_{15} = 1/5,\quad b_3 - b_9 + b_{15} = 1/3,$$

and $$b_1 - b_3 + b_5 - b_7 + b_9 - b_{11} + b_{13} - b_{15} = 1.$$

Hence we get the following table.

Harmonics	Computed values	True values	Harmonics	Computed values	True values
b_1	1·224	1·273	b_9	0·111	0·141
b_3	0·378	0·424	b_{11}	0·091	0·116
b_5	0·267	0·255	b_{13}	0·077	0·098
b_7	0·143	0·182	b_{15}	0·067	0·085

We see that the error in the computed value of the first harmonic due to the method is greater than 4 per cent. and in b_7 and higher harmonics it is greater than 21 per cent.

Let us now consider the wave shown in Fig. 77. In this case the difficulty arises as to the true value of the ordinate when the abscissa has the values $\pi/3$, $2\pi/3$, $4\pi/3$, ... If the value of x is a little less than $\pi/3$, y is 1 and if it is greater than $\pi/3$, y is 0. It would seem reasonable, therefore, to take $(0 + 1)/2$ as the value of y when x is $\pi/3$, and a rigorous mathematical proof of this procedure can be given. From (14) it follows that $b_m = 0$, and from (13) we get that $a_{3m} = 0$. Hence neglecting the twenty-fifth and higher harmonics we get the following table.

Harmonics	Computed values	True values	Harmonics	Computed values	True values
a_1	1·121	1·103	a_{13}	0·077	0·085
a_5	-0·200	-0·221	a_{17}	-0·059	-0·065
a_7	0·143	0·158	a_{19}	0·053	0·058
a_{11}	-0·091	-0·100	a_{23}	-0·044	-0·048

The error in the value of a_1 is about two per cent. and in the value of a_5 and the higher harmonics it is greater than ten per cent.

Comparison of the series methods.

The waves analysed above are discontinuous and the analysis shows that there is an infinite series of harmonics the amplitudes of which are smaller the higher their order. The amplitudes of the harmonics however diminish slowly as their order increases. In both the examples considered the amplitude of the hundred and first harmonic is about one per cent. of the amplitude of the fundamental. They therefore put the series method to a severe test. We shall now consider waves which although they are discontinuous and have an infinite number of harmonics yet approach roughly in shape to a sine wave.

Let us consider the wave which has the trapezoidal shape shown in Fig. 78. If the length of the straight line forming the top of the wave be $\lambda/6$, it is easy to show by Fourier's method (see Chapter XV) that its equation is

$$y = \frac{6\sqrt{3}}{\pi^2}\left[\sin\frac{2\pi x}{\lambda} - \frac{1}{5^2}\sin 5\frac{2\pi x}{\lambda} + \frac{1}{7^2}\sin 7\frac{2\pi x}{\lambda} - \dots\right].$$

As the wave approximates in shape much more closely to a sine wave than the cases previously considered, the amplitudes of the harmonics are smaller and we shall therefore consider the effect that neglecting the seventh and higher harmonics has on the accuracy of our results.

Using the area method we get by formula (12)

$$b_1 + b_3/3 + b_5/5 = \pi/3, \quad b_3 = 0, \text{ and } b_5 = -\pi/(3 \cdot 5^2).$$

Hence

$$b_1 = 1{\cdot}056, \quad b_3 = 0, \text{ and } b_5 = -0{\cdot}0419.$$

The true values are 1·053, 0, and − 0·0421 respectively.

If we take the ordinates method we get by (14)

$$b_1 - b_3 + b_5 = 1, \quad b_3 = 0, \text{ and } b_5 = -1/5^2.$$

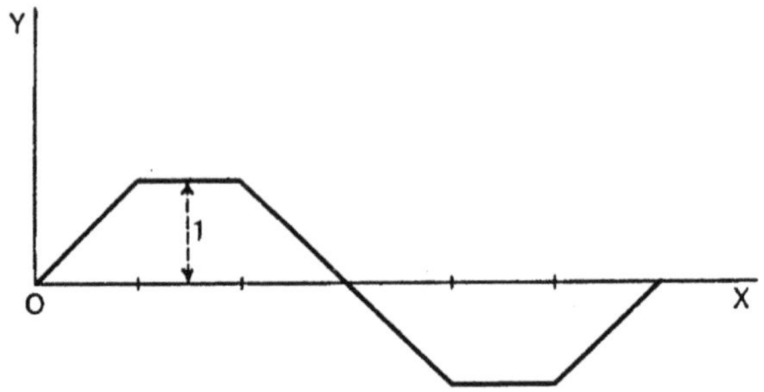

Fig. 78. Trapezoidal wave.

Hence

$$b_1 = 1{\cdot}040, \quad b_3 = 0, \text{ and } b_5 = -0{\cdot}040.$$

Both methods show that all coefficients the orders of which are multiples of three vanish. In this case by taking higher harmonics into account the accuracy of the ordinates method could be made quite satisfactory, but a very large number of ordinates would have to be measured with high accuracy.

As a further example let us consider the triangular wave shown in Fig. 79. By Fourier's method (see Chapter XV) we find that the equation to this wave is

$$y = \frac{8}{\pi^2}\left[\sin\frac{2\pi x}{\lambda} - \frac{1}{3^2}\sin 3\frac{2\pi x}{\lambda} + \frac{1}{5^2}\sin 5\frac{2\pi x}{\lambda} - \dots\right].$$

The values of the ordinates and the areas required in formulae (12) and (14) can easily be written down. The results of computing the harmonics by the two methods given above, when the twenty-fifth and higher harmonics are neglected, are given in the following table.

Harmonics	True values	Formula (12) (Areas)	Formula (14) (Ordinates)
b_1	0·8106	0·8106	0·8129
$-b_3$	0·0901	0·0901	0·0921
b_5	0·0324	0·0326	0·0356
$-b_7$	0·0165	0·0166	0·0181
b_9	0·0100	0·0097	0·0123
$-b_{11}$	0·0067	0·0065	0·0083
b_{13}	0·0048	0·0046	0·0059
$-b_{15}$	0·0036	0·0035	0·0044
b_{17}	0·0028	0·0027	0·0035
$-b_{19}$	0·0022	0·0022	0·0028
b_{21}	0·0018	0·0018	0·0023
$-b_{23}$	0·0015	0·0015	0·0019

In obtaining these results we assume that no less than 49 areas and 49 ordinates have been accurately measured. It will be seen that in this case for all practical purposes the area method, although laborious, is sufficiently accurate.

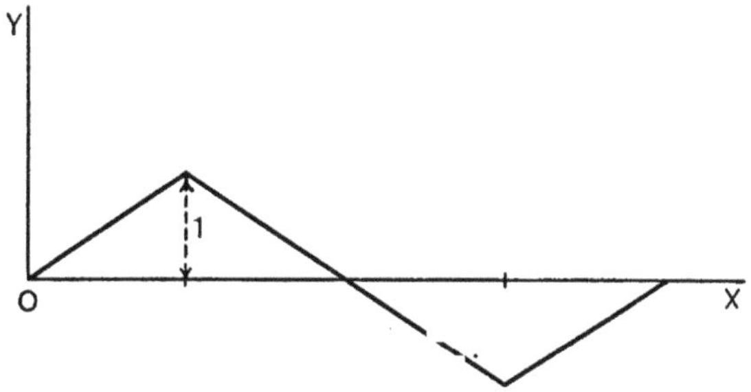

Fig. 79. Triangular wave.

Approximate formulae. As a general rule it is best to employ methods which aim directly at finding the values of the Fourier integrals. When the highest accuracy is desired it is essential to use these methods. One advantage they possess is that the amplitude and phase angle of each harmonic are determined independently of the others. In this case we can use the formulae (11) to (14) to check the accuracy of our calculations.

To determine a_1 and b_1 accurately it is necessary to divide the wave length λ into 24 equal parts and measure the lengths of the ordinates $y_0, y_1, y_2, \dots y_{23}$ at the ends and at the points of division, the ordinate y_{24} being equal to y_0 as the curve is periodic. By (3) we have

$$a_1 = \frac{2}{\lambda}\int_0^\lambda y \cos\frac{2\pi x}{\lambda}\,\partial x$$
$$= \frac{2}{\lambda}\left[\int_0^{\lambda/4} + \int_{\lambda/4}^{\lambda/2} + \int_{\lambda/2}^{3\lambda/4} + \int_{3\lambda/4}^{\lambda}\right] y\,\partial x.$$

Hence applying Weddle's rule to each integral separately, we get

$$a_1 = \frac{1}{40}\left[y_0 \cos 0 + y_2 \cos 2\frac{2\pi}{24} + y_4 \cos 4\frac{2\pi}{24} + y_6 \cos 6\frac{2\pi}{24}\right.$$
$$+ 5\left(y_1 \cos\frac{2\pi}{24} + y_5 \cos 5\frac{2\pi}{24}\right) + 6y_3 \cos 3\frac{2\pi}{24}$$
$$\left. + y_6 \cos 6\frac{2\pi}{24} + y_8 \cos 8\frac{2\pi}{24} + \ldots\right]$$
$$= \tfrac{1}{80}\left[4y_0 + y_4 + y_{20} - (4y_{12} + y_8 + y_{16})\right]$$
$$+ \frac{\sqrt{3}}{80}\left[y_2 + y_{22} - (y_{10} + y_{14})\right]$$
$$+ \tfrac{1}{8}\left[\cos 15^\circ\,(y_1 + y_{23} - y_{11} - y_{13})\right.$$
$$\left. + \sin 15^\circ\,(y_5 + y_{19} - y_7 - y_{17})\right]$$
$$+ \frac{3\sqrt{2}}{40}\left[y_3 + y_{21} - y_9 - y_{15}\right] \quad\ldots\ldots\ldots(15).$$

Similarly we find that

$$b_1 = \tfrac{1}{80}\left[y_2 + 4y_6 + y_{10} - (y_{14} + 4y_{18} + y_{22})\right]$$
$$+ \frac{\sqrt{3}}{80}\left[y_4 + y_8 - (y_{16} + y_{20})\right]$$
$$+ \tfrac{1}{8}\left[\sin 15^\circ\,(y_1 + y_{11} - y_{13} - y_{23})\right.$$
$$\left. + \cos 15^\circ\,(y_5 + y_7 - y_{17} - y_{19})\right]$$
$$+ \frac{3\sqrt{2}}{40}\left[y_3 + y_9 - y_{15} - y_{21}\right] \quad\ldots\ldots\ldots(16).$$

Applying these formulae to the rectangular wave shown in Fig. 76, we get

$$a_1 = 0,$$

and

$$b_1 = \frac{12}{80} + \frac{\sqrt{3}}{20} + \frac{1}{2}(\sin 15^\circ + \cos 15^\circ) + \frac{3\sqrt{2}}{10}$$
$$= 1{\cdot}27324,$$

which is correct to the last figure.

The accuracy of the formulae (15) and (16) is thus of a high order. When using these formulae in everyday work it is customary to write 0·259 for sin 15° and 0·966 for cos 15°.

Similarly taking the same 24 ordinates we get

$$a_2 = \tfrac{1}{80}[4(y_0 + y_{12} - y_6 - y_{18}) + y_2 + y_{10} + y_{14} + y_{22} - (y_4 + y_8 + y_{16} + y_{20})]$$
$$+ \frac{\sqrt{3}}{16}[y_1 + y_{11} + y_{13} + y_{23} - (y_5 + y_7 + y_{17} + y_{19})] \quad \ldots\ldots (17),$$

$$b_2 = \frac{\sqrt{3}}{80}[y_2 + y_4 + y_{14} + y_{16} - (y_8 + y_{10} + y_{20} + y_{22})]$$
$$+ \tfrac{1}{16}[y_1 + y_5 + y_{13} + y_{17} - (y_7 + y_{11} + y_{19} + y_{23})]$$
$$+ \tfrac{3}{20}[y_3 + y_{15} - (y_9 + y_{21})] \quad \ldots\ldots (18),$$

$$a_3 = \tfrac{1}{40}[2y_0 + y_8 + y_{16} - (y_4 + 2y_{12} + y_{20})]$$
$$+ \frac{\sqrt{2}}{16}[y_1 + y_7 + y_{17} + y_{23} - (y_5 + y_{11} + y_{13} + y_{19})]$$
$$+ \frac{3\sqrt{2}}{40}[y_9 + y_{15} - (y_3 + y_{21})] \quad \ldots\ldots (19),$$

$$b_3 = \tfrac{1}{40}[y_2 + y_{10} + 2y_{18} - (2y_6 + y_{14} + y_{22})]$$
$$+ \frac{\sqrt{2}}{16}[y_1 + y_{11} + y_{17} + y_{19} - (y_5 + y_7 + y_{13} + y_{23})]$$
$$+ \frac{3\sqrt{2}}{40}[y_3 + y_9 - (y_{15} + y_{21})] \quad \ldots\ldots (20).$$

To obtain the same accuracy for a_n as that given by formula (15) for a_1 we should generally have to measure $24n$ ordinates. Hence the accuracy of (19) is usually not as high as that of (15) as we have only measured the same number of ordinates and the curve $y = f(x) \sin 3(2\pi x/\lambda)$ is in general much more irregular than the curve $y = f(x) \sin(2\pi x/\lambda)$. Applying (19) and (20) to the rectangular wave, Fig. 76, we find that $a_3 = 0$ and

$$b_3 = \frac{1}{40}(0) + \frac{\sqrt{2}}{16}(0) + \frac{3\sqrt{2}}{10} = 0{\cdot}4243\ (0{\cdot}4244),$$

the number in the bracket being the correct value. The formulae for the higher harmonics can be written down by Weddle's rule without difficulty, but in this case (see p. 140) a larger number of ordinates must be measured.

As a further example let us take the case of the semicircular wave which is shown in Fig. 80 (a). The wave length is taken equal to 2.

If we divide the wave length into twelve equal parts and erect ordinates, we get by Weddle's rule

$$10b_1 = 5y_{1/6} + \sqrt{(3)} \cdot y_{2/6} + 6y_{3/6},$$

the formula simplifying owing to the symmetry of the wave. Hence since $y = \sqrt{(x - x^2)}$, we get $b_1 = 0{\cdot}568$. We shall show later that the true value is $0{\cdot}567$. To get b_3 with an accuracy of the same order we must measure the lengths of eight ordinates, the formula being

$$30b_3 = 5\,(y_{1/18} + y_{5/18} - y_{7/18}) + \sqrt{(3)} \cdot (y_{2/18} + y_{4/18} - y_{8/18}) + 6\,(2y_{3/18} - y_{9/18}).$$

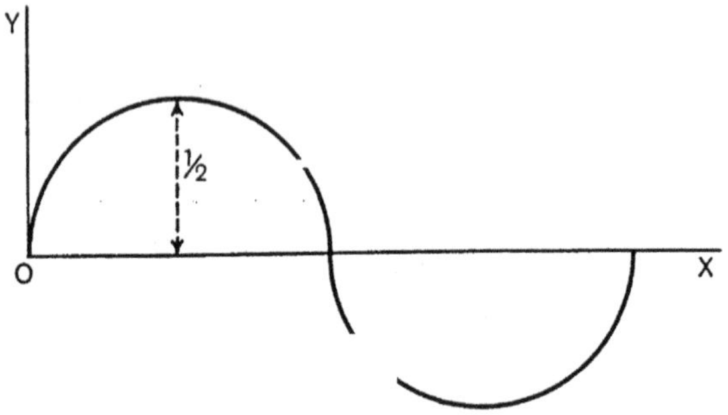

Fig. 80 (a). Semicircular wave.

In general we have

$$\begin{aligned} 10\,(2n+1)\,b_{2n+1} = {} & 5\,\{y_1 + y_5 - y_7 - y_{11} + \dots + (-)^n y_{6n+1}\} \\ & + (\sqrt{3}) \cdot \{y_2 + y_4 - y_8 - y_{10} + \dots + (-)^n y_{6n+2}\} \\ & + 6\,\{2y_3 - 2y_9 + \dots + (-)^n y_{6n+3}\}, \end{aligned}$$

where we have written y_p for $y_{p/6(2n+1)}$.

To find the value of b_{2n+1} therefore by this method we should have to measure the lengths of $5n + 3$ ordinates. Proceeding in this way we get $b_3 = 0{\cdot}0942$, $b_5 = 0{\cdot}0423$, $b_7 = 0{\cdot}0252$ and $b_9 = 0{\cdot}0171$. The values of b_9 and b_7 are correct but the true values of b_5 and b_3 are $0{\cdot}0422$ and $0{\cdot}0939$ respectively.

Semicircular wave.

We shall now find the value of b_m for a semicircular wave by means of Fourier's formula. An inspection of Fig. 80 (a) will show that the values of the ordinates corresponding to x and $-x$ are equal in value but of opposite sign. Hence there are no cosine terms in the series for y.

Let the equation to the first loop of the wave shown in Fig. 80 (a) be $y^2 = x(1-x)$. Then by Fourier's formula we get

$$b_m = 2\int_0^1 \{x(1-x)\}^{\frac{1}{2}} \sin m\pi x \partial x.$$

To evaluate this integral put $x = \frac{1}{2}(1 - \cos\xi)$, so that

$$\{x(1-x)\}^{\frac{1}{2}} = \tfrac{1}{2}\sin\xi \text{ and } \partial x = \tfrac{1}{2}\sin\xi\partial\xi.$$

Hence, by substituting,

$$b_m = \frac{1}{2}\int_0^\pi \sin^2\xi \sin\left(\frac{m\pi}{2} - \frac{m\pi\cos\xi}{2}\right)\partial\xi.$$

When m is an even integer, the value of the integral from 0 to $\pi/2$ is equal but of opposite sign to the value from $\pi/2$ to π, and hence b_m is zero. When m is an odd integer

$$\begin{aligned} b_m &= \frac{1}{2}\sin\frac{m\pi}{2}\int_0^\pi \sin^2\xi \cos\left(\frac{m\pi\cos\xi}{2}\right)\partial\xi \\ &= \frac{1}{2}\sin\frac{m\pi}{2}\int_0^\pi \sin^2\xi\left\{1 - \frac{(m\pi\cos\xi)^2}{2^2 \,.\, 2!} + \frac{(m\pi\cos\xi)^4}{4^2 \,.\, 4!} - \ldots\right\}\partial\xi \\ &= (-)^{(m-1)/2}\frac{\pi}{4}\left\{1 - \frac{1}{2!\,1!}\left(\frac{m\pi}{4}\right)^2 + \frac{1}{3!\,2!}\left(\frac{m\pi}{4}\right)^4 - \ldots\right\}. \end{aligned}$$

Thus the equation to the circular wave is

$$\begin{aligned} y = \frac{\pi}{4}\Bigg[&\left\{\sum_1^\infty \frac{(-)^{n+1}}{n!\,(n-1)!}\left(\frac{\pi}{4}\right)^{2(n-1)}\right\}\sin x \\ &- \left\{\sum_1^\infty \frac{(-)^{n+1}}{n!\,(n-1)!}\left(\frac{3\pi}{4}\right)^{2(n-1)}\right\}\sin 3x + \ldots\Bigg]. \end{aligned}$$

Now if $J_1(x)$ be the Bessel's function of the first order

$$J_1(2x) = x\sum_1^\infty \frac{(-)^{n+1}}{n!\,(n-1)!}x^{2(n-1)},$$

and hence

$$\begin{aligned} y &= J_1(\pi/2) \,.\, \sin\pi x - \tfrac{1}{3}J_1(3\pi/2) \,.\, \sin 3\pi x \\ &\quad + \tfrac{1}{5}J_1(5\pi/2) \,.\, \sin 5\pi x - \ldots \\ &= 0{\cdot}5668 \sin\pi x + 0{\cdot}0939 \sin 3\pi x \\ &\quad + 0{\cdot}0422 \sin 5\pi x + 0{\cdot}0252 \sin 7\pi x \\ &\quad + 0{\cdot}0171 \sin 9\pi x + \ldots. \end{aligned}$$

In Fig. 80 (b), for which the author is indebted to A. E. Kennelly,

(1), (3), (5) and (7) are the graphs of the first, third, fifth and seventh harmonics respectively. The curve 3′ shows the addition of (1) and (3). The curve 5′ shows the addition of (1), (3) and (5), and the curve 7′ shows the addition of (1), (3), (5) and (7).

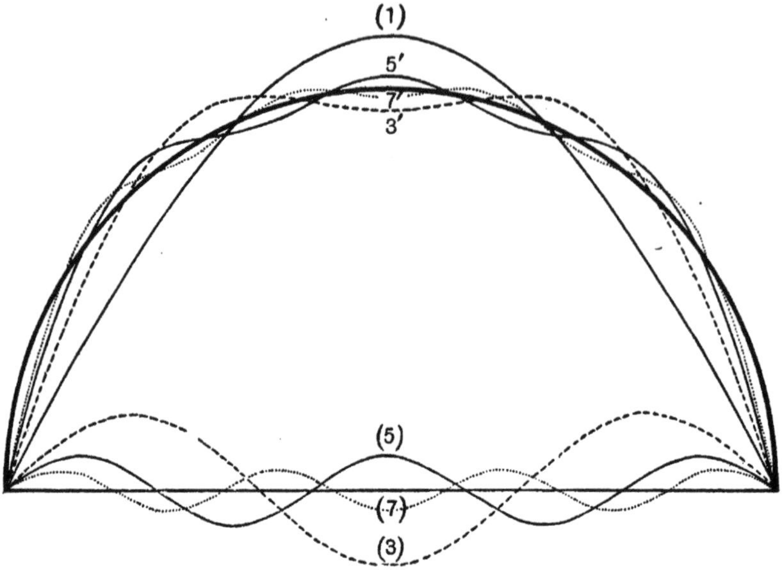

Fig. 80 (b). The curves (1), (3), (5) and (7) are the first, third, fifth and seventh harmonics of a semicircular wave.
3′ = (1) + (3), 5′ = (1) + (3) + (5), 7′ = (1) + (3) + (5) + (7).

Special formulae.

In certain cases the harmonics can be determined very simply by means of series formulae. In alternating current work the positive half of the wave is generally exactly similar to the negative half, and so $a_0 = 0$ and all the even harmonics vanish. If in addition $f(x) = f(\lambda/2 - x)$ so that the first half of the positive loop is the image of the second half, formula (1) becomes

$$y = f(x) = b_1 \sin (2\pi/\lambda)\, x + b_3 \sin 3\, (2\pi/\lambda)\, x + \dots.$$

Hence we easily prove that

$$b_1 - b_{11} + b_{13} - b_{23} + \dots + b_{12n-11} - b_{12n-1} + \dots$$
$$= (y_{\lambda/12} + y_{\lambda/4})/3 + y_{\lambda/6}/\sqrt{3} \;\dots\dots(21),$$

and

$$b_5 - b_7 + \dots + b_{12n-7} - b_{12n-5} + \dots$$
$$= (y_{\lambda/12} + y_{\lambda/4})/3 - y_{\lambda/6}/\sqrt{3} \;\dots\dots(22).$$

Similarly we can show that

$$b_1 + \frac{b_{11}}{11} + \frac{b_{13}}{13} + \dots + \frac{b_{12n-11}}{12n-11} + \frac{b_{12n-1}}{12n-1} + \dots$$
$$= \frac{2\pi}{3\lambda}\left[2\int_0^{\lambda/4} - \int_0^{\lambda/6}\right] y\,\partial x + \frac{2\pi}{\lambda\sqrt{3}}\left[\int_0^{\lambda/4} - \int_0^{\lambda/12}\right] y\,\partial x \quad \dots(23),$$

and

$$\frac{b_5}{5} + \frac{b_7}{7} + \dots + \frac{b_{12n-7}}{12n-7} + \frac{b_{12n-5}}{12n-5} + \dots$$
$$= \frac{2\pi}{3\lambda}\left[2\int_0^{\lambda/4} - \int_0^{\lambda/6}\right] y\,\partial x - \frac{2\pi}{\lambda\sqrt{3}}\left[\int_0^{\lambda/4} - \int_0^{\lambda/12}\right] y\,\partial x \quad \dots(24).$$

We also have by (14) and (12)

$$b_3 - b_9 + b_{15} - \dots = \tfrac{1}{3}\left[2y_{\lambda/12} - y_{\lambda/4}\right] \quad \dots\dots\dots(25),$$

and
$$b_3 + \frac{b_9}{3} + \frac{b_{15}}{5} + \dots = \frac{2\pi}{\lambda}\left[2\int_0^{\lambda/6} - \int_0^{\lambda/4}\right] y\,\partial x \quad \dots\dots(26).$$

These equations sometimes enable us to find b_1, b_3, ... with very little trouble. For example in the case of the trapezoidal wave (Fig. 78) we see at once that

$$\int_0^{\lambda/4} y\,\partial x = \lambda/6, \int_0^{\lambda/6} y\,\partial x = \lambda/12 \text{ and } \int_0^{\lambda/12} y\,\partial x = \lambda/48.$$

Hence by (23), $b_1 + b_{11}/11 + \dots = 1{\cdot}053$, and thus neglecting small fractions of the amplitudes of higher harmonics of the orders $12n - 1$ and $12n - 11$ we get $b_1 = 1{\cdot}053$ which is correct to the last figure.

The equation to the positive half of the semicircular wave Fig. 80 (a) is $y = \sqrt{(x - x^2)}$, and thus, noticing that λ equals 2, we get

$$y_{\lambda/12} = 0{\cdot}3727, \quad y_{\lambda/6} = 0{\cdot}4714 \text{ and } y_{\lambda/4} = 0{\cdot}5000.$$

We also have

$$\int_0^{\lambda/12} y\,\partial x = 0{\cdot}0430, \int_0^{\lambda/6} y\,\partial x = 0{\cdot}1146 \text{ and } \int_0^{\lambda/4} y\,\partial x = 0{\cdot}1964.$$

Hence by only measuring three ordinates and three areas we get by formulae (21) to (26)

$$b_1 - b_{11} + b_{13} - \dots = 0{\cdot}563,$$
$$b_5 - b_7 + b_{17} - \dots = 0{\cdot}019,$$
$$b_1 + b_{11}/11 + b_{13}/13 + \dots = 0{\cdot}569,$$
$$b_5/5 + b_7/7 + b_{17}/17 + \dots = 0{\cdot}013,$$
$$b_1 - b_3 + b_5 - \dots = 0{\cdot}500,$$

and
$$b_3 - b_9 + b_{15} - \dots = 0{\cdot}088.$$

From these equations we might deduce as a first approximation that $b_1 = 0{\cdot}569\ (0{\cdot}567)$, $b_3 = 0{\cdot}088\ (0{\cdot}094)$, etc. where the numbers in the brackets give the true values. With the possible exception of the value of b_1 we should have no guarantee of the accuracy of our results, as the equations show that the higher harmonics are not negligible; and since for a circular wave $f'(x) =$ infinity when x is zero, there must be an infinite number of terms in the Fourier series for $f(x)$.

We have seen (p. 134) that the values found by Weddle's rule are

$$b_1 = 0{\cdot}568, \quad b_3 = 0{\cdot}0942, \quad b_5 = 0{\cdot}0423, \quad b_7 = 0{\cdot}0252, \quad \ldots.$$

As only ordinates need to be measured when using Weddle's rule the accuracy of the necessary data is higher than in those methods in which the measurement of areas has to be made.

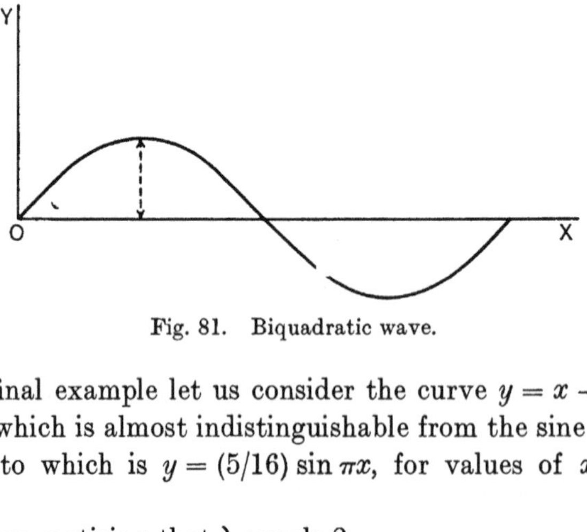

Fig. 81. Biquadratic wave.

As a final example let us consider the curve $y = x - 2x^3 + x^4$ (Fig. 81) which is almost indistinguishable from the sine curve the equation to which is $y = (5/16) \sin \pi x$, for values of x between 0 and 1.

We have, noticing that λ equals 2,

$$y_{\lambda/4} = 5/16, \quad y_{\lambda/6} = 22/81 \quad \text{and} \quad y_{\lambda/12} = 205/1296.$$

Hence, if we neglect the eleventh and higher harmonics, we get by (21)

$$b_1 = (y_{\lambda/12} + y_{\lambda/4})/3 + y_{\lambda/6}/\sqrt{3}$$
$$= 0{\cdot}313704,$$

the true value being $0{\cdot}313705$.

We also have

$$\int_0^{\lambda/4} y\,\partial x = 1/10, \quad \int_0^{\lambda/6} y\,\partial x = 61/1215 \quad \text{and} \quad \int_0^{\lambda/12} y\,\partial x = 263/19440,$$

and thus, by (23), we get

$$b_1 = 0{\cdot}313705,$$

which is correct to the last figure.

By Fourier's method we can show that the equation to the curve is

$$y = \frac{96}{\pi^5}\left[\sin\frac{2\pi x}{\lambda} + \frac{1}{3^5}\sin 3\frac{2\pi x}{\lambda} + \ldots\right],$$

and so the values of the higher harmonics are extremely small. This explains the high accuracy attainable in this case by using the series formulae.

If the alternating current wave be such that $f(x) = -f(\lambda/2 - x)$ its equation must be of the form

$$y = f(x) = a_1 \cos(2\pi/\lambda)\,x + a_3 \cos 3\,(2\pi/\lambda)\,x + \ldots.$$

In this case we have

$$a_1 - \frac{a_{11}}{11} + \frac{a_{13}}{13} - \frac{a_{23}}{23} + \ldots = \frac{2\pi}{3\lambda}\left[\int_0^{\lambda/12} + \int_0^{\lambda/4}\right] y\,\partial x + \frac{2\pi}{\lambda\sqrt{3}}\left[\int_0^{\lambda/6} y\,\partial x\right] \quad \text{.........(27)},$$

and

$$\frac{a_5}{5} - \frac{a_7}{7} + \frac{a_{17}}{17} - \frac{a_{19}}{19} + \ldots = \frac{2\pi}{3\lambda}\left[\int_0^{\lambda/12} + \int_0^{\lambda/4}\right] y\,\partial x - \frac{2\pi}{\lambda\sqrt{3}}\left[\int_0^{\lambda/6} y\,\partial x\right] \quad \text{.........(28)}.$$

For instance if we take the wave form shown in Fig. 77 and neglect $a_{11}/11$, etc. we get

$$a_1 = \frac{\pi}{6} + \frac{\pi}{9}\sqrt{3} = 1{\cdot}13\ (1{\cdot}10),$$

and

$$\frac{a_5}{5} - \frac{a_7}{7} + \ldots = -\,0{\cdot}081\ (-\,0{\cdot}067),$$

where the true values are given in the brackets.

The large errors in this case are due to the very distorted shape of the wave analysed. For waves approaching cosine shape the accuracy attainable would be far higher.

Formulae for alternating waves.

The examples given above prove that for a smooth wave we need to measure the lengths of twelve ordinates at least in order to find a_1 and b_1 with a probable inaccuracy of only a few per cent. To determine a_n and b_n with the same accuracy we ought to measure $12n$ ordinates. With alternating waves all the even harmonics are zero. Let us take the origin at a point where y is zero and divide the wave length λ into twelve equal parts, erecting ordinates $y_1, y_2, \ldots y_{11}$ at the points of division. It is to be noticed that $y_0 = -y_6 = y_{12} = 0$. We readily find by Weddle's rule that

$$20a_1 = y_2 - y_4 + 5\sqrt{(3)}\,.\,(y_1 - y_5),$$
$$20b_1 = \sqrt{(3)}\,.\,(y_2 + y_4) + 5\,(y_1 + y_5) + 12y_3.$$

To find a_3 and b_3 we divide the wave length into 36 equal parts and we shall denote the new ordinates at the points of division by $y_1, y_2, \ldots y_{35}$. In this case $y_0 = -y_{18} = y_{36} = 0$.

Applying the rule we get

$$60a_3 = y_2 - y_4 - y_8 + y_{10} + y_{14} - y_{16} + 4\,(-y_6 + y_{12}) + 5\sqrt{(3)}\,.\,(y_1 - y_5 - y_7 + y_{11} + y_{13} - y_{17}),$$
$$60b_3 = \sqrt{(3)}\,.\,(y_2 + y_4 - y_8 - y_{10} + y_{14} + y_{16}) + 5\,(y_1 + y_5 - y_7 - y_{11} + y_{13} + y_{17}) + 12\,(y_3 - y_9 + y_{15}).$$

In general to find a_{2n+1} and b_{2n+1} we divide the wave length into $12\,(2n+1)$ equal parts and erect ordinates at the points of division. The formulae are

$$20\,(2n+1)\,a_{2n+1} = y_2 - y_4 - y_8 + y_{10} + \ldots + y_{12n+2} - y_{12n+4} + 4\,(-y_6 + y_{12} - \ldots + y_{12n}) + 5\sqrt{(3)}\,.\,(y_1 - y_5 - y_7 + y_{11} + \ldots + y_{12n+1} - y_{12n+5}),$$

and

$$20\,(2n+1)\,b_{2n+1} = \sqrt{(3)}\,.\,(y_2 + y_4 - y_8 - y_{10} + \ldots + y_{12n+2} + y_{12n+4}) + 5\,(y_1 + y_5 - y_7 - y_{11} + \ldots + y_{12n+1} + y_{12n+5}) + 12\,(y_3 - y_9 + y_{15} - \ldots + y_{12n+3}).$$

For example for a rectangular wave (Fig. 76) we find at once that

$$a_{2n+1} = 0,$$

and

$$b_{2n+1} = \frac{11 + \sqrt{3}}{10\,(2n+1)} = \frac{1{\cdot}27321}{2n+1}.$$

The true value of b_{2n+1} is $1{\cdot}27324/(2n+1)$. In this case therefore the accuracy is highly satisfactory.

If the value of the first harmonic is desired to a higher accuracy we may use (15) and (16), p. 132, noticing that for alternating currents $y_1 = -y_{13}$, $y_2 = -y_{14}$, etc.

Method of finding the form factor.

Harmonic analysis can be usefully employed to find the form factor of an E.M.F. wave when the wave is approximately sine shaped. Let the equation to the wave be

$$y = A_1 \sin\{(2\pi/\lambda)x - \epsilon_1\} + A_3 \sin\{3(2\pi/\lambda)x - \epsilon_3\} + \dots .$$

Then by squaring and taking mean values we find at once that

$$2y^2_{\text{eff.}} = A_1^2 + A_3^2 + A_5^2 + \dots .$$

Also if A be the area of the positive loop of the wave, y_{mean} is equal to $2A/\lambda$ and hence the form factor $y_{\text{eff.}}/y_{\text{mean}}$ can be found.

As an example let us suppose that the positive loop of the wave from $x = 0$ to $x = \lambda/2 = 1$, is the parabola $y = x - x^2$.

Then by the ordinates method

$$b_1 - b_3 + b_5 - \dots = y_{1/2} = \tfrac{1}{4},$$
$$b_3 - b_9 + b_{15} - \dots = (\tfrac{1}{3})(2y_{1/6} - y_{1/2}) = 1/108, \text{ etc.}$$

Since the curve is very like a sine wave we shall neglect the fifth and higher harmonics. Hence we get

$$b_1 = 0{\cdot}2593 \text{ and } b_3 = 0{\cdot}0093.$$

The true values are 0·2580 and 0·0096. We also have $y_{\text{mean}} = 1/6$ and thus

$$y_{\text{eff.}}/y_{\text{mean}} = 6\{(0{\cdot}2593)^2 + (0{\cdot}0093)^2\}^{\frac{1}{2}}/\sqrt{2}$$

which equals 1·100 approximately. It may be easily proved that the true value is $6/\sqrt{30}$ which equals 1·096....

Conclusion.

The results given above prove that it is best to determine the values of the Fourier constants $a_0, a_1, a_2, \dots b_1, b_2, \dots$ of a periodic curve directly by mechanical quadrature. In particular the rule for quadrature given by Weddle will be found most useful. The various series formulae given above enable us when the curves are roughly sine shaped to find the values of the lower harmonics very readily. Their main use however lies in checking the values of the Fourier constants computed by other methods.

REFERENCES.

JOSEPH FOURIER, *Théorie Analytique de la Chaleur*, 1822. Translated by A. Freeman, 1878.

C. F. GAUSS, 'Theoria Interpolationis Methodo Nova Tractata,' *Werke*, Band 3, p. 265.

THOMAS WEDDLE, 'On a New and Simple Rule for approximating to the Area of a Figure by means of 7 Equidistant Ordinates,' *The Cambridge and Dublin Mathematical Journal*, p. 79, February, 1854.

JAMES O'KINEALY, 'Fourier's Theorem.' (Suggested method of proof by the Theory of Differential Equations.) *Phil. Mag.*, Vol. 48, p. 95, August, 1874.

I. TODHUNTER, *Integral Calculus*, Chap. XIII.

R. STRACHEY, 'On the Computation of the Harmonic Components of a series representing a Phenomenon, occurring in Daily and Yearly Periods,' *Proc. Roy. Soc.*, Vol. 42, p. 61, 1887.

C. RUNGE, 'Über die Zerlegung empirisch gegebener periodischer Funktionen in Sinuswellen,' *Zeitschrift für Mathematik und Physik*, Vol. 49, p. 443, 1903.

H. S. CARSLAW, *Introduction to the Theory of Fourier's Series and Integrals etc.*

E. W. HOBSON, *The Theory of Functions of a Real Variable and the Theory of Fourier's Series.*

A. E. KENNELLY, 'The Harmonic Analysis of the Semicircle and of the Ellipse,' *Annals of Mathematics*, Sec. Ser. Vol. 7, No. 2, January, 1906.

J. FISCHER-HINNEN, 'Über die Zerlegung periodischer Kurnen in ihre harmonischen Wellen,' *Elektrotechnik und Maschinenbau*, Vol. 27, p. 335, 1909.

SILVANUS THOMPSON, 'A new Method of Approximate Harmonic Analysis by Selected Ordinates,' *Proc. of the Phys. Soc.*, Vol. 26, p. 275, 1911.

F. W. GROVER, 'Analysis of Alternating Current Waves by the Method of Fourier, with special Reference to Methods of Facilitating the Computations,' *Bulletin of the Bureau of Standards*, Vol. 9, p. 567, 1914.

A. RUSSELL, 'Practical Harmonic Analysis,' *Proc. of the Phys. Soc.*, Vol. 27, p. 149, 1915.

For the connection between the Method of Least Squares and Fourier Coefficients *see* CHARLES H. LEES, *Proc. of the Phys. Soc.*, Vol. 26, p. 275, 1914, and for a simple introduction to Runge's method of harmonic analysis *see* GIBSON'S *Introduction to the Calculus*, Chap. XI.

For the description of an accurate mechanical harmonic analyser *see* A. A. MICHELSON and S. W. STRATTON, 'A New Harmonic Analyser,' *Phil. Mag.*, Vol. 45, p. 85, 1898.

CHAPTER IV

Dangers from the harmonics in the E.M.F. waves. Methods of analysing waves. Blondel's method. Pupin's resonance method. Armagnat's method. Analysis of electromotive force waves. Resonance of the first harmonic. Resonance of the fifth harmonic. Resonance of the seventh harmonic. Resonance of the eleventh harmonic. Interference of two resonating harmonics. Measuring irregularities in the speed of alternators. Causes of the harmonics in electromotive force waves. Harmonics caused by slots. Harmonics in the E.M.F. waves of three phase machines. Annulling harmonics by special windings. Methods of preventing the slots in the armature from producing harmonics. References.

Dangers from harmonics in the E.M.F. waves.

IN many distributing systems we have mains of high electrostatic capacity in circuit with transformers having considerable inductance, and the armature of the alternator itself has also considerable inductance. We have seen in Vol. I, p. 138, that when we have a condenser of capacity K in series with an inductive coil whose inductance is L, resonance of the nth harmonic in the applied potential difference wave ensues when

$$LKn^2 (2\pi f)^2 = 1,$$

where f is the frequency of the first harmonic. In this case, very high potential differences are established between various parts of the circuit, sometimes causing sparks which break down the insulation of the cables, or of the armature or transformer windings. As a rule $LK(2\pi f)^2$ is much smaller than unity, so that there is little danger of resonance with the first harmonic. The danger arises when there is a pronounced high harmonic. The formula given above shows that the nth harmonic will resonate with only the $1/n^2$ part of the capacity required for resonance by the first harmonic. It would be dangerous to run up an alternator, which had a jagged electromotive force wave, to its normal speed with its field excited if its terminals were connected with mains having considerable electrostatic capacity. There would be a serious risk of the resonance of some of the harmonics at particular speeds. It is therefore essential to consider the causes of these harmonics in the electromotive force wave of the machine, and whether there are any methods of preventing their formation.

Methods of analysing waves.

The wave form of an alternator can be found directly by means of an oscillograph, ondograph, or rheograph. We can then apply any of the various analytical methods described in Chapter III, to analyse this curve into its harmonics. It is found, however, in practice that in some cases owing to the irregularities in the speed of the engines driving the alternators, etc., the curves cannot be traced with sufficient accuracy to make graphical methods useful, and so recourse is had to experimental methods of finding the amplitudes of the harmonics and their phases relative to the fundamental harmonic.

Let $f(t)$ denote the electromotive force wave of the machine, then by Fourier's Theorem we have

$$f(t) = \Sigma a_n \cos n\omega t + \Sigma b_n \sin n\omega t \quad \ldots\ldots\ldots\ldots(1)$$

$$= \Sigma \sqrt{(a_n^2 + b_n^2)} \,.\, \sin (n\omega t + \phi_n) \quad \ldots\ldots(2),$$

where $\tan \phi_n = a_n/b_n$.

There is no constant term in the series, as $f(t)$ is a purely alternating function. In order to find $f(t)$ we have to determine the amplitudes $\sqrt{(a_n^2 + b_n^2)}$ of the various harmonics and their time angles ϕ_n.

Blondel's method.

If we apply the potential difference wave we wish to analyse to the terminals of a non-inductive resistance R, then the current wave in this resistance will be similar to the applied potential difference wave, since by Ohm's law $i = f(t)/R$, and therefore the curves of i and $f(t)$ only differ in the scale of the ordinates. Thus, if we can analyse the current wave in the non-inductive resistance into its various harmonics, we can find the corresponding harmonics in $f(t)$. The time angles of the harmonics in the potential difference and current waves will be the same in the two cases, but the amplitudes in the former case will be R times the amplitudes in the latter case.

Let the current i pass through the fixed coil of an electro-dynamometer, the reactance of this coil being negligible compared with R, and let the current from one phase of an auxiliary two phase alternator which produces a sine shaped electromotive force wave pass through a non-inductive resistance and through the movable coil. When the speed of the auxiliary machine is varied, we get large deflections of the dynamometer at particular frequencies. Let us suppose that the auxiliary current is $I \sin(n\omega t - \alpha)$ so that the frequency is $n\omega/2\pi$, then we have

$$\frac{1}{T}\int_0^T \frac{f(t)}{R} \cdot I \sin(n\omega t - \alpha)\, dt = \frac{I}{R} \cdot \frac{1}{T}\int_0^T f(t) \sin(n\omega t - \alpha)\, dt$$
$$= (I/2R)(b_n \cos\alpha - a_n \sin\alpha),$$

where b_n and a_n are the coefficients of $\sin n\omega t$ and $\cos n\omega t$ in (1).

Hence, by Vol. I, p. 123, if k be the constant of the electro-dynamometer and D_1 be the deflection, we have

$$(I/2R)(b_n \cos\alpha - a_n \sin\alpha) = k^2 D_1.$$

Similarly by sending the current $I \cos(n\omega t - \alpha)$ from the other two terminals of the auxiliary machine through the movable coil we find that

$$(I/2R)(b_n \sin\alpha + a_n \cos\alpha) = k^2 D_2,$$

where D_2 is the new reading of the dynamometer.

Therefore

$$(a_n^2 + b_n^2)^{\frac{1}{2}} = 2Rk^2 (D_1^2 + D_2^2)^{\frac{1}{2}}/I,$$

and thus the amplitude of the nth harmonic is found. We have assumed that α remains constant during the time of taking the readings D_1 and D_2. It would be advisable therefore that the auxiliary machine be connected, through a variable speed gearing, directly with the shaft of the alternator.

Pupin's resonance method.

In Pupin's method the potential difference wave, which is to be analysed, is applied to the terminals of a circuit consisting of a condenser K in series with a choking coil L which contains no iron and the eddy current loss in which is negligible. Pupin placed an electrometer across the condenser K, but in practice an electrostatic voltmeter is now more convenient. The capacity or the inductance is varied continuously, and the values of K and L for which the voltmeter readings attain maximum values are noted. These voltmeter readings enable us to determine the amplitudes of the various harmonics in the potential difference wave.

Let i_n denote the nth harmonic in the current wave; then we know (see Vol. I, p. 137) that

$$i_n = \frac{A_n \sin (n\omega t + \phi_n - \psi_n)}{R\{1 + (L - 1/Kn^2\omega^2)^2 (n\omega/R)^2\}^{\frac{1}{2}}},$$

where $\qquad A_n = \sqrt{a_n^2 + b_n^2},$

and $\qquad \tan \psi_n = (LKn^2\omega^2 - 1)/(Kn\omega R).$

Now the amplitude of i_n is a maximum when $n\omega\sqrt{LK}$ equals unity, and it is given by

$$i_n' = (A_n/R) \sin (n\omega t + \phi_n).$$

Hence i_n' is in phase with the nth harmonic of the applied potential difference wave and is a simple sine wave.

If e_n denote the nth harmonic of the potential difference wave at the terminals of the condenser, then

$$e_n = \frac{1}{K}\int i_n' \, dt = -(A_n/Kn\omega R) \cos (n\omega t + \phi_n)$$
$$= -(Ln\omega/R) A_n \cos (n\omega t + \phi_n).$$

Therefore, if V_n be the effective value of e_n, we have

$$V_n = Ln\omega A_n/(R\sqrt{2}), \text{ and } A_n = \sqrt{2}RV_n/(Ln\omega).$$

Now, if R be small compared with $L\omega$, the amplitudes of the other harmonics in the applied P.D. wave are very small compared with V_n when n equals $1/2\pi f\sqrt{LK}$. Hence V_n practically equals the reading of the electrostatic voltmeter, and thus A_n can be found. The amplitudes of the other harmonics are found in the same way. It is to be noticed that this method does not determine the phase differences of the various harmonics.

Armagnat's method.

Armagnat uses an oscillograph instead of an electrostatic voltmeter. In this case we can easily arrange to get a picture of the potential difference wave and of its resonating nth harmonic on the screen at the same time. We arrange a condenser in series with a choking coil and find the current wave in the circuit by means of an oscillograph. The wave of potential difference is found by means of a second oscillograph the circuit of which, in series with a large non-inductive resistance, is placed in parallel with the circuit of the first oscillograph, which is in series with the condenser and the choking coil. Exact resonance ensues when the amplitude of the nth harmonic, shown by the vibrations of the mirror of the first oscillograph, has its maximum value. If the oscillograph be standardised so that we know the value of the current which produces the observed maximum deflection, we can find A_n, for $A_n = RI_n$ nearly, where R is the combined resistance of the oscillograph circuit and of the choking coil. The observed current is practically equal to i_n.

Again, since there is resonance, i_n and e_n are in phase with one another. By noting, therefore, the angle of lag between i_n and the applied potential difference wave on the screen, we can find ϕ_n. We have assumed, hitherto, that, when resonance ensues with a particular harmonic, the other harmonics produce currents which are negligible in comparison. In many cases this assumption is not permissible. We shall now find the values of the other currents when there is resonance of the nth harmonic. In this case, we have

$$n\omega\sqrt{LK} = 1;$$

and since

$$i_{n+m} = \frac{A_{n+m}}{R} \cdot \frac{\sin\{(n+m)\omega t + \phi_{n+m} - \psi_{n+m}\}}{[1 + \{L - 1/K(n+m)^2\omega^2\}^2\{(n+m)^2\omega^2/R^2\}]^{\frac{1}{2}}},$$

we have

$$I_{n+m} = \frac{(n+m)A_{n+m}}{\{(n+m)^2R^2 + (m^2+2mn)^2L^2\omega^2\}^{\frac{1}{2}}}.$$

In practice, the interference of the first harmonic is usually much the most troublesome. In this case,

$$I_1 = \frac{A_1}{\{R^2 + (n^2-1)^2L^2\omega^2\}^{\frac{1}{2}}},$$

while

$$I_n = \frac{A_n}{R}.$$

Hence the greater the ratio of $L\omega$ to R the smaller will be the effect of this interference. It is therefore important to arrange that the resistance of the resonant circuit shall be as small as possible. The minimum value of this resistance, if the applied voltage cannot be varied, is limited by the maximum permissible readings of the oscillograph. The presence of harmonics other than the nth has generally only the effect of producing a slight curvature of the median line without appreciably altering either the amplitude or the phase of the nth harmonic. This will be seen in Figs. 83 and 84 below. Sometimes however nodes and loops are produced (Fig. 86).

Again we have

$$\tan \psi_{n+m} = \frac{[L - 1/\{K(n+m)^2\omega^2\}](n+m)\omega}{R}$$
$$= \{m + mn/(m+n)\}(L\omega/R).$$

When $L\omega/R$ is large, it will be seen that ψ_{n+m} is practically equal to $+ 90$ degrees when there is resonance of the nth harmonic if m be positive. If m be negative, so that $n + m$ is less than n, then ψ_{n+m} will be nearly $- 90$ degrees.

Suppose that a small error is made in adjusting the resonance so that

$$n^2\omega^2 LK - 1 = \epsilon,$$

where ϵ is a small fraction, then, in this case

$$i_n = \frac{A_n}{R} \cdot \frac{\sin(n\omega t + \phi_n - \psi_n)}{\{1 + \epsilon^2/(K^2n^2\omega^2R^2)\}^{\frac{1}{2}}}$$
$$= (A_n \cos\psi_n/R)\sin(n\omega t + \phi_n - \psi_n),$$

and $\tan\psi_n = \epsilon/Kn\omega R.$

These formulae show that, if the regulation of the resonance is not quite exact, both the amplitude and phase of the nth harmonic are affected. It is necessary, therefore, that the variation of LK be done in a manner that is practically continuous. A variable inductance of the Ayrton and Perry type is suitable, or a large drum on which flexible wire can gradually be coiled so as to increase continually the value of the self-inductance of the circuit.

Unless the speed of the alternator is almost perfectly constant it is practically impossible to photograph the resonance curves, as a slight variation of speed makes the resonance no longer perfect. Also, the greater the ratio of L to R the more difficult it is to get exact resonance. Hence it is sometimes necessary to increase the resistance of the resonant circuit so as to diminish the effects of the irregularities in the speed of the generator.

Analysis of electromotive force waves.

The following experimental analysis of an electromotive force wave was made by Armagnat. The machine experimented on was a small rotary converter, with two distinct windings on its armature, so arranged that the applied direct current potential difference was equal to the effective value of the alternating voltage. The pressure of the

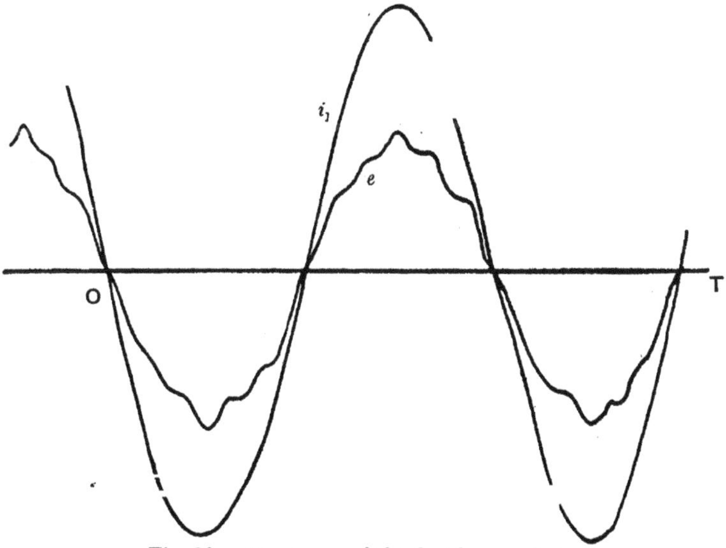

Fig. 82. Resonance of the first harmonic i_1.

direct current supply circuit was unsteady, and, owing to this cause, it was almost impossible to get photographs of the curves at the moment of exact resonance. The frequency of the alternating current was about 26.

Resonance of the first harmonic.

In Fig. 82, e gives the shape of the wave of electromotive force which is analysed, and i_1 gives the phase of the principal harmonic. In this case the value of L was 2·006 henrys, of K, 9 microfarads, and the resistance of the circuit was 204 ohms. The value of ϕ_1 (Fig. 82) is zero, and a_1 measured in a certain scale is 6500.

Resonance of the fifth harmonic.

In Fig. 83 the fifth harmonic is approximately in resonance, but the interference of the first harmonic is evident. The equation to the curve marked i in the figure is approximately

$$i = I_1 \sin(\omega t - \pi/2) + I_5 \sin(5\omega t + \phi_5).$$

Putting ωt equal to zero we get

$$i' = -I_1 + I_5 \sin\phi_5.$$

When ωt is $2\pi/5$, we have

$$i'' = -I_1 \cos(2\pi/5) + I_5 \sin\phi_5.$$

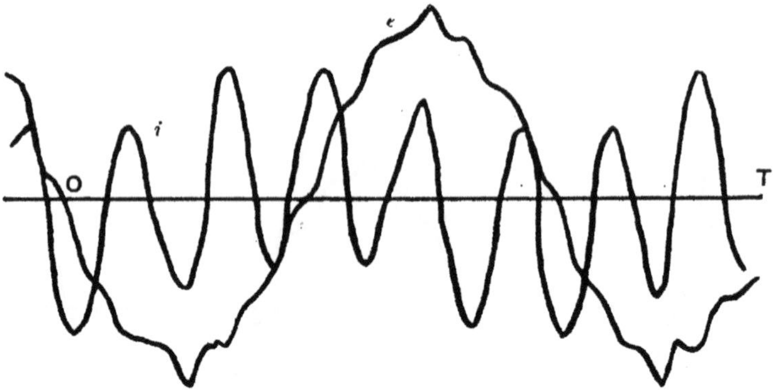

Fig. 83. Resonance of the fifth harmonic.

Similarly we can write down the values of the ordinates when ωt is $4\pi/5$, $6\pi/5$ and $8\pi/5$. Thus, we get by addition

$$i' + i'' + \ldots = -I_1\{1 + \cos(2\pi/5) + \ldots + \cos(8\pi/5)\} + 5I_5 \sin\phi_5$$
$$= 5I_5 \sin\phi_5.$$

Therefore

$$I_5 \sin\phi_5 = (i' + i'' + \ldots)/5 = l.$$

Also when ωt is $\pi/2$ we have

$$I_5 \cos\phi_5 = i_1' = m.$$

Therefore

$$I_5 = \sqrt{l^2 + m^2} \text{ and } \tan\phi_5 = l/m.$$

The values of L, K and R in this case were 0·240, 3 and 13 respectively. The sum of the five ordinates $i' + i'' + \ldots$ is practically zero while i_1' is large. Thus ϕ_5 is zero and a_5 is $13I_5$, that is, $13i_1'$. On the scale in which A_1 is represented by 6500, A_5 is represented by 136.

Approximate resonance of the seventh harmonic is shown in Fig. 84. The interference of the first harmonic is again in evidence. We can find I_7 and ϕ_7 from the curve in practically the same way as we found I_5 and ϕ_5. Starting from t equal to zero and taking the ordinates at distances $2\pi/7$ apart, we have

Resonance of the seventh harmonic.

$$I_7 \sin \phi_7 = (i' + i'' + \dots)/7,$$

and
$$-I_7 \cos \phi_7 = i_1',$$

where i_1' is the value of i when ωt is $\pi/2$.

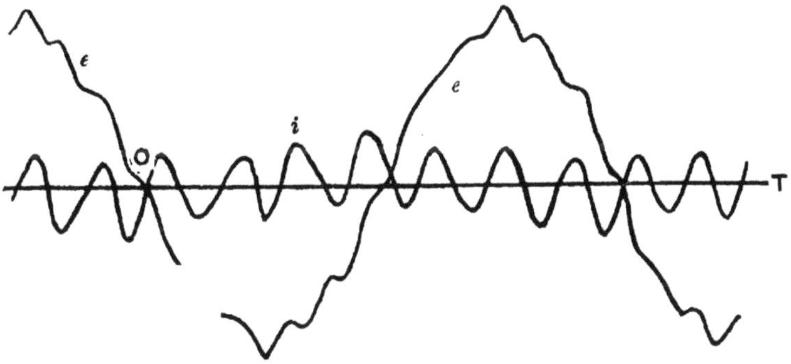

Fig. 84. Resonance of the seventh harmonic.

The values of L, K and R were 0·240, 1·6 and 6·8 and the values of a_7 and ϕ_7 are 57 and $+ 3\pi/4$.

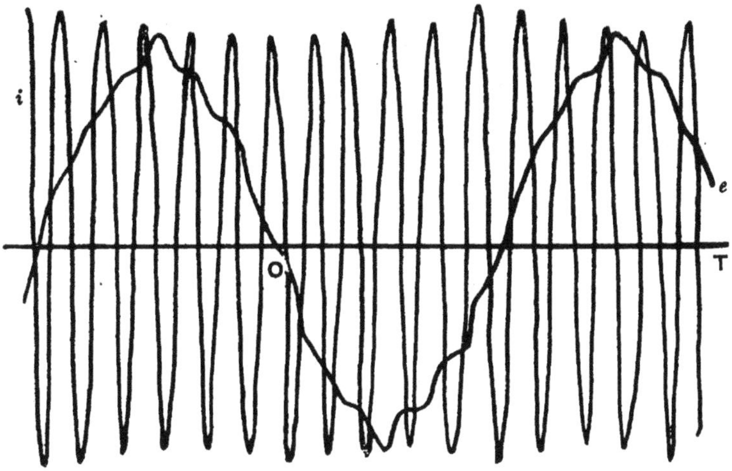

Fig. 85. Resonance of the eleventh harmonic.

In both figures 83 and 84 the amplitude of the component produced by the first harmonic is about a quarter of that produced by the resonating harmonic.

Resonance of the eleventh harmonic.

In Fig. 85 the eleventh harmonic is nearly in resonance, and the interference of the first harmonic is now much less. In this case, the values of L, K and R were 0·240, 0·63 and 13 respectively and the approximate values of a_{11} and ϕ_{11} are 357 and $-\pi$. The variations of the angular velocity of the machine are now in evidence, and the amplitude of the curve i is continually varying.

Interference of two resonating harmonics.

The curve i shown in Fig. 86 illustrates the curious effect produced by the interference of the eleventh and thirteenth harmonics.

The values of L, K and R which produced this effect were ·0106, 0·90 and 18·1 respectively. By this experi-

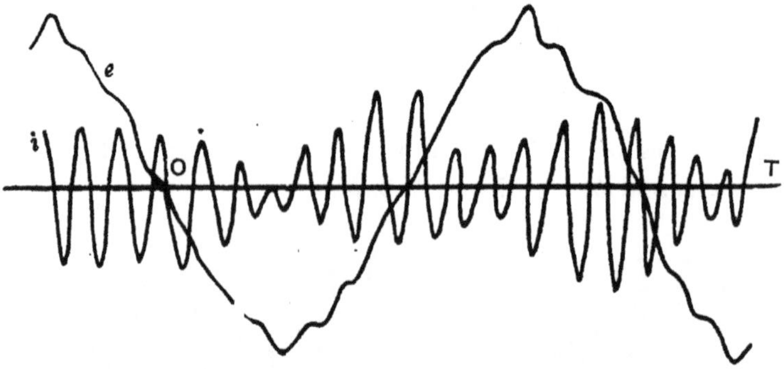

Fig. 86. The thirteenth harmonic interfering with the eleventh harmonic.

mental analysis Armagnat finds that the equation to the curve e is approximately of the form

$$e = 6500 \sin \omega t + 136 \sin 5\omega t + 75 \sin (7\omega t + 3\pi/4)$$
$$+ 357 \sin (11\omega t - \pi) + 90 \sin 13\omega t.$$

The electromotive force wave contains other harmonics, the seventeenth for example, but their amplitude is quite negligible when compared with the amplitudes of the harmonics given above.

Measuring irregularities in the speed of alternators.

Resonance methods can also be applied to measure irregularities in the speed of an alternator. If, for example, in an oscillograph we suppress the displacement of the spot of light in the direction of the time axis we get a luminous straight line. The length of this line is proportional to the amplitude of the wave of current passing through the oscillograph. If this luminous line be primarily due to a resonating high harmonic of the electromotive force wave, the variations in its length will indicate slight variations in the speed of the machine. Now if we let it fall on a strip of sensitive paper wound on a drum which is made to move synchronously with the axis of the alternator, we get a trace on the strip the breadth of which varies with the speed of the alternator.

Causes of the harmonics in electromotive force waves.

If we have an irregular distribution of the magnetic flux in the air-gap of an alternator, then, since the electromotive force is due to the armature conductors being cut by or cutting lines of force, it is obvious that the E.M.F. wave will be irregular. In order, therefore, that the electromotive force wave on open circuit may be a sine curve it is necessary that the distribution of the flux density round the air-gap at every instant follow the harmonic law. With smooth core armatures this can be attained approximately by making the pole pieces of a suitable shape. In modern alternators the field magnets are of cast steel, but the pole pieces are laminated and are cut back so that the distribution of the flux in the air-gap is often approximately sine shaped.

When there are slots in the armature it is evident that the flux density in the air-gap cannot follow the harmonic law, and so in this case we should expect to find harmonics in the E.M.F. wave. Also, when an alternator is working on a load, the reaction of the currents in the armature will distort the field, and harmonics will be introduced into the electromotive force wave of the machine (see Chapter I, p. 59). This latter effect could only be got rid of by constructing a machine with negligible armature reaction.

Harmonics caused by slots.

Let us consider the case of a polyphase alternator, the field magnets of which rotate. Let us suppose that the armature has slots so that there will be a continual variation of the reluctance of the air-gaps as the field magnets rotate. If there are n slots in the polar step, then, during the time that a point on the circumference passes over the nth part of this distance, the flux will go through all its values. This time is the $2n$th part of the period of the alternating current. Now, if $4\pi\mathcal{R}/10$ be the reluctance of the path of the field flux, we have, on open circuit,

$$\Phi = n'C/\mathcal{R},$$

where Φ is the flux traversing the path and $n'C$ the ampere-turns of the exciting current producing this flux. Since $\mathcal{R}$ fluctuates with a frequency $2n/T$, Φ will vary, and an electromotive force will act on the exciting circuit, inducing in it an alternating component. The effect of this induced current is to diminish the amplitude of the variation of Φ from its mean value Φ_m. It has no effect on the frequency $2nf$ of the fluctuations of Φ, where f is the frequency of the alternating current. We may therefore write

$$\Phi = \Phi_m \{1 + \lambda F (2n\omega t)\},$$

where $F (2n\omega t)$ is a periodic alternating function the maximum value of which is unity, and λ is generally a very small fraction.

Now, if ϕ be the instantaneous value of the flux embraced by an armature coil, we can write

$$\begin{aligned} \phi &= F_1 (\omega t) \Phi \\ &= F_1 (\omega t) \Phi_m \{1 + \lambda F (2n\omega t)\}. \end{aligned}$$

By Fourier's Theorem we may write

$$F_1 (\omega t) = A_1 \sin (\omega t - a_1) + A_3 \sin (3\omega t - a_3) + \dots,$$

and $$F (2n\omega t) = B_1 \sin (2n\omega t - \beta_1) + B_3 \sin (6n\omega t - \beta_3) + \dots .$$

Hence a typical term in the series for $2\Phi_m F_1 (\omega t) F (2n\omega t)$ is

$$\begin{aligned} &2\Phi_m A_p B_s \sin (p\omega t - a_p) \sin (s 2n\omega t - \beta_s) \\ &= \Phi_m A_p B_s \cos \{(2ns - p) \omega t + a_p - \beta_s\} \\ &- \Phi_m A_p B_s \cos \{(2ns + p) \omega t - a_p - \beta_s\}, \end{aligned}$$

where p and s can only be odd integers.

The orders of the harmonics in ϕ due to this typical term are therefore

$$2ns + p \text{ and } 2ns - p.$$

Now the electromotive force is proportional to the rate at

which ϕ varies with the time, hence the order of the harmonics in the electromotive force wave will also be $2ns + p$ and $2ns - p$. Therefore the lowest harmonics due to the slots which are introduced into the electromotive force wave of a polyphase machine on open circuit are $2n + 1$ and $2n - 1$ respectively.

The same reasoning applies also to single phase machines on open circuit, the lowest harmonics introduced by the action of the slots being $2n - 1$ and $2n + 1$.

Harmonics in the E.M.F. waves of three phase machines.

The lowest possible orders of the harmonics introduced into the electromotive force waves of three phase machines can easily be written down by the formulae given above. Consider, for example, a three phase machine having one slot per pole and per phase, and therefore having three slots in the polar step. The lowest harmonics introduced by the action of the armature slots will be the $(2 \times 3 - 1)$th and the

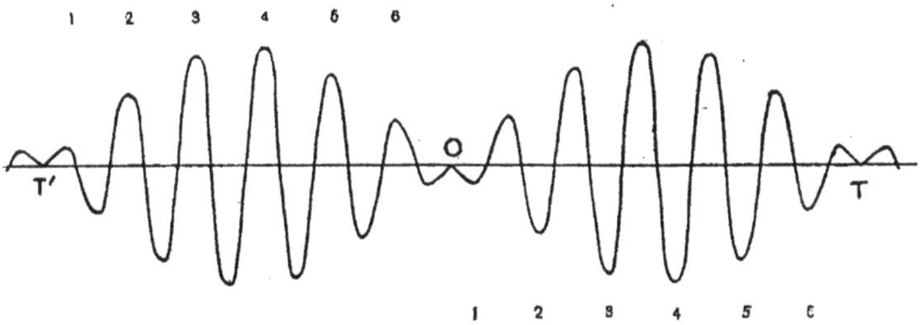

Fig. 87. The jagged component of an E.M.F. wave produced by six slots.

$(2 \times 3 + 1)$th, that is, the fifth and the seventh. If the machine had two slots per pole and per phase then the lowest harmonics would be the eleventh and the thirteenth, and if it had three slots per pole and per phase they would be the seventeenth and the nineteenth.

In Fig. 86 we saw the effect produced by the interference of the eleventh and thirteenth harmonics. To a first approximation we can assume that the equation to this curve is

$$y = I_{11}' \sin(11x - \pi/2) + I_{13}' \sin(13x + \pi/2).$$

If we make the further assumption that I_{11}' and I_{13}' are each equal to unity, we get

$$y = -2 \sin x \sin 12x$$

as the equation to the curve. This curve is shown in Fig. 87, and it will be seen that it is not unlike the curve in Fig. 86. Blondel has suggested that when considering resonance in net-works in practice, when eleventh and thirteenth harmonics are involved, it is sufficient merely to consider this curve, which we may regard as a twelfth harmonic with a periodically varying amplitude. The curve shows the effect of the alternate increase and diminution of the reluctance caused by the six teeth in the polar step. It is to be noticed that the disturbing effect changes sign in each half of the period of the first harmonic. This change of sign is well shown at T', O and T (Fig. 87). In practice we are only able to calculate very roughly the lowest frequency at which resonance will ensue in a net-work. It is sufficient therefore to make sure that this frequency does not approach $2nf$ in value, where n is the number (odd or even) of the slots in the polar step and f is the frequency of the alternating current.

Annulling harmonics by special windings.

Let us consider the electromotive force generated in one phase of a polyphase generator with a distributed winding in the armature. In order to fix our ideas let us take the case of an ordinary ring armature with a Gramme winding and suppose that the two slip rings to which the phase winding is connected have p coils between them, the angle between the planes of two adjacent coils on the armature being a. Let $e_1, e_2, \ldots e_p$ be the electromotive forces generated in the coils, then

$$e_1 = A_1 \sin(\omega t - \phi_1) + \ldots + A_n \sin(n\omega t - \phi_n),$$

and

$$e_p = A_1 \sin\{\omega t - \phi_1 - (p-1)a\} + \ldots + A_n \sin\{n\omega t - \phi_n - n(p-1)a\}.$$

Therefore, since

$$e = e_1 + e_2 + \ldots + e_p,$$

we have

$$e = A_1 \frac{\sin(pa/2)}{\sin(a/2)} \sin\left(\omega t - \phi_1 - \frac{p-1}{2}a\right) + \ldots + A_n \frac{\sin(pna/2)}{\sin(na/2)} \sin\left(n\omega t - \phi_n - \frac{p-1}{2}na\right).$$

Hence if $\sin(pna/2)/\sin(na/2)$ is less than $\sin(pa/2)/\sin(a/2)$ the ratio of the amplitude of the nth to that of the first harmonic is less in the resultant wave than in the wave generated in a single coil.

We see also that if $\sin(pna/2)$ is zero and $\sin(na/2)$ is not zero, the nth harmonic in the resultant wave vanishes. In a single phase machine pa is π, and as n is always an odd number in practice the nth harmonic is not annulled.

Consider now the case of a three phase machine with a rotating armature. In this case we have pa equal to $2\pi/3$, hence

$$pna/2 = n\pi/3, \text{ and } na/2 = n\pi/3p.$$

If n be a multiple of 3 but not of $3p$, there is no nth harmonic in the resultant electromotive force wave.

Methods of preventing the slots in the armature from producing harmonics.

The variation of the reluctance of the magnetic circuit due to the slots in the armature can be prevented by two methods. In the first method the slots (Fig. 88) are inclined at a certain angle to the axis of the rotor. This angle is chosen so that a line drawn parallel to the axis through the middle point of a slot directly

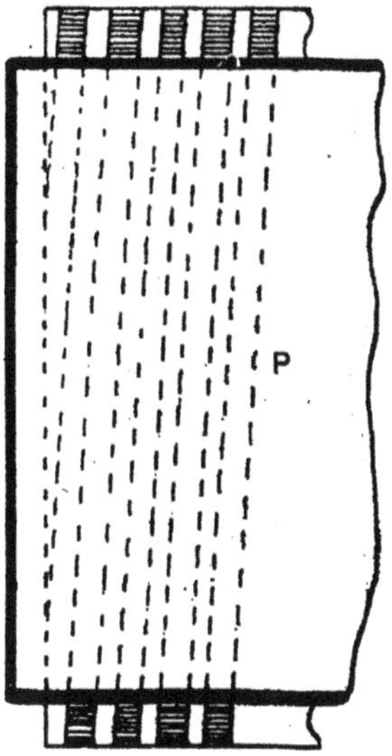

Fig. 88. Inclined slots. P is the pole piece.

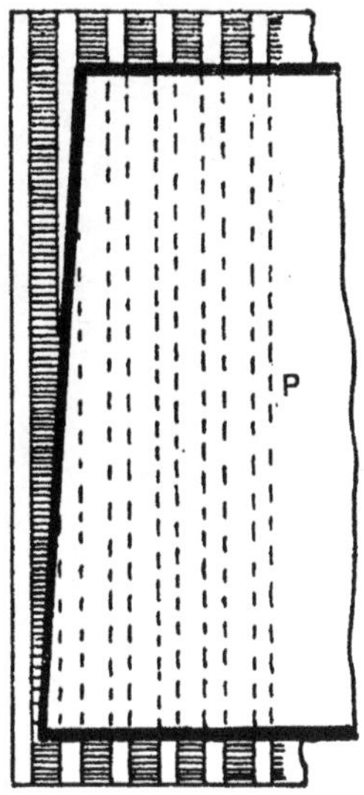

Fig. 89. Inclined pole piece. The slots are parallel to the axis of the rotor.

under a side edge of the pole piece will pass through the middle point of an adjacent slot directly under the opposite edge of the pole piece. Whether the field or armature rotates, the reluctance of the magnetic circuit of the field will in this case be constant.

Fig. 90. Field magnets with inclined pole pieces to obtain a pure sine wave.

An alternative constructional method of eliminating the harmonics caused by slots is to make the slots parallel to the axis and to incline the pole pieces so that if we project the edge of a pole piece on to the armature, this projection will be parallel to the line joining the middle point of a slot directly under the side edge of a pole piece to the middle point of an adjacent slot directly under the opposite edge (Fig. 89). This method of getting a sine wave of E.M.F. was adopted as early as 1892 by the Oerlikon Company in the generator they constructed for the historic Frankfort-Lauffen power-transmission experiments. In Fig. 90 is shown the rotor of an Oerlikon generator in which this device is employed.

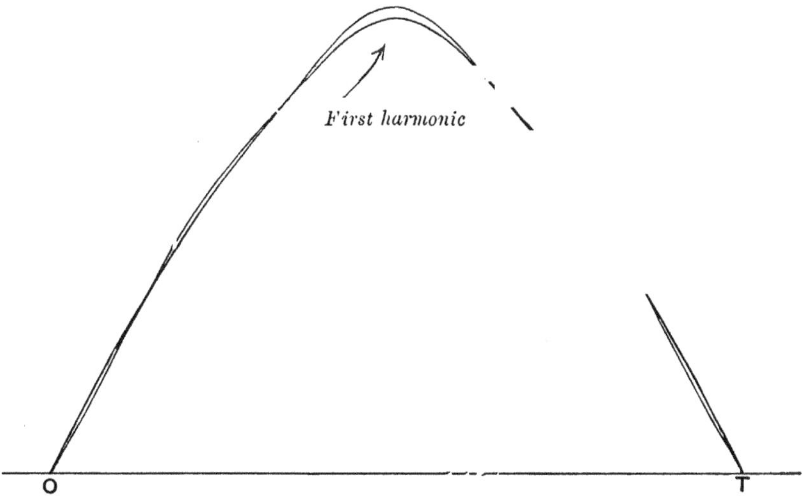

Fig. 91. Positive half of the wave of the mesh E.M.F. of a three phase Oerlikon alternator.
Type 6295. 2350 Kws. 315 Revs. 42 ~.

Both these methods are found very useful in practice; and if, in addition, the pole pieces are bent slightly back so as to make the air-gap of variable depth, it is possible to construct machines which on open circuit will give an electromotive force wave practically indistinguishable from a sine wave.

In Fig. 91 the mesh E.M.F. of a three phase alternator with inclined pole pieces is shown. It will be seen that the first harmonic is very nearly coincident with the wave. The alternator, which has sixteen poles, was built by the Oerlikon Company and has an output of 2350 kilowatts at 315 revolutions per minute. Its frequency is therefore 42.

Owing, however, to the fact that the effects of hysteresis are always appreciable, it would be practically impossible, even on open circuit, to get a perfect sine wave. On closed circuit owing to armature reaction, harmonics, the orders of which are odd, are introduced into the E.M.F. wave generated, and ripples the frequencies of which are even multiples of the frequency of the machine are introduced into the exciting current.

It has been suggested that a sine distribution of flux might be obtained by arranging the windings on the field magnets so as to get this effect. In most cases this would present practical difficulties, and it would be uneconomical owing to the leakage of the field flux.

REFERENCES.

M. I. PUPIN, 'Resonance Analysis of Alternating Currents,' *The American Journal of Science*, Vol. 48, p. 379, May, 1894.

H. ARMAGNAT, 'Application des Oscillographes à la Méthode de Résonance,' *L'Éclairage Électrique*, Vol. 30, p. 373, March, 1902.

C. F. GUILBERT, 'Sur l'élimination des Harmoniques dans les Alternateurs,' *L'Éclairage Électrique*, Vol. 31, p. 377, June, 1902.

F. GUÉRY, 'Sur la Production des Harmoniques dans les Machines à Courants Alternatifs,' *L'Éclairage Électrique*, Vol. 36, p. 51, July, 1903.

R. V. PICOU, 'Oscillations Électriques et Surélévations de Tension Correspondantes,' *Bulletin de la Société Internationale des Électriciens*, Vol. 4, p. 267, May, 1904.

CHAPTER V

Synchronous motors. Bipolar alternator. Bipolar synchronous motor. Equation for the torque. Graphical proof that the equilibrium is stable. Multipolar synchronous motor. Polyphase synchronous motor. The armature reaction of synchronous motors. Generator and synchronous motor. Diagram of the armature electromotive forces. Formula for the potential difference at the terminals. Shape of the resultant electromotive force wave. The current vector. Formulae for the output of the generator and the intake of the motor. Condition for stable running. Fundamental equation. Effect of varying the excitation of the motor and the generator. Graphical solution. Limiting values of the motor electromotive force. Efficiency of the transmission. Method of increasing the efficiency. Variation of the current with the load on the motor. Variation of the current with the excitation of the generator. Variation of the current with the excitation of the motor. Variation of the power factor with the load. Variation of the power factor with the excitation of the motor. References.

Synchronous motors.

WHEN an alternating current dynamo is supplying current to any circuit, then, owing to the electrical losses in the machine, the mean power given to the rotor by the prime mover must be greater than the electrical output. A torque has to be applied to the pulley of the rotor to overcome the magnetic attractions and repulsions between the field poles and the armature poles, as these forces, in accordance with Lenz's law, tend to prevent the rotation. Now the polarity of the armature coils alternates with the same frequency as the current. If, therefore, the values of the currents in the armature coils at any instant were the same as when the machine is acting as an alternator, but if they were flowing in the opposite direction, the attractions and repulsions would become repulsions and attractions, and so the induced torque would be in the direction of rotation and the machine would act as a motor. In order, therefore, to turn an alternator into a motor we need to supply it with alternating current the frequency of which is exactly equal to the frequency of the current it would give when running as an alternator at the same speed. A motor of this kind is called a synchronous motor. It is found in practice that its efficiency is high, and that a considerable mechanical load can be put on the pulley without pulling it out of step with the pulsations of the supply current. In order to understand the action of this type of motor, let us consider the working of a single phase alternator which has an armature rotating in a bipolar field.

Bipolar alternator.

Let us first consider the simple alternator illustrated in Fig. 92. We suppose that the armature is simply a bundle of iron stampings wrapped round with a coil of insulated wire the ends of which are connected with two slip rings. We may suppose that the field is produced either

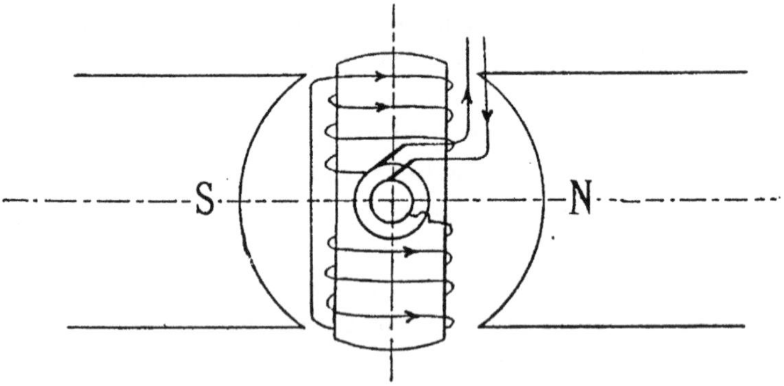

Fig. 92. Single phase alternator or synchronous motor.

by permanent magnets or by electromagnets excited by direct current. If we rotate the armature at a constant speed, the electromotive force generated in the coil will be a maximum when it is in the position shown in the figure, and it will be zero when the axis of the coil is horizontal, that is, when it embraces the maximum magnetic flux. If SN is the position of the axis of the coil at the moment when we begin to measure time, we may, on making certain assumptions, express the electromotive force generated in the coil by $E \sin \omega t$, where ω is the angular velocity of the armature of the two pole machine and $\omega/2\pi$ is therefore the frequency of the alternating E.M.F. generated.

If the external circuit be closed through a large non-inductive resistance, the current in the armature will be in phase with the armature electromotive force. Hence the current will be a maximum in the position of the armature shown in Fig. 92, and if the direction of rotation is with the hands of a watch, that is, if it rotates against the brushes shown in the figure, the arrow heads will indicate the direction of the current. The top part of the armature will, therefore, be a north pole. It will be seen that work has to be done against the magnetic attractions and repulsions of the field poles in order to maintain the mean angular velocity. If g denote the instantaneous value of the moment, about the axis of the armature, of the mechanical forces which have to be applied to it, so that its angular velocity may not vary, $g\omega$ will be the rate at which work is given to it, and if this be expressed in watts, we shall have

$$g\omega = ei,$$

where e and i are the instantaneous values of the electromotive force and current respectively; we neglect the losses due to friction, eddy currents and hysteresis. Since e is zero twice in every revolution, and i is in phase with e, we see that in this case g is also zero twice in every revolution. When e and i have not the same time lag, g must vanish four times every revolution and it is sometimes negative and sometimes positive. If the machine has $2p$ poles, the frequency of the alternating currents generated will be $p\omega/2\pi$, and the torque will vanish $4p$ times every revolution, provided that the current and electromotive force do not vanish simultaneously. If they do vanish simultaneously the torque will vanish $2p$ times every revolution.

Bipolar synchronous motor.

Let us suppose that when the angular velocity of the armature of the above machine is ω, the slip rings are put in circuit with mains supplying alternating current of frequency f, and suppose that ω is $2\pi f$. Then, if the armature is rotating in the direction against the hands of a watch as indicated by the brushes in Fig. 92, and if the current is a maximum in the position illustrated and flows in the direction of the arrow heads, there will be a torque in the direction of the motion, as the top part of the armature is a north pole. A quarter of a period later the axis of the armature coil will be horizontal. The current reverses in the armature at this instant, and we see that the side of the armature which is uppermost is always a north pole. Similarly, the lower side of the armature is always a south pole, and hence the torque is always in the same direction. The effect of the alternating current, therefore, in this case, is to produce a mechanical torque which tends to accelerate the angular velocity of the armature.

Equation for the torque.

Let us now consider the case when the alternating current supplied can be represented by $I \sin \omega t$. We suppose that t is zero when the axis of the armature coil is in the position SN (Fig. 92). Owing to the direction of rotation being opposite to the direction it has when the machine acts as an alternator, the electromotive force developed will be always in opposition to the current, and hence work will be given to the armature. This electromotive force, developed in the armature, is generally referred to as the back E.M.F. of the armature. Since, in our case, it is proportional to $\sin \omega t$, we may write

$$g\omega = EI \sin^2 \omega t,$$

hence

$$g = G \sin^2 \omega t$$
$$= \tfrac{1}{2}G - \tfrac{1}{2}G \cos 2\omega t,$$

where G, which equals EI/ω, is the maximum torque on the armature. We see at once that the mean torque over a whole revolution is $\tfrac{1}{2}G$ in this case.

In general, when the alternating current supplied is

$$I \sin (\omega t - a),$$

we have

$$g = G \sin \omega t \sin (\omega t - a)$$
$$= \tfrac{1}{2}G \cos a - \tfrac{1}{2}G \cos (2\omega t - a).$$

The mean torque is therefore $\tfrac{1}{2}G \cos a$ and it only vanishes when a is + or − ninety degrees. For all values of a between these limits the mean work done on the armature during a revolution is positive. If, however, a be greater than ninety degrees, the mean torque is negative, and the armature is giving work to the electric circuit. In this case the machine is acting as a generator.

In Fig. 93 the mean torque is shown graphically for all values of α. BOB' is a vertical line through the axis of the armature, and NOS is the horizontal line. The lines OB and OB' are each equal to $\frac{1}{2}G$ and circles are described with these lines for diameters. We can suppose that lines, like OP, drawn to points on the circumference of the upper circle are positive, and that lines, like OP', drawn to points on the circumference of the lower circle are negative. If

Graphical proof that the equilibrium is stable.

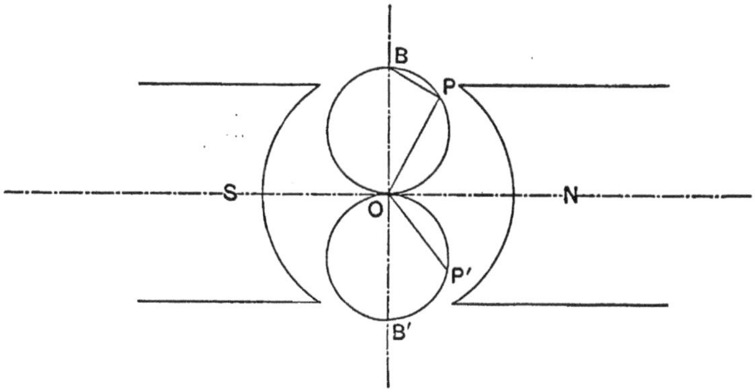

Fig. 93. *OP* gives the mean value of the accelerating torque when the phase difference between the current and the back E.M.F. is the angle *BOP*.

the angle BOP is α, OP equals $BO \cos \alpha$, that is, $\frac{1}{2}G \cos \alpha$, and hence this line gives the value of the mean accelerating torque when the current lags by an angle α behind the counter E.M.F. generated in the armature. If the angle BOP' is α', then, since OP' is drawn to the lower circle it is negative and equals $\frac{1}{2}G \cos \alpha'$, the mean retarding torque when the angle of lag is α'.

Let us consider the case when the current supplied to the slip rings lags ninety degrees in phase behind the back electromotive force of the armature. In this case, α will be ninety degrees, and so OP will be zero. If the armature now slows down, the difference in phase between the back E.M.F. and the current will diminish, and OP will rapidly increase, tending to drive the armature more quickly. On the other hand, if the armature quickens when α is ninety degrees, a retarding torque OP' will be applied to the rotor by the electrical forces.

When a is less than ninety degrees and $\frac{1}{2}G \cos a$ is the mean value of the retarding torque of the mechanical forces applied to the armature, the machine will run as a motor, and, if the moment of inertia of the armature be considerable, there will be only slight fluctuations in its speed due to the fluctuations in the value of g. When the armature slows down the mean torque increases, and when it quickens the mean torque diminishes. Hence an alternator used in this fashion makes a very satisfactory motor. It is called a synchronous motor because it is always exactly in step with the alternating current supplied.

Multipolar synchronous motor.

If the machine have $2p$ poles, and we make the same assumptions as before, then

$$g = G \sin p\omega t \sin (p\omega t - a)$$
$$= \tfrac{1}{2}G \cos a - \tfrac{1}{2}G \cos (2p\omega t - a),$$

where G is the maximum value of the torque when the current and the back electromotive force are in phase with one another. We see that, in general, g vanishes $4p$ times during one revolution of the armature. The angular velocity of the armature is $2\pi f/p$, and it therefore makes $60f/p$ revolutions per minute. A twenty pole machine, for example, supplied with alternating current, having a frequency of 50, would make 300 revolutions per minute.

Polyphase synchronous motor.

Consider a polyphase alternator with its terminals connected with three mains supplying three phase currents, of frequency f, and suppose that the angular velocity of the rotor is $2\pi f/p$ where $2p$ is the number of poles. If we make the assumption that the currents and the back electromotive forces in the armature coils follow the harmonic law, then, if g be the instantaneous value of the torque exerted by the magnetic forces on the armature, we have

$$g\omega = EI \{\sin p\omega t \sin (p\omega t - a)$$
$$+ \sin (p\omega t + 2\pi/3) \sin (p\omega t - a + 2\pi/3)$$
$$+ \sin (p\omega t + 4\pi/3) \sin (p\omega t - a + 4\pi/3)\}$$
$$= EI \{(3/2) \cos a\},$$

and, therefore,

$$g = (3EI/2\omega) \cos a.$$

In this case, therefore, the torque on the armature is absolutely constant. We can also easily show that when the armature quickens the torque is diminished, and when it slows down the torque is increased. Hence a polyphase alternator, provided that it generates a sine wave of E.M.F. and is supplied with harmonic currents, will run very smoothly as a synchronous motor.

The armature reaction of synchronous motors.

In Chapter I we explained Blondel's theory of two reactions. The current in the armature was resolved into two components, one of which was in phase with the armature electromotive force, and the other was in quadrature with it. The former merely produced a transverse magnetisation of the field, and the latter partially demagnetised or magnetised the field magnets, according as the current was lagging or leading. Formulae were found for these effects. In a synchronous motor the same effects will be produced, but since, when an alternator is acting as a motor, the currents in the armature are flowing in the opposite direction to that in which they flow when the machine is acting as a generator, the magnetic effects will be reversed. That is to say, the transverse magnetisation will be in the opposite direction to that in which it is in a generator and a lagging current will now tend to magnetise the field magnets whilst a leading current will demagnetise them. The formulae for these effects are given on pages 43 and 52.

Generator and synchronous motor.

We shall now discuss what happens when the load on a synchronous motor is varied. In this case, as a rule, the phase difference between the current and the applied potential difference alters, and this produces a change in the magnetic field of the generator. It is therefore essential, when discussing the working of a synchronous motor, to take into consideration also the generator or generators supplying it with electric power.

In order to simplify the problem as much as possible, we will consider the case of two alternating current machines which are similar to one another in all respects. We shall suppose that they are running at the same speed and that their terminals are in metallic connection. If their field magnets be excited, there will be in general an electromotive force round the circuit of the

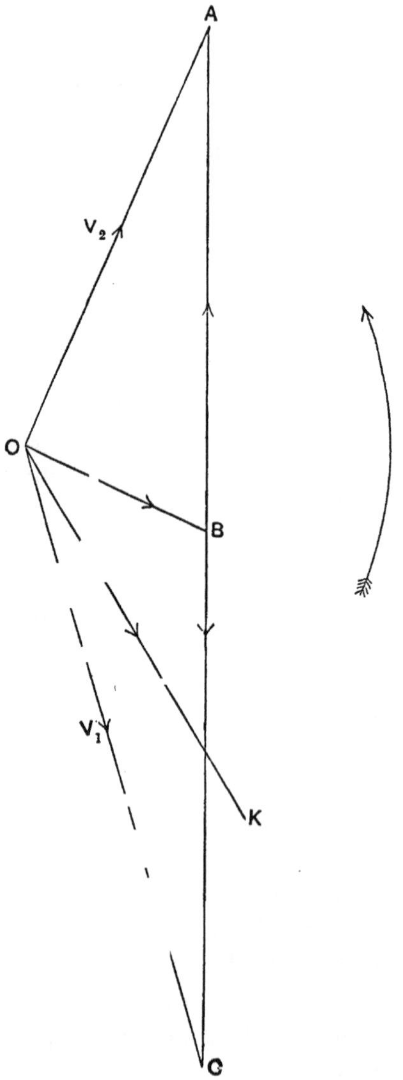

Fig. 94. Vector diagram of a generator and a synchronous motor.

two armatures and the same current will be flowing in each. In the case when the electromotive forces generated in the two machines are equal at every instant but act in opposite directions round the circuit of the two armatures, there will be no current, and hence no power will be conveyed from one machine to the other. In order that power may be transmitted, a current must flow, and hence the two electromotive forces cannot be in exact opposition in phase. The simplest method of discussing this problem is to represent the electromotive forces generated in each machine by vectors. We can suppose that these vectors take account of armature reaction.

Diagram of the armature electromotive forces.

Let OC in Fig. 94 represent the effective value V_1 of the armature electromotive force of the generator, and let OA represent the effective value V_2 of the armature electromotive force of the motor. Join AC and bisect it in B, then twice OB represents the resultant of V_1 and V_2 in magnitude and phase, and is the effective value of the electromotive force that drives the current round the circuit. Since by the triangle of vectors V_1 is equivalent to the vectors OB and BC, and V_2 is equivalent to OB and BA, it follows that BA and BC are each equal to the voltage V between the connecting mains. The lower part of the diagram OBC refers to the generator and the upper part to the motor. Hence, although BA and BC represent the same voltage, yet we have drawn them as if they were in opposition in phase. The phase of the potential difference voltage must be drawn in opposite directions when looked at from the generator or the motor end of the circuit. In one case the current in a circuit bridging the mains would appear at a particular instant to be going from left to right, whilst in the other case it would appear at the same instant to be going from right to left.

Formula for the potential difference at the terminals.

If V be the potential difference between the connecting mains, and θ the phase difference between the electromotive forces of the two machines, then

$$V = \tfrac{1}{2}(V_1^2 + V_2^2 - 2V_1V_2\cos\theta)^{\frac{1}{2}} \quad \text{......(1)}.$$

The maximum value of V is therefore $\frac{1}{2}(V_1 + V_2)$, and it has this value when $\cos\theta$ is -1, that is, when the phase difference between the two electromotive forces is 180 degrees. In this case, the electromotive forces are in opposition so far as the circuit of the armatures is concerned, but they would be in phase as regards their action on a circuit bridging the two mains joining the terminals of the machines. We shall see in Chapter VII that this

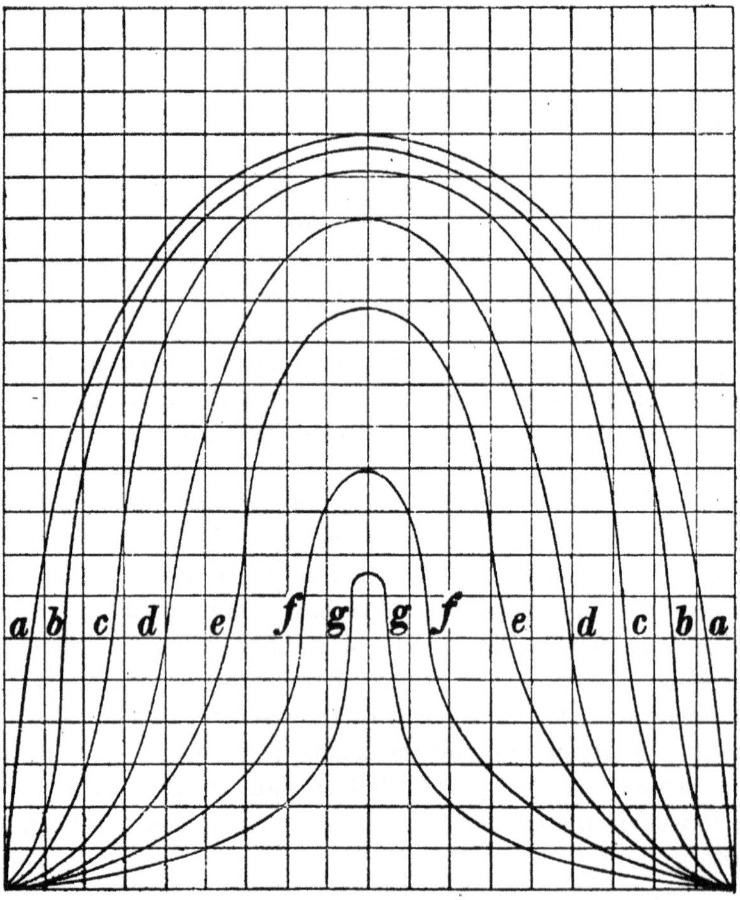

Fig. 95. Resultant of two rounded electromotive force waves when their angles of time lag are (*a*) 0°, (*b*) 15°, (*c*) 30°, (*d*) 60°, (*e*) 90°, (*f*) 120°, (*g*) 150°.

is approximately the case when the two alternators are running in parallel.

If the machines give electromotive force waves of different shapes, then θ can never be as great as 180 degrees. For example, if the electromotive force wave of the generator were sine shaped and had an effective value of 1000 volts whilst the motor wave were rectangular and had an equal effective value, the maximum value of V would be 974·8 volts.

Shape of the resultant electromotive force wave.

The effective value of the electromotive force wave producing the current round the circuit of the two armatures is represented in magnitude by twice OB, and in phase by OB (Fig. 94). This, however, tells us nothing about the shape of the resultant wave. If the waves were sine curves then, whatever value the angle of lag between them might have, their resultant would also be a sine curve. In the general case, however, the shape of the resultant wave is quite different from the shape of either of its components. In Fig. 95 the variation in the shape of the resultant of two equal semicircular shaped waves is shown for the case of angles of time lag equal to 0, 15, 30, 60, 90, 120, and 150 degrees respectively. It will be seen that when the time lag is small we get a rounded wave, but when it is nearly 180 degrees we get a very peaky one.

If both the component waves are peaky, then, as a rule, the shape of their resultant is rounded when they are nearly in opposition, and peaky when they are nearly in phase. This change of shape of the wave of the electromotive force which produces the current in the armatures makes the phase difference between it and the current produced a variable quantity, and so we are not justified in making the assumption that the impedance of the circuit of the two armatures is constant. Again the armature reaction of each machine depends on the magnitude and phase of the current, and so the shape of the electromotive force waves must also vary from this cause.

The current vector.

If the circuit of the armatures of the alternator and the synchronous motor acted like a non-inductive resistance, the current vector would be represented in phase by OB (Fig. 94). In general, however, the phase difference between the current and the resultant electromotive force is large. Suppose that OK in Fig. 94 represents this vector, and let the angle BOK equal γ. If we suppose that OK is in the plane OAC, the angle KOA would be the phase difference between the current and the motor electromotive force, and the angle KOC would be the phase difference between the current and the generator electromotive force. This would be true if the waves which OA, OC and OK represent were all sine waves. In practice it is not true, and hence for a rigorous theory we should need to have recourse to solid geometry (see Vol. I, Chapter XII). The formulae obtained by making the assumption that the vectors are all in one plane are useful and instructive, but it has to be remembered that they are only approximate.

Formulae for the output of the generator and for the intake of the motor.

The formulae for the power generated in the alternator and received by the motor can easily be deduced from Fig. 94. In this figure

OC is the vector of the alternator E.M.F., V_1,
OA is the vector of the motor E.M.F., V_2,
OK is the vector representing the current, A_1,
AOC is the phase difference θ between OA and OC,
and the angle BOK is γ.

The angle BOK represents the angle of lag of the current behind the resultant E.M.F. round the circuit of the armatures. This resultant E.M.F. is represented by twice OB.

We shall also denote the impedance of the circuit of the two armatures and their connecting mains by Z, the electrical power generated by the alternator by W_1 and the electrical power given to the motor by W_2. Since KOC is the phase difference between the vectors OK and OC, that is, between A and V_1, we get

$$W_1 = A \,.\, V_1 \cos KOC.$$

Now, since Z is the impedance of the circuit of the armatures, and $2 \,.\, OB$ is the resultant E.M.F., we have

$$A = 2 \,.\, OB/Z.$$

Again

$$\begin{aligned} 2 \,.\, OB \cos KOC &= 2 \,.\, OB \cos (BOC - \gamma) \\ &= 2 \,.\, OB \cos BOC \cos\gamma + 2 \,.\, OB \sin BOC \sin\gamma \\ &= (V_1 + V_2 \cos\theta) \cos\gamma + V_2 \sin\theta \sin\gamma \\ &= V_1 \cos\gamma + V_2 \cos(\theta - \gamma). \end{aligned}$$

Hence, substituting for $A \cos KOC$ in the formula for W_1, we get

$$W_1 = (V_1/Z)\,\{V_1 \cos\gamma + V_2 \cos(\theta - \gamma)\} \;\;\ldots\ldots(2).$$

Similarly $$W_2 = -V_2 A \cos AOK$$
$$= -(V_2/Z)\{V_2 \cos\gamma + V_1 \cos(\theta+\gamma)\} \quad \ldots(3).$$

When the running is steady (2) and (3) give us the relations between the various quantities involved. We see from (3) that W_2 is a maximum when θ is $\pi - \gamma$. It is then equal to

$$(V_2/Z)\{V_1 - V_2 \cos\gamma\}.$$

Hence the smaller the impedance Z of the circuit, and the nearer γ is to 90 degrees, the greater is the load that can be put on the motor.

If we write $\pi - \gamma + \alpha$ for θ in equations (2) and (3), we get

Condition for stable running.

$$W_1 = (V_1/Z)\{V_1 \cos\gamma - V_2 \cos(2\gamma - \alpha)\}$$

and

$$W_2 = (V_2/Z)\{V_1 \cos\alpha - V_2 \cos\gamma\}.$$

If we suppose that α varies owing to irregularities in the speed of the motor or generator, then

$$\frac{dW_1}{d\alpha} = -\frac{V_1 V_2}{Z}\sin(2\gamma - \alpha)$$

and

$$\frac{dW_2}{d\alpha} = -\frac{V_1 V_2}{Z}\sin\alpha.$$

Hence if α be positive W_2 diminishes when α increases. We see that when the motor quickens the power given to it diminishes, and similarly when it slows down the power given to it increases, and so the electric forces called into play tend to keep the speed constant. Also, in practice, 2γ is always greater than α, and hence W_1, the load on the generator, diminishes as α increases, and so this also has the effect of tending to restore α to its original value. Therefore positive values of α correspond to stable positions of running.

When α is zero, W_2 has its maximum value

$$(V_2/Z)\{V_1 - V_2 \cos\gamma\}.$$

Hence the smaller the impedance Z of the circuit, and the nearer γ is to 90 degrees, the greater is the load that can be put on the motor. We see also that if power is to be given to the motor, V_2 must be less than $V_1/\cos\gamma$.

Again, since

$$\cos \alpha = \frac{W_2 Z}{V_1 V_2} + \frac{V_2}{V_1} \cos \gamma,$$

it follows that for every load W_2 on the motor there is a positive and a negative value of α which satisfies this equation. We have already seen that the positive value of α corresponds to the stable position of running, and we can show in an exactly similar way that the negative value corresponds to an unstable position. When W_2 is zero, α is $\cos^{-1} \{(V_2/V_1) \cos \gamma\}$. Hence the stable positions of running are given by

$$\theta = \pi - \gamma + \alpha,$$

where α can have any value between 0 and $\cos^{-1} \{(V_2/V_1) \cos \gamma\}$. If V_2 be less than V_1, θ may be greater than π. In this case we may regard the generator as the leading machine. For different loads, α has different values, but in all cases the mean angular velocity of the rotor is exactly the same, namely $2\pi f/p$, where f is the frequency of the alternating current and $2p$ is the number of poles of the motor.

When a load is put on a synchronous motor gradually, α slowly diminishes. When the load is so great that α vanishes, then the angular velocity of the rotor diminishes, and the applied potential difference being no longer in step with the back electromotive force of the armature, a large pulsating current is set up, which blows the fuses or opens the magnetic cut-outs which are used to protect the machine.

Fundamental equation.

We know that the square of the effective value of the total electromotive force round the circuit of the armatures is $V_1^2 + V_2^2 + 2V_1 V_2 \cos \theta$, where θ is the phase difference between V_1 and V_2. Hence, if Z be the impedance of this circuit, and A the effective value of the current flowing in it, we have

$$\begin{aligned} A^2 Z^2 &= V_1^2 + V_2^2 + 2V_1 V_2 \cos \theta \\ &= V_1^2 + V_2^2 + 2V_1 V_2 \cos \{(\theta + \gamma) - \gamma\} \\ &= V_1^2 + V_2^2 + 2V_1 V_2 \cos (\theta + \gamma) \cos \gamma \\ &\qquad + 2V_1 V_2 \sin (\theta + \gamma) \sin \gamma. \end{aligned}$$

Substituting for $\cos(\theta+\gamma)$ and $\sin(\theta+\gamma)$ their values obtained from (3) and noticing, since the minimum value of θ for steady running is $\pi-\gamma$, that $\sin(\theta+\gamma)$ is either zero or negative, we get

$$A^2Z^2 = V_1^2 - 2W_2Z\cos\gamma - V_2^2\cos 2\gamma$$
$$- 2\sin\gamma\,\{V_1^2V_2^2 - (W_2Z + V_2^2\cos\gamma)^2\}^{\frac{1}{2}} \quad \ldots(4).$$

This is the fundamental equation of the synchronous motor. If we square this equation and simplify we get

$$(A^2Z^2 - V_1^2 + V_2^2 + 2W_2Z\cos\gamma)^2 = 4\sin^2\gamma\,(A^2Z^2V_2^2 - W_2^2Z^2).$$

This equation is sometimes given as the fundamental equation, but (4) is more useful in practice as the values of the variables found from it correspond to stable positions of running only, and the current is given directly in terms of the other variables.

We shall first consider the effect of varying the excitation of the motor or the generator on the current in the circuit, and on the power factor of the motor load.

Effect of varying the excitation of the motor and the generator.

In equation (4), if we regard V_1 as variable and V_2, W_2, Z and γ as constants, then, for each value of V_1, we get a definite value of A. Equating the first differential coefficient of A with regard to V_1 to zero, solving the resulting equation for V_1 and substituting in (4), we find that the minimum value of A is W_2/V_2. Hence the minimum value of the current obtained by varying the excitation of the generator is W_2/V_2. It is, therefore, in exact opposition in phase to V_2.

Similarly, when we vary the excitation of the motor, the minimum value of the current is given by

$$A = \frac{V_1}{2Z\cos\gamma} - \left\{\frac{V_1^2}{4Z^2\cos^2\gamma} - \frac{W_2}{Z\cos\gamma}\right\}^{\frac{1}{2}}.$$

In this case we can show that the current is in phase with V_1, so that the minimum value of the current is W_1'/V_1, where W_1' is the electric power generated when V_1 and A are in phase.

Graphical solution.

These theorems can be proved more easily as follows. When we vary V_1 by altering the excitation of the generator, the power W_2 given to the motor circuit is $-AV_2 \cos\theta_2$, where θ_2 is the phase difference between A and V_2. Since this power is constant, A will be a minimum when $-\cos\theta_2$ is a maximum, that is, when θ_2 is 180°, and in this case A is W_2/V_2.

In Fig. 96, (a) gives the diagram for the minimum value of the current when the excitation of the generator is varied.

Again, from equations (2) and (3), or directly from the fact that the electrical power generated by the alternator equals the power

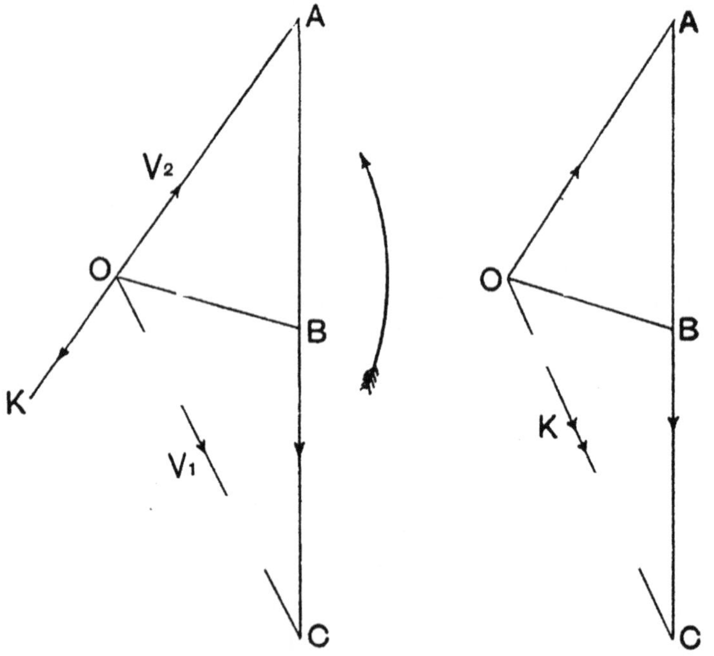

Fig. 96. (a) The minimum value of the current when the excitation of the generator is varied. (b) The minimum value of the current when the excitation of the motor is varied.

given to the motor together with the power expended by the driving electromotive force, we have

$$W_1 = W_2 + AZ \,.\, A\cos\gamma.$$

Now W_1 equals $AV_1 \cos\theta_1$, where θ_1 is the phase difference between A and V_1. Substituting this value for W_1 in the equation, and solving the resulting quadratic equation for A, we find that

$$A = \frac{V_1 \cos\theta_1}{2Z\cos\gamma} - \left\{\frac{V_1^2 \cos^2\theta_1}{4Z^2\cos^2\gamma} - \frac{W_2}{Z\cos\gamma}\right\}^{\frac{1}{2}}.$$

For every load W_2 there are two possible values of the current, but the larger one corresponds to the unstable position of running, and so we have prefixed the negative sign to the radical in this equation.

Since the differential coefficient of A with respect to $\cos\theta_1$ is a negative quantity, it follows that A diminishes as $\cos\theta_1$ increases. Hence it has its minimum value when $\cos\theta_1$ is unity, and this gives us the same value of A as before. Also, in this case, W_1' equals AV_1, and hence the minimum value of A is W_1'/V_1. This case is illustrated in (*b*) Fig. 96. The vector OK of the current coincides in direction with OC, the vector of the generator electromotive force.

Limiting values of the motor electromotive force.

In order that the value A of the current given by equation (4) may be real, $W_2Z + V_2^2\cos\gamma$ must be less than V_1V_2. It follows that the value of V_2 must lie between

$$\frac{V_1}{2\cos\gamma} + \left\{\frac{V_1^2}{4\cos^2\gamma} - \frac{W_2Z}{\cos\gamma}\right\}^{\frac{1}{2}}$$

and

$$\frac{V_1}{2\cos\gamma} - \left\{\frac{V_1^2}{4\cos^2\gamma} - \frac{W_2Z}{\cos\gamma}\right\}^{\frac{1}{2}}.$$

We see that if $\cos\gamma$ be small, the back electromotive force of the motor may be considerably greater than the electromotive force of the generator which is driving it. The maximum possible value of W_2 is, however, $V_1^2/4Z\cos\gamma$.

Suppose, for example, that γ is 60 degrees, then V_2 must lie between

$$V_1 + (V_1^2 - 2W_2Z)^{\frac{1}{2}} \text{ and } V_1 - (V_1^2 - 2W_2Z)^{\frac{1}{2}}.$$

The maximum value of W_2 in this case is $V_1^2/2Z$.

Efficiency of the transmission.

If η denote the ratio of the power given to the motor to the total power generated by the alternator, then η is the fractional efficiency of the transmission. With our usual notation

$$\eta = \frac{W_2}{W_2 + A^2Z\cos\gamma},$$

and hence for a given load W_2 the efficiency is a maximum when the current is a minimum. If we vary the excitation of the generator, V_2 remaining constant, the minimum value of the current is W_2/V_2. Hence the maximum efficiency in this case is

$$\frac{1}{1+(W_2 Z/V_2^2)\cos\gamma},$$

and this diminishes as W_2 is increased.

Again, when we vary the excitation of the motor, keeping V_1 constant, the maximum efficiency occurs when A has its minimum value W_1/V_1, and in this case

$$\eta = \frac{W_2}{W_1} = \frac{W_2}{AV_1},$$

and
$$A = \frac{V_1}{2Z\cos\gamma} - \left\{\frac{V_1^2}{4Z^2\cos^2\gamma} - \frac{W_2}{Z\cos\gamma}\right\}^{\frac{1}{2}}.$$

Hence the maximum efficiency when the load is W_2 can be found.

Method of increasing the efficiency.

The results given above indicate the following method of procedure as being theoretically desirable, when we wish to increase the efficiency by raising the voltage. First adjust the excitation of the motor until the current is a minimum. Then increase the excitation of the generator until the current is again a minimum. Then go back to the motor and increase its excitation until the current is reduced again to its smallest value, and so on backwards and forwards between the two machines until the desired efficiency is attained. In practice a limit to the possible excitation is soon reached. It would save time to over-excite the motor in the first instance, but the theoretical method is worth remembering.

Variation of the current with the load on the motor.

As the fundamental equation (4) is complicated, we shall illustrate it graphically by drawing curves for various particular cases. It is to be remembered that we have made the assumption that the current vector and the electromotive force vectors are in one plane, and we now make the further assumptions that the impedance Z and the angle of lag γ remain constant as the excitations vary. The curves arrived at are very similar to those obtained by actual experiments, and show that the main phenomena connected with the working of synchronous motors could have been predicted from the properties of triangles. On the other hand the anomalous results sometimes obtained, when the electromotive force waves are very distorted from the sine shape, show that our assumptions are not justifiable in these cases.

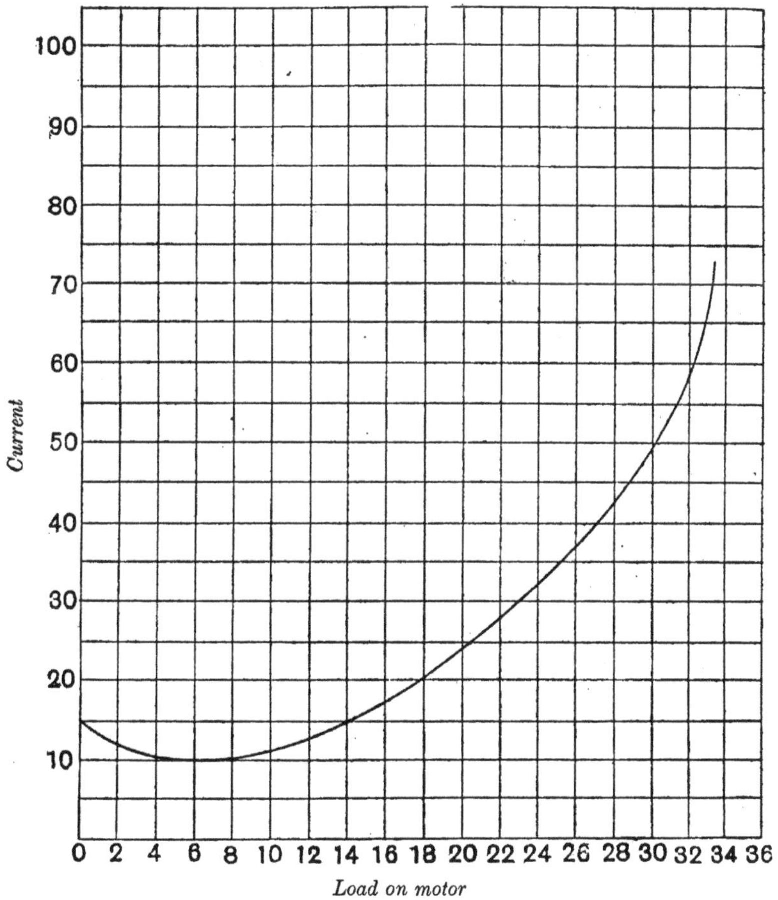

Fig. 97. Variation of current with the load on a synchronous motor.

In Fig. 97 a curve is shown illustrating how the current A varies with the load W_2 on the motor. The angle of lag γ of the current behind the resultant or driving electromotive force round the circuit of the armatures has been taken equal to 45°. In practice it is usually greater than this.

It will be noticed that when W_2 is zero the current is 15 amperes, but when the load is 7·5 kilowatts the current is only 10 amperes. It will be seen that the current varies very little for a small increase or diminution of this load. On the other hand, when the load is heavy, a slight increase of it will cause a large increase of the current.

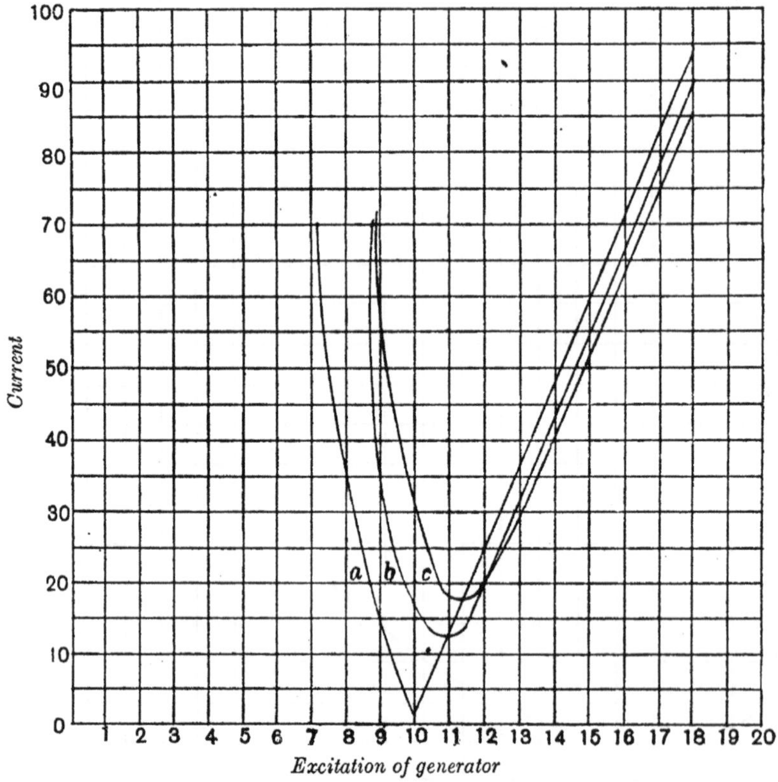

Fig. 98. Effect of varying the excitation of the generator on the current taken by a synchronous motor at three different loads '*a*,' '*b*' and '*c*' when γ is 45 degrees

Fig. 98 shows how the current varies with the excitation of the generator for three different loads which are to one another in the ratios 1 : 16 : 25. The angle of lag γ has again been taken equal to 45 degrees. When running at a high voltage increasing the load diminishes the current, but at a low voltage increasing the load increases the current.

Variation of the current with the excitation of the generator.

The curves in Fig. 99 show how the current varies with the excitation of the motor when γ is 90 degrees. The curve '*a*' shows the machine running on a zero load. In this case the curve is simply two lines meeting one another at the point 10 on the axis of x. The curves '*b*' and '*c*' show the machine running on a light and a moderate load respectively. The curves do not intersect one another in this case.

Variation of the current with the excitation of the motor.

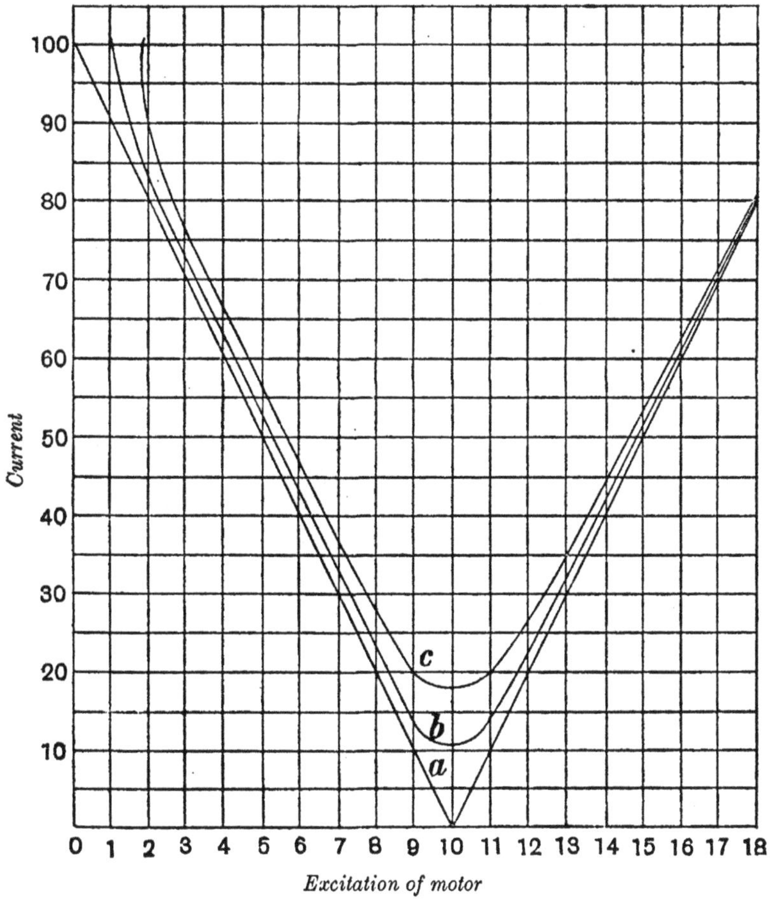

Fig. 99. The curves '*a*,' '*b*' and '*c*' show how the current varies with the excitation of a synchronous motor at light loads when γ is 90°.

In Fig. 100 we have taken γ equal to 45 degrees, all the other data remaining the same as in the preceding illustration. It is to be noticed that the curves now cross one another. Comparing them with the curves shown in the preceding diagram it will be seen that the new curves are much narrower than the old, and that there is a superior as well as an inferior limit to the excitation of the motor.

Curves similar to those shown in Figs. 98, 99 and 100 were first obtained experimentally by Mordey. They are generally called V curves.

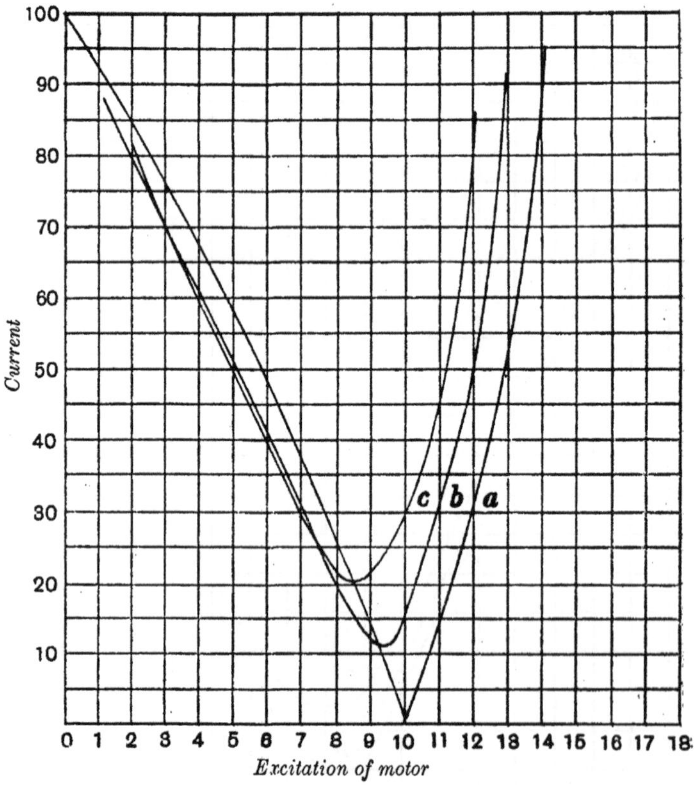

Fig. 100. The curves '*a*,' '*b*' and '*c*' are the V curves of a synchronous motor on light loads when γ is 45°.

Variation of the power factor with the load.

The power factor of the motor circuit is the cosine of the angle between the current vector OK and the line joining the extremities of the two vector electromotive forces OA and OC shown in Fig. 94. When the motor is feebly excited, V_2 is small, and hence the angle BOC

is small. In practice the angle BOK is nearly 90 degrees; hence the current OK lags behind the applied potential difference BC in phase. As we increase the excitation, V_2 increases, and OK becomes parallel to BC for a particular excitation. It would apparently follow that the power factor must always be unity for

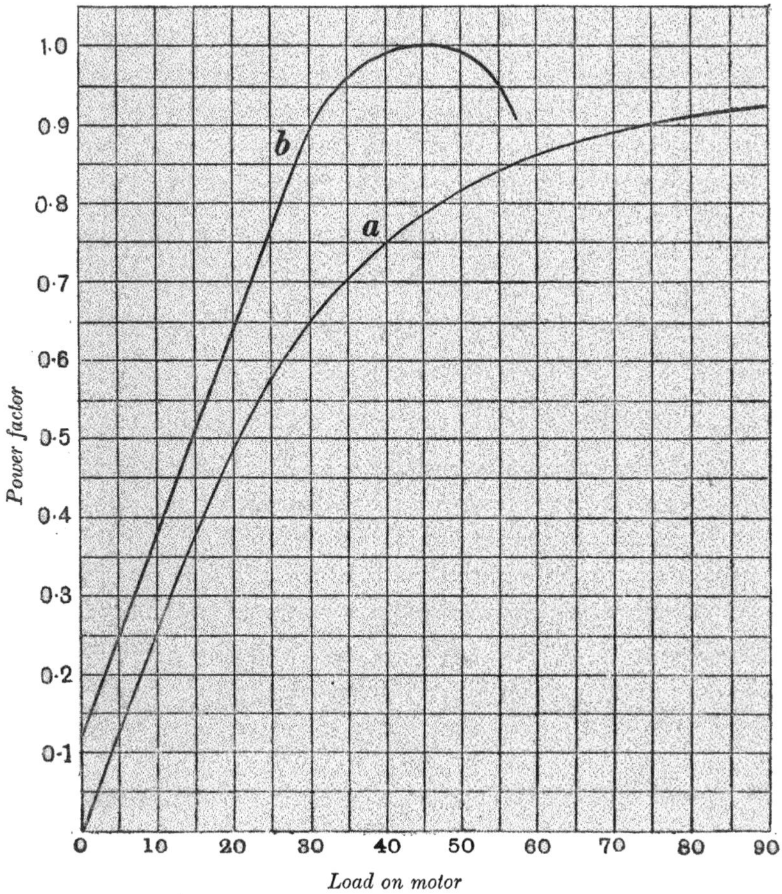

Fig. 101. The curve 'a' gives the relation between the power factor and the load when γ is 90° and 'b' gives the relation when γ is 45°.

a particular excitation. We have, however, to remember that we have made the assumption that OK and BC are in one plane, and this is never exactly true in practice.

In Fig. 101 'a' shows the relation between the power factor and the load when γ is 90 degrees, and 'b' the relation when γ is

45 degrees. In '*a*' the power factor increases with the load, and attains its maximum value 0·96 when the load has its maximum value of 120 kilowatts. In '*b*' the power factor is unity when the load is 42 kilowatts, and it then diminishes. The maximum permissible value of the load is now only 56 kilowatts.

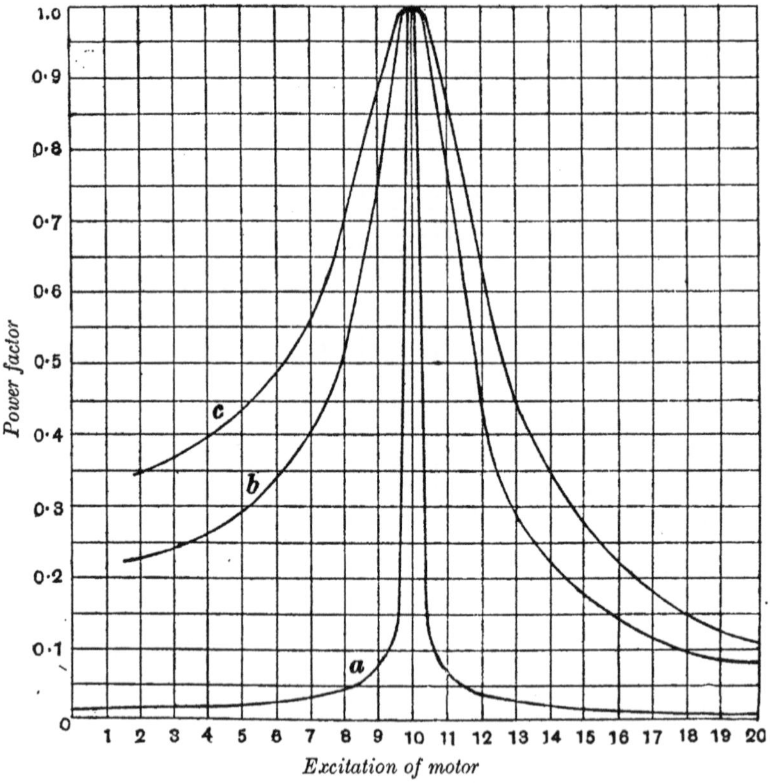

Fig. 102. How the power factor of a synchronous motor varies with the excitation at constant loads.

Variation of the power factor with the excitation of the motor.

In Fig. 102 the data, with the exception of the curve '*a*' which represents a very light load, are the same as for the curves in Fig. 99, so that the two sets of curves can be compared. For each curve the power factor equals unity for an excitation denoted by 10. For values of the excitation less than this, the current lags behind the applied potential difference by an angle ϕ, where $\cos\phi$ is the power factor, and for values of the excitation greater than 10, the phase difference is leading. From Fig. 96 (*b*) we see that when we gradually increase the excitation of the motor, the current attains its minimum value after the power factor becomes unity.

The ease with which large lagging or leading currents can be obtained by under-exciting or over-exciting synchronous motors sometimes makes them useful in general testing work when large choking coils or condensers are not available (see page 96).

REFERENCES.

H. WILDE, 'On a Property of the Magneto-electric Machine to Control and Render Synchronous the Rotations of the Armatures of a number of Electromagnetic Induction Machines,' *Phil. Mag.*, Vol. 37, p. 54, Jan. 1869. (Read at a meeting of the Literary and Philosophical Society of Manchester, Dec. 15, 1868.)

J. HOPKINSON, 'On the Theory of Alternating Currents particularly in Reference to two Alternate-current Machines connected to the same circuit,' *Journ. of the Soc. Tel. Eng.*, Vol. 13, p. 496, 1884.

W. G. ADAMS, 'The Alternate-current Machine as a Motor,' *Journ. of the Soc. Tel. Eng.*, Vol. 13, p. 515, 1884.

W. M. MORDEY, 'Alternate-current Working,' *Journ. of the Inst. of El. Eng.*, Vo. 18, p. 592, 1889.

J. SWINBURNE, 'Transformer Distribution,' *Journ. of the Inst. of El. Eng.*, Vol. 19, p. 163, Feb. 1891.

A. RUSSELL, 'Notes on the Theory of Synchronous Motors and of Alternators in Parallel,' *The Electrical Review*, Vol. 49, p. 88, 1901.

CHAPTER VI

Blondel's bipolar diagram. Lines of equal power when the excitation of the generator is varied. Lines of equal phase when the motor excitation is constant. The excitation of the generator required to give a power factor of unity. The circle limiting the current vector. Example. Synchronous motor supplied from constant potential mains. Use of synchronous motors for raising the power factor. Rotary condenser. Reactance motors. Synchronous motors with alternating fields. The starting of single phase synchronous motors. Polyphase synchronous motors. The starting of polyphase motors. Determination of the moment of inertia of the rotor. Methods of determining the efficiency of a motor. Brake tests. Experimental results. Advantages of synchronous motors. References.

Blondel's bipolar diagram.

THE following graphical method of studying the working of a synchronous motor is instructive and is useful in practice. We make the assumptions that the vectors of the electromotive forces and the currents can be represented by lines in one plane, and that the impedance of the circuit of the armatures is constant. The effects of armature reaction are also neglected. In the diagram (Fig. 103) OP represents the armature electromotive force V_1 of the generator and OO_1 the armature electromotive force V_2 of the synchronous motor. The angle POO_1 is the supplement of the phase difference between V_1 and V_2 and hence, by the triangle of vectors, O_1P is the effective value of the resultant electromotive force round the circuit of the armatures. Let the line O_1B give the phase of the current and draw PB perpendicular to O_1B, then O_1B will represent the watt electromotive force acting (Vol. I, p. 286) round the circuit. If we multiply the effective value A of the current by O_1B we get the power expended in heating the armature windings. Owing to eddy currents this power will be greater than $R \cdot A^2$, where R is the resistance of the armature coils. The value of O_1B will be therefore greater than $R \cdot A$. In practice, it is customary to assume that O_1B equals $nR \cdot A$, where n is a number greater than unity. Usual values for n are 1·5 and 2.

If Z be the impedance of the circuit of the armatures, O_1P will be equal to $Z \cdot A$, and if γ be the phase difference PO_1B we have $\cos\gamma$ equal to nR/Z.

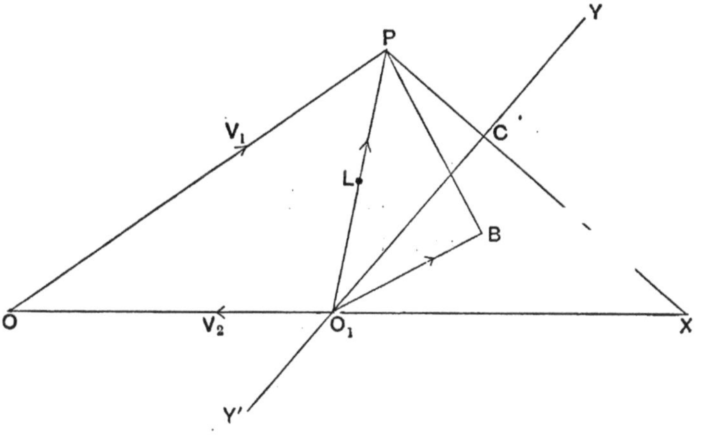

Fig. 103. Blondel's bipolar diagram.

YO_1Y' is inclined at an angle γ to O_1X. O_1P represents on a certain scale the current vector, when the phase difference between it and V_2 is measured from O_1Y'.

Since O_1P equals $Z \cdot A$ we can make O_1P represent the current in magnitude by assuming that the length representing one ampere is Z times the length representing one volt. In other words the scale in which the amperes are measured must be Z times the scale in which the volts are measured. We can also make O_1P represent the current in phase by assuming that the phase difference between this vector and V_2 is measured by the angle it makes with a line $Y'O_1Y$ which is inclined to OX at an angle γ.

To prove this, let us suppose that the line $Y'O_1Y$ makes an angle γ with OX. Since the angle PO_1B is also equal to γ, it follows that the angle PO_1Y equals the angle BO_1X and therefore the angle BO_1O equals the angle PO_1Y'. Hence, if we measure the phases of the electromotive force vectors by the inclinations of these lines to OX we can measure the phase of the current by the inclination of O_1P to O_1Y'. We see, therefore, that if the scale of the amperes is Z times the scale of the volts and the phase of O_1P is measured by its inclination to O_1Y', then O_1P will represent the current vector completely.

This diagram is known as the bipolar diagram. It enables us to see easily how the current and electromotive force vectors vary with the excitations of the machines.

If PCX (Fig. 103) be drawn at right angles to O_1Y, then, in the scale in which the currents are measured, O_1C will represent the watt current with respect to V_2, and PC will give the wattless current. If the power given to the motor is W_2, we have

$$W_2 = V_2 \,.\, A \cos BO_1X$$
$$= V_2 \,.\, A \cos PO_1C$$
$$= V_2 \,.\, O_1C,$$

where O_1C represents the watt current. Its length must be measured in the ampere scale.

Lines of equal power when the excitation of the generator is varied.

Let the excitation of the motor and the load on it be kept constant, whilst the excitation of the generator is varied. Then, since W_2 and V_2 remain constant, the watt current O_1C must also be constant. For all excitations of the generator, therefore, under the given conditions, P must lie on the line PCX. This line may be called, therefore, a line of equal power. In general, all lines drawn perpendicular to O_1Y are lines of equal power.

If P and O are on the same side of O_1Y, PC the wattless component of the current will be lagging with respect to V_2. In this case the current will be leading with respect to the P.D. applied at the motor terminals, and so the armature reaction will weaken the field of the motor. If, however, P and O are on opposite sides of O_1Y, the wattless component PC will be leading with respect to V_2 and the field of the motor will be strengthened by the armature reaction. It has to be remembered that in obtaining the diagram we have neglected these reactions.

In the last chapter we saw that, when the phase difference between V_1 and V_2 is $\pi - \gamma$, the running is unstable. This can be seen also from the bipolar diagram. When OP (Fig. 103) is perpendicular to PX, the angle POO_1 is equal to the angle YO_1X. It is therefore equal to γ. Hence the angle between V_1 and V_2 is $\pi - \gamma$. In this case OP, which represents V_1, is a minimum for the given load corresponding to the watt current O_1C. If the load were to diminish, O_1C would diminish, and there would be a stable position of running, but if it were to increase, OP would not reach the new power line drawn through C, there would be no position for stable running, and the machine would drop out of step.

When V_1 equals OC, the wattless component of the current with respect to V_2 vanishes, and when V_1 is greater than OC the wattless component is leading. For values of V_1 greater than OX we may consider that the generator is the leading machine.

Lines of equal phase when the motor excitation is constant.

The phase difference between a current vector O_1P (Fig. 103) and the vector OO_1, representing the armature electromotive force of the motor, will be PO_1Y'. Hence this angle represents the phase difference between any of the current vectors, which point in the direction O_1P, and the motor E.M.F. The line O_1P may be called, therefore, a line of equal phase. In general, every line drawn through O_1 is a line of equal phase.

When the current is in opposition in phase to V_2, O_1Y will be the line of equal phase. If θ be the phase difference between V_1 and V_2 in this case, we see from the diagram that

$$- V_1 \cos\theta - V_2 = ZA \cos\gamma$$

and hence

$$V_1 A \cos(\pi - \theta) = V_2 A + ZA^2 \cos\gamma.$$

This could also have been written down directly, since the electric power generated must always be equal to the power given to the motor together with the power expended in heating the circuit.

The excitation of the generator required to give a power factor of unity.

In Fig. 103 the angle PO_1Y is not the phase difference between the current and the applied potential difference at the motor terminals. If L (Fig. 104) be the middle point of O_1P and the lines have the same meanings as in Fig. 103, then, if we make the assumption that the motor and generator are exactly similar machines, OL will represent in magnitude and phase the potential difference applied at the motor terminals. If the angle OLO_1 equals the angle LO_1B (γ), then OL and O_1B will be parallel and the phase difference between them will vanish. Whenever, therefore, the angle OLO_1 is γ, the power factor of the motor circuit will be unity. In this case, the locus of L is a circle having OO_1 for a chord and touching O_1Y at O_1. The locus of P will also be a circle, for if we draw through P a line PO' parallel to LO this line will always cut O_1O produced

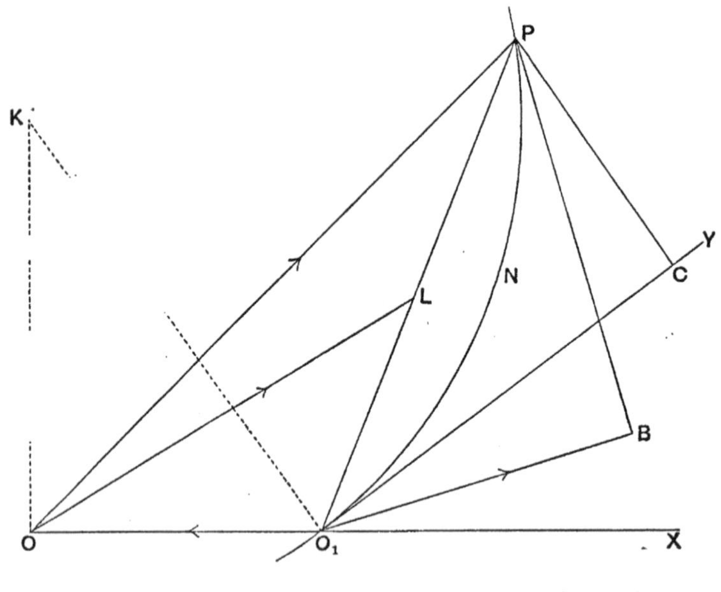

Fig. 104. When P lies on the circle O_1NP, the power factor of the motor circuit is unity.

in a point O' so that $O'O$ equals OO_1 and the angle O_1PO' will equal OLO_1 and will therefore be constant. This circle will also touch O_1Y at O_1 and its radius will be equal to $V_2 \operatorname{cosec} \gamma$. Whenever P lies on this circle the power factor of the motor circuit is unity.

When O_1P represents the current, we have seen that PC represents its lagging wattless component and O_1C represents its power component. From Fig. 104 it is obvious that, when the power factor of the motor circuit is unity, the current must have a component which lags relatively to the counter electromotive force of the motor armature. In this case, the armature reaction always tends to weaken the field of the motor and to strengthen the field of the generator.

It will be seen at once from Fig. 104 that as O_1C increases, OP increases. Hence as the load on the motor is increased we must increase the excitation of the generator if the power factor of the motor circuit is to be kept at its maximum value. We also see that when the load on the motor remains constant and we gradually increase the excitation of the generator from a low value, the power factor of the motor circuit attains its maximum value before the current attains its minimum value.

The circle limiting the current vector.

If $A_{max.}$ be the maximum permissible value of the current in the armature, then O_1P in Figures 103 and 104 must be less than $Z \cdot A_{max.}$. Hence, if we describe a circle (Fig. 105) with centre O_1 and radius equal to $Z \cdot A_{max.}$ P must be somewhere within this circle, for all possible positions of running.

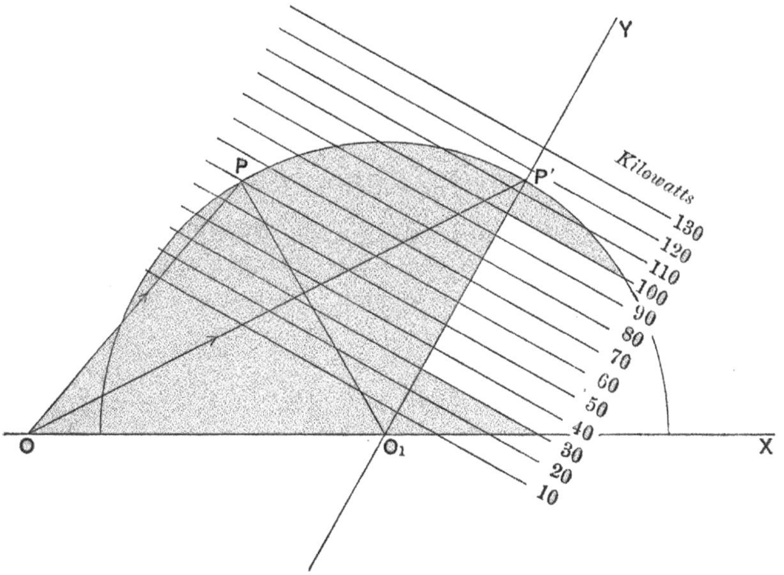

Fig. 105. The radius of the circle PP' is the maximum current the armature can carry. P lies within this circle in all practical cases.

Example.

Suppose that we have two similar and equal machines, one acting as a generator and the other as a synchronous motor, and suppose that they are coupled together through a long transmission line. Let (r_1, l_1) be the constants for the line and (nR, L) be the approximate constants for the armatures, then, making the usual assumptions, we get

$$\tan\gamma = \omega\,(l_1 + 2L)/(r_1 + 2nR),$$

and

$$Z^2 = (r_1 + 2nR)^2 + \omega^2\,(l_1 + 2L)^2.$$

Hence γ and Z can be determined approximately.

Draw a line OO_1 (Fig. 105) equal to the armature electromotive force of the motor. With centre O_1 and radius $Z \,.\, A_{max.}$ describe a circle. Along $O_1 Y$ mark off points at equal distances apart and through these points draw lines perpendicular to $O_1 Y$. These lines will give the lines of equal power and the distance between them can be chosen, so that each represents a load which is a multiple of a kilowatt. Now suppose that the excitation of the generator gives an electromotive force on open circuit of V_1 volts. With centre O and radius V_1 describe a circle, and let it cut the circle which limits the current in P. We see that, if we draw a line from P perpendicular to $O_1 Y$, the maximum power that the transmission line can transmit for this excitation of the generator can be read off at once along $O_1 Y$. The maximum possible power that the line could transmit would be $O_1 P'$ and to transmit this power the generator would have to be capable of producing an electromotive force on open circuit equal to OP'.

Synchronous motor supplied from constant potential mains.

We suppose that the excitation of the synchronous motor is constant and that γ also remains constant. With centre O (Fig. 106) and radius V_1, equal to the constant potential difference between the mains, desçribe a circle PP'. If the load on the motor be W_2, mark off, on the scale of the volts, a length $O_1 D$ along $O_1 Y$, so that $V_2 \,.\, O_1 D/Z$ equals W_2. In the case we are considering, we can neglect the impedance of the rest of the circuit in comparison with that of the motor armature. Hence Z will be practically equal to the impedance of the motor armature. If we draw DP perpendicular to $O_1 Y$, then OP will be the vector of the applied potential difference.

We see from the diagram that, if the perpendicular through M is the tangent to the circle at P', O_1M corresponds to the maximum load on the motor. This point, however, corresponds to an unstable position of working, as the slightest increase of the load would make the motor fall out of step. Since OP' is parallel to O_1Y, the angle POO_1 equals γ and, at the maximum load, W_2 equals $OO_1 . O_1M/Z$, that is, $V_2(V_1 - V_2 \cos\gamma)/Z$, which agrees with the formula we found on page 173.

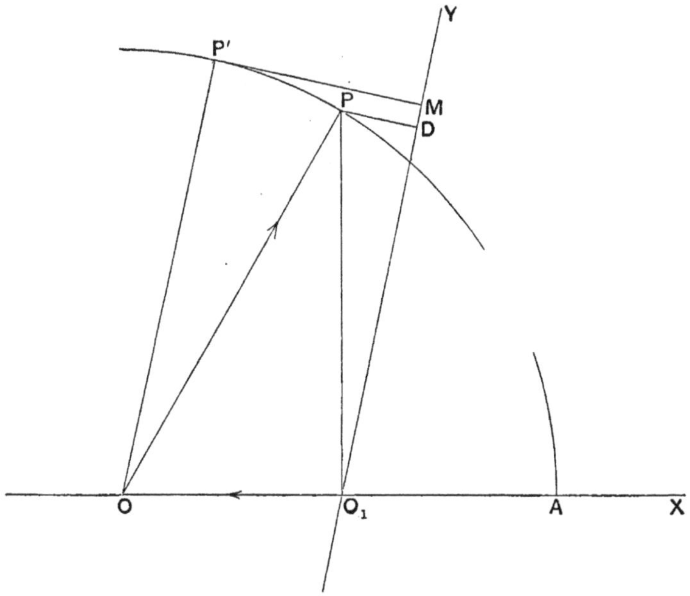

Fig. 106. Diagram for synchronous motor working on constant potential mains.

Use of synchronous motors for raising the power factor.

For a given distribution of power from a Central Station, the higher the power factor of the load, provided that the current is lagging, the more economical will be the distribution, as the current, and therefore also the copper losses in the mains, and armature, diminish as the power factor increases, the voltage and the load remaining constant. Also, since the current is lagging, the excitation has to be increased in order to neutralise the demagnetising effect of the armature currents on the field magnets, and this increases the total loss. Swinburne suggested in 1891 that the power factor of a distributing system might be increased by using over-excited synchronous motors at the supply station to neutralise the wattless component of the load current. We saw in Vol. I, p. 140, how a condenser shunt could be utilised to furnish the necessary magnetising current for a choking coil. In a similar way a comparatively small synchronous motor can be employed to raise appreciably the power factor of a distributing system.

Let us suppose that the effective value of the current in the main is A_1 and that $A_1 \sin \psi_1$ is its wattless component. Let also $-A_2 \sin \psi_2$ be the wattless component of the auxiliary motor current. If $\cos \psi$ be the value of the power factor when the synchronous motor is running, we have

$$\tan \psi = (A_1 \sin \psi_1 - A_2 \sin \psi_2)/(A_1 \cos \psi_1 + A_2 \cos \psi_2).$$

If we wish a power factor of unity, $A_2 \sin \psi_2$ would need to be equal to $A_1 \sin \psi_1$. Hence the smaller the value of ψ_1 the smaller the synchronous motor required. Now it is easy to see from the theorems given in Vol. I, Chap. x, that a power factor of unity could never be actually obtained. Let us suppose that $\cos \psi$ is the power factor required for satisfactory working. Then writing the equation given above in the form

$$A_2 \sin \psi_2 = A_1 \sin \psi_1 (1 - \tan \psi/\tan \psi_1) - A_2 \cos \psi_2 \tan \psi,$$

and regarding ψ as constant, we see that the smaller the value of ψ_1, for a given wattless current $A_1 \sin \psi_1$, the smaller will be the value of $A_2 \sin \psi_2$. Hence the greater the power factor of the load, for a given value of the wattless current, the smaller will be the size of the synchronous motor required to raise the power factor of the station to a desired value. This can also be easily seen graphically.

Rotary condenser. When a synchronous motor is used merely for regulating the power factor of the load on a power station it is sometimes called a rotary condenser. It is of especial use in connection with long transmission lines working at very high pressures, as a small motor can supply both the lagging current required at light loads and the leading current required at heavy loads. For example, when a 6000 horse power plant in India, which transmitted power 90 miles, reached the limit of its capacity, the installation of a 1000 kilovolt-ampere rotary condenser enabled 50 per cent. more power to be transmitted for the same voltage drop.

It has to be remembered however that in most cases it is better and cheaper to raise the power factor of consumers' motors rather than to use a rotary condenser. If the induction motors, for instance, can be replaced by suitable self-synchronising synchronous motors the power factor of which is unity or if they can be fitted with some device for raising the power factor then, as a rule, a more economical system is obtained.

Reactance motors.

When the number of turns in the windings of the armature is large, the wattless component $A \sin \psi$ of the current appreciably magnetises the field. Hence a motor wound in this manner will be self-exciting. It is not possible, however, to get much power from it, as the power factor needs to be low in order that the wattless component of the current may be sufficiently large to magnetise the field. Industrially, these motors have not hitherto proved successful.

Synchronous motors with alternating fields.

Ferraris pointed out that if we excite the field magnets of a synchronous motor with alternating current from the supply mains, then, in certain cases, the machine will still act as a synchronous motor but its speed will be double that at which it runs when its field magnets are excited with direct current. The windings of the field magnets may be connected either in parallel or in series, with the armature windings.

The working principle of the machine can easily be understood from Fig. 92, p. 162. Let us suppose that the current is a maximum in the position shown in the figure. At this instant the torque will be in the direction against the hands of a watch. A quarter of a period later the axis of the armature coil will be again vertical, but, the current being zero, there will be little magnetism left either in the field magnets or in the armature. A quarter of a period later, the polarity of both field magnets and armature will have changed, but the armature being in its initial position the torque will still be in the same direction. Hence, the mean torque over a whole period will not be zero but will act in the direction against the hands of a watch. The machine, therefore, will act as a motor when the speed is double that of synchronism. The objections to this type of motor are the high speed requisite and its low efficiency. It has the great advantage, however, of not requiring separate excitation. It could be started by means of an ordinary small asynchronous motor having less than half the number of poles of the large motor which is excited by the alternating current.

If we neglect the effects of armature reaction we may discuss the working of this motor by means of the theory of rotary fields, explained in Vol. I, Chap. XVIII. The alternating field due to the field magnets can be replaced by two fields of half the maximum strength rotating in space with angular velocities ω/p and $-\omega/p$ respectively. The oscillatory field due to the armature current rotates with angular velocity $2\omega/p$, and hence it may be considered as made up of two rotary fields rotating with angular velocities $3\omega/p$ and ω/p respectively. The mean value of the torque produced by the action of the field which rotates with the angular velocity $3\omega/p$ on the fields rotating with angular velocities ω/p and $-\omega/p$ will be zero. Similarly, the mean value of the torque due to the fields rotating with angular velocities ω/p and $-\omega/p$ will be zero. The fields, however, which both rotate with an angular velocity ω/p will produce a steady torque and so the armature will rotate.

The starting of single phase synchronous motors.

Synchronous motors are not self-starting, and hence some device has to be employed for this purpose. When the power station is not very far from the generating station, direct current from the exciter of the generator may be transmitted by special mains to the exciter of the motor, and the latter may be driven by its own exciter, acting as a motor, until it attain synchronous speed. As about ten per cent. of the total power of the synchronous motor may be required and as the direct current is transmitted at a low pressure, this method can only be applied economically in very special cases.

When a small auxiliary direct current motor or a small

asynchronous induction motor (Chap. XIII) is available the machine may be started on a loose pulley. In order that the small motor may not get overheated before the large machine gets up speed, we have to use some device that permits the small motor to run at approximately constant speed during the whole process. One method of doing this is to mount the small motor on slide rails, and to put a small conical friction wheel at one end of its shaft. This wheel presses on a large friction disc keyed to the shaft of the synchronous motor. At first, the conical wheel rotates near the circumference of the disc, and thus makes many turns for one turn of the machine armature. As the machine speeds up, the small motor is moved on its slide rails by means of a hand-screw until the conical pulley reaches the centre of the disc when synchronism is attained.

When no direct current or alternating current motor is available to start the machine, its armature must be provided with a special winding which starts the motor by producing, in conjunction with the other windings, a rotary field. When the machine gets up to synchronous speed the starting winding is cut out and the load is put on the pulley, the machine now running as a synchronous motor. We have seen in Vol. I, Chap. XVIII, how a rotary field can be produced by two currents which are not in phase with one another. In order to obtain a powerful rotary field we need to have the phase difference between the currents approximately equal to ninety degrees. One way of doing this is to put an electrolytic condenser in series with one of the circuits, so that the current in it may be in advance in phase of the applied potential difference, while the current in the other winding lags behind the phase of the applied P.D. An electrolytic condenser generally consists of iron plates placed in a solution of soda contained in an iron vessel. It will be seen that, during the start, the motor is provided with currents in different phases just like a two phase motor.

Polyphase synchronous motors.

The theory of polyphase synchronous motors is practically identical with that of single phase machines. We can regard the armature of the generator as consisting of three single phase armatures all keyed together, the phase difference between any two of the three applied potential differences being 120°. We can make a similar supposition with regard to the motor. Since the components of the torque due to the currents in the three windings generally vanish at different instants, the torque on the armature of the motor will be much steadier than in the case of single phase machines. In the case of sine waves and a balanced load, we have seen (p. 166) that

$$g\omega = (3/2)\, EI \cos \alpha$$
$$= 3VA \cos \alpha,$$

where g is the instantaneous value of the torque and ω is the instantaneous value of the angular velocity of the armature. On the given assumptions, therefore, the power given to the armature is the same at every instant and so the torque is absolutely constant. Although owing to armature reactions, hysteresis, eddy currents, etc., it is exceedingly unlikely that the given assumptions could ever strictly be justified, yet, as the frequency of the variations of the torque must be at least three times as rapid as the frequency of the alternating current, we see that its variations will have little effect on the angular velocity of the rotor.

The action that takes place between the currents in the armature and the magnetic field in single phase machines is different from the corresponding action in polyphase machines. Let us suppose, for instance, that the field magnets form the rotor. In single phase machines the magnetic field produced by the armature currents is an oscillatory one and pulsates with a frequency $\omega/2\pi$. If there are $2p$ poles the angular velocity of the rotor is ω/p. Now the fixed oscillatory field due to the armature currents may be replaced by two magnetic fields gliding in opposite directions with angular velocities ω/p and $-\omega/p$ respectively, the intensity of each of the gliding magnetic fields being equal to half that of the fixed pulsating field. The action of the field gliding in the opposite sense to the rotation of the rotor adds nothing to the total torque, and thus, the torque, produced on the rotor by a fixed pulsating field, will only be equal to half that produced by a gliding magnetic field of equal intensity.

In polyphase machines the armature produces a rotating magnetic field, and hence there will be a torque on the rotor, even when the latter is at rest, tending to turn it in the direction of the rotation of the field. We have seen, in Vol. I, Chap. XVIII, that if H be the amplitude of the magnetic field produced in one pole of the armature by a current in a phase winding, then $3H/2$ is the strength of the rotating magnetic field produced by the poles of the three phases. Hence, when the machine is running at synchronous speed, the torque produced is three times as great as the mean torque produced when it runs as a single phase machine with only one of the phase windings in circuit. The fluctuations of the torque, however, when run as a single phase machine would be violent. The torque would vanish in this case at least $2p$ times, and in general $4p$ times, every revolution of the rotor.

Starting polyphase motors.

To start a polyphase motor we open the field magnet circuit and connect the armature with the polyphase mains through starting resistances. When the rotor attains synchronous speed the field circuit is closed. At the moment of switching in the armature windings and until the motor gets up speed, the rapidly reversing flux in the field magnet windings sets up very high electromotive forces which may give rise to a spark and so break down the insulation. Hence the field magnet coils are generally wound in sections which are on open circuit during the start. When a direct current motor is available it is generally best to use it to start the synchronous motors and so avoid all the risks of a breakdown in the insulation.

Determination of the moment of inertia of the rotor.

The moment of inertia of a rotor may be determined by noting the time that it takes to slow down after both the alternating current supply for the armature and the direct current supply for the field magnet coils have been switched off. Let us suppose that the rotor slows down from an angular velocity ω_1 to an angular velocity ω_2 in t_1 seconds and that the retarding torque in dyne-centimetres is g. It is found by experiment that g is very nearly constant. Let Mk^2 be the moment of inertia of the rotor about its axis, and let θ be the angle which a radius of the rotor makes with the horizontal, t seconds after the era of reckoning. The equation of motion is

$$Mk^2 \frac{\partial^2\theta}{\partial t^2} = \text{the moment of the applied forces}$$
$$= -g,$$

and thus, by integrating,

$$Mk^2 \frac{\partial\theta}{\partial t} = A - gt,$$

where A is a constant.

If $\partial\theta/\partial t$ is ω_1 when t is zero and is ω_2 after t_1 seconds, we have

$$Mk^2\omega_1 = A$$

and

$$Mk^2\omega_2 = Mk^2\omega_1 - gt_1,$$

and thus

$$gt_1 = Mk^2 (\omega_1 - \omega_2).$$

If we now apply a constant torque g_1 to the rotor by means of a mechanical or an electrical brake, we get, in a similar manner,

$$(g + g_1) t_2 = Mk^2 (\omega_1 - \omega_2),$$

where t_2 is the number of seconds the rotor takes to slow down from the angular velocity ω_1 to the angular velocity ω_2 when the brake is applied. We have, therefore,

$$(g + g_1) t_2 = gt_1$$

and

$$g = g_1 t_2/(t_1 - t_2).$$

Hence finally

$$Mk^2 = g_1 t_1 t_2/\{(t_1 - t_2)(\omega_1 - \omega_2)\}$$
$$= g_1 t_1 t_2/\{2\pi (t_1 - t_2)(n_1 - n_2)\},$$

and n_1 and n_2—the revolutions per second of the rotor at the two given speeds—can be measured by a tachometer.

Methods of determining the efficiency of a motor.

The losses in a synchronous motor are due to heating of the armature coils, bearing and brush friction, hysteresis, eddy currents, wind friction and excitation losses. It is found in practice that the retarding torque due to the bearing and brush friction is nearly independent of the speed, and so also is the retarding torque due to hysteresis. Now experiment shows that the torque due to wind friction is approximately proportional to the angular velocity of the rotor, and it is generally assumed that the torque due to eddy currents is also proportional to it. This latter assumption is, however, often inadmissible. When the frequency of the alternating current is high, when the eddy current losses in the pole faces caused by the fluctuations in the value of the flux density due to the slots in the armature are appreciable, or when eddy currents are induced in the copper conductors or unlaminated masses of metal, this assumption must not be made.

On the given assumptions the torque due to hysteresis, bearing and brush friction can be denoted by a constant, B. The sum of the two torques due to wind friction and eddy currents may be denoted by $D\omega$ where D is a constant and ω is the angular velocity of the rotor. Let R be the resistance of the armature of the motor and suppose that we run it at two different speeds ω_1 and ω_2, on open circuit, with the field excited. If W_1 and W_2 be the watts, measured by a wattmeter, supplied to the motor at these speeds, we have

$$W_1 - RA_1^2 = (B + D\omega_1)\,\omega_1 = B\omega_1 + D\omega_1^2 \quad \text{...........(1)},$$

and $$W_2 - RA_2^2 = (B + D\omega_2)\,\omega_2 = B\omega_2 + D\omega_2^2 \quad \text{...........(2)},$$

where A_1 and A_2 are the readings of the ammeter, in series with the armature, in the two cases. From equations (1) and (2) B and D can be determined readily. Knowing the values of B and D, the efficiency η of the motor can be found approximately by the formula

$$\eta = (W - RA^2 - B\omega - D\omega^2)/(W + X),$$

where W is the power taken by the motor from the alternating current mains and X is the power expended in exciting the field magnets. This method is due to Swinburne. In practice as the load is increased so also is the excitation, and thus X in the above equation is not a constant. When we can neglect the alternating current component, due to armature reaction or to fluctuations of the reluctance of the magnetic circuit, in the field magnet windings, the value of X, in watts, is the product of the reading of the ammeter in the circuit of the field magnet coils by the reading of the voltmeter placed across the terminals of the exciting circuit.

The above method is purely electrical. When a transmission dynamometer, that is, a transmission coupling which indicates the torque which it transmits, is available, the test is best made as follows. Couple the motor directly with the shaft of the generator, by means of the dynamometer, in such a way that the motor helps to drive the generator. The torque g on the shaft can be measured by the dynamometer, and multiplying this by the angular velocity ω we get the load on the motor. This also represents the power returned to the generator. If now we measure the power W, in watts, supplied to the motor and also the power X required for excitation, we have

$$\eta = g\omega/(W + X).$$

In this formula, $g\omega$ must be measured in watts, so that the unit in which g is measured must equal 10^7 dyne-centimetres. It is to be noticed that this is an economical method of testing, as the power taken from the generator $W - g\omega$ represents merely the losses in the motor.

Brake tests.

The efficiency of small synchronous motors can be determined to an accuracy of about one per cent. by means of an absorption brake. By this apparatus a retarding torque, which can be easily measured, is applied to the circumference of the pulley of the rotor by means of friction. As all the useful power of the motor is expended in heating the pulley and the surfaces in contact with it, special water cooling arrangements have to be devised when the power expended at the rubbing surfaces cannot be radiated away quickly enough. If g_1 be the torque in dyne-centimetres applied by the brake, we have

$$\eta = g_1\omega 10^{-7}/(W + X),$$

where ω is $2\pi n$, and n is the number of revolutions of the shaft per second. If the torque g be measured in kilogramme-metres, we have

$$\eta = 9{\cdot}81 g\omega/(W + X).$$

An ordinary direct current dynamo can be employed very usefully as an absorption brake. The dynamo must have all its losses carefully measured in the first instance, so that we know approximately the power expended in hysteresis, eddy currents and friction. For a twenty kilowatt dynamo the sum of these losses generally lies in value between ten and fifteen per cent. of the maximum rated output of the machine. An error, therefore, of ten per cent. in determining, for example, the eddy current losses will only introduce an error of about one per cent. in the calculated value of the total power absorbed by the dynamo at full load. The electrical output of the dynamo can be measured to an accuracy of about the half of one per cent. by means of a carefully calibrated ammeter and voltmeter. The electrical power generated is usually expended in a water resistance. Lead plates connected with the terminals of the dynamo are placed in a vertical position and at some distance apart from one another in a tank containing salt water. The adjustment of the load is made by varying the distance apart of the plates, by raising or lowering the plates so as to vary the area of the immersed portion of the plate, or by both these methods.

Experimental results.

The results of tests on a three phase synchronous motor made by the Oerlikon Company are given in Figures 107 and 108. The machine is designed for a frequency of 50 and for an output of 525 horse power when the potential difference applied to each pair of slip rings is 3500. The number of poles is 16 so that the armature makes $60 \times 50/8$, that is, 375 revolutions per minute.

In Fig. 107, O.C.C. is the open circuit characteristic and S.C.C. is the short circuit characteristic. $V_{1/1}$ is the V curve at full load, $V_{1/2}$ is the V curve at half load and V_0 is the V curve at no load. It will be seen that these curves closely resemble the theoretical curves shown in Fig. 99, p. 181.

In Fig. 108, $\cos \psi_{1/1}$ is the power factor curve at full load when the excitation is varied, and $\cos \psi_{1/2}$ is the power factor curve at half load. The curve η gives the efficiency at various loads, the power factor being unity in every case. The curve $P_{ex.}$ gives the excitation losses, the curve P_H gives the combined losses due to hysteresis and eddy currents, the armature losses RA^2 are given by the curve P_R and the curve P_F gives the losses due to solid and air friction. The solid friction is the friction of the bearings and the friction of the brushes pressing on the collector rings.

The high power factor obtained shows that the counter electromotive force wave of the motor and the electromotive force wave of the generator are approximately sine shaped.

The advantages of synchronous motors are that they are simple to construct mechanically, they can easily be wound for high pressures and, as a rule, their power factor is high. Their distinguishing peculiarity is that they run at exactly the same speed at all loads. The moment the armature gets out of step with the field large alternating currents flow in it which cause the fuses to melt

The advantages of synchronous motors.

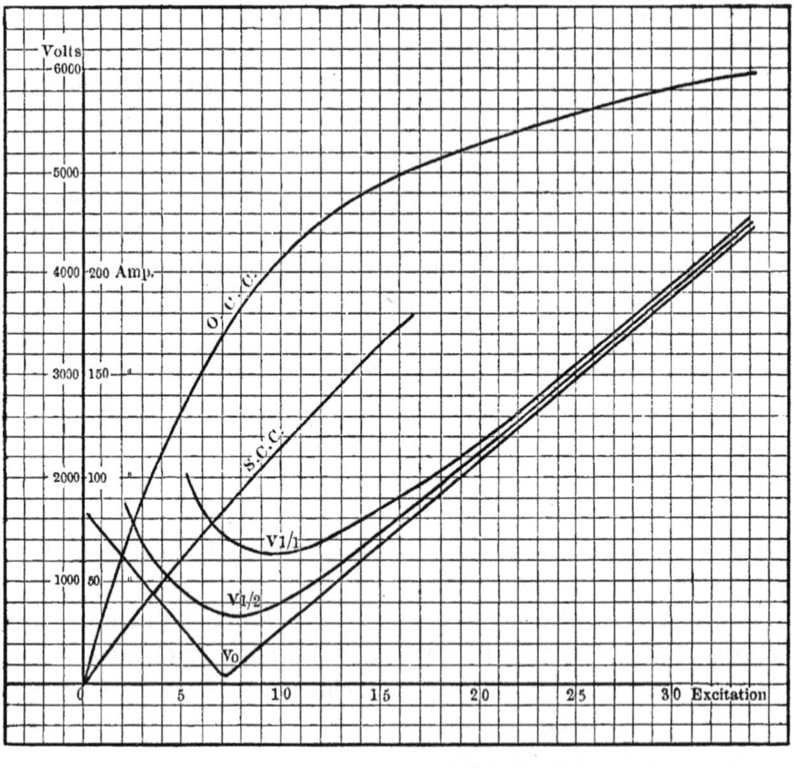

Fig. 107. Characteristics and *V* curves of a 525 H.P. Oerlikon motor. The curves are for no load, half load, and full load.

or the magnetic cut-outs to open the circuit. The only way of altering the speed is by altering the frequency of the supply current. This constancy of speed is invaluable for some purposes in connection with oscillographs, ondographs, rectifiers, etc. It is also useful sometimes when we wish to drive a dynamo at a constant speed.

If the power factor of a synchronous motor be high and the resistance of the armature windings be small, the efficiency is also high. In order to get a high power factor the wave of the resultant electromotive force in the armature circuit must be approximately sine shaped, and thus both the applied potential difference and the back E.M.F. of the armature must be approximately sine shaped. Particular attention is paid to this point by designers of synchronous motors. Motors which work well when supplied with alternating current from the supply mains at

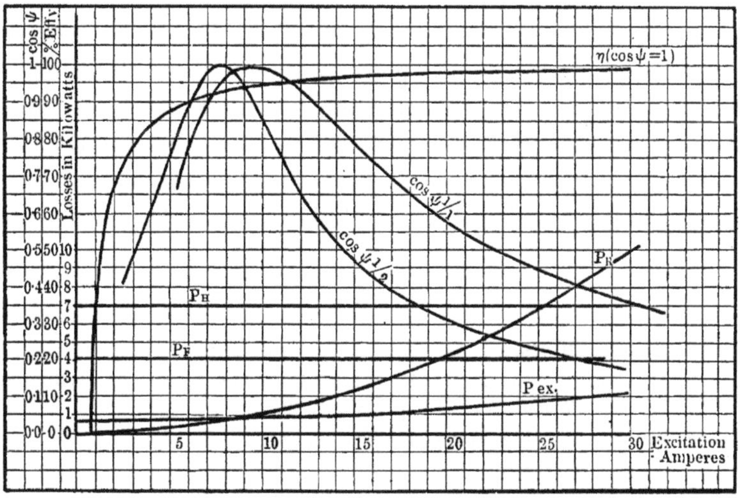

Fig. 108. The efficiency curve and the curves showing the losses in a 525 horse power three phase synchronous motor, 3500 volts, 50 frequency, 375 revs. per minute.

$\cos \psi_{1/1}$ = Power factor at full load.
$\cos \psi_{1/2}$ = Power factor at half load.

certain times of the day are sometimes found to take an excessive current even on a light load at other times of the day. This is due to variations in the wave shape of the supply. When the power station is some distance from the supply station and the mains connecting them have considerable electrostatic capacity, the distortion of the wave shape of the pressure of the supply is often excessive at light loads.

Synchronous motors are only of limited use for ordinary power work from supply mains. The speed cannot be regulated and a supply of direct current is wanted for excitation. Special starting devices have also to be used. They sometimes set up phase swinging (Chap. VII) which causes serious oscillations of the pressure between the mains of the supply circuit and a consequent blinking of the lamps supplied from these mains.

A synchronous motor is often coupled directly to a direct current generator, both the machines being mounted on the same bedplate. The combination is called a synchronous motor generator. There are several of these motor generators in the sub-stations connected with the Charing Cross Company's City of London Works. The synchronous motors are supplied with current at 10,000 volts by means of three core mains connected with the three phase generators at the power station. Each motor drives either one dynamo or two balancing dynamos, the distribution being on the three wire direct current system. To diminish the risk of a breakdown, the motor generator sets are of very solid construction and the high tension windings are heavily insulated.

REFERENCES.

ANDRÉ BLONDEL, *Moteurs Synchrones à Courants Alternatifs.*

A. HAY, 'Determining the Moments of Inertia of the Rotors of Dynamos,' *The Electrical Review*, Vol. 47, p. 287, 1900.

W. E. SUMPNER, 'Testing Motor Losses,' *Journ. of the Inst. of El. Eng.*, Vol. 31, p. 632, 1902.

W. H. PATCHELL, 'The Charing Cross Company's City of London Works,' *Journ. of the Inst. of El. Eng.*, Vol. 36, p. 66, 1905.

For a description of many forms of absorption brake, see J. BUCHETTI, *Guide pour l'essai des Moteurs*, Chapter VIII. The English translation by A. Russell is entitled, 'Engine Tests and Boiler Efficiencies.'

E. ROSENBERG, 'Self-Synchronising Machines,' *Journ. of the Inst. of El. Eng.*, Vol. 51, p. 62, 1913.

G. KAPP, 'On Phase-Advancing,' *Journ. of the Inst. of El. Eng.*, Vol. 51, p. 243, 1913.

CHAPTER VII

Phase swinging. B. Hopkinson's method. The stability of the motion in special cases. General case. Products of the pairs of the roots of a biquadratic. The criterion for stability. The period of the phase swing. Effect produced by the distortion of the field. Amortisseurs. The theory of damping coils. References.

Phase swinging.

WHEN a synchronous motor is running on a load, small periodic pulsations of the supply current can nearly always be noticed whenever there is any change in the resisting torque due to the load. In some cases these pulsations are damped out rapidly. In others they are very persistent and lead to instability of the motion, so that the machine falls readily out of step. As the periodic time of these oscillations may be as long as two or three seconds it can sometimes be measured easily. The pulsations are due to variations of the angular velocity of the rotor. If we consider a fixed radius of the rotor we can imagine that the motion of this radius consists of isochronous vibrations about a mean position which rotates with constant angular velocity. The phase difference between the applied potential difference and the back E.M.F. of the motor will therefore be a periodic function of the time. It is customary to refer to the period of the pulsations of the current as the period of the phase swing, and the phenomenon is called phase swinging. It is to be noticed that the period of the phase swing is large compared with the period of the current, and so, in finding an approximate formula for it, we can neglect the forces due to the relative velocity of the rotor and the field. We shall take this into account later on, but the following elementary discussion leads to a formula for the period of the phase swing which is of practical value.

We suppose that the generator is directly coupled to its engine and that it runs at a speed which is uninfluenced by slight variations in the load. Let W_2 be the power given to the motor, then, by formula (3) on p. 173, we have

$$W_2 = -(V_2^2/Z)\cos\gamma - (V_1V_2/Z)\cos(\theta+\gamma)$$
$$= g\omega + W_0 \quad \text{...(1)},$$

where g is the retarding couple due to the load, ω the angular velocity of the rotor, and W_0 the power expended in heating the armature of the motor, overcoming friction, etc. We shall suppose that the generator is large compared with the motor, so that its speed is unaffected by the small fluctuations in the power taken by the motor. We shall also suppose that the motor has only two poles, and that the electromotive forces follow the harmonic law, so that if $\theta + x$ be the disturbed value of θ, x is the angle between the actual position of a radius of the rotor in space and the position it would have if the motor were running steadily.

Let W_2' be the new value of W_2, then, neglecting for the present the forces due to the relative angular velocity of the rotor and the field, we have

$$W_2' = -(V_2^2/Z)\cos\gamma - (V_1V_2/Z)\cos(\theta+\gamma+x) \quad \text{...(2)}$$
$$= g'\omega' + W_0',$$

and therefore, by (1),

$$g'\omega' + W_0' = g\omega + W_0 + 2(V_1V_2/Z)\sin(\theta+\gamma+x/2)\sin(x/2).$$

In practice ω' and ω are practically equal. We may also assume that W_0' is equal to W_0, and thus, since x is a small angle, we have

$$(g'-g)\,\omega = (V_1V_2/Z)\sin(\theta+\gamma)\,.\,x, \text{ approximately.}$$

Now, for steady running, $\theta+\gamma$ is greater than π (p. 174). Let us assume that θ is $\pi - \psi$, so that ψ must be less than γ. Let Mk^2 be the moment of inertia of the rotor, then

$$Mk^2\frac{\partial^2 x}{\partial t^2} = \text{the moment of the effective forces about the axis of rotation} = g' - g$$
$$= -(V_1V_2/Z\omega)\sin(\gamma-\psi)\,.\,x.$$

Since, in practice, ω is nearly constant, it follows that the acceleration of x is approximately proportional to x. Hence the motion is simple harmonic, and if T be the period of an oscillation,

$$T = 2\pi\,(\text{displacement}/\text{acceleration})^{\frac{1}{2}}$$
$$= 2\pi\,[(Mk^2Z\omega)/\{V_1V_2\sin(\gamma-\psi)\}]^{\frac{1}{2}}.$$

When the machine has two poles, the electromotive force vectors rotate at the same rate as the rotor, but, when it has $2p$ poles, they rotate p times faster. In the latter case,

$$\frac{Mk^2}{p}\frac{\partial^2 x}{\partial t^2} = -\frac{V_1V_2}{Z\omega}\sin(\gamma-\psi)\,.\,x,$$

and therefore

$$T = 2\pi\,[(Mk^2Z\omega)/\{pV_1V_2\sin(\gamma-\psi)\}]^{\frac{1}{2}}.$$

If the rotor make n revolutions per second, ω equals $2\pi n$, and the frequency f equals pn. Hence, we can also write

$$T = (2\pi/p)\,[(2\pi Mk^2Zf)/\{V_1V_2\sin(\gamma-\psi)\}]^{\frac{1}{2}}.$$

If V_1 and V_2 are expressed in volts and Z in ohms, then V_1V_2/Z is given in watts. Now one watt is 10^7 ergs per second, and if M be measured in kilogrammes and k in metres, $Mk^2\,.\,10^7$ will be numerically equal to Mk^2, when M is measured in grammes and k in centimetres. If M, therefore, be measured in kilogrammes and k in metres, the formula given above will give T in seconds.

For a two phase machine with two separate windings the formula is

$$T = (2\pi/p)\,[(\pi Mk^2Zf)/\{V_1V_2\sin(\gamma-\psi)\}]^{\frac{1}{2}},$$

and for a three phase machine we have

$$T = (2\pi/p)\,[(2\pi Mk^2Zf)/\{3V_1V_2\sin(\gamma-\psi)\}]^{\frac{1}{2}},$$

where V_1 is the voltage in one phase of the generator winding, and V_2 is the counter-electromotive force in one phase of the motor winding.

If M be measured in pounds and k in feet, then, for machines with q phases, the formula is

$$T = (0{\cdot}32/p)\,[(Mk^2Zf)/\{qV_1V_2\sin(\gamma-\psi)\}]^{\frac{1}{2}}.$$

We see that, when V_1 and V_2 are equal, the periodic time of the swings varies inversely as V_2, and therefore inversely as the excitation. We also see that the frequency varies inversely as the square root of the moment of inertia of the rotor.

In finding the formulae given above we have made the assumption that the torque depends only on the relative positions of the rotors of the generator and the motor, and, therefore, that it is independent of their relative angular velocities. Hence the motion would be similar to that of an undamped pendulum, and the oscillations once started would continue until the external forces were altered. The solution, therefore, although it gives us a formula of practical importance, is only an approximation, and leaves unexplained many of the troublesome phenomena noticed in everyday work. To obtain a deeper insight into the practical problem we must take into account the damping forces, and thus introduce into our equation of motion a term which is proportional to the angular velocity of the rotor. We shall make the assumption that the potential difference between the terminals of the supply mains always obeys the sine law.

The equation of motion will be of the form

$$Mk^2 \frac{\partial^2 \theta}{\partial t^2} + b \frac{\partial \theta}{\partial t} + c\theta = 0 \quad .$$

approximately; and when b/Mk^2 is small, an approximate solution of this equation is

$$\theta = \theta_0 \epsilon^{-bt/2Mk^2} \sin \{(c/Mk^2)^{\frac{1}{2}} t + a\}, \quad .$$

where θ_0 and a are constants depending on the initial conditions. Hence, if b is positive, the amplitude of the oscillations continually diminishes and the motion is stable. When, however, b is negative the motion is unstable. In this case, when once the oscillations are started they will get greater and greater, until finally the machine falls out of step, and the cut-outs act. In order, therefore, to discuss the stability of the motion, we must find an expression for the damping term. For a full account of the nature of the motion which can be represented by linear equations and the conditions of stability, the student is referred to Chapter VI of E. J. Routh's *Advanced Rigid Dynamics*. The discussion of the motor problem given below is founded on B. Hopkinson's solution.

B. Hopkinson's method.

Let the flux of induction linked with the armature and the field coils of a two pole synchronous motor be denoted by Φ_A, and let

$$\phi = N\Phi_A \sin \omega t = \Phi \sin \omega t,$$

where N is the number of turns in series on the armature and ωt is the angle which defines the position of a radius of the rotor at the time t. We neglect, for the present, the armature reaction, and we assume that the reluctance of the paths of the field flux is the same in all positions of the armature. We assume, therefore, that Φ is constant. Now the instantaneous value of the electromotive force generated in the motor armature is $\omega\Phi \cos \omega t$, and thus, with our usual notation, we shall have $\omega\Phi$ equal to $\sqrt{2}V_2$.

Let R be the resistance of the motor circuit, and Li, where i is the instantaneous value of the current in the armature, the flux of induction, round the armature wires and the connecting wires, which is not linked with the field coils. We assume that R is constant, and, in getting an approximate result, we can assume that L is constant also. It is to be noticed that R includes the resistance of the connecting mains and L includes their inductance. The equation which determines the steady motion of the motor is

$$e_1 = Ri + L\frac{\partial i}{\partial t} + \frac{\partial}{\partial t}(\Phi \sin \omega t),$$

where e_1 is the instantaneous value of the applied potential difference. If V_1 and V_2 be the effective values of the applied potential difference and of the motor E.M.F. respectively, and if $\pi - \psi$ is the phase difference between them, we may write the equation in the form

$$Ri + L\frac{\partial i}{\partial t} = \sqrt{2}V_1 \cos(\omega t + \psi) - \sqrt{2}V_2 \cos \omega t,$$

if we choose the origin of time at the instant when the field flux linked with the armature is zero. We have seen that for steady running ψ must be less than γ where $\tan \gamma$ equals $L\omega/R$. Solving the equation we get

$$i = i_1 \sin \omega t + i_2 \cos \omega t,$$

where
$$\left.\begin{aligned} i_1 &= (\sqrt{2}\,V_1/Z)\sin(\gamma - \psi) - (\sqrt{2}\,V_2/Z)\sin\gamma \\ i_2 &= (\sqrt{2}\,V_1/Z)\cos(\gamma - \psi) - (\sqrt{2}\,V_2/Z)\cos\gamma \end{aligned}\right\} \quad \ldots\ldots(a),$$
and
$$Z = (R^2 + L^2\omega^2)^{\frac{1}{2}}.$$

The component $i_1 \sin \omega t$ is wattless with respect to the motor E.M.F. and its amplitude i_1 may have a positive or negative value.

For a motor the amplitude i_2 of the watt component is always positive.

Let us now suppose that the steady running of the motor is slightly disturbed. Let i_1, i_2 and $\Phi \sin \omega t$ become $i_1 + x$, $i_2 + y$, and $\Phi \sin (\omega t + \xi)$ respectively. We can suppose that x, y and ξ are small quantities, so that we can neglect their squares or products. The equation for the disturbed motion is

$$\begin{aligned}\sqrt{2} V_1 \cos (\omega t + \psi) &= Ri + L \frac{\partial i}{\partial t} + \frac{\partial}{\partial t} \{\Phi \sin (\omega t + \xi)\} \\ &= R \{(i_1 + x) \sin \omega t + (i_2 + y) \cos \omega t\} \\ &\quad + L\omega (i_1 + x) \cos \omega t - L\omega (i_2 + y) \sin \omega t \\ &\quad + L \sin \omega t \frac{\partial x}{\partial t} + L \cos \omega t \frac{\partial y}{\partial t} \\ &\quad + \Phi \cos (\omega t + \xi) \left(\omega + \frac{\partial \xi}{\partial t}\right).\end{aligned}$$

For steady motion x, y and ξ are all zero, and thus, equating the coefficients of $\cos \omega t$ on each side of the equation, we get

$$\left.\begin{aligned}\sqrt{2}\, V_1 \cos \psi &= Ri_2 + L\omega i_1 + \omega \Phi. \\ \text{Similarly} \quad -\sqrt{2}\, V_1 \sin \psi &= Ri_1 - L\omega i_2.\end{aligned}\right\} \quad \ldots\ldots\ldots(b).$$

Solving these equations for i_1 and i_2, and noticing that $\omega \Phi$ equals $\sqrt{2} V_2$, we get the equations (a) given above. Equating the coefficients of $\cos \omega t$ on each side of the equation for the disturbed motion, we get

$$L \frac{\partial y}{\partial t} + L\omega x + Ry + \Phi \frac{\partial \xi}{\partial t} \cos \xi = 0.$$

Similarly
$$L \frac{\partial x}{\partial t} - L\omega y + Rx - \Phi \left(\omega + \frac{\partial \xi}{\partial t}\right) \sin \xi = 0.$$

Since ξ is a small angle, we may write 1 and ξ for $\cos \xi$ and $\sin \xi$ respectively. In practice also, $\partial\xi/\partial t$ is small compared with ω, and thus we may write ω for $\omega + \partial\xi/\partial t$. Hence these equations become

$$L \frac{\partial y}{\partial t} + L\omega x + Ry + \Phi \frac{\partial \xi}{\partial t} = 0 \quad \ldots\ldots\ldots\ldots\ldots(1),$$

and
$$L \frac{\partial x}{\partial t} - L\omega y + Rx - \Phi \omega \xi = 0 \quad \ldots\ldots\ldots\ldots\ldots(2).$$

Let g be the instantaneous value of the accelerating torque. Then, writing θ for $\omega t + \xi$, we get, by equating the two expressions for the power given to the rotor,

$$g\frac{\partial\theta}{\partial t} = i\frac{\partial}{\partial t}(\Phi \sin\theta).$$

Thus $$g = i\frac{\partial}{\partial\theta}(\Phi\sin\theta)$$

$$= \{(i_1 + x)\sin\omega t + (i_2 + y)\cos\omega t\}\,\Phi\cos(\omega t + \xi)$$

$$= \tfrac{1}{2}\Phi(i_1 + x)\{\sin(2\omega t + \xi) - \sin\xi\}$$
$$+ \tfrac{1}{2}\Phi(i_2 + y)\{\cos(2\omega t + \xi) + \cos\xi\},$$

and since ξ is small, we may write

$$g = \tfrac{1}{2}\Phi i_2 + \tfrac{1}{2}\Phi(y - \xi i_1)$$
$$+ \text{ periodic terms of frequency } \omega/\pi$$
$$+ \text{ small quantities.}$$

Now, when the motion is steady, y and ξ are zero, and thus $\Phi i_2/2$ is a measure of the constant resisting torque, and $(\Phi/2)(y - \xi i_1)$ is the torque which accelerates the rotor. If Mk^2 be the moment of inertia of the rotor, we have, therefore,

$$2Mk^2\frac{\partial^2\xi}{\partial t^2} + \Phi i_1\xi - \Phi y = 0 \quad ....(3).$$

The solution of the equations (1), (2) and (3) will approximately determine the motion when disturbed. To solve these equations, let us suppose that

$$x = A\epsilon^{mt}, \quad y = B\epsilon^{mt}, \quad \text{and} \quad \xi = C\epsilon^{mt},$$

where A, B and C are constants. Substituting these values of x, y and ξ in (1), (2), and (3), and dividing out the exponential terms, we have

$$L\omega A + (Lm + R)B + \Phi m C = 0,$$
$$(Lm + R)A - L\omega B - \Phi\omega C = 0,$$
$$-\Phi B + (\Phi i_1 + 2Mk^2m^2)C = 0.$$

Eliminating A, B and C from these equations, we have

$$\begin{vmatrix} L\omega, & Lm + R, & \Phi m \\ Lm + R, & -L\omega, & -\Phi\omega \\ 0, & -\Phi, & \Phi i_1 + 2Mk^2m^2 \end{vmatrix} = 0.$$

Expanding and simplifying, this reduces to

$$(\Phi i_1 + 2Mk^2m^2)\{(Lm+R)^2 + L^2\omega^2\} + L\Phi^2(\omega^2+m^2) + R\Phi^2 m = 0 \quad\text{.........(4)},$$

or

$$am^4 + bm^3 + cm^2 + dm + e = 0 \quad\text{.........(5)},$$

where

$$a = 2Mk^2L^2,$$

$$b = 4Mk^2LR,$$

$$c = 2Mk^2Z^2 + L\Phi(\Phi + Li_1),$$

$$d = R\Phi(\Phi + 2Li_1),$$

and

$$e = L\Phi^2\omega^2 + \Phi i_1 Z^2,$$

where

$$Z^2 = R^2 + L^2\omega^2;$$

and thus, by equations (a) given above,

$$c = 2Mk^2Z^2 + (2V_2L/Z^2\omega^2)\{V_2R^2 + V_1ZL\omega\sin(\gamma-\psi)\},$$

$$d = (2RV_2/Z^2\omega^2)\{V_2(R^2 - L^2\omega^2) + 2V_1ZL\omega\sin(\gamma-\psi)\},$$

and $e = (2/\omega)V_1V_2Z\sin(\gamma-\psi)$.

An inspection of the constants will show that a, b, c and e are always positive, since in the cases we are considering ψ is less than γ. We also see that d must necessarily be positive if R be greater than $L\omega$.

The stability of the motion in special cases.

Before finding the general criterion for the stability of the motion, it will be instructive to consider the special cases in which the equation (5) can be solved easily. We shall first consider the special case when R is negligible. Putting R equal to zero in equation (4), so that $Z = L\omega$ and $\gamma = \pi/2$, we have

$$L(\Phi i_1 + 2Mk^2m^2)(\omega^2 + m^2) + \Phi^2(\omega^2+m^2) = 0.$$

The roots of this equation are

$$\pm\omega\sqrt{-1} \quad\text{and}\quad \pm m_2\sqrt{-1},$$

where $m_2{}^2 = \Phi(\Phi + Li_1)/(2Mk^2L) = (V_1V_2\cos\psi)/(Mk^2\omega^2L)$.

The values of x and y in this case are

$$x = A_1\cos(\omega t + a_1) + A_2\cos(m_2t + a_2)$$

and

$$y = B_1\cos(\omega t + \beta_1) + B_2\cos(m_2t + \beta_2),$$

where A_1, a_1, ... are constants. We see, therefore, since

$$i = (i_1 + x)\sin\omega t + (i_2 + y)\cos\omega t,$$

that the components of the current when the motion is disturbed have frequencies $\omega/2\pi$, ω/π, $(\omega + m_2)/2\pi$ and $(\omega - m_2)/2\pi$ respectively. Now, in practice, m_2 is much smaller than ω, so that in the time that $\sin \omega t$ takes to go through all its values, $\sin m_2 t$ and $\cos m_2 t$ will have altered by a very small amount only. Assuming that $\sin m_2 t$ and $\cos m_2 t$ are constant during the time $2\pi/\omega$, we find that the effective value A of the current is given by

$$A^2 = C^2 + i_1 A_2 \cos(m_2 t + a_2) + i_2 B_2 \cos(m_2 t + \beta_2) + \tfrac{1}{2}A_2^2 \cos^2(m_2 t + a_2) + \tfrac{1}{2}B_2^2 \cos^2(m_2 t + \beta_2),$$

where C^2 is a constant.

Thus A^2, and therefore also A, goes through all its values in the time $2\pi/m_2$. Hence when the resistances of the armature and connecting mains of a synchronous motor are negligible, we see that the variations of the reading of the ammeter have a period given by

$$2\pi \{(Mk^2L\omega^2)/(V_1 V_2 \cos\psi)\}^{\frac{1}{2}}.$$

In practice, when the steady running of a synchronous motor is disturbed by a sudden variation in the resisting or the driving torque, the ammeter pointer sometimes gives a periodic series of readings the period of which may be a few seconds. It is found that, when R is negligible, the square of this periodic time is approximately directly proportional to the moment of inertia of the rotor and inversely proportional to its excitation, and this is in agreement with the formula given above. It is to be noticed, however, that, in addition to the components of the current of slow period which are set up by the disturbance, there may be also components having a period approximately equal to the period of the applied potential difference. The pointer of the motor ammeter cannot follow these rapid variations of the current, and so their effect is merely to increase the ammeter reading.

When we neglect the resistance R of the motor armature and the leads, the solution obtained shows that, once oscillations are set up about the position of steady running, the amplitude of these oscillations remains constant, and there is no cause tending either to increase or diminish them. In order to show how oscillations are damped out, let us consider the case when there is no magnetic leakage, that is, when L is zero.

Putting L equal to zero in equation (4), we find that

$$R\{\Phi i_1 + 2Mk^2m^2\} + \Phi^2 m = 0.$$

Therefore

$$m = -(\Phi^2/4Mk^2R) \pm (1/4Mk^2R)\{8Mk^2R^2\Phi i_1 - \Phi^4\}^{\frac{1}{2}}\sqrt{-1}.$$

Thus, if $8Mk^2R^2i_1$ is greater than Φ^3, oscillations of the ammeter pointer will ensue when the steady running is disturbed; the successive amplitudes of the swings, however, will diminish in geometrical progression. The greater the value of $\Phi^2/4Mk^2R$, the more effective will be the damping. Hence the damping effect increases with Φ, but diminishes if the moment of inertia of the rotor or the resistance of the armature be increased. If $8Mk^2R^2i_1$ is less than Φ^3, there will be no oscillations. We see, therefore, that when L is negligible the running is stable.

General case.

We shall now consider the general case. The four roots of equation (5) may be real, or two may be real and two imaginary, or the whole four may be imaginary. It has to be remembered that imaginary roots occur in pairs. If $p + n\sqrt{-1}$ is a root of the equation, $p - n\sqrt{-1}$ is also a root. The term in the solutions of the differential equations corresponding to this pair of imaginary roots is $A\epsilon^{pt}\cos(nt + \alpha)$, where A and α are constants. If p is positive we see that the amplitude of the swing is increasing, and this corresponds to an unstable oscillation. If p is negative the oscillation is diminishing and the oscillation represented by this term is stable. Similarly we can show that, for stable motion, equation (5) must have no real positive root, as this would introduce a term in our solutions which would increase with the time. We conclude, therefore, that in order that the running of the synchronous motor be stable the real roots and the real parts of the complex roots of equation (5) must be negative.

Products of the pairs of the roots of a biquadratic.

In order to find the required criterion we shall first, by Routh's method, find the products of the pairs of all the roots of equation (5). Writing $x \pm y$ for m in the equation, so that $x + y$ is one root, $x - y$ is another root and x is therefore the arithmetic mean between two roots, we get

$$a(x \pm y)^4 + b(x \pm y)^3 + c(x \pm y)^2 + d(x \pm y) + e = 0.$$

Thus $ay^4 + (6ax^2 + 3bx + c)y^2 + ax^4 + bx^3 + cx^2 + dx + e = 0,$

and $$(4ax + b)y^3 + (4ax^3 + 3bx^2 + 2cx + d)y = 0.$$

Rejecting the solution, y equal to zero, and eliminating y between the two equations, we get

$$64a^3x^6 + \dots + bcd - ad^2 - eb^2 = 0.$$

Now each value of x is the arithmetic mean between two values of m, and thus the product of the roots of this sextic equation

$$= \tfrac{1}{64}(m_1 + m_2)(m_1 + m_3)(m_1 + m_4)(m_2 + m_3)(m_2 + m_4)(m_3 + m_4)$$
$$= (bcd - ad^2 - eb^2)/(64a^3).$$

Thus, if we denote $bcd - ad^2 - eb^2$ by X, the product of the pairs of roots of (5) will be X/a^3.

Let us suppose first of all that the biquadratic has two pairs of imaginary roots $p_1 \pm n_1\sqrt{-1}$ and $p_2 \pm n_2\sqrt{-1}$. Then, by considering the sum of the roots of (5), we have

$$2(p_1 + p_2) = -b/a = \text{a negative quantity},$$

and

$$X/a^3 = 4p_1p_2\{(p_1 + p_2)^2 + (n_1 + n_2)^2\}\{(p_1 + p_2)^2 + (n_1 - n_2)^2\}.$$

Since, for stability, p_1 and p_2 must both be negative, and their sum is always negative, so that both cannot be positive, we see that the criterion in this case is that X must be positive. This is also the criterion when p_1 equals p_2.

Let us now suppose that two of the roots are real and two imaginary. Writing $n_2'\sqrt{-1}$ for n_2 in the preceding paragraph, we see that the roots are now $p_1 \pm n_1\sqrt{-1}$ and $p_2 \pm n_2'$. Thus, we have

$$2(p_1 + p_2) = -b/a,$$
$$X/a^3 = 4p_1p_2[\{(p_1 + p_2)^2 + n_1^2 - n_2'^2\}^2 + 4n_1^2n_2'^2],$$

and, by equation (5), the product of the roots is given by

$$e/a = (p_1^2 + n_1^2)(p_2^2 - n_2'^2).$$

As before, we see that p_1 and p_2 are both negative when X is positive, and since e/a is positive, p_2 is numerically greater than n_2'; and thus the two real roots are both negative. The criterion for stability in this case also is that X must be positive. Finally, when all the roots are real, none of them can be positive when a, b, c, d and e are positive. In our equation a, b, c and e are necessarily positive. If d be zero or negative, X is negative, and thus we see that when X is positive d must be positive, and the real roots are all negative.

The criterion for stability.

The criterion for the stability of the running of the synchronous motor is therefore that $bcd - ad^2 - eb^2$ must be greater than zero. Substituting for the coefficients their values, this criterion becomes

$$\Phi\,(\Phi + 2Li_1) + (4Mk^2/L)\,(R^2 - L^2\omega^2) > 0.$$

This inequality may be written in the form

$$R^2 > L^2\omega^2 - 2ZV_1V_2L^2\omega \sin(\gamma - \psi)/(LV_2^2 + 2Mk^2 \,.\, Z^2\omega^2).$$

Hence, if R is greater than $L\omega$, the motion is stable, since $\sin(\gamma - \psi)$ is positive.

In the particular case when X is zero the sum of one pair of the roots of equation (5) must be zero. Hence it easily follows that these roots must be $\pm\sqrt{d/b}\sqrt{-1}$, and the other roots, if real, are negative, and, if imaginary, they have their real parts negative. Thus the equilibrium in this case is neutral for one type of free oscillations and is stable for other displacements.

The sum of the squares of the roots of equation (5) is, by Newton's theorem, $b^2/a^2 - 2c/a$. If this expression be negative, some of the roots of the equation must be imaginary. We see, therefore, that when $2ac$ is greater than b^2 we must have at least one pair of imaginary roots, and these correspond to stable or unstable oscillations. This condition may be written

$$Mk^2\,(L^2\omega^2 - R^2) + (V_2L/Z^2\omega^2)\,\{V_2R^2 + V_1ZL\omega \sin(\gamma - \psi)\} > 0.$$

Hence, if $L\omega$ is greater than R, which is generally the case in practice, there will be at least one type of free oscillations set up. From the criterion for stability we see that these oscillations will be, in general, unstable in the ideal case we are considering.

The period of the phase swing.

Since the period of the phase swing is very long compared with the period of the applied potential difference, we shall consider the case when n/ω is a small quantity, $p \pm n\sqrt{-1}$ being a pair of the roots of the biquadratic (5). We shall assume that p is a very small quantity, otherwise the swings would be damped out or would increase so rapidly that the phase swing would not be a noticeable phenomenon. We shall also assume that $a(p+n\sqrt{-1})^4$, which equals $2Mk^2L^2(p+n\sqrt{-1})^4$, may be put equal to zero in equation (5).

Substituting $p+n\sqrt{-1}$ for m in equation (5), and noticing that, on our assumptions, we may write

$$a(p+n\sqrt{-1})^4=0, \qquad b(p+n\sqrt{-1})^3=-bn^3\sqrt{-1},$$

$$c(p+n\sqrt{-1})^2=2cpn\sqrt{-1}-cn^2, \text{ and } d(p+n\sqrt{-1})=dp+dn\sqrt{-1},$$

we get, by equating the real terms in the resulting equation to zero,

$$\begin{aligned} cn^2 &= dp+e \\ &= e \quad\quad (6), \end{aligned}$$

approximately, since, in practice, dp is small compared with e. Similarly, by equating the coefficient of $\sqrt{-1}$ to zero, we get

$$2cp = bn^2 - d \quad\quad (7).$$

Neglecting the small terms in the value of c in (6), we can write $c = 2Mk^2Z^2$ and hence we find that

$$2Mk^2Z^2n^2 = (2/\omega)\, V_1 V_2 Z \sin(\gamma - \psi),$$

and thus $T = 2\pi\,[(Mk^2Z\omega)/\{V_1V_2\sin(\gamma-\psi)\}]^{\frac{1}{2}}$,

which agrees with the result given on p. 209.

Similarly from (7) we find that

$$2Mk^2p = -\,V_2{}^2R\,(R^2 - L^2\omega^2)/(\omega^2Z^4),$$

approximately.

We see that, if R is greater than $L\omega$, p is negative and so the motion is stable, but if R is less than $L\omega$, which is the usual case in practice, the motion is unstable. In the latter case, the amplitude of the phase swing begins to increase according to the law ϵ^{pt}. On the given assumptions, therefore, R must be greater than $L\omega$ if the running is to be steady. In other words, γ must be less than 45° for steady running. In this case, the smaller the moment of inertia of the rotor, and the greater the excitation of the field, the more effective will be the damping.

If we keep the excitation of the field constant, then V_2/ω, which equals $\Phi/\sqrt{2}$, will also be constant, and thus, for all values of the frequency, we have

$$2Mk^2p = -(\Phi^2/2)(R^3 - RL^2\omega^2)/(R^2 + L^2\omega^2)^2,$$

and hence

$$\frac{\partial p}{\partial \omega} = (RL^2V_2^2/Mk^2\omega)\{(3R^2 - L^2\omega^2)/(R^2 + L^2\omega^2)^3\}.$$

If, therefore, $L\omega$ is greater than R but less than $R\sqrt{3}$, we see that p, which in this case we may call the coefficient of instability, increases as the frequency increases. If, however, $L\omega$ is greater than $R\sqrt{3}$, the coefficient of instability diminishes as the frequency is increased. Let us now suppose that R is greater than $L\omega$, so that p is negative. Then, the greater the numerical value of p the smaller will be the value of ϵ^{-pt} for a given value of t, and the more rapidly will the free oscillations of long period be damped out. Since, when R is greater than $L\omega$, $\partial p/\partial\omega$ is always positive, and p is negative, it follows that, in this case, increasing the frequency diminishes the numerical value of p, and therefore the damping.

It must be noticed that we have neglected the damping effect produced by the resistance of the air. In addition, since we have made the assumption that the field of the motor is unaffected by the oscillations of the armature current, we have neglected the damping effects caused by the eddy currents induced in the iron and the copper.

Effect produced by the distortion of the field.

When the pulsations of the current are small, the modification of the formulae given above introduced by the distortion of the field due to these pulsations can be taken into account without much difficulty. B. Hopkinson has considered this case (*Proc. Roy. Soc.*, Vol. 72, p. 235). He proves that the distortion of the field slightly increases the instability.

On our assumption we see that, when $L\omega$ is greater than R, slow oscillations are always set up when the motion is disturbed. They gradually increase in amplitude until finally the machine falls out of step. Phenomena similar to this are often noticed in practical working. They may, however, be primarily due to other causes. For instance, periodic fluctuations in the driving torque of the engine of the generator or in the retarding torque due to the load may synchronise with the electrical forces tending to maintain the free oscillations, and thus cause the machines to set up phase swinging. The theory we have outlined shows that, when the running is disturbed, there are electrical forces called into play which tend to make the machine fall out of step.

Amortisseurs.

In order to prevent phase swinging, Hutin and Leblanc provided the field magnets with 'amortisseurs,' or 'dampers,' which tend to prevent any relative change between the positions of the magnetic field due to the armature and the field due to the field magnets. These dampers sometimes consist of heavy copper circuits surrounding the poles, or of copper rods embedded in the poles and having their ends joined by copper rings. Since, in polyphase machines, under normal conditions, both magnetic fields are fixed relatively to these circuits, no currents will be induced in them. When, however, phase swinging is set up, the alteration of the magnetic flux in these circuits produces a torque which generally tends to prevent any departure from the normal running.

For polyphase machines running synchronously (see the next chapter) these dampers are useful, as the magnetic reactions produced tend to prevent the machines from falling out of step. Another effect of the dampers is to reduce the potential drop at the terminals on heavy inductive loads, as they prevent the armature reaction from appreciably demagnetising the field magnets.

For single phase machines dampers are not so useful. The magnetomotive force due to the currents in the armature of a single phase machine sets up a pulsating magnetic field. This may be resolved into two magnetic fields gliding in opposite directions. One of these has no effect on the dampers, when the running is steady, as it is fixed relatively to them, but, when the running is disturbed, the currents induced in them by this component help to damp out the oscillations. The other produces in the damping coils an alternating current of double the frequency of the supply current. Owing to the high inductance of the damper circuits, the currents induced in them by this field are rarely large, and the retarding torque due to it is generally small.

It has been observed in practical work that the humming noise often made by single phase machines when running is reduced considerably when damping circuits are used. This is due to a diminution in the amplitude of the flux variations.

The theory of damping coils.

B. Hopkinson, in the paper quoted above, has found an approximate solution for the disturbed motion of a synchronous motor provided with damping coils. The effect of these coils is generally to increase the stability of the motion. It is proved that if the period of the phase swing be decreased, the damping will be increased. For instance, if the moment of inertia of the flywheel be increased, the regulation will be improved. The interesting result is also proved that it is possible to use too much copper in constructing the damping coils.

The ordinary field magnet coils must act to a certain extent like damping coils. The alternating currents induced in them tend to prevent sudden variations in the value of the field flux. If we neglect the cross flux and the leakage, the damping effect is the same, whether we utilise the extra copper required for the damping coils in making these coils, or whether we utilise it in reducing the resistance of the exciting circuit. The latter method has the incidental advantage of reducing appreciably the excitation losses.

If the exciting circuit had no resistance, there could be no variation of the induction linked with it, and consequently no damping effects would ensue. Similarly, if it had infinite resistance, there would be no damping. Hence there must be a particular value of the time constant of the exciting circuit for which the damping effects are a maximum.

B. Hopkinson also proves that, in order that the damping coils may increase the stability of the running, the watt component i_2 of the current, with reference to the back E.M.F. of the motor, must be greater than a current which is approximately equal to $\omega R\Phi/Z^2$. Hence, increasing the load on the motor may make the running stable; a result which he has verified experimentally.

All the conclusions given in this chapter have been obtained on the supposition that the applied potential difference is sine shaped and that L is a constant. In practice, L may vary by 50 per cent. for different positions of the rotor. It is sufficient, however, when making rough calculations in connection with synchronous motors, to take its mean value. Accurate quantitative results would be exceedingly difficult to obtain, and would be too complicated for practical use.

It is to be noticed that i_2 is positive for a motor and negative for a generator. B. Hopkinson's method, therefore, can also be applied to the case of a generator running in parallel with other generators.

REFERENCES.

PAUL BOUCHEROT, 'La Théorie des Alternateurs,' *La Lumière Électrique*, Vol. 45, p. 201, July, 1892.

ANDRÉ BLONDEL, 'Couplages et Synchronisation des Alternateurs,' *La Lumière Électrique*, Vol. 45, p. 351, Aug. 1892.

GISBERT KAPP, 'Das Pendeln parallel geschalteter Maschinen,' *Elektrotechnische Zeitschrift*, Vol. 20, p. 134, 1899.

HANS GÖRGES, 'Ueber das Verhalten parallel geschalteter Wechselstrommaschinen,' *Elektrotechnische Zeitschrift*, Vol. 21, p. 188, 1900.

BERTRAM HOPKINSON, 'The "Hunting" of Alternating Current Machines,' *Proceedings of the Royal Society*, Vol. 72, p. 235, June, 1903.

CHAPTER VIII

The parallel running of alternators. Circuit breakers in parallel running. Reverse power relays. The theory of parallel running. Improving the electric regulation. Effects of wave shape. Inductive loads. Condenser loads. Armature reaction. Free oscillation of long period. Practical running. Forced oscillations. Electro-mechanical resonance. The stresses on the shafts coupling dynamos and engines. The whirling of shafts. Connecting a machine with the bus bars. Methods of synchronising. Two transformer method. Phase indicating transformer. High potential voltmeter method. Rotating field device. Optical methods. Synchronising device for three phase plant. Synchroscope for three phase plant. References.

The parallel running of alternators.

THE efficiency of a steam engine or a steam turbine is much higher when it is running on a heavy load than when it is running lightly loaded. It is necessary, therefore, for the engineer of a central station to arrange that his engines never run for long periods on light loads. To illustrate the importance of this point, let us consider how the efficiency of a high speed steam engine, for example, varies with the load. Let W be the number of pounds of steam consumed by the engine per hour, and let P be the brake horse power developed. A linear equation of the form

$$W = a + bP$$

will express very approximately the relation between W and P; the constants a and b in this equation being different for different engines. This equation is known as Willans's law, and is true whether the engine is used with or without a steam condenser. The constant b is the same in both cases; the effect of the condenser is merely to diminish the value of a. In a Willans and Robinson high speed engine, when working without a steam condenser, the steam consumed per hour at no load is about a quarter of that consumed per hour at full load. If w denote the number of pounds of steam consumed per hour, per brake horse power developed, we have

$$w = W/P = b + a/P.$$

In a Willans and Robinson engine, therefore, if a be the number of pounds consumed per hour at no load and P_m be the full load brake horse power, we have, since $3a = bP_m$,

$$w = a\,(1/P + 3/P_m),$$

approximately. Thus, at one-fifth full load, for example, the value of w would be twice as great as at full load. As the coal consumed is roughly proportional to the number of pounds of steam that leave the boiler, we see that the coal bill for the units generated at one-fifth load will be about twice as great as the coal bill for an equal number of units generated at full load. In addition the efficiency of the alternators is less at a fifth load than at full load. For economical working, therefore, it is essential never to have the machines running for long periods on light loads.

In central stations, each engine is generally coupled to its own alternator so as to avoid the losses consequent on the use of gearing. It would not conduce to economical working to have each generator supplying a set of mains connected with no other generator, as the pressure between every pair of supply mains has always to be maintained, whatever may be the load, and thus we should often have several engines and alternators running on light loads. All the alternators, therefore, are connected in parallel to two mains called 'bus bars' with which the mains supplying the transformers are also connected, and care is taken to ensure that the number of machines running at any time is only sufficient to carry the load.

We saw in Chapter v that, if two alternators have the same frequency, and if they are connected in series, the running is stable when the phase difference between the armature electromotive forces is nearly 180°. In this case, the terminals which are connected with the same bus bar are practically at the same potential, and so the machines are working in parallel so far as a circuit joining the two bus bars is concerned. Hence, when two alternators are connected in this manner, the stable position of running on no load occurs when their armature E.M.F.s are nearly in opposition round the circuit formed by the armatures, and consequently when both the E.M.F.s are acting nearly in phase with one another, and tending to produce a potential difference between the bus bars. To a first approximation, therefore, the electric forces tend to make two alternators run in parallel when they are connected with the bus bars, provided that the effective values of their electromotive forces lie between certain limits.

Circuit breakers in parallel running.

When fuses or magnetic circuit breakers are placed in the leads connecting the terminals of an alternator with the bus bars, then, if the device in one only of the connecting leads acts, the insulation of the machine may be subjected to excessive stresses. The electrical forces no longer constrain the alternator to run in parallel with the others, and so it will sometimes be running in series with them. In this case, the effective value of the P.D. between the terminals of the circuit-breaking device will have double its normal value. The P.D., also, between the armature of the machine and the field poles may have nearly double its working value, and this may start an arc between the armature windings and the poles which may ruin the machine. For this reason, amongst others, therefore, fuses and 'excess current circuit breakers' are now rarely placed in the circuits of the connecting leads.

Reverse power relays.

If for any reason the prime mover driving an alternator which is running in parallel with the bus bars fail to give the requisite power to the alternator or if a partial failure of the alternator such as a short circuit of part of the armature windings occur, then it will be driven as a synchronous motor and this fact may escape the attention of the attendant. To prevent this happening reverse power relays are used. The principle on which they act is the same as that on which watt hour meters act. So long as the alternator is supplying power to the bus bars a meter connected across the leads joining it to the bus bars will rotate in the positive direction. When however the alternator is running as a synchronous motor the meter will rotate in the opposite direction. In the reverse power relay, when the alternator begins to run as a motor the forces slowly called into play move a contact piece which closes a tripping coil circuit, and this opens a switch which puts the machine out of circuit. It is necessary to make the device act sluggishly, so that any transient cause, like a momentary sticking of the governor of the steam engine, will not cause it to act.

In the manufacture of reverse power relays the principle of the power factor indicator (Vol. I, p. 407) can also be utilised, as it is easy to make a device of this nature operate the tripping coil when the angle of lag or lead exceeds ninety degrees, that is, when the alternator is acting as a motor. In the event of a break in the circuit of the field magnet windings the machine does not fall out of step, but in practically every case would still run as an alternator, the field being excited by the armature reaction. There would, however, be an excessive increase in the armature current. This would be detected at once by the attendant, who would then take the necessary steps to locate the break.

The theory of parallel running.

In order to simplify the theory of parallel running we shall assume that the electromotive force and current waves are sine shaped, so that the current vector is in the same plane as the electromotive force

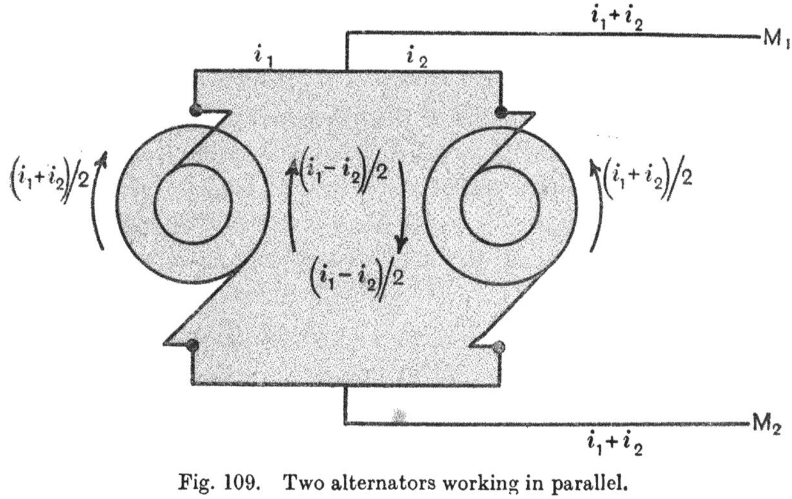

Fig. 109. Two alternators working in parallel.

vectors, and we shall also assume that the effective values of the electromotive force of each machine are the same. We shall suppose that the two machines are similar and equal, and that the load is constant.

If i_1 and i_2 be the instantaneous values of the currents in the armatures, we can always write

$$i_1 = \tfrac{1}{2}(i_1 + i_2) + \tfrac{1}{2}(i_1 - i_2),$$

and

$$i_2 = \tfrac{1}{2}(i_1 + i_2) - \tfrac{1}{2}(i_1 - i_2).$$

Hence, we may consider that each machine (Fig. 109) is supplying a current $(i_1 + i_2)/2$ to an external circuit, and that there is a synchronising current $(i_1 - i_2)/2$ in the armatures. Suppose that the load is inductive and that β is the phase difference between the external current and the external potential difference. Let (Fig. 110) OC and OA represent the two armature electromotive forces, each of which has an effective value V. Let B be the middle point of AC, then, as in Fig. 94, p. 168, BA or BC will represent the voltage V_1 in the external circuit. Let OK represent the synchronising current, and let OD and OE be each equal to half the current C in the external circuit. If we draw OA' parallel to BA, the angle $A'OD$ will be equal to β. Let the angles BOC and BOA be each equal to $\theta/2$, and let W_1 and W_2 be the loads on the machines, the vector electromotive forces of which are OC and OA respectively. Then, since the electric power generated is the product of the apparent watts multiplied by the cosine of the phase difference, we get from Fig. 110

$$W_1 = V \,.\, \tfrac{1}{2}C \,.\, \cos EOC + V \,.\, A \,.\, \cos KOC,$$

where A is the effective value of the synchronising current. Now A is $2 \,.\, OB/Z$, where Z is the impedance of the circuit of the armatures. The angle BOK is the angle of lag of the synchronising current behind the electromotive force driving it. We shall denote this angle by γ. Hence, noticing that $2 \,.\, OB$ is $2V\cos(\theta/2)$, we get

$$W_1 = (1/2)\, VC \cos(\beta + \pi/2 - \theta/2) + 2\,(V^2/Z)\cos(\theta/2)\cos(\theta/2 - \gamma)$$
$$= (1/2)\, VC \sin(\theta/2 - \beta) + (V^2/Z)\,\{\cos\gamma + \cos(\theta - \gamma)\}.$$

Similarly

$$W_2 = (1/2)\, VC \sin(\theta/2 + \beta) + (V^2/Z)\,\{\cos\gamma + \cos(\theta + \gamma)\},$$

where W_2 is the power generated by the second machine. It is easy to see that when C is zero these formulae agree with the formulae for a generator coupled to a synchronous motor given in Chapter v, p. 172.

The difference between the loads on the two machines is given by the equation

$$W_1 - W_2 = -VC\cos(\theta/2)\sin\beta + (2V^2/Z)\sin\theta\sin\gamma.$$

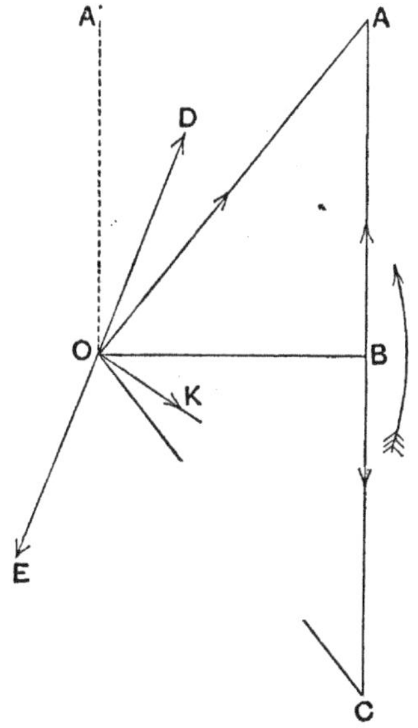

Fig. 110. Vector diagram for two alternators in parallel.

If θ equal π, that is, if the machines are in phase with regard to the external circuit, the right-hand side of this equation equals zero, and the load on one machine equals that on the other for all values of the external load.

In practice, θ is generally less than π. Let us suppose that θ is $\pi - x$, where x is a small angle. Substituting in the above equation, we find that

$$\begin{aligned} W_1 - W_2 &= -VC\cos\{(\pi - x)/2\}\sin\beta + 2(V^2/Z)\sin(\pi - x)\sin\gamma \\ &= V\{(2V\sin\gamma)/Z - (O\sin\beta)/2\}x, \end{aligned}$$

since we may write x for $\sin x$, and $x/2$ for $\sin(x/2)$, when x is small. In this formula it is immaterial whether we write V or V_1, since $V_1 = V \sin(\theta/2) = V \cos(x/2) = V$ approximately. If the external circuit were non-inductive, β would be zero, and $W_1 - W_2$ would be independent of the load, since in this case

$$W_1 - W_2 = \{(2V^2 \sin\gamma)/Z\}\, x.$$

It follows that, when $\pi - x$ diminishes, that is, when x increases, the difference between the load on the leading and lagging machine increases and this tends to good regulation.

Improving the electric regulation.

From the formula given above for $W_1 - W_2$ it follows that increasing the value of the electromotive force V greatly increases the accelerating and braking effects called into play by the mutual electric forces generated round the circuit of the armatures. Hence, on the assumptions we are now making, and neglecting the question of the stability of the free oscillations, we see that the greater the excitation the better is the electric regulation of the parallel running of the machines.

Effects of wave shape.

When the shapes of the electromotive force waves of the two machines are different they can never be in exact opposition in phase, and so, as we have seen on p. 169, the bus bar voltage is less than the voltage of either machine, and the electromotive force V' round the circuit of the armatures will be large. If the wave shapes are very unlike one another, the circulating current may be so large that parallel running is impossible.

Inductive loads.

Since, on our assumptions, we have

$$W_1 - W_2 = V\,\{(2V \sin\gamma)/Z - (C \sin\beta)/2\}\, x,$$

we see that if $(C \sin\beta)/2$ equals $(2V \sin\gamma)/Z$ there is no electric regulating effect, and if $(C \sin\beta)/2$ is greater than this value, the machines tend to run in series. When the machines, therefore, are working on a heavy inductive load, that is, when C and β are large, the machines will have a tendency to fall out of step.

If, on the other hand, the external load acts like a condenser, β will be negative, and hence the regulating effect will be better than for a non-inductive circuit. In this case also, as the current increases, the retarding and accelerating effects will increase, provided that $C \sin \beta$ increases.

Condenser loads.

In the preceding investigation we have not considered the effects of armature reaction on the parallel running of the machines. The magnetomotive force of the currents in the armature when they lag behind the armature electromotive force tends to demagnetise the field magnets. Hence the electromotive forces generated are reduced, and this tends to bad regulation. Similarly, with leading currents the armature reaction tends to increase the electromotive forces generated, and thus improves the running of the machines.

Armature reaction.

We should expect, therefore, when alternators are working in parallel on a heavy inductive load, that the running would be unsteady and that breakdowns would be frequent. The stability could be improved by the use of an over-excited synchronous motor (a rotary condenser) connected between the bus bars so as to raise the power factor of the circuit. A battery of static condensers would have a similar effect when each is connected in parallel across the circuit. If the potential difference be too high for the condensers they could be connected in parallel groups across the mains or a transformer might be used to reduce the pressure, the condensers being connected across the low pressure terminals.

We shall now find a formula for the free oscillation of long period or 'phase swing' which practically always occurs if an alternator is 'paralleled' slightly out of step, that is, if it is switched on to the bus bars at an instant when its electromotive force is not exactly in phase with the potential difference between the bars. In order to simplify the problem we shall assume that, initially, there is only one machine connected with the bars. We shall also assume that the oscillation is so slow that we may use vector diagrams. We assume, therefore, that the periodic time of the free oscillation is great compared with that of the alternating current. In

Free oscillation of long period.

approximate working this assumption may safely be made. We shall also neglect all the damping forces.

Let g_1' and g_2' be the instantaneous values of the torques applied to the first and second machines respectively, and let g_1'' and g_2'' be the torques required to overcome the mechanical retarding forces. Then, the torques employed in developing electrical energy will be $g_1' - g_1''$ and $g_2' - g_2''$ respectively. We shall make the assumption that these applied torques are constant. We shall suppose that each machine has the same number, $2p$, of poles, so that the mean angular velocities of the rotors are the same.

By the fundamental equations we have, when the running is steady,

$$g_1\omega = \tfrac{1}{2}VC \sin(\theta_0/2 - \beta) + (V^2/Z)\{\cos\gamma + \cos(\theta_0 - \gamma)\},$$

and $$g_2\omega = \tfrac{1}{2}VC \sin(\theta_0/2 + \beta) + (V^2/Z)\{\cos\gamma + \cos(\theta_0 + \gamma)\},$$

where g_1 and g_2 are equal to $g_1' - g_1''$ and $g_2' - g_2''$ respectively, and θ_0 is the phase difference between the vectors representing the electromotive forces.

Let us now suppose that, owing to a momentary variation of the driving torque or the load, θ_0 becomes $\theta_0 + x$ at a particular instant, then since ω remains practically constant, we get the following equations for the differences Δg_1 and Δg_2 between the new and the old values of the torque

$$\omega\Delta g_1 = VC\cos(\theta_0/2 + x/4 - \beta)\sin(x/4)$$
$$- 2(V^2/Z)\sin(\theta_0 - \gamma + x/2)\sin(x/2),$$

and $$\omega\Delta g_2 = VC\cos(\theta_0/2 + x/4 + \beta)\sin(x/4)$$
$$- 2(V^2/Z)\sin(\theta_0 + \gamma + x/2)\sin(x/2).$$

Now, when θ is $\theta_0 + x$, let us suppose that a given radius of the rotor of the first machine makes an angle θ_1 with the horizontal, and let also a radius of the rotor of the second machine make an angle θ_2 with it. Then, the radius of the second machine may be chosen, so that $\theta_1 - \theta_2$ is equal to θ/p, where θ or $\theta_0 + x$ is the angle between the vectors of the electromotive forces. Let $M_1k_1^2$ be the moment of inertia of the first machine together with the moment of inertia of the shaft and the flywheel of the engine to which it is coupled. Let $M_2 k_2^2$ be the corresponding moment of inertia for the second machine. Then, since the moment of inertia of a rotating body multiplied by its angular acceleration equals the moment of the forces about the axis of the rotor, we have

$$M_1 k_1^2 \frac{\partial^2 \theta_1}{\partial t^2} = \Delta g_1,$$

and

$$M_2 k_2^2 \frac{\partial^2 \theta_2}{\partial t^2} = \Delta g_2.$$

Noticing that $p(\theta_1 - \theta_2)$ equals $\theta_0 + x$, we get

$$\frac{\partial^2 x}{\partial t^2} = \frac{p \Delta g_1}{M_1 k_1^2} - \frac{p \Delta g_2}{M_2 k_2^2}.$$

If x be small we can write x for $\sin x$, and hence we find that

$$\frac{\partial^2 x}{\partial t^2} = -\mu x,$$

where

$$\mu = (pV/4\omega)\,[\{(4V/Z)\sin(\theta_0 - \gamma) - C\cos(\theta_0/2 - \beta)\}/M_1 k_1^2 \\ - \{(4V/Z)\sin(\theta_0 + \gamma) - C\cos(\theta_0/2 + \beta)\}/M_2 k_2^2].$$

If μ be positive, the motion is therefore simple harmonic, and the period is $2\pi/\sqrt{\mu}$.

If we suppose that $M_1 k_1^2$ is very large compared with $M_2 k_2^2$ and that C is zero, we have

$$T = 2\pi\,[(M_2 k_2^2 Z\omega)/\{-pV^2 \sin(\theta_0 + \gamma)\}]^{\frac{1}{2}},$$

and putting $\pi - \psi$ for θ_0, we get

$$T = 2\pi\,[(M_2 k_2^2 Z\omega)/\{pV^2 \sin(\gamma - \psi)\}]^{\frac{1}{2}}.$$

This agrees with the formula for the time of the slow free oscillation of a synchronous motor given on p. 209. It has been proved by several experimenters that the period of the phase swing varies directly as the square root of the moment of inertia $M_2 k_2^2$ and inversely as the excitation. Now the excitation is proportional to V, and thus the formula has been partially verified experimentally.

Let us suppose that $M_1 k_1^2$ equals $M_2 k_2^2$. The formula for the period of the phase swing is now given by

$$T = 2\pi\,[\{(2M_1 k_1^2 \omega)/pV\}/\{4(V/Z)\cos\psi \sin\gamma - C\sin\beta\cos(\psi/2)\}]^{\frac{1}{2}}.$$

On a non-inductive load, β is zero, and thus

$$T = 2\pi\,[(M_1 k_1^2 Z\omega)/(2pV^2 \cos\psi \sin\gamma)]^{\frac{1}{2}}.$$

Hence the time of swing is a minimum when ψ is zero, and increases as ψ increases, that is, as the phase difference between the electromotive force vectors of the two machines diminishes.

The time of swing is practically independent of β in most cases, for $4\,(V/Z)\cos\psi\sin\gamma$ is generally much greater than $C\sin\beta\cos(\psi/2)$. It makes, therefore, little difference to the period of the phase swing whether the load acts like a condenser or a choking coil, provided that the armature reaction of the alternators is negligible.

When phase swinging is set up between two machines, we have seen that to a first approximation the motion is simple harmonic. It follows, therefore, that, when the phase difference between the electromotive force vectors of the two machines is a maximum or a minimum, their rotors are moving with the same angular velocity, and when they pass through the positions which they have when the running is stable, the difference between their angular velocities is a maximum.

Practical running.

In practice, the problems connected with parallel running are much more complicated than those considered above. Not only has armature reaction to be taken into account, but we have also to consider the stability of the motion. We saw, in Chapter VII, that when the steady motion of a synchronons motor is disturbed, then, in some cases, the ensuing motion is unstable. Similarly when an alternator is running in parallel with other alternators the steady motion may be unstable.

Let us suppose that we have several alternators connected with the bus bars and working in parallel. If we assume that the potential difference between the bars is sine shaped and is practically undisturbed by oscillations of the current in the circuit of one of the machines, then the analytical work given in the preceding chapter applies, the only difference being that the watt component i_2 of the current is negative. We see, therefore, that it is possible for two types of free oscillations of different periods to be set up in the circuit of each alternator. Thus, if there are n alternators, we may have $2n$ principal free oscillations, and these oscillations may all be taking place at the same time. As $L\omega$ is generally greater than R for each alternator circuit, we see that, on the usual assumptions, there must be at least n types of free oscillations. If the damping forces due to armature reaction and eddy currents were negligible, the running would be inherently unstable.

We saw in the last chapter that, so far as the free oscillations are concerned, they can be damped effectively by means of suitably chosen damping coils. We saw also that, in some cases, the same effect could be produced by diminishing the resistance of the exciting coils of the field magnets. In general the effect of the eddy currents generated when the steady motion is disturbed is to damp out the ensuing disturbances.

We can see, also, that machines which produce electromotive force waves differing widely in shape are not well adapted for running in parallel as the circulating currents are large. Even when machines giving a sine shaped wave of E.M.F. on open circuit are used, the circulating currents are large when the machines are very unequally loaded, as the shapes of their E.M.F. waves are then different. The damping effect of the inductances of the armatures of the alternators on the high harmonics in the current disturbances will however be considerable, and thus the effect of the fundamental harmonic will be the most important.

Forced oscillations.

In what precedes we have merely considered the free oscillations that are set up when the steady running is disturbed. When the disturbing force is periodic we get forced oscillations as well. For example, when an alternator is driven by a single crank reciprocating engine, the fluctuations in the driving torque are large, and this torque vanishes at least twice in every revolution. Even in an engine with three cranks, the torque is not absolutely steady, and forced oscillations will be set up in the running of the alternator. These oscillations, in practice, are often sufficiently large to produce current oscillations which can be observed by noticing the continual oscillations of the pointer of the machine ammeter. As the variations of the torque give rise to free oscillations also, we should expect that the ammeter pointer would vibrate in an irregular manner, but that in general it would go through all its values during the time the rotor takes to make a complete revolution, and when the forced oscillations are appreciable, this is found to be the case.

Electro-mechanical resonance.

The magnitudes of the forced oscillations set up, when the alternators are acted on by periodic disturbing forces, depend not only on the magnitude of the amplitudes of the disturbing forces but also, in a very special manner, on the periodic times of these forces. If the period of the disturbing force is nearly the same as that of one of the free oscillations, the resulting forced oscillation will be very large. In particular, when the period of the disturbing force equals the period of a free oscillation, electro-mechanical resonance ensues, and, unless the damping be very powerful, the oscillations will increase until the large currents cause the circuit breakers to act or the machines have to be switched out of circuit owing to the large periodic rushes of current through their armatures.

Many dynamical illustrations can be given of this kind of resonance. A heavy pendulum, for instance, can be set into violent oscillation by a series of little pushes, provided that they are properly timed. Similarly the 'rolling' of ships at sea is explained. When the period of the waves synchronises with the period of the free oscillation of the ship, it may roll very heavily even although the height of the waves be small.

When the period of the disturbing force is not approximately equal to any of the periods of the free oscillations, the effect produced is practically always small. If the period of the disturbing force be much smaller than the period of the quickest of the free oscillations, the resulting disturbance will, in general, be quite negligible. This is illustrated by noticing the apparently absolutely steady deflections of the pointers attached to the movable coils of several types of electric measuring instruments when traversed by alternating currents, even when the frequency of these currents is very low.

In practice, therefore, we have to arrange that none of the periods of the free oscillations is approximately equal to the period of any of the disturbing forces. In modern stations each alternator is directly coupled to a steam-engine. The disturbing forces are generally due to the variations in the driving torque, but in some cases they are due to the oscillations of the governors of the steam-engines. It is well known (see Routh's *Advanced Rigid Dynamics*, p. 73) that the oscillations of the balls in a Watt's governor are unstable. For this reason various damping devices are sometimes employed in connection with steam-engine governors. If these devices are inefficient, periodic fluctuations will be set up. When the balls are at their greatest distance apart the lever acting on the throttle valve will diminish or cut off the supply of steam, and when they are at their minimum distance the valve may be fully opened. These pulsations will therefore produce a periodic fluctuation of the pushing force on the piston, and therefore also a fluctuation in the driving torque. This will give rise to a forced oscillation of the current in the armature. If the period and real exponential of the disturbing force in this case are nearly the same as the period and real exponential of a free vibration of the current, a very large forced oscillation may be set up. The remedy for the resonance due to this cause is to use efficient dampers for the governors. They may be fitted, for example, with a dash pot, that is, a loosely fitting piston working in a small closed cylinder containing air. The piston, whilst offering practically no resistance to slow changes of its position, offers a great resistance to sudden changes.

If the periodic times of the disturbing forces are known, care must be taken that none of them equals the period of any of the free oscillations. Now the period of the free oscillations of an alternator can be varied by increasing or diminishing the moment of inertia of the flywheel, and this would be the best remedy to apply in practice. The periods of the free oscillations can also be increased or diminished by varying the excitation of the alternator.

The stresses on the shafts coupling dynamos with engines.

In designing the shaft necessary to couple an electric generator to its prime mover, the stresses which it will have to withstand in actual working must be studied carefully. Heavy shafts, quite free from flaws, have fractured when rotating at moderate speeds although they were only transmitting a small fraction of the working torque for which they were designed. When the shaft has been replaced by a new one of the same dimensions, it has been noticed, on several occasions, that it fractures at the same critical speed as the shaft which it replaced. As the forces applied to the shaft are small compared with the static forces which it can safely withstand, the fracture may possibly be due to mechanical resonance. One of the applied periodic forces, due, for instance, to the pulsations of the driving torque of the engine or to the pulsations of the resisting torque of the load, may have the same period as one of the free torsional oscillations of the rotating shaft. It is of importance, therefore, to be able to calculate the frequency of these free torsional oscillations.

If we have a thin rod of circular cross section clamped at one end and if the length of the free part of the rod be l centimetres, the frequencies of the free torsional oscillations are given by $\{(2m + 1)/4l\}\sqrt{\mu/\rho}$, where m is zero or a positive integer, μ the rigidity, and ρ the density of the metal forming the shaft. For steel $\sqrt{\mu/\rho}$ is about 330,000, and for wrought iron it is not much smaller. It will be seen, therefore, that the frequency of these oscillations is very high. In practice, however, when calculating the free torsional oscillations, we must consider the shaft, rotor, crank arms and flywheel as forming a simple body, and this makes the exact calculation of the periods of these oscillations very difficult. There are apparently, in this case, only a limited number of possible periods, and the frequency need not be high. The curve showing the driving torque of the engine is generally very different from a sine curve, and so the periodic torque may be supposed to be the resultant of a series of periodic torques some of which have appreciable amplitudes. The frequencies of these harmonic torques are multiples of n where n is the number of the revolutions of the crank per second. If the field of the alternator be excited as the rotor is driven up to the normal speed, then, if the machine be an inductor machine or if the number of slots in the armature be few, an appreciable pulsating torque due to the eddy current and hysteresis losses caused by the variations of the reluctance in the path of the field flux will be produced. It will be seen, therefore, that there are many forces of different frequencies applied to the shaft, and it is highly probable that, when the rotor is being run up to speed, a component of the applied forces of appreciable amplitude will pass through synchronism with a free torsional vibration, and so there will be a risk of the shaft being fractured. If the amplitude of the applied resonating forces be sufficiently great to overcome the damping due to the friction of the bearings, etc., the risk will be serious. In this connection, we must remember that alternating stresses of high frequency produce metallic fatigue in the shaft, and so, for this reason alone, they are more likely to cause it to fracture than alternating stresses of the same amplitude but of a lower frequency.

Torsional vibrations are not the only type of vibrations which can be set up in a shaft with a straight axis fixed in direction. When torsional vibrations are started in a shaft at rest, we have one or more sections of the shaft absolutely at rest, whilst the other sections are in motion. When the shaft is rotating, one or more sections of the shaft are moving with uniform angular velocity whilst the other sections move relatively to them. In a second type of vibrations (ortho-radial) the angular velocities of all points equidistant from the axis are the same. Any line in the shaft parallel to the axis always remains a straight line, but its angular velocity varies in a periodic manner. If the axis of a circular cylinder were fixed we could start a vibration of this type by applying equal tangential forces to every point on the circumference of the cylinder, and then removing them simultaneously. In some cases vibrations of this type are more likely to be set up than torsional vibrations. They have been studied by Chree, who finds that, in the case of a solid circular cylinder, the frequencies are given by the equation $J_2\{2\pi fa(\rho/\mu)^{\frac{1}{2}}\} = 0$, where f is the frequency, a the radius of the cross section and J_2 denotes the Bessel's function of the second order. The three smallest values of $2\pi fa(\rho/\mu)^{\frac{1}{2}}$ which satisfy this equation are approximately equal to 5·14, 8·42, and 11·6 respectively. Comparing the lowest frequency f_2 of this type of vibration with the lowest frequency f_1 of the torsional vibrations, we see that $f_2/f_1 = 3{\cdot}27l/a$. In general $3{\cdot}27l$ is greater than a, and thus the frequency of the second type of vibration is usually greater than that of the first type. In the case, however, of a flywheel considered apart from the shaft the vibrations of the second type would be less rapid than the torsional vibrations.

The whirling of shafts.

Another possible explanation of the fracture of shafts is that it is due to 'whirling.' When the length of the shaft is considerable, this explanation seems the more probable. The phenomenon of whirling has been investigated theoretically by Greenhill and Chree, and both theoretically and experimentally by Dunkerley. The following simple explanation of the cause of whirling was first given by Chree. Let us consider the case of a thin rod of circular section, firmly clamped at one end to a shaft which is capable of rotation about its axis. If we pull the free end slightly to one side and let it go, the rod will execute a number n of complete vibrations per second; the time taken by the extremity of the axis of the rod to pass from one position of maximum amplitude to the next being $1/(2n)$. If we now make the rod rotate about its axis as well as vibrate, it will be found that the time taken by the extremity of the axis of the rod to pass from one position of maximum amplitude to the next is greater than $1/(2n)$. We shall call this time half the period of the transverse vibration of the rod when rotating. If the velocity of rotation of the rod be increased, the period of the transverse vibration gets slower and slower until, finally, when it makes n revolutions per second, whirling ensues. The transverse vibrations get slower, as the angular velocity increases, owing to the centrifugal forces acting in the opposite direction to the elastic stresses tending to restore the rod to its initial position, and thus the resultant stress is diminished.

Experiments made by the author show that, when the critical angular velocity is reached, the free end of the rod describes rapidly widening loops round the axis of rotation of the revolving clamp, and the rod either fractures near the clamped end or bends round until it rotates with its free end practically perpendicular to its initial direction. When a rod whirls, it acts apparently in much the same way as a piece of fairly stiff rope would act when rotated under similar conditions. The rope, however, whirls at a much lower speed.

From the equations for the vibration of thin rotating rods given by the theory of elasticity, it follows, at once, that

$$(2\pi f)^2 + \omega^2 = (2\pi F)^2,$$

where ω is the actual angular velocity, F the number of vibrations per second when there is no rotation, and f the number of vibrations per second when the angular velocity is ω. The condition for instability is that f is zero, and hence the corresponding value Ω of the angular velocity is given by

$$\Omega^2 = (2\pi F)^2.$$

Instability arises when the frequency of the transversal vibrations is *nil*, as there is then no righting force.

The same reasoning applies when the rod, which we suppose to be unloaded, is supported by two bearings. If the frequency of the transverse vibrations when the rod is not rotating be F, the rod will whirl when the angular velocity is $2\pi F$. Chree has shown that when a loaded shaft is rotating, the frequency equation, in many of the cases considered by Dunkerley and himself, is of the form

$$(2\pi f)^2 + a\omega^2 = (2\pi F)^2 \quad \text{.....................}(a),$$

where a is approximately constant, and F is the frequency of the transverse vibrations of the loaded shaft in the absence of rotation. The whirling velocity Ω is now given by the equation

$$\Omega^2 = (2\pi F)^2/a.$$

If f_1 and ω_1 be simultaneous values of f and ω, we have

$$(2\pi)^2 (F^2 - f_1^2) = a\omega_1^2,$$

and thus we find that

$$\Omega^2 = \omega_1^2 F^2/(F^2 - f_1^2).$$

Hence, by determining F, f_1 and ω_1, we can find Ω. In order to check our result it would be advisable to find Ω from other simultaneous values of ω and f.

It must be remembered that whirling is a phenomenon of instability and not of resonance. It is not a case of synchronism between a free vibration of a system and one of the applied periodic disturbing forces. When whirling begins the centrifugal forces overpower the righting forces and the shaft tends to fly outwards. It is possible, however, that for speeds less than that at which whirling ensues, we may have equality of period between the variations in the thrust and pull of the connecting rod of the reciprocating engine, on the crank pin, and the transverse vibrations of the rotating shaft. Owing to the rotation, the period of these vibrations is diminished, and care must be taken that the period of none of the component disturbing forces, which set up transverse vibrations, coincides with this diminished period. This kind of resonance might produce breaking stresses in the shaft.

Connecting a machine with the bus bars.

When several alternators connected with the bus bars of a central station are running in parallel and it is desired to put a new machine in circuit, the procedure is as follows. The first operation is to run the machine up to the proper speed and excite the field magnets until the electromotive force is equal to, or preferably a little greater than, the voltage between the bus bars. We then connect some form of synchroniser, several of which are described below, between the machine and the bus bars so that we can find when they are in step. When the synchroniser indicates the proper moment we close the main switch and gradually increase the driving power of the engine, by adjusting the governor or otherwise, so as to open wider the throttle or expansion valve until the engine takes its due share of the load on the station. Altering the excitation of the field increases or diminishes the current, and hence we adjust the excitation until the power factor of the load on the machine is the same as that of the loads on the other machines. The excitation is adjusted by means of a rheostat in the circuit of the field magnet windings of the exciter. Altering the excitation makes very little difference in the load taken by the machine.

Methods of synchronising.

In order to tell when the electromotive force of the incoming machine is exactly in step with the potential difference between the bus bars, various devices are employed. One of them consists of an iron core transformer with three windings. One of these windings is connected across the terminals of the machine and another is connected across the bus bars. When the two applied potential differences are in phase with respect to the load, the magnetomotive forces acting on the core of the transformer balance one another. At this instant, there is no electromotive force in the third coil, and a lamp connected across its terminals is dark. When the potential difference at the terminals of the machine is in opposition with the potential difference between the mains, the magnetomotive forces acting on the core of the transformer are in phase, and hence the alternating magnetic flux generated produces an electromotive force in the third coil, and the lamp glows. The proper moment for switching on is when the lamp is dark. When the speed of the incoming machine is near its proper value, the pulsations of the light given out by the lamp can easily be noticed. When the period of the pulsation is five or six seconds, the switch is closed in the middle of a period of darkness. It is advisable not to have the lamp bright, when the voltage is a maximum, as otherwise the eyes get dazzled. It will be found that a dull red is generally quite sufficient.

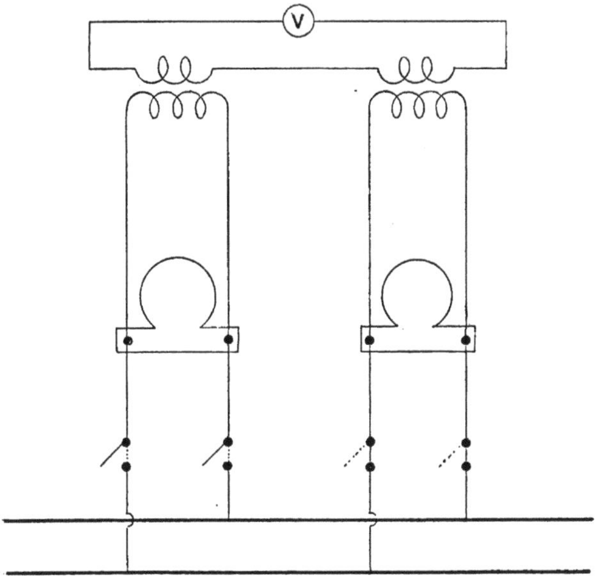

Fig. 111. Method of synchronising with two transformers.

In Fig. 111 the connections are given for the two transformer method of synchronising. In this method the two transformers have their secondaries connected in series through a voltmeter. They may be connected so that the voltmeter has either its maximum or its minimum reading at the proper moment for closing the switch. In practice, however, it is better to arrange so that the voltmeter has its maximum reading when the voltages of the two machines are in phase, as this instant is more definitely indicated by the instrument.

Two transformer method.

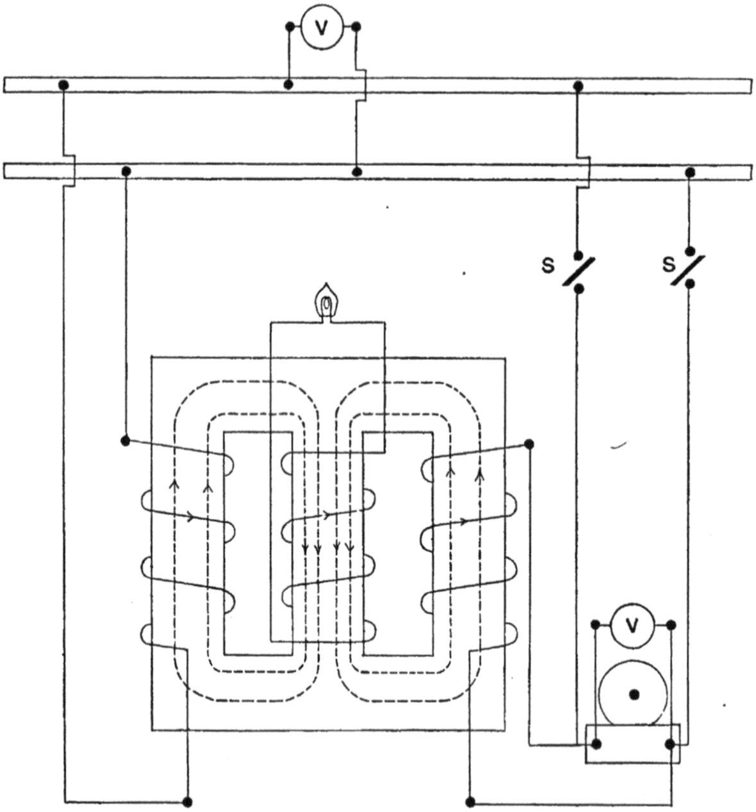

Fig. 112. Method of synchronising by means of a phase indicating transformer. The switch *S* of the incoming machine is closed when the lamp has its maximum brightness.

Phase indicating transformer. A special transformer is sometimes used to indicate the moment when the voltage of the incoming machine is in phase with the voltage across the bus bars. Its action will be understood from Fig. 112. When the voltage of the machine is not in phase with the voltage between the bus bars, the magnetising currents in the coils round the outer cores of the transformer will flow for a fraction of a period in opposite directions. The resultant flux in the middle core is a maximum when the magnetising forces due to the currents in the outer coils are in phase with one another. The electromotive force induced in the coil round the middle core will therefore be a maximum, and the lamp in series with it will be brightest when the potential differences between the bus bars and the terminals of the machine are in phase with one another. The switch *S* is closed at this instant. A voltmeter may be used instead of a lamp.

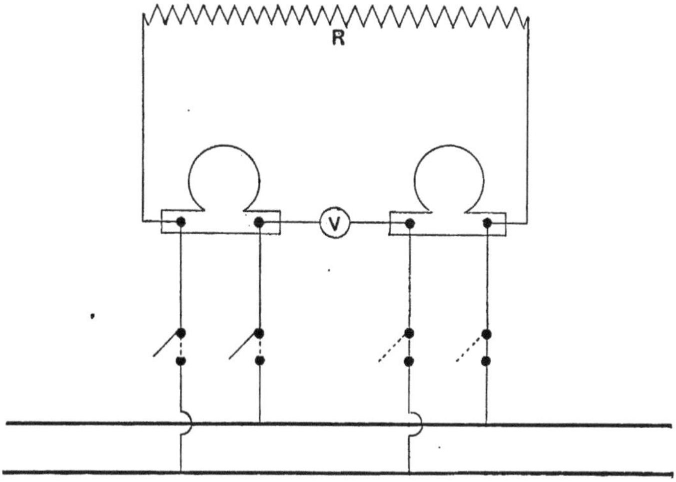

Fig. 113. Method of synchronising by means of a high potential voltmeter.

High potential voltmeter method. The connections for synchronising by means of a high potential voltmeter are shown in Fig. 113. The voltmeter is connected to a terminal of each machine and the circuit is completed by means of a high resistance *R*. If the voltmeter be an electrostatic one, the resistance *R* may be very large. A thick pencil line drawn on a piece of ground glass will answer the purpose of completing the circuit. When the voltmeter has its maximum reading the switch is closed.

Rotating field device.

In commercial work, devices depending on rotating magnetic fields are very often used. The diagram (Fig. 114) illustrates the connections of a rotating field synchroniser for single phase machines. The armature represented by the circle is free to move about an axis through O and has a pointer OP rigidly attached to it. The coil AA' is excited directly by the alternator, which has to be connected with the bus bars, and the coil BB', the axis of which is at right angles to that of AA', is in series with a choking coil, the combination being

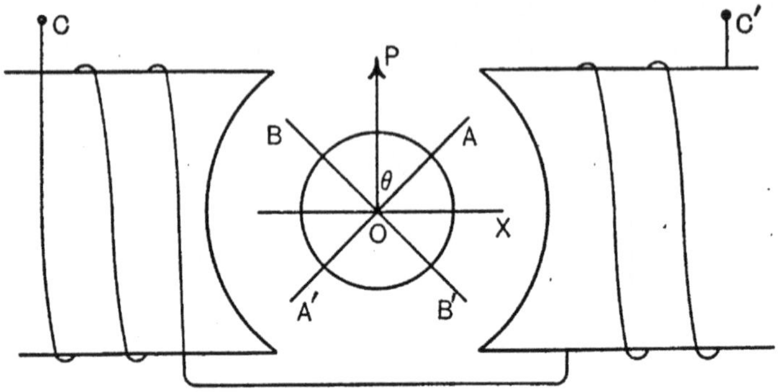

Fig. 114. Rotating field device.

excited by the machine. In the simple theory, therefore, we may assume that the currents in these coils are in quadrature. The terminals C and C' of the field magnet windings are connected directly with the bus bars.

Let $\omega/2\pi$ be the frequency of the potential difference between the bus bars and let $\omega'/2\pi$ be the frequency of the alternator. Then if we suppose that the currents in AA' and BB' are $I \sin \omega' t$ and $I \cos \omega' t$ respectively, so that they are equal in magnitude, and if we suppose that the current in the field magnet winding CC' is $I \sin (\omega t - \epsilon)$, so that ϵ is the phase difference between the potential difference of the machine and the bus bars, the torque g in the position shown in the diagram is given by

$$g = k \sin \omega' t \sin (\omega t - \epsilon) \sin \theta + k \cos \omega' t \sin (\omega t - \epsilon) \cos \theta,$$

where k is a constant depending on the strength of the field, the number of turns, the magnitude of the coils, etc. This follows because the torque on AA' is proportional to the current it carries and the strength of the field, and we are assuming that the latter factor is proportional to the current in the field magnet windings. Hence, we have

$$g = k \sin(\omega t - \epsilon) \cos(\omega' t - \theta)$$
$$= (k/2) [\sin\{(\omega + \omega') t - \epsilon - \theta\} + \sin\{(\omega - \omega') t + \theta - \epsilon\}].$$

We see that the torque consists of two alternating components the amplitudes of which are equal and the frequencies of which are $(\omega + \omega')/2\pi$ and $(\omega - \omega')/2\pi$ respectively. When ω' is nearly equal to ω the effect produced by the rapidly alternating torque due to the first component can be neglected, and hence we may write, without appreciable error, that

$$g = (k/2) \sin\{(\omega - \omega') t + \theta - \epsilon\} \quad \ldots\ldots\ldots\ldots (a).$$

When $\omega' = \omega$ we have

$$g = (k/2) \sin(\theta - \epsilon),$$

and hence, since a very small torque will move the rotor, the position of equilibrium is given by $\theta = \epsilon$ approximately. If therefore the pointer be fixed in the direction OA, the instant for synchronising will occur when it is vertical. If when the pointer is vertical the incoming machine is slow, ω' will be less than ω and we see from (a) that g will be positive; but if it be too fast $\omega - \omega'$ will be negative and the torque will therefore be negative. Hence when too 'fast' it rotates in one direction and when too 'slow' in the other, and these words can therefore be engraved on the scale so as to remind the operator.

Optical methods. If a number of white spots be painted round the rotor of an alternator and be illuminated by light from an arc lamp supplied with alternating current of frequency f, then, in certain cases, the white spots appear to be stationary. Suppose, for example, that there are m white spots painted at equal angular distances apart round the circumference of the rotor. Since the light from the arc is pulsating with a frequency $2f$, it follows that if a spot make the mth part of a complete revolution in the time $1/2f$, the spots will appear to be stationary as they will have their maximum illuminations always in the same m places and their minimum illuminations at points midway between these places. Now, if there are $2p$ poles on the rotor and it makes n revolutions per second, each pole will make the $2p$th part of a complete revolution in the time $1/2pn$. Thus, if m equals $2p$, the frequency of the alternating current supplied by the machine when the spots appear stationary will equal f provided that n equals f/p.

If the lamp be supplied with alternating current taken from the bus bars, then, as the alternator speeds up, the spots present the appearance of a ring of a uniform grey colour, owing to the persistence of luminous impressions on the retina. At a certain speed they appear to be rotating rapidly, but this apparent velocity diminishes as the speed is increased, and finally when synchronism is attained they appear to be stationary. For higher speeds they appear to rotate in the opposite direction. If the alternator be a flywheel alternator with a ring of field poles round its circumference, the spokes of the alternator sometimes answer the purpose of the white spots and appear to be stationary when the alternator is running in synchronism with the others.

It has to be noticed, however, that this method only tells us when the speed is right. It gives no indication of the phase. When the windings of the armature of the alternators are embedded in slots, then, if the incoming machine have the proper speed, the switch may be safely closed, since the high inductance of the armature prevents any excessive rush of current and the machine is pulled into step by the magnetic attractions and repulsions of the armature and field poles. In machines with small armature inductance this cannot be done, and so transformers, with pilot lamps or voltmeters, must be used in addition to the optical device.

Synchronising device for three phase plant.

In Fig. 115 the connections are shown for the Siemens and Halske synchronising device for three phase plant. 1, 2, and 3 are three lamps, which can be connected across the terminals of the incoming machine and the three bus bars by means of the switches A and

B. The contact studs marked a_1, a_2, and a_3 in each switch are connected with the terminals a_1, a_2, and a_3 of the machines. Suppose now that we turn the handle of the switch *A* until the studs a_1, b_1, and c_1 make connection with the segmental contact pieces by means of the radial conductors. Let us also turn *B* round in the same manner until the studs a_2, b_2, and c_2 make contact with the segmental pieces. Now, following out the connections in

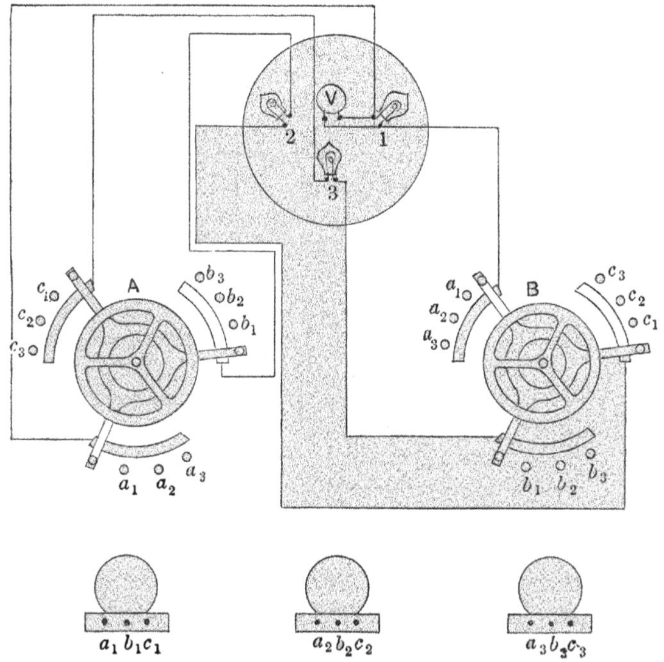

Fig. 115. Synchronising device for three phase alternators.

Fig. 115, we see that a_1 and a_2 are connected through the lamp 1; b_1 and c_2 are connected through the lamp 2; and c_1 and b_2 are connected through the lamp 3. Notice the want of symmetry of these connections. If the machines are in phase with one another, 1 will be out and 2 and 3 will be bright. If the frequency of the machines be not quite the same, the lamps will be bright in turn, the direction of the apparent rotation of the light depending on whether the incoming machine is faster or slower than the other. Hence we can tell whether the speed of the incoming machine is too high or too low. When the apparent rotation is very slow we close the main switch when the lamp 1 is dark and the lamps 2 and 3 are bright.

The number of studs round the segmental contact pieces of the switches depends on the number of machines in the station. For more accurate adjustment, voltmeters like V (Fig. 115) can be placed across the lamps. In practice, the three phase machines are wound for high voltages, and hence step-down transformers must be used, the lamps being placed in their secondary circuits.

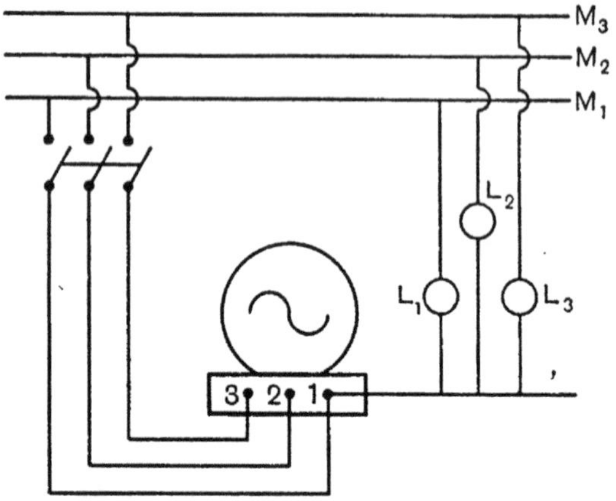

Fig. 116. Synchroscope for three phase plant.

Synchroscope for three phase plant.

When the three phase machines are star wound and have their neutral points connected with the earth either directly or, which is more customary, through a choking coil, a simpler method of synchronising (Fig. 116) can be employed. If the connections are made as shown in this diagram, then it will be seen that when the lamp L_1 is dark and L_2 and L_3 equally bright the potential difference of the machine is in synchronism and in phase with that between the bus bars, and the switch may be closed. When the speed is slightly too low the lamps attain their maximum brightness successively, thus producing an appearance of rotation in one direction, and when the speed is too high the rotation appears to be in the other direction.

REFERENCES.

L. ANDREWS, *Electricity Control.*

A. RUSSELL, 'Notes on the Theory of Synchronous Motors and of Alternators in Parallel,' *The Electrical Review*, Vol. 48, p. 919, 1901.

LEONARD WILSON, 'The Effect of Governors on the Parallel Running of Alternators,' *Journ. of the Inst. of El. Eng.*, Vol. 28, p. 389, 1899.

M. R. GARDNER and R. P. HOWGRAVE-GRAHAM, 'The Synchronising of Alternators,' *Journ. of the Inst. of El. Eng.*, Vol. 28, p. 658, 1899.

'Synchroniser for Three Phase Plant,' *The Electrical Review*, Vol. 44, p. 318, 1899.

BERTRAM HOPKINSON, 'The Parallel Working of Alternators.' Paper read before Section G of the British Association at Southport. *The Electrician*, Vol. 51, p. 886, 1903.

H. H. BARNES, 'Notes on Fly-wheels,' *Trans. of the Am. Inst. of El. Eng.*, Vol. 21, p. 343, 1904.

For torsional vibrations, see Lord RAYLEIGH, *Theory of Sound*, Vol. 1, § 159.

For transverse cylindrical vibrations of the second type, see C. CHREE, 'The Equations of an Isotropic Elastic Solid in Polar and Cylindrical Coordinates, their Solution and Application,' *Trans. Camb. Phil. Soc.*, Vol. 14, p. 355, 1887.

C. CHREE, H. R. SANKEY and W. E. M. MILLINGTON, 'The Strength of Shafts subject to small Forces rhythmically applied,' *Proc. Inst. Civ. Eng.*, Vol. 162, p. 371, Nov. 1905.

For the whirling of shafts, see A. G. GREENHILL, 'On the Strength of Shafting when exposed both to Torsion and to End Thrust,' *Proc. Inst. of Mech. Eng.*, p. 182, 1883.

S. DUNKERLEY, 'On the Whirling and Vibration of Shafts,' *Phil. Trans.*, A., p. 279, 1894.

C. CHREE, 'The Whirling and Transverse Vibrations of Rotating Shafts,' *Phil. Mag.*, [6], Vol. 7, p. 504, May 1904.

CHAPTER IX

The alternating current transformer. Raising or lowering the pressure. Transformer ratio. Magnetising current. Magnetising power. Power factor at no load. Closed and open iron circuit transformers. Core and shell transformers. Constant potential and constant current transformers. Floating coil transformers. Formulae for transformers. Air core transformer. Maximum power factor. Formulae for the air core transformer. The theory of the floating coil transformer. Inductive load on the secondary. Condenser load on the secondary. No magnetic leakage. General solution.

The alternating current transformer.

From the mechanical point of view the construction of the alternating current transformer is very simple. If a bundle of iron wires be bent into the form of a ring (Fig. 117) and two coils, *PP* and *SS*, of insulated copper wire be wound round it, we may use this piece of

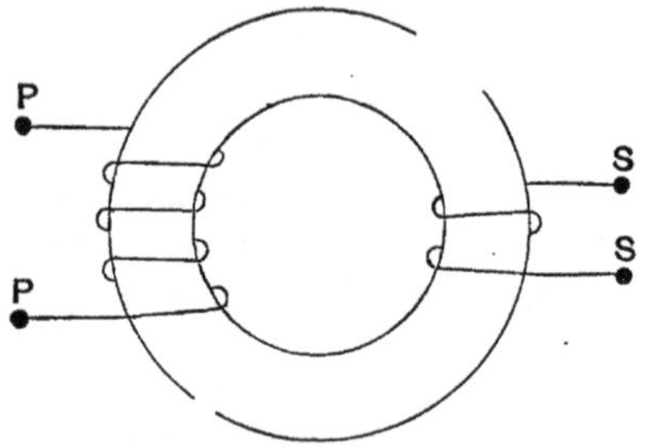

Fig. 117. Alternating current transformer having a closed iron circuit.

apparatus as an alternating current transformer. It will be seen that it has three fundamental parts, two coils of insulated copper wire and an iron core linking them together magnetically. We may apply the alternating potential difference to either of the coils and take power from the other. The coil to which the P.D. is applied is called the primary coil, and the other the secondary coil.

Let there be n_1 turns of wire in the primary coil PP and n_2 turns in the secondary coil SS. Then, if the relative proportions of copper and iron have been properly chosen, we have

$$V_1/V_2 = n_1/n_2 \text{ very approximately,}$$

where V_1 is the effective value of the voltage applied at the terminals of the primary coil, and V_2 is the voltage between the secondary terminals. For instance, if n_1 be 100 and n_2 be 10, and if the applied potential difference be 200 volts, V_2 will be 20 volts. When the transformer has been properly designed it is found that an appreciable amount of current and, therefore, of electric power can be taken from the secondary without the voltage V_2 being lowered by more than one or two per cent. of its initial value and without excessive heating of the primary or secondary coils.

Raising or lowering the pressure.

From the formula given above it is obvious that a transformer can be used for either raising or lowering the pressure of the supply. If we apply 20 volts to the secondary terminals of the transformer described above, we get 200 volts across the primary terminals. When it is used for reducing the pressure it is called a step-down transformer, and when it is used for raising the pressure it is called a step-up transformer. A transformer does for electric pressures what a lever does for mechanical forces. In the one case, the ratio of the electric pressures remains constant; in the other case, the ratio of the mechanical forces remains constant. If the electric pressures are too great the transformer burns out, and if the mechanical forces are too great the lever breaks. If copper had infinite conductivity and iron infinite resistivity, and if there were no hysteresis loss in it, a transformer would be a perfect machine, absorbing power, at one pressure, at the primary terminals, and giving out the same amount, at another pressure, at the secondary terminals. In an analogous manner a lever would be mechanically perfect if it were absolutely rigid and frictionless.

Transformer ratio.

The ratio of the effective value of the applied potential difference to the effective value of the potential difference at the secondary terminals, on open circuit, is called the transformer ratio. We can prove, as follows, that this ratio is approximately equal to the ratio of the number of primary to the number of secondary turns of the transformer when the magnetic leakage is negligible, that is, when practically all the magnetic flux generated in the primary is linked with the secondary, and when, also, the resistance of the primary coil is negligible. If Φ be the total flux in the core at any instant, the value of the potential difference e_2 across the secondary terminals at this instant is given by

$$e_2 = -n_2 \frac{\partial \Phi}{\partial t} \qquad \text{..........................} (\alpha),$$

where n_2 is the number of turns in the secondary winding.

Since the flux Φ embraced by the n_1 turns of the primary winding is continually altering, the electromotive force induced in the primary windings by this varying flux in the core is $-n_1 \partial\Phi/\partial t$. It follows that, if R_1 be the resistance of the primary coil, we have by Ohm's law

$$i_1 = (e_1 - n_1 \partial\Phi/\partial t)/R_1,$$

and therefore $$e_1 = R_1 i_1 + n_1 \partial\Phi/\partial t.$$

In practice, the resistance of the primary circuit is very small. In addition, the reluctance of the magnetic circuit is very small, and hence a small change in the value of i_1 when the secondary is on open circuit produces a large change in the value of the flux. With transformers at ordinary frequencies, therefore, the maximum value of $R_1 i_1$ is very small compared with the maximum value of $n_1 \partial\Phi/\partial t$. We can write, therefore, during practically all the period

$$e_1 = n_1 \partial\Phi/\partial t \qquad \text{..........................} (\beta),$$

and thus, making these assumptions, we get from (α) and (β)

$$n_2 e_1 + n_1 e_2 = 0,$$

and hence $$V_1/V_2 = n_1/n_2.$$

When the secondary terminals are connected through a resistance, a current will flow in the secondary coil, and the equations become more complicated. We shall find and discuss these equations later on. For the present it is sufficient to notice that by Lenz's law the secondary current will flow in the direction which tends to prevent any change taking place in the value of the magnetic flux in the core. The magnetomotive force due to it therefore will oppose the magnetomotive force due to the primary current.

Magnetising current.

The effective value A_0 of the current in the primary coil of a transformer, when the secondary is on open circuit, and a potential difference V_1 of given value and at a given frequency is maintained between the primary terminals, is called the magnetising current of the transformer. Now the primary of a transformer when the secondary is open circuited acts like an inductive coil, and we saw in Vol. I, p. 133, that the current taken by such a coil varies considerably with the shape of the wave of the applied potential difference. We should therefore expect, for this reason alone, that the magnetising current of a transformer would vary with the shape of the wave of the applied potential difference, and this is found to be the case in practice. Potential difference waves which are approximately sine shaped generally produce the maximum magnetising currents. In order to give a definite meaning to the magnetising current of a transformer it is customary to specify that the applied wave of potential difference must be sine shaped.

Magnetising power.

The power W_0, in watts, taken by the primary coil when the secondary is on open circuit, and a potential difference V_1 of specified frequency is maintained between the primary terminals, is called the magnetising power taken by the transformer. The power taken varies with the shape of the applied potential difference wave. Hence, when ordering transformers, it is necessary to specify the shape of the wave of the applied potential difference that is to be used in making the test. It is customary to specify that the wave of the applied P.D. must be approximately sine shaped.

The power factor $\cos\psi_0$ at no load is the power factor of the primary circuit when the secondary is on open circuit. The following relation is always true,

Power factor at no load.

$$\cos\psi_0 = W_0/V_1A_0;$$

but it is found by experiment that, like W_0 and A_0, $\cos\psi_0$ varies with the shape of the applied potential difference wave, although V_1 and the frequency are kept constant.

Closed and open iron circuit transformers.

A transformer which consists merely of an iron core wound with primary and secondary coils, like the one shown in Fig. 117, is called a closed iron circuit transformer. The path of the magnetic flux in this type of transformer is practically confined to the iron, and hence its reluctance is small. It follows that very small changes in the value of the current produce very large back electromotive forces, and therefore the magnetising current in a closed iron circuit transformer is small.

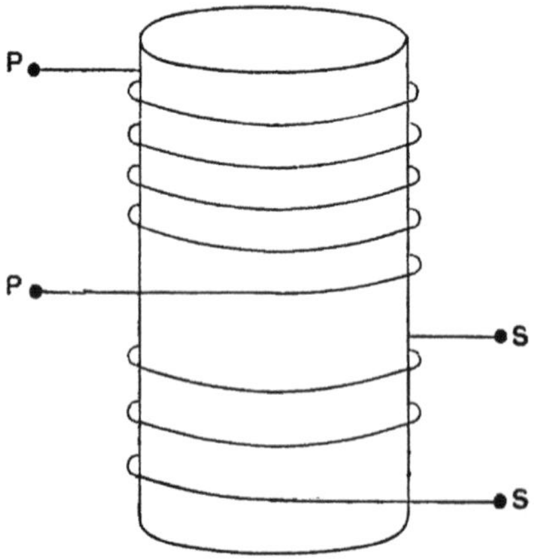

Fig. 118. Open iron circuit transformer.

If the iron core of a closed iron circuit transformer be sawn across and the ends pulled apart, we get an open iron circuit transformer. The reluctance of the path of the magnetic flux is considerably increased (Fig. 118), and so it will take a larger magnetising current, and therefore the losses due to the heating $R_{\mathrm{I}}A_0^2$ of the primary coil will be increased. In practice, however,

$R_1 A_0^2$ is only a small fraction of the no load losses, that is, of the losses when the primary is connected with the live mains and the secondary is on open circuit. By using more copper in the primary coil and less iron in the core it is easy to make the no load losses for an open iron circuit transformer less than for a closed iron circuit transformer, but the magnetising current is much greater, cos ψ_0 being consequently much smaller. The large magnetising current taken by open iron circuit transformers is a serious objection to their use in practice, and hence nearly all modern transformers have a closed iron circuit.

Core and shell transformers.

The transformers we have considered hitherto are core transformers. In a shell transformer the primary and secondary coils are placed one over the other and are encased in a sheath formed of iron plates insulated from one another. In Fig. 119 the cross section of a transformer of this

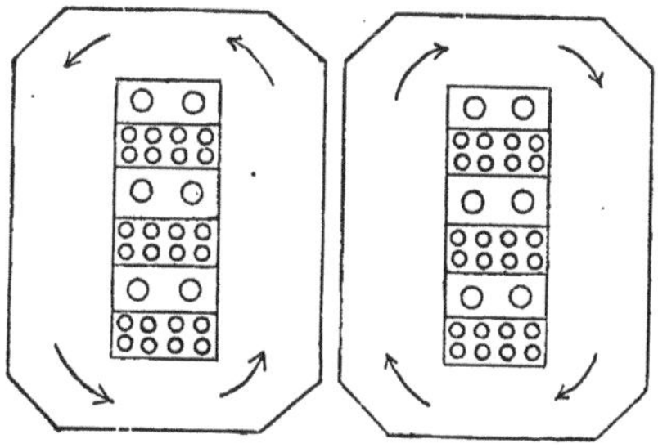

Fig. 119. Section of a shell transformer. The arrow heads indicate the directions of the flux in the iron plates. The circles represent the sections of the primary and secondary wires.

type is shown. The inner core contains the primary and secondary copper coils, which are sandwiched together in such a way that the number of lines of force common to both coils is a maximum. The sheath is built up of centre-hole iron stampings, each of which has a slit from the centre hole to the boundary, so that the iron strip can be bent and easily slipped round the copper coils. The strip is then straightened so that the two edges of the slit touch one another. These stampings form paths of small reluctance for the flux of induction which embraces both coils. They are generally pressed tightly together by the ends of the frame in which they are held and, as they heat considerably during the working of the transformer, air spaces are left for ventilating purposes. Shell transformers are all practically of the closed iron circuit type and have very small magnetising currents.

Constant potential and constant current transformers.

Constant potential transformers are those which are intended to be used with a constant potential difference applied across their primary terminals. If they are to be used on a lighting circuit, it is essential that the potential difference drop on the secondary between no load and full load, that is, the difference between the secondary potential differences at no load and full load, should not be more than about two per cent. If a transformer has been economically constructed, then, when there is the maximum potential difference drop at the secondary terminals, there ought to be the maximum permissible heating of the transformer itself.

Transformers which are constructed so that, whatever the resistance in the secondary circuit may be, the current in the primary will only alter by a fraction of its open circuit value, that is, of its value when the secondary is on open circuit, are called constant current transformers. The leakage of magnetic lines from the iron circuit linking the primary to the secondary coil in this case must be made large. When the secondary coil is short circuited the primary current is always larger than when the secondary is open circuited, but the power expended is approximately the same in the two cases. For a particular value of the resistance of the secondary the power given to it is a maximum.

Floating coil transformers.

It is desirable sometimes, as for example in arc lamp series lighting, to maintain the current in the secondary constant whatever the load on it may be, although the potential difference applied to the primary terminals is always kept constant. This can be managed by suspending the secondary coil over the primary and counterbalancing its weight. The principle on which this transformer is constructed is illustrated in Fig. 120. *PP* and *SS* are sections of the primary and secondary coils of a closed iron circuit transformer. *W* almost counterbalances the weight of *SS* so that on no load it rests lightly on the fixed primary. When the secondary circuit is closed the induced secondary current, by Lenz's law, repels the current in the primary. The force of repulsion separates the two coils, and thus the magnetic leakage between them is increased and the mutual inductance diminished. The induced electromotive force

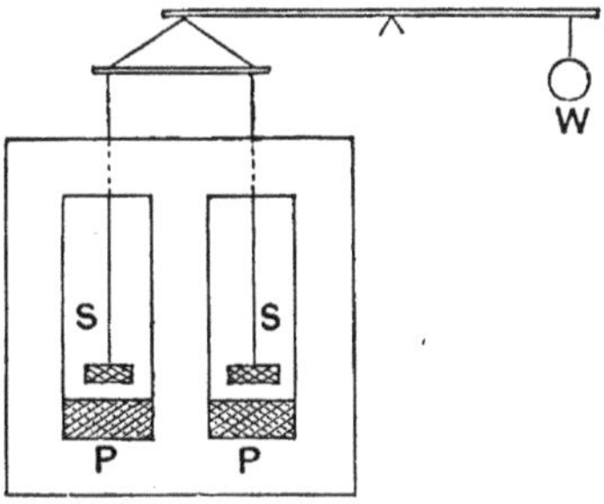

Fig. 120. Floating coil transformer.

and the current in the secondary, therefore, are diminished, and the position of equilibrium is determined by the relative values of the weights of *W* and the coil *SS*. A properly designed transformer of this type will maintain the secondary current very approximately constant at all loads.

Formulae for transformers.

Although the fundamental principle of the ordinary alternating current constant potential transformer is so simple, yet the best way of utilising the iron and copper required for its construction is a problem of considerable complexity. If there is too much magnetic leakage between the coils in any given design, then this will very considerably increase the expense of making the transformer. It is therefore essential to know the effects produced by varying the relative amounts of the copper and iron, and also the effects produced by varying the magnetic leakage on the potential difference drop at the terminals of the secondary. We will first consider the case of the air core transformer, for although we are not always justified in deducing the formulae for the iron core transformer from the formulae for the air core transformer, yet the converse process is always permissible and serves as a valuable check on the accuracy of our results.

We saw in Vol. I, Chap. XIV, that the equations to the air core transformer are

Air core transformer.

$$e_1 = R_1 i_1 + L_1 \frac{\partial i_1}{\partial t} + M \frac{\partial i_2}{\partial t}$$

and

$$0 = R_2 i_2 + M \frac{\partial i_1}{\partial t} + L_2 \frac{\partial i_2}{\partial t},$$

where R_1, L_1 and R_2, L_2 are the resistance and inductance of the primary and secondary coils respectively, and M is the mutual inductance between them. These equations can be written in the form

$$\left.\begin{aligned} e_1 &= R_1 i_1 + L_1 \frac{\partial}{\partial t}\left(i_1 + \frac{M}{L_1} i_2\right) \\ \text{and} \quad - M \frac{\partial}{\partial t}\left(i_1 + \frac{M}{L_1} i_2\right) &= R_2 i_2 + L_2 \sigma \frac{\partial i_2}{\partial t} \end{aligned}\right\} \quad \text{.........(1)},$$

where

$$\sigma = 1 - M^2/L_1 L_2$$

$$= \text{the leakage factor.}$$

When the resistance of the primary coil is negligible, the problem is greatly simplified. In this case the secondary current is determined by the equation

$$-\frac{M}{L_1} e_1 = R_2 i_2 + L_2 \sigma \frac{\partial i_2}{\partial t} \quad \text{..................(2).}$$

Hence the secondary current is equal to that produced in a coil $(R_2, L_2\sigma)$ by a potential difference $-(M/L_1)\, e_1$ applied to its terminals.

Again, since R_1 is zero, we get from (1)

$$\frac{\partial}{\partial t}(L_1 i_1 + M i_2) = e_1.$$

If i_0 be the instantaneous value of the primary current at no load, that is, when i_2 is zero, we have

$$\frac{\partial}{\partial t}(L_1 i_0) = e_1.$$

Hence

$$L_1 i_1 + M i_2 = L_1 i_0 + \text{constant.}$$

Since the mean value of the left-hand side of this equation over a whole period must be zero and the mean value of i_0 is also zero, the constant must be zero, and thus we have

$$L_1 i_1 + M i_2 = L_1 i_0 \quad \text{........................(3).}$$

It follows that the vectors of A_1, $(M/L_1)\,A_2$ and A_0, the effective values of i_1, $(M/L_1)\,i_2$ and i_0, can be represented by lines drawn in a plane.

If we multiply equation (2) by i_2 and integrate over a whole period, we get

$$-(M/L_1)\,V_1 A_2 \cos\theta' = R_2 A_2^2,$$

and therefore $$\cos\theta' = -(L_1/M)\,(R_2 A_2/V_1) \quad \text{..............(4)},$$

where θ' is the phase difference between e_1 and i_2. If θ equals $\pi - \theta'$, θ will be the phase difference between e_1 and $-i_2$, and this is an acute angle.

When the secondary coil is short circuited, that is when R_2 is zero, we have

$$-\frac{M}{L_1}e_1 = L_2\sigma\frac{\partial i_2}{\partial t}$$

$$= \frac{L_2\sigma}{M}\frac{\partial}{\partial t}(L_1 i_0 - L_1 i_1).$$

We have, also, $$e_1 = L_1\frac{\partial i_0}{\partial t},$$

and therefore, in this case,

$$-Mi_0 = (L_1 L_2\sigma/M)\,i_0 - (L_1 L_2\sigma/M)\,i_1$$

and $$i_1 = \{1 + M^2/(L_1 L_2\sigma)\}\,i_0 = i_0/\sigma.$$

It follows that the shape of the wave of the current in the primary when the secondary is short circuited is the same as the shape of the wave of the primary current when the secondary is on open circuit. Also if A_s denote the vector of the primary current when the secondary coil is short circuited, then A_s will be at right angles to V_1 the vector of the applied potential difference, and it will be in phase with A_0. We also have

$$\sigma = A_0/A_s.$$

Again from (3) we have

$$(M/L_1)\,i_2 = i_0 - i_1,$$

and if i_2 be the short circuit current in the secondary,

$$-(M/L_1)\,i_2 = i_1 - i_0 = i_0(1-\sigma)/\sigma = i_1(1-\sigma).$$

Hence the phase difference between the primary and secondary currents when the secondary is short circuited is 180 degrees.

It is to be noticed, however, that it is impossible in practice to make the resistance of the secondary circuit absolutely zero, as the resistance of the secondary coil itself is always appreciable. By properly designing the transformer, exact opposition of phase of the currents on short circuit can very nearly be obtained.

If A_s'' denote the effective value of the short circuit current in the secondary, we have

$$(M/L_1)\,A_s'' = A_s - A_0.$$

In Fig. 121, if OY represents the vector of the applied potential difference V_1, and if OA and OB are the vectors of the open circuit current A_0 and the current A_s in the primary when the secondary

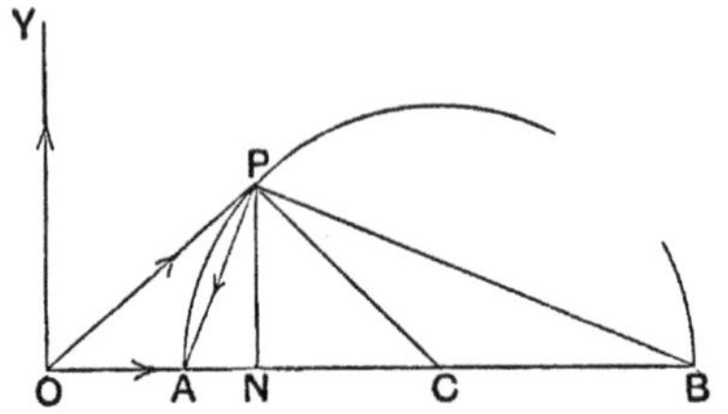

Fig. 121. Diagram of the primary and secondary currents in the ideal air core transformer. OY gives the phase of the applied potential difference, OP is the primary current vector and PA is M/L_1 times the secondary current vector. For sine waves the locus of P is a circle.

is short circuited, OAB will be a straight line at right angles to OY. Also BA will be equal to $(M/L_1)\,A_s''$ and will give the phase of the short circuit current in the secondary.

Again let OP (Fig. 121) represent A_1, then, since OA represents A_0 we see by equation (3) that PA represents $(M/L_1)\,A_2$. If we suppose that the applied potential difference wave is not sine shaped and that its shape is invariable, then from (2) we see that the shape of the current wave i_2 depends on the relative values of R_2 and $L_2\sigma$ and is continually altering as R_2 varies. Since

$$L_1 i_1 = \int e_1 \partial t - M i_2,$$

we see that i_1 also varies in shape as R_2 varies. It follows that a linear relation cannot connect the variables i_0, e_1 and i_1, since i_1 varies with R_2 but e_1 and i_0 are invariable. Therefore OP cannot lie in the same plane as OY and OA. Hence the point P does not necessarily lie in the plane *YOB*, except in the special case when the applied potential difference wave is sine shaped. Since *P* coincides with the points *A* and *B* when the secondary is open circuited and short circuited respectively, we see that as the current in the secondary increases from zero to its maximum value *P* describes a curve in space starting from the point *A* and finishing at the point *B*.

Let us now suppose that $e_1 = \sqrt{2}V_1 \sin \omega t$, then from (2) we have

$$i_2 = - \{(M/L_1)\sqrt{2}V_1 \sin(\omega t - \theta)\}/(R_2^2 + L_2^2\sigma^2\omega^2)^{\frac{1}{2}},$$

where $\tan\theta = L_2\sigma\omega/R_2$. The angle θ is thus the phase difference between e_1 and $-i_2$. The inclination of *PA* to *OY* (Fig. 121) is therefore θ, and the angle *PAB* is $\pi/2 - \theta$. We also have

$$A_2 = (M/L_1)V_1/(R_2^2 + L_2^2\sigma^2\omega^2)^{\frac{1}{2}} = (M/L_1)\, V_1 \sin\theta/(L_2\sigma\omega).$$

If we draw *PB* at right angles to *AP*, then since

$$AB = AP/\sin\theta = (M/L_1)\, A_2/\sin\theta = (M/L_1)^2\, V_1/(L_2\sigma\omega),$$

we see that *AB* is independent of the value of θ. Thus, since the angle *APB* is a right angle, we see that when the applied P.D. and the frequency are constant, the locus of *P* is a circle described on *AB* as diameter.

Again when A_2 is zero, V_1 is $L_1\omega A_0$. Thus substituting this value for V_1 in the formula given above, we find that

$$AB = M^2A_0/(L_1L_2\sigma) = \{(1-\sigma)/\sigma\}\, A_0.$$

Maximum power factor. It is easy to see from Fig. 121 how the currents in the ideal air core transformer vary as the load on the secondary increases. When the resistance of the secondary is infinite, the magnetising current is *OA*, and the power factor is zero. As the load increases, the primary current *OP* continually increases. The angle ψ which *OP* makes with *OY* is the phase difference between the primary current and the applied potential difference. When ψ has its smallest value, $\cos\psi$ the power factor of the primary circuit has its maximum value. Hence the power factor of the primary is a maximum when *OP* is a tangent to the circle *APB*.

In this case, we know, by geometry, that

$$OP^2 = OA \,.\, OB.$$

Now, we have already shown that

$$OA = A_0, \quad OB = A_s, \quad A_0 = \sigma A_s$$

and $$AB = A_s - A_0 = (1/\sigma - 1)\, A_0.$$

Hence, if A_m be the value of the primary current when the power factor is a maximum and $\cos\psi_m$ denote this maximum value, we have

$$A_m^2 = A_0^2/\sigma = \sigma A_s^2,$$

and therefore $$A_m = A_0/\sqrt{\sigma} = \sqrt{\sigma} A_s.$$

Since OP is a tangent to the circle in this case, the angle OPC is a right angle, and therefore

$$\cos\psi_m = \cos OCP = CP/(OA + AC) = AB/(2 \,.\, OA + AB)$$
$$= \{(1/\sigma - 1)\, A_0\}/\{2A_0 + (1/\sigma - 1)\, A_0\} = (1 - \sigma)/(1 + \sigma).$$

Similarly, $$\sin\psi_m = 2\sqrt{\sigma}/(1 + \sigma)$$

and $$\tan\psi_m = 2\sqrt{\sigma}/(1 - \sigma).$$

We also have $$\tan(\psi_m/2) = \sqrt{\sigma}.$$

The maximum power factor $\cos\psi_m$ of the primary may also be expressed by a series, for

$$\cos\psi_m = 1 - 2\sigma + 2\sigma^2 - 2\sigma^3 + \ldots.$$

In many practical applications σ is small and we can write

$$\cos\psi_m = 1 - 2\sigma.$$

We see from the diagram that after OP attains the value $A_0/\sqrt{\sigma}$ the power factor $\cos\psi$ continually diminishes and is zero when the secondary is short circuited.

Formulae for the air core transformer.

We shall now give a list of the formulae for the ideal air core transformer, that is, the air core transformer the resistance of the primary coil of which is zero. As these formulae are frequently used by practical men as a foundation on which to base rules for designing both transformers and induction motors, the student is recommended to make himself thoroughly familiar with them. Most of the formulae follow at once from the simple diagram shown in Fig. 121.

Let V_1, A_1 and $\cos\psi$ be the applied potential difference, the primary current, and the primary power factor respectively. Let A_0 be the magnetising current, σ the leakage factor $1 - M^2/L_1L_2$, and R_2 the resistance of the secondary coil. Let also V_2 and A_2 be the secondary potential difference and current respectively.

In Fig. 121, OP is A_1, OA is A_0, OB is A_s, AP is $(M/L_1) A_2$ and the angle POY is ψ. If we denote the phase difference between the applied potential difference V_1 and the secondary current A_2 by $\pi - \theta$, then, if all the vectors are in one plane, we see, since PB and AB are perpendicular to AP and OY respectively, that the angle PBA equals θ. As the secondary current, which is proportional to AP, increases from zero to its maximum value, θ increases from 0 to $\pi/2$.

We have already shown that A_0 equals σA_s; it therefore follows that

$$AB = A_s - A_0 = (1/\sigma - 1) A_0 = 2 . CP.$$

If we draw PN at right angles to AB, we have

$$BN = BP \cos\theta = AB \cos^2\theta$$

and

$$AN = AP \sin\theta = AB \sin^2\theta.$$

Also since PN^2 equals $AN . NB$, we have

$$\begin{aligned} PN &= AB \sin\theta \cos\theta \\ &= (1/\sigma - 1) A_0 \sin\theta \cos\theta. \end{aligned}$$

Now

$$\begin{aligned} OP^2 &= OB^2 + PB^2 - 2 . OB . PB \cos\theta \\ &= (A_0/\sigma)^2 + AB^2 \cos^2\theta - 2 (A_0/\sigma) AB \cos^2\theta \\ &= (A_0/\sigma)^2 + (1/\sigma - 1) A_0^2 \cos^2\theta \{(1/\sigma - 1) - 2/\sigma\} \\ &= (A_0/\sigma)^2 \sin^2\theta + A_0^2 \cos^2\theta, \end{aligned}$$

and therefore

$$A_1 = (A_0/\sigma) \{\sin^2\theta + \sigma^2 \cos^2\theta\}^{\frac{1}{2}} \quad \text{...............(5)}.$$

Hence

$$\sin\theta = (\sigma/A_0) \{(A_1^2 - A_0^2)/(1 - \sigma^2)\}^{\frac{1}{2}} \quad \text{.........(6)},$$

and

$$\cos\theta = (1/A_0) \{(A_0^2 - \sigma^2 A_1^2)/(1 - \sigma^2)\}^{\frac{1}{2}} \quad \text{.........(7)}.$$

Again, we have $\cos\psi = PN/OP$,

and therefore

$$\cos\psi = \{(1 - \sigma) \sin\theta \cos\theta\}/\{\sin^2\theta + \sigma^2 \cos^2\theta\}^{\frac{1}{2}} \quad \text{......(8)}.$$

If $\cos\psi_m$ denote the maximum value of the primary power factor, we have

$$\cos\psi_m = (1 - \sigma)/(1 + \sigma) \quad \text{..................(9)}.$$

If W_1 be the power given to the primary, we have

$$W_1 = V_1 A_1 \cos\psi \quad \text{.........(10)}$$

$$= V_1 A_0 \{(1-\sigma)/\sigma\} \sin\theta \cos\theta \quad \text{.........(11).}$$

If θ vary, then W_1 has its maximum value when θ is 45 degrees, and we then have

$$W_1 = V_1 A_0 (1-\sigma)/2\sigma \quad \text{.........(12).}$$

Since the power expended in the secondary when the primary resistance is zero equals the power given to the primary, we get

$$W_1 = R_2 A_2^2 \quad \text{.........(13),}$$

and from (2) $\quad W_1 = (M/L_1) V_1 A_2 \cos\theta \quad$ ....(14).

From (11), (14) and (6) we also get

$$A_2 = (L_1/M) \{(1-\sigma)/(1+\sigma)\}^{\frac{1}{2}} (A_1^2 - A_0^2)^{\frac{1}{2}} \quad \text{...(15).}$$

Hence the difference of the squares of A_1 and A_0 is always directly proportional to A_2^2. This result could also be proved directly from the geometry of the figure.

If r_2 and x be the resistances of the secondary coil and of the external non-inductive load respectively, $x + r_2$ will equal R_2 the resistance of the secondary circuit, and V_2 equals xA_2. Hence, from (13) and (14), we have

$$V_2 = \{x/(x+r_2)\} R_2 A_2 = \{x/(x+r_2)\} (M/L_1) V_1 \cos\theta \text{ ...(16).}$$

Let V_2' denote the secondary voltage on open circuit, then V_2' equals $(M/L_1) V_1$ and thus

$$V_2 = V_2' \cos\theta - r_2 A_2 \quad \text{.........(17).}$$

We may also write this equation in the form

$$V_2 = V_2' - r_2 A_2 - 2V_2' \sin^2(\theta/2) \text{ ....(18).}$$

If the phase difference between the currents i_1 and i_2 be $\pi - a$, we see from Fig. 124 that a is the angle OPA, and therefore

$$a = \psi - \theta \text{ ....(19)}$$

and $\quad \cos a = (OP^2 + PA^2 - AO^2)/(2 \, . \, OP \, . \, PA)$

$$= \{L_1^2 (A_1^2 - A_0^2) + M^2 A_2^2\}/(2ML_1 A_1 A_2) \text{ ...(20).}$$

Again, we have

$$L_1 A_1 \sin\psi - L_1 A_0 = L_1 (ON - OA)$$

$$= L_1 \, . \, AN$$

$$= L_1 (M/L_1) A_2 \sin\theta$$

$$= MA_2 \sin\theta \quad \text{.........(21).}$$

Similarly $$L_1 A_1 \cos \psi = M A_2 \cos \theta \qquad \text{.................(22).}$$

Formulae (21) and (22) may also be deduced from the equation

$$L_1 i_1 + M i_2 = L_1 i_0.$$

We also have

$$L_1 A_1 \cos \alpha = M A_2 + L_1 A_0 \sin \theta \qquad \text{...........(23).}$$

Now from (11) and (14) we get

$$A_0 \sin \theta = \{\sigma/(1-\sigma)\} (M/L_1) A_2 \qquad \text{...........(24),}$$

and hence $$A_1 \cos \alpha = \{1/(1-\sigma)\} (M/L_1) A_2 \qquad \text{...........(25).}$$

If η be the efficiency of this ideal transformer,

$$\eta = \frac{\text{useful power}}{\text{total power}} = \frac{V_2 A_2}{V_1 A_1 \cos \psi} = \frac{x}{r_2 + x}.$$

The greater, therefore, the value of x, that is, the smaller the load, the higher will be the efficiency.

The theory of the floating coil transformer.

In the floating coil transformer (see Fig. 120) the secondary coil has its plane parallel to the plane of the primary and the axes of the coils are coincident. We shall now investigate the law according to which the magnetic leakage must vary with the relative positions of the coils, so that the mean value of the repulsive force between them and the effective value of the current in the secondary may be constant at all distances. Let i_1 and i_2 be the instantaneous values of the currents in the coils and let M be their mutual inductance. The instantaneous value f of the repulsion between them is given by the equation

$$f = i_1 i_2 \, \partial M / \partial x,$$

where x is measured along the axis of the coils. Hence, if the mean value of this force be F, we have

$$F = - A_1 A_2 \cos \alpha \, \partial M / \partial x,$$

and therefore by (25)

$$F = - \frac{1}{1-\sigma} \frac{M}{L_1} A_2^2 \frac{\partial M}{\partial x}.$$

Now since σ equals $1 - M^2/L_1 L_2$, we have

$$\partial \sigma / \partial x = - (2M/L_1 L_2) \, \partial M / \partial x,$$

and thus $$F = L_2 A_2^2 (\partial \sigma / \partial x)/(1-\sigma).$$

But by hypothesis F and A_2 are to be constant at all distances, we have therefore

$$(\partial\sigma/\partial x)/(1-\sigma) = F/L_2A_2^2 = \text{a constant} = k,$$

and thus
$$\sigma = 1 - (1-\sigma_0)\,\epsilon^{-kx},$$

where σ_0 is the value of σ when x is zero. Hence the leakage factor σ must increase with x according to this logarithmic law if the effective value of the secondary current is to remain absolutely constant.

Inductive load on the secondary.

If we put an inductive load (x, N) across the secondary terminals, the formulae become

$$e_1 = L_1\frac{\partial i_1}{\partial t} + M\frac{\partial i_2}{\partial t},$$

$$0 = (r_2 + x)\,i_2 + M\frac{\partial i_1}{\partial t} + (L_2 + N)\frac{\partial i_2}{\partial t}.$$

Hence we see that the effect is to increase L_2 in the ratio of $L_2 + N$ to L_2, and also to increase the leakage factor to σ', where

$$\sigma' = 1 - M^2/\{L_1(L_2+N)\} = 1 - L_2(1-\sigma)/(L_2+N)$$
$$= (N + L_2\sigma)/(N + L_2).$$

The short circuit current in the primary is now A_0/σ' and is less than when N is zero. The diameter of the circle in Fig. 123 is therefore diminished. Also, since by (6) and (15)

$$\sin\theta = \sigma M A_2/\{(1-\sigma)\,L_1A_0\},$$

we see that, for a given value of A_2, θ is increased by increasing σ. Now, by (18), we have

$$V_2 = V_2' - r_2A_2 - 2V_2'\sin^2(\theta/2).$$

The voltage drop, $V_2' - V_2$, for a given current is therefore greater the more inductive the load.

Condenser load on the secondary.

When the curve of potential difference is sine shaped, we may replace a condenser of capacity K by an inductive coil $\{0, -1/(K\omega^2)\}$. We see, therefore, from the preceding section, that the effect of a condenser in the secondary circuit is to diminish the resultant self-inductance of the secondary circuit. It can even make it negative. The leakage factor σ' is given by

$$\sigma' = 1 - M^2/\{L_1 (L_2 - 1/K\omega^2)\} = 1 - L_2 (1 - \sigma)/(L_2 - 1/K\omega^2)$$
$$= \sigma (K - 1/L_2\sigma\omega^2)/(K - 1/L_2\omega^2).$$

As the capacity K is increased from zero to $1/(L_2\omega^2)$, σ' increases from unity to infinity. When K increases from $1/(L_2\omega^2)$ to $1/(L_2\sigma\omega^2)$, σ' increases from negative infinity to zero, and finally, when K is greater than $1/(L_2\sigma\omega^2)$, σ' is positive and equals σ when K is infinite. It is easy to see that an infinite condenser would act exactly like a non-inductive coil of zero resistance.

Let us first suppose that the value of K lies between zero and $1/(L_2\omega^2)$ so that σ' is positive and greater than unity. If we now suppose that the condenser is in series with a non-inductive load x,

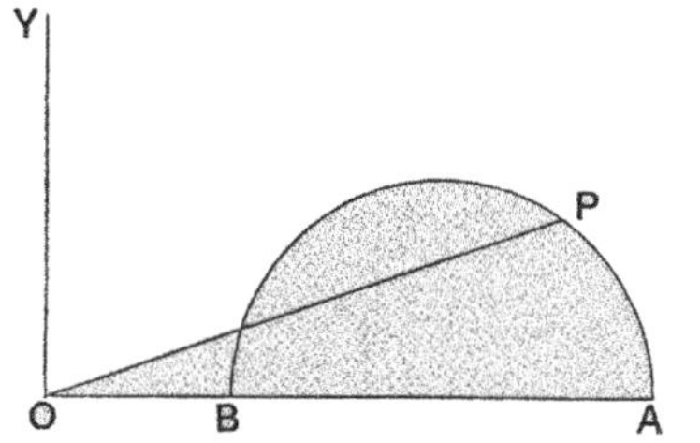

Fig. 122. Condenser K in series with the secondary, K being less than $1/(L_2\omega^2)$.

then, since the short circuit current A_0/σ' is less than A_0, B in Fig. 122 will be to the left of A, and the locus of the extremity

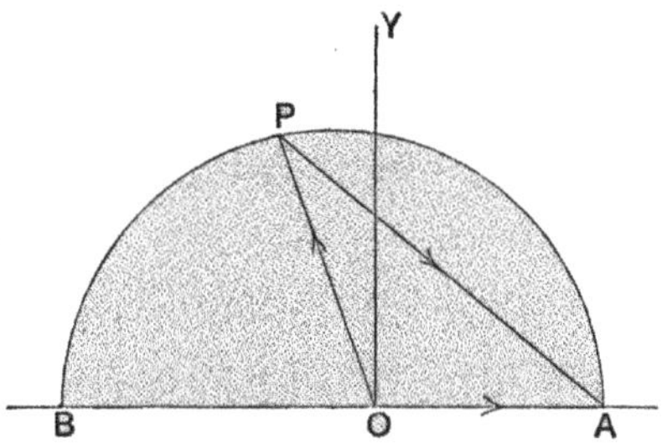

Fig. 123. Air core transformer with a condenser K in series with the secondary, K being greater than $1/(L_2\omega^2)$ but less than $1/(L_2\sigma\omega^2)$.

of the primary current vector OP will be the semicircle described on AB as diameter. If K is made equal to $1/(L_2\omega^2)$, then B

coincides with O, and the current in the primary continually diminishes as the resistance of the secondary is diminished. When the resistance of the secondary is zero, the primary current is zero, although the secondary current is now a maximum. The primary circuit therefore acts like a non-conductor when resonance takes place in the secondary.

When K is greater than $1/(L_2\omega^2)$, but less than $1/(L_2\sigma\omega^2)$, σ' is negative and B is to the left of O (Fig. 123). In the particular case, when σ' is -1 and K is therefore $2/\{L_2(1+\sigma)\omega^2\}$, we see that the primary current is constant in magnitude whatever may be the load on the secondary.

When K equals $1/(L_2\sigma\omega^2)$, σ' is zero, and the transformer acts exactly as if it had no magnetic leakage (Fig. 124). The locus of P

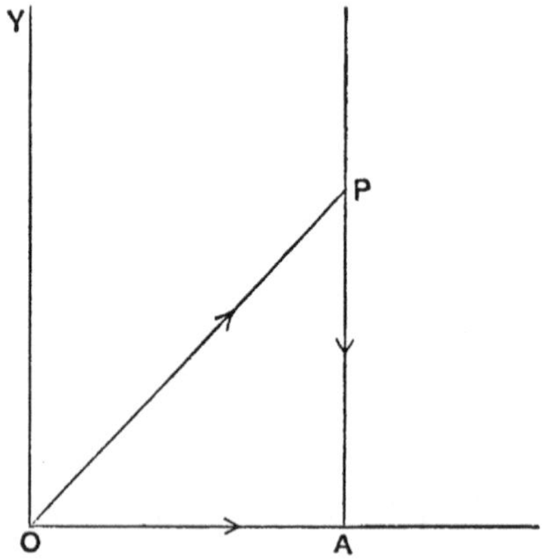

Fig. 124. Air core transformer with no magnetic leakage. $M^2 = L_1L_2$.

in this case is a straight line, as the centre of the circle is at infinity, and we have

$$A_1^2 = A_0^2 + \{(M/L_1)A_2\}^2.$$

The primary and secondary potential differences are also always in exact opposition in phase.

When K is greater than $1/(L_2\sigma\omega^2)$ then σ' is positive and less than unity. In this case (Fig. 125) OB is A_0/σ', and is very large when σ' is small. Finally, when K is infinite OB is A_0/σ, and we get the ordinary transformer diagram.

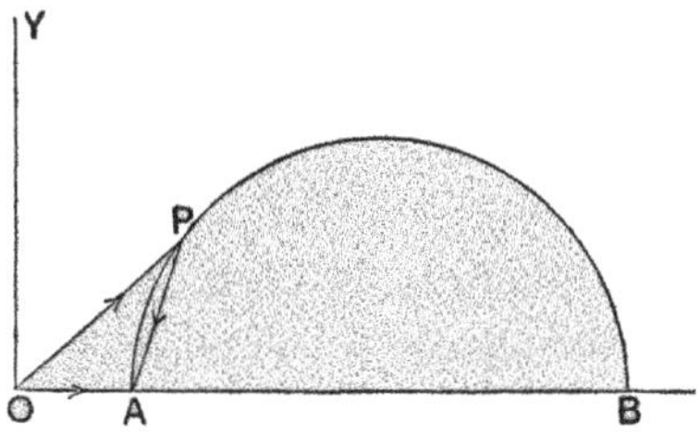

Fig. 125. Air core transformer with a condenser K in series with the secondary when K is greater than $1/(L_2\sigma\omega^2)$.

When there is no magnetic leakage σ is zero, and the problem becomes much simpler. In this case we will take the resistance of the primary into account. Let us suppose that the load is non-inductive. Replace the transformer by its equivalent net-work (see Vol. I, p. 339). The choking coil L_1 (Fig. 126) is shunted by the non-inductive resistance $(L_1{}^2/M^2)\,R_2$, and is in series with the resistance R_1. The current in

No magnetic leakage.

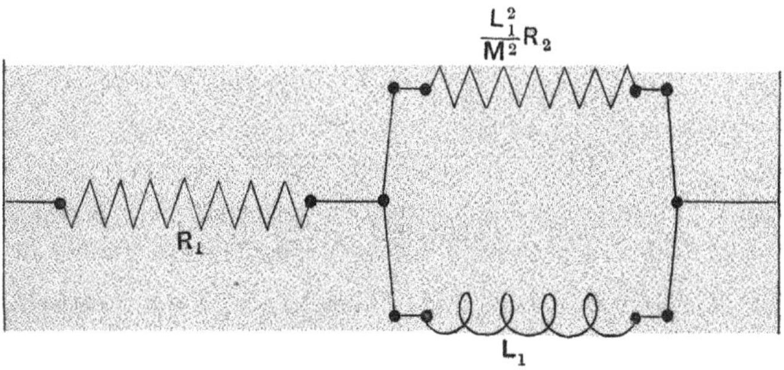

Fig. 126. Equivalent net-work of a transformer with no magnetic leakage.

the secondary is in opposition in phase to the current i' in the non-inductive branch $(L_1{}^2/M^2)\,R_2$, and its magnitude is L_1/M times this current. Our equations are

$$e_1 = R_1 i_1 + (L_1{}^2/M^2)\,R_2 i' \qquad (a),$$

$$(L_1{}^2/M^2)\,R_2 i' = L_1 \frac{\partial i}{\partial t} \qquad (b),$$

and

$$i_1 = i + i' \qquad (c),$$

where i is the current in the choking coil L_1.

Now, we see from (*b*) that whatever the shape of the applied wave of potential difference, the currents i and i' are in quadrature, and thus we have

$$A_1^2 = A^2 + \{(M/L_1)\, A_2\}^2,$$

since $i' = -(M/L_1)\, i_2$, where i_2 is the secondary current.

In Fig. 127 let OY be equal to V_1 and let OP represent $R_1 A_1$. Describe a semicircle on OY as diameter, and let YP produced

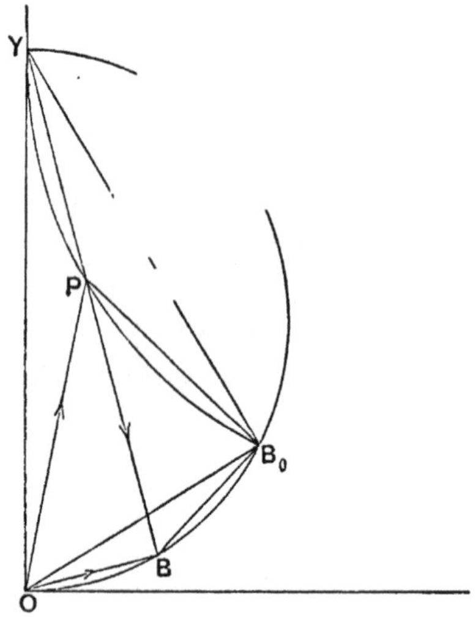

Fig. 127. Transformer diagram when the magnetic leakage is zero. $OP = R_1 A_1$, $BP = (M/L_1)\, R_1 A_2$, $OB = R_1 A$, $OB_0 = R_1 A_0$, and $OY = V_1$.

meet this circle in B. Join OB. Then since, by hypothesis, OP represents $R_1 A_1$, therefore from equation (*a*) we see that YP will represent $(L_1/M)\, R_2 A_2$ in magnitude and phase.

Now equation (*c*) is

$$i_1 = i + i',$$

and since i and i' are in quadrature we get, on multiplying each side of the equation by i', and taking mean values,

$$A_1 A' \cos a = A'^2,$$

but $$A' = (M/L_1)\, A_2,$$

and therefore $$\cos a = (MA_2)/(L_1 A_1),$$

where a is the phase difference between A_1 and A', and $\pi - a$ is therefore the phase difference between A_1 and A_2. In Fig. 127 PB gives the phase of A_2, and the angle OPB equals a.

Hence $\cos OPB = (MA_2)/(L_1A_1) = BP/OP$,

and thus $BP = (M/L_1)\,R_1A_2$.

We have also

$$OB^2 = OP^2 - BP^2 = R_1^2\,\{A_1^2 - (M/L_1)^2\,A_2^2\} = R_1^2A^2,$$

and hence $OB = R_1A$.

We shall now make the assumption that the applied potential difference wave is sine shaped. In this case PY equals ωL_1A, since this line represents the voltage across the choking coil in Fig. 126.

Thus $OB/PY = R_1A/\omega L_1A = R_1/\omega L_1 =$ a constant.

Let B_0 (Fig. 127) be the position of B when the secondary is on open circuit. Then we have

$$OB/PY = OB_0/B_0Y,$$

and therefore $OB/OB_0 = PY/B_0Y$.

Also, since the angle $BOB_0 =$ the angle B_0YP, it follows that the triangles BOB_0 and B_0YP are therefore similar, and the angle B_0PY equals the angle B_0BO and is therefore constant. Hence the locus of P is a circle passing through B_0 and Y.

The secondary electromotive force is R_2A_2, and this equals M/L_1 times PY. The value of the magnetic flux also is proportional to the current in L_1 (Fig. 126), that is, to A. Hence the magnetic flux is proportional to OB. We have seen that the secondary current is proportional to PB. The magnetic flux therefore continually diminishes and the secondary current continually increases as the resistance of the secondary is diminished. We can see from the diagram that the primary current which is proportional to OP diminishes slightly at first (Vol. I, p. 344).

When the secondary is short circuited, P coincides with Y and $V_1 = R_1A_1 = (M/L_1)\,R_1A_2$. Hence the power $R_1A_1^2$ given to the transformer is entirely expended in heating the primary coil.

When the magnetic leakage is not negligible the formulae are more complicated. This case is completely discussed in Vol. I, Chap. XIV.

General solution.

We saw in Vol. I, p. 342, that when a potential difference $E_1 \sin \omega t$ is applied to the primary terminals of a transformer, the primary and secondary currents are given by

$$i_1 = E_1 \sin(\omega t - a_1)/\{(R_1 + m_1^2 R_2)^2 + (L_1 - m_1^2 L_2)^2 \omega^2\}^{\frac{1}{2}}$$
$$= I_1 \sin(\omega t - a_1),$$

and $$i_2 = -\{M I_1 \omega \cos(\omega t - a_1 - a_2)\}/(R_2^2 + L_2^2 \omega^2)^{\frac{1}{2}}.$$

In these equations

$$m_1^2 = M^2\omega^2/(R_2^2 + L_2^2\omega^2),$$
$$\tan a_1 = (L_1 - m_1^2 L_2)\,\omega/(R_1 + m_1^2 R_2),$$

and $$\tan a_2 = L_2\omega/R_2.$$

If the applied wave e_1 be given by the equation

$$e_1 = E_1 \sin(\omega t - \beta_1) + E_3 \sin(3\omega t - \beta_3) + \ldots,$$

then, by writing down the values of i_1 and i_2 for each term separately and adding them up, we get the complete solution. The square of the effective value of the primary current would be equal to

$$V_1^2/\{(R_1 + m_1^2 R_2)^2 + (L_1 - m_1^2 L_2)^2\}$$
$$+ V_3^2/\{(R_1 + m_3^2 R_2)^2 + (L_1 - m_3^2 L_2)^2\} + \ldots,$$

where $m^2_{2n-1} = M^2(2n-1)^2\omega^2/\{R_2^2 + L_2^2(2n-1)^2\omega^2\}$.

The complete analytical solution of the air core transformer can thus be written down by Maxwell's method.

CHAPTER X

The alternating current transformer. Losses on open circuit and under load. Difficulty of the sine curve assumption. Flux and applied potential difference wave. Magnetising current. Shape of the magnetising current wave. Magnetising current obtained on the sine wave assumption. Core losses. Copper losses. Values of the resistances. Constant potential transformer with no magnetic leakage. $B_{\text{max.}}$ is nearly constant at all loads. The hysteresis and eddy current losses in the core. Resultant ampere turns Example. The secondary voltage. Example. Output. Efficiency. Examples. Regulation on inductive load. The short circuit test. Numerical example. Parallel operation. Equivalent net-work. Inductive and condenser loads. Resonance with transformers. References.

The alternating current transformer.

WHEN iron sheets are placed in the path of the flux of an air core transformer, then, for the same power in the secondary circuit, the primary current is considerably reduced. The magnetising current, in particular, is very much smaller. We see, therefore, that unless the induced currents and the hysteresis losses in the iron sheets are excessive, it is more economical to use an iron core transformer, as not only the losses due to the heating of the copper in the primary coil, but also the losses in the mains and in the armatures of the generators due to the primary current are much smaller. The initial cost also of iron core transformers is much less, and so they are practically always employed.

In order to reduce the losses due to eddy currents, the core is generally built up of plates of thin sheet iron insulated from one another. In Vol. I, Chap. xx, we saw that these eddy currents dissipate power directly by heating the iron in which they flow. They cause losses by screening the interior of the iron sheets from the magnetic forces, and thus make the primary current larger than that required to produce the same magnetic flux if it were uniformly distributed throughout the core. We saw also that the irregular distribution of the magnetic flux increases the hysteresis loss. We can, however, make the eddy current losses very small by using very thin sheet iron.

Since there is no known formula that gives the magnetic force H as a function of the magnetic induction B which it produces, the problem of finding the relations between the currents and the voltages in an alternating current transformer does not admit of an exact analytical solution. Approximate solutions, however, can be obtained which are of value in practical work. We shall first consider from a general point of view the various losses that take place in the copper and iron used in the construction of the transformer.

Losses in a transformer when the secondary is on open circuit.

The principle of the action of the iron core transformer is the same as that of the air core transformer. When the secondary is on open circuit, we have a current in the primary coil magnetising the core and producing a magnetic flux the bulk of which is linked with the secondary coil. The losses in this case are mainly due to the heating $R_1A_0^2$ of the primary coil by the primary current, and to eddy current and hysteresis loss in the core. In addition there may be eddy current losses in the copper of the secondary winding or even in the copper of the primary winding itself. Sometimes also, when the transformer is enclosed in a cast iron case, leakage flux from the primary may cause eddy currents in the case. When the frequency is high the current density over the cross section of the primary winding is not uniform, and this increases the value of R_1 and therefore the $R_1A_0^2$ losses. In practice $R_1A_0^2$, where R_1 is the resistance of the primary coil, gives the minimum possible value of the copper losses.

Losses under load.

As a non-inductive load on the secondary circuit increases, and therefore as the secondary current increases, the primary current increases also. If $4\pi\mathcal{R}/10$ be the reluctance of the path of the magnetic flux ϕ, common to both primary and secondary coils, we have at every instant, by the fundamental magnetic equation,

$$\phi = (n_1 i_1 + n_2 i_2)/\mathcal{R},$$

where i_1 and i_2 are the instantaneous values of the primary and secondary currents, and n_1 and n_2 are the numbers of turns of the primary and secondary coils respectively. Some of the magnetic lines linked with the secondary current do not pass through the primary circuit, and, as in the case of the leakage flux from the primary, these lines may give rise to eddy currents and so increase the losses. We may divide the losses in the loaded transformer into iron and copper losses. The iron losses are due mainly to hysteresis and eddy currents in the core, but the losses in the iron case are sometimes appreciable. The copper losses, $R_1 A_1^2 + R_2 A_2^2$, are caused by the primary and secondary currents heating the coils, and in addition there are losses due to eddy currents in the coils themselves. For frequencies higher than fifty it is advisable to use stranded conductors for the primary and secondary windings if they have to carry large currents, as otherwise the eddy current losses are appreciable.

Difficulty of the sine curve assumption.

In order to simplify the theory, we assume that the applied potential difference wave is sine shaped. Even in this case, however, the current wave will not be sine shaped owing to the fact that the flux in the iron is not proportional to the magnetising force. To simplify the problem, therefore, we must assume not only that the P.D. is sine shaped but that the current and the magnetic flux also obey the harmonic law. We shall show that this virtually amounts to assuming that the shape of the hysteresis loop of the iron in the core of our imaginary transformer is an ellipse.

If the current in the primary winding of the transformer when the secondary is on open circuit is $I \sin \omega t$, we may write

$$h = H_m \sin \omega t,$$

where h denotes the instantaneous value of the magnetising force and H_m is its maximum value. If the magnetic flux also obey the harmonic law we can write

$$b = - B_r \cos \omega t + B \sin \omega t,$$

where b is the instantaneous value of the flux density, B_r the remanence, and B the flux density when the magnetising force is H_m.

If we eliminate the trigonometrical functions from the above equation we get

$$\frac{(b - Bh/H_m)^2}{B_r^2} + \frac{h^2}{H_m^2} = 1.$$

Plotting out the curve represented by this equation we get the ellipse shown in Fig. 128. In this figure a real B, H curve for iron sheets is superposed on the ellipse. The remanence is the same in each case, but the coercive force is a little greater for the

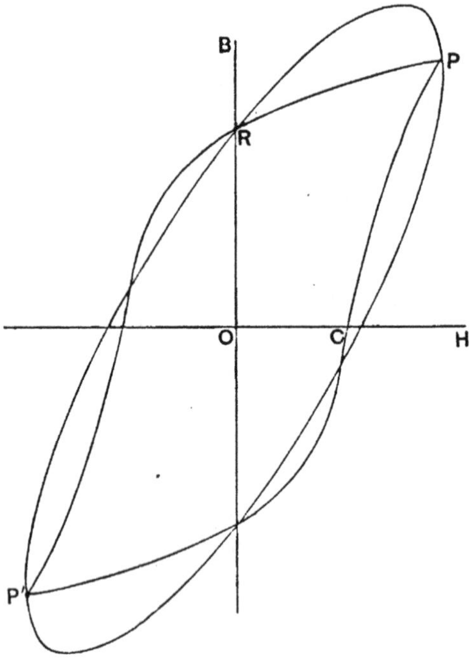

Fig. 128. Real hysteresis loop $PRP'CP$ and hypothetical elliptic hysteresis loop.

ellipse. In the hypothetical iron the induction density goes on increasing for some time after the magnetising force has begun to diminish, whilst in the real iron H and B attain their maximum values at the same instant. This is the main difference between the real and the hypothetical hysteresis loop.

It is proved in treatises on Conic Sections that if we transform an equation of the form

$$ax^2 + 2hxy + by^2 = 1$$

from one set of rectangular axes to another, the quantity $ab - h^2$ remains unaltered. It follows, by referring the ellipse to its principal axes, that its area is equal to $\pi/\sqrt{ab - h^2}$.

Hence the area of the ellipse in Fig. 128 is

$$\pi H_m B_r.$$

Now the work done in taking a cubic centimetre of iron which obeys the elliptic law through a cycle is

$$(1/4\pi) \int h \partial b = \pi H_m B_r / 4\pi = H_m B_r / 4 \text{ ergs.}$$

When b is zero h equals H_c the coercive force, hence

$$H_c = H_m B_r / (B_r^2 + B^2)^{\frac{1}{2}}.$$

If B_m denote the maximum value of the induction density, it is not difficult to show that

$$B_m = (B_r^2 + B^2)^{\frac{1}{2}}.$$

Hence $$H_c B_m = H_m B_r, \text{ or } H_c / H_m = B_r / B_m.$$

We have already in Chapters I and II made the assumptions that the flux and the current follow the harmonic law; we have therefore assumed that the hysteresis loop of the iron is an ellipse. We can see that the area of the ellipse is greater than that of the real hysteresis loop, and hence it may be supposed to take into account some of the eddy current losses.

Flux and applied potential difference wave.

The maximum value of the flux produced in the core of a transformer cannot be predetermined unless we know the shape of the applied potential difference wave. The voltmeter reading on the primary side only gives us the effective value of the voltage. It gives no indication of the wave shape. Let e_1 and i_1 be the instantaneous value of the primary voltage and current respectively. If n_1 be the number of primary turns, we may write

$$e_1 = R_1 i_1 + n_1 \frac{\partial \phi}{\partial t} + n_1 \frac{\partial \phi_a}{\partial t},$$

where ϕ equals the mean value per turn of the instantaneous flux linking the primary with the secondary circuit, and ϕ_a equals the mean value per turn of the instantaneous flux linked with the primary alone. The path of the flux ϕ we may consider to be entirely in the iron, whilst the path of the flux ϕ_a is partly in the iron and partly in the air and copper or entirely in the air and copper. The fluxes ϕ and ϕ_a are therefore not in phase with one another, and the complete problem is very complex. In practice, however, the maximum values of the terms $R_1 i_1$ and $n_1 \partial\phi_a/\partial t$ are quite negligible compared with the maximum value of $n_1 \partial\phi/\partial t$. To a first approximation, therefore, we have

$$n_1 \frac{\partial\phi}{\partial t} \cdot 10^{-8} = e_1,$$

where ϕ is the resultant flux in C.G.S. units and e_1 is in volts. When e_1 is zero, $\partial\phi/\partial t$ vanishes, that is, the rate of increase or decrease of ϕ is zero, and therefore ϕ must have a maximum or a minimum value. Owing to the maximum positive value of the alternating current obtained from an alternator being exactly equal to its maximum negative value, the maximum and minimum values of ϕ are equal numerically but have opposite signs. Let $\Phi_{max.}$ and $-\Phi_{max.}$ be these values respectively, and let e_1 vanish when t is t_1, then we have

$$\int_{t_1}^{t_1+\frac{T}{2}} n_1 \frac{\partial\phi}{\partial t} \partial t \cdot 10^{-8} = \int_{t_1}^{t_1+\frac{T}{2}} e_1 \partial t,$$

and therefore $\int_{-\Phi_{max.}}^{\Phi_{max.}} n_1 \partial\phi \cdot 10^{-8} =$ the area of the applied potential difference wave

$$= A',$$

and thus $$2n_1 \Phi_{max.} = A' \cdot 10^8.$$

If we write $\Phi_{max.} = S \cdot B_{max.}$ where S is the mean cross sectional area of the core, we get

$$B_{max.} = 10^8 \cdot A'/(2n_1 S).$$

In calculating A' in this formula, the ordinates must be measured in volts, and the abscissae in seconds. The maximum induction density is therefore directly proportional to the area of the wave of the applied potential difference.

It is proved in Vol. I, Chap. IV that, when the effective value of the applied voltage is maintained constant, the more peaky the wave the less will be its area. The more peaky, therefore, the wave, the smaller will be the value of $B_{max.}$, and hence, by Steinmetz's law, the less will be the hysteresis loss.

If the frequency f vary, the shapes of the applied waves being always similar and their effective values equal, the areas of the waves will be directly proportional to the period, and thus will be inversely proportional to the frequency. For instance, if we were to increase the frequency to nf, the maximum value of the induction density would diminish to $B_{\max.}/n$.

Let the instantaneous value of the applied potential difference be denoted by $E \sin \omega t$, then

$$A' = \int_0^{T/2} E \sin \omega t \partial t = (E/\omega)\Big[-\cos \omega t\Big]_0^{T/2} = 2E/\omega = \sqrt{2} V_1/\pi f,$$

where A' is the area of the positive half of the wave, and V_1 is the effective value, of the applied potential difference.

Therefore $\sqrt{2} V_1/\pi f = 2n_1 S B_{\max.} 10^{-8}$,

and
$$V_1 = \pi \sqrt{2} n_1 f S B_{\max.} 10^{-8}$$
$$= 4 . 443 n_1 f S B_{\max.} 10^{-8}.$$

Let the form factor (p. 16), that is, the ratio of the effective value V_1 to the mean value v_m of the applied P.D. wave, be k, then we have

$$V_1 = k v_m = k\, 2A'/T = 2kfA' = 4kn_1 f S B_{\max.} 10^{-8}.$$

Values of k are given on p. 18. For very peaky waves k can be very large, and therefore a mere knowledge of the value of V_1 only determines the maximum possible value that $B_{\max.}$ can have, namely, $V_1 10^8/(4n_1 f S)$. It has this value when k is unity, that is, for a rectangular wave.

Magnetising current.

When b has its maximum value so also has h, and we have therefore $B_{\max.} = \mu H_{\max.}$, where μ is the permeability of the iron when the magnetising force is $H_{\max.}$. Now if $I_{\max.}$ be the maximum value of the primary current, $H_{\max.} = 4\pi n_1 I_{\max.}/10l$, where l is the mean length of the path of the flux in the iron. We thus find that

$$I_{\max.} = (10l/4\pi n_1)\, H_{\max.}$$
$$= (10l/4\pi n_1)\{V_1 10^8/(4\mu k n_1 f S)\}$$
$$= l V_1 10^9/(16\pi\mu k n_1^2 f S),$$

where the symbols have the same meaning as in the preceding paragraph.

Let k' be the amplitude factor of the current wave, that is, the ratio of A_0 to $I_{max.}$, then, we have $A_0 = k'I_{max.}$, and thus A_0 can be found when k and k' are known.

Let A_0 and A_0' be the magnetising currents of the high tension and low tension coils of a transformer when used as a step-down and step-up transformer respectively. Since the maximum value of the flux in the core will be the same in the two cases, the magnetising ampere turns will also be the same, provided that the wave shapes are the same, and hence $A_0/A_0' = n_2/n_1$.

The following test of a five kilowatt hundred volt to five volt transformer illustrates how the magnetising current of a transformer varies with the frequency and also shows the practical limitations of the formula given above. In the first test (Fig. 129)

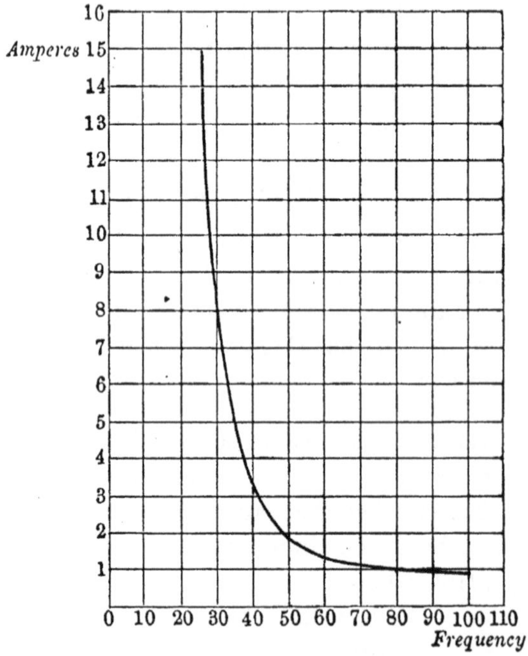

Fig. 129. How the magnetising current of the high tension coil varies with the frequency.

the high tension P.D. is maintained at 100 volts at all frequencies from 100 to 25. The magnetising current varies from 0·85 to 15 amperes. The frequency for which the transformer is constructed is 80, and so its magnetising current is one ampere. For frequencies below 25, a very slight change of the frequency produces a very great change in the current, and for frequencies above 100 the current is practically independent of the frequency.

A potential difference of five volts was now maintained across the low tension terminals, the primary being on open circuit, and the frequency was varied between 110 and 25. The current varied from 18·5 to 180 amperes (Fig. 130). At high frequencies

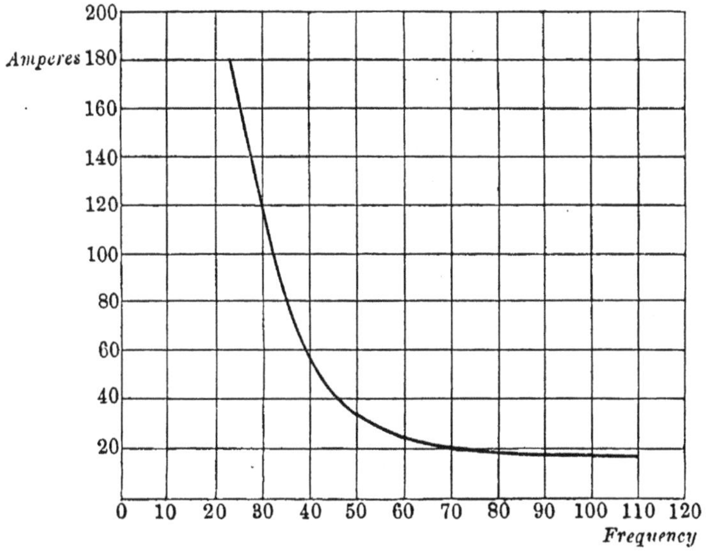

Fig. 130. How the magnetising current of the low tension coil varies with the frequency.

the current tends to attain a constant value and at low frequencies it increases very rapidly. It does not increase as rapidly as the high tension magnetising current at low frequencies, owing to the relatively greater value of its resistance. The curves show that to a first approximation we have $A_0/A_0' = n_2/n_1$.

Shape of the magnetising current wave. If we neglect the primary resistance and suppose that there is no magnetic leakage, we have, when the variables are measured in C.G.S. units, $e_1 = n_1 \partial\phi/\partial t$. By the differential calculus we see that the flux ϕ has a maximum or a minimum value when e_1 is zero. If we suppose that e_1 is zero when t is zero, we get

$$\phi = -\frac{1}{2n_1}\int_0^{T/2} e_1 \partial t + \frac{1}{n_1}\int_0^t e_1 \partial t,$$

the flux having its maximum and minimum values when t is $T/2$ and zero respectively. The current will also have its maximum and minimum values at these instants. The flux ϕ vanishes at the instant when the ordinate e_1 divides the area of the positive half of the wave into two equal portions. If the curve be symmetrical, this will be when t equals $T/4$. In the time 0 to $T/4$, ϕ increases from $-\Phi_{max.}$ to zero and therefore (see Fig. 131)

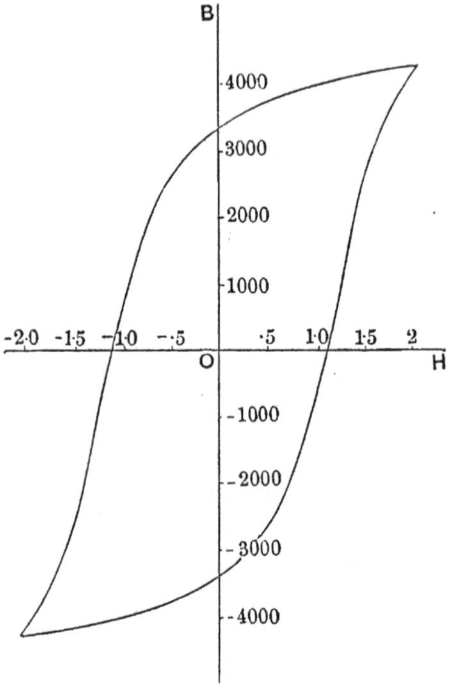

Fig. 131. Hysteresis loop for steel strips.

i increases from $-I_{max.}$ to I_c, where I_c is the current which produces the coercive force. In the time $T/4$ to $T/2$, i increases from I_c to $I_{max.}$. We see, therefore, that, even when the applied wave of potential difference is symmetrical, the current wave is unsymmetrical, varying more rapidly in the first quarter of a period than in the second, and similarly it varies more rapidly in the third quarter of a period than in the fourth.

When we know the shape and the magnitude of the hysteresis loop of the iron forming the core of the transformer and also the shape of the applied P.D. wave, we can construct the current wave as follows. At the times 0, $T/2n$, $2T/2n$, ... $nT/2n$, erect ordinates to the P.D. curve e_1, and calculate the value of ϕ by means of the formula given above, the integrals being evaluated by means of a planimeter. We then find from the hysteresis loop the values of the currents corresponding to these values of ϕ, and, choosing any convenient scale, mark off these values along the corresponding ordinates of the curve e_1. Drawing a curve through these points, we get the wave of magnetising current. It is to be noticed that we have neglected the effects of magnetic leakage and of eddy currents, and we have supposed that the primary resistance is negligible. In many practical cases this is permissible.

The curve in Fig. 132 shows the shape of the magnetising current wave when the applied P.D. wave is sine shaped and when

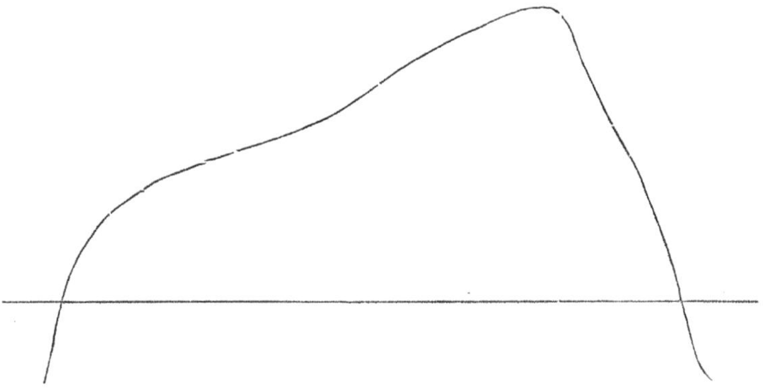

Fig. 132. Shape of the wave of the magnetising current, when the core is built up of steel strips the hysteresis loop of which is shown in Fig. 131 and the applied P.D. wave is sine shaped.

the hysteresis loop of the iron in the core is as given in Fig. 131. In this case, the shape of the current wave is not unlike the shape of the tooth of a carpenter's saw. Hence it is described sometimes as being shaped like a saw-tooth.

F. J. Dykes has made an harmonic analysis of the curve shown in Fig. 132. He finds that the equation to the curve is

$$\begin{aligned} i = 0{\cdot}59 \sin \omega t &- 0{\cdot}71 \cos \omega t \\ + 0{\cdot}05 \sin 3\omega t &- 0{\cdot}22 \cos 3\omega t \\ - 0{\cdot}02 \sin 5\omega t &- 0{\cdot}05 \cos 5\omega t \\ - 0{\cdot}02 \sin 7\omega t &- 0{\cdot}02 \cos 7\omega t \\ + \ldots . \end{aligned}$$

It will be seen that the amplitude of the third harmonic is approximately equal to a quarter of the amplitude of the first harmonic. The presence of this large third harmonic in the wave of the magnetising current often produces appreciable effects in practice, especially in polyphase working.

We can also construct the wave of applied potential difference necessary to produce a sine shaped wave of magnetising current. We first of all construct the flux wave by means of Fig. 115, and then draw the required wave of potential difference by means

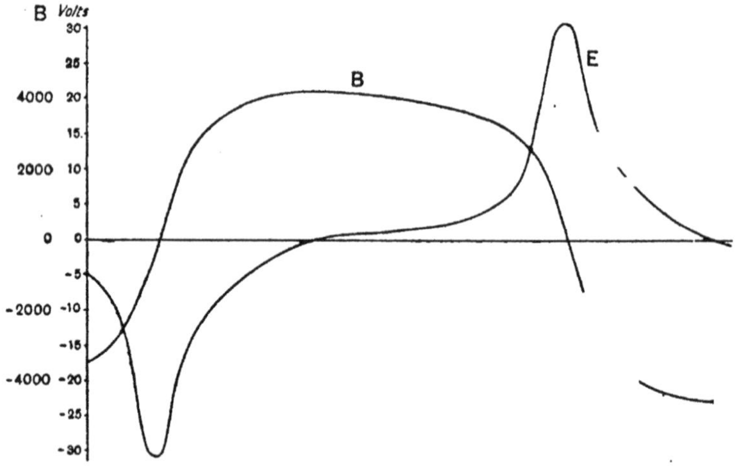

Fig. 133. Shape of the wave of the applied P.D. required to produce a sine shaped magnetising current.

of the formula $e_1 = n_1 \partial\phi/\partial t$. The ordinate e_1 is therefore equal to n_1 times the slope of the flux wave. The curves shown in Fig. 133 were constructed in this manner by F. J. Dykes.

Core losses.

In Vol. I, Chap. xx, formulae were found for the eddy current losses in metal sheets when subjected to harmonic magnetising forces. In order to produce these forces in the core of a transformer built up of iron plates, the applied potential difference wave must have the shape shown in Fig. 133. We assumed, however, that the permeability of the metal was constant, and therefore that the flux wave also obeyed the harmonic law. We see from Fig. 133 that this assumption is not justified. If the permeability were constant, the hysteresis loop would be a straight line and the hysteresis loss would be zero. The formulae, therefore, when applied to transformer cores, can only be regarded as roughly approximate.

We proved, however, that the power expended in the secondary coil of an air core transformer, when the primary resistance is negligible, is given by (Vol. I, p. 489)

$$(M/L_1)^2\, V_1^2 R_2/(R_2^2 + a^2 L_2^2 \sigma^2 \omega^2),$$

where a is a constant which has its minimum value unity when the applied wave of P.D. is sine shaped. Now, since we may regard the path of a filament of eddy current as a secondary circuit of a transformer, we may consider, when the screening effect of the eddy currents is negligible, that the eddy current loss can be represented by the sum of a series of terms of the form

$$R_2\{(M/L_1)\, V_1\}^2/(R_2^2 + a^2 L_2^2 \sigma^2 \omega^2).$$

If the amplitudes of the higher harmonics of the applied wave of potential difference be small compared with the amplitude of the first harmonic, we see from the formula given in Vol. I, p. 136, that a is nearly unity. When the applied waves have the shapes shown in Vol. I, fig. 33, p. 128, and R_2 is small compared with $L_2 \sigma \omega$, a is

$$(2/\pi)\,[\{(n+2)\,(2n+3)\}/\{2\,(2n+1)\}]^{\frac{1}{2}}.$$

For a triangular wave n is 1 and a is 1·007 nearly. Hence the values of a for a sine wave and a triangular wave differ from one another by less than one per cent. We should therefore expect that the difference between the eddy current losses in the core produced by a triangular shaped wave and a sine shaped wave of equal effective voltage and having the same frequency, when applied to the primary terminals of the transformer, would be too small to be measurable.

The formula also shows us that, if $L_2\sigma\omega$ be small compared with R_2, the eddy current losses in the core will be practically independent of the shape of the wave of the applied potential difference. It has been proved experimentally that in several types of transformer, the eddy current loss is approximately independent of the shape of the wave of the applied potential difference, when the effective primary voltage and frequency are constant.

As a rule, the hysteresis loss in the core is much larger than the eddy current loss. We see by Steinmetz's law that it depends practically only on the value of $B_{max.}$, and therefore, when the primary resistance of the transformer is negligible, on the value of the area of the applied wave of potential difference. In Vol. I, Chapter IV, many illustrations are given showing how waves of equal effective voltage may vary in shape. It is proved that peaky waves have a smaller area for a given effective voltage than rounded waves, and so, although they cause practically the same eddy current loss, they cause smaller hysteresis losses.

In practice, transformers for use on low frequency circuits work at higher induction densities than those for use with higher frequencies. For instance, in transformers constructed for use in circuits where the frequency is 25, $B_{max.}$ may be 8000 or even 10,000 C.G.S. units. On the other hand, if they are constructed for a frequency of about 100, 4000 C.G.S. units would be a usual value for $B_{max.}$. The iron sheets used in the construction of the core are generally from 10 to 20 mils, that is, from 0·025 to 0·05 centimetres in thickness. In commercial transformers, therefore, we may regard the screening effect of the eddy currents as negligible. By Steinmetz's formula, it can easily be shown that the hysteresis loss per kilogramme of the core is practically the same when the frequency is 100 and $B_{max.}$ is 4000, and when the frequency is 25 and $B_{max.}$ is 10,000. The eddy current losses per kilogramme, however, would generally be less in the latter case.

In a choking coil with no iron in the core, the sine shaped wave produces the maximum magnetising current (Vol. I, p. 136). In a transformer, with the secondary on open circuit, the current has to do work, owing to hysteresis and eddy current losses in the core. The wattless component of the current, however, produces a flux which is practically the same as that which would be produced if the iron were absent. For a given applied effective voltage therefore the wattless component of the current is a maximum for a sine shaped wave of P.D. The hysteresis losses will be a maximum for the wave of given effective voltage that has the maximum area, that is, for the rectangular wave. In this case, if we assume that the eddy current losses are the same whatever the shape of the wave, the watt component of the current will be a maximum. The watt component is therefore a maximum for the rectangular wave and the wattless component for a sine wave. We should therefore expect that the magnetising current of a transformer would be a maximum for a wave shape a little more rounded than a sine curve, and this is found to be the case in practice.

Copper losses.

The copper losses at any load are given by $R_1A_1^2 + r_2A_2^2$, where R_1 and r_2 are the resistances of the windings. If the frequency be high, then the real losses will be greater than those calculated by this formula owing to the current density being greater near the circumference of the conductors than along their axes (Vol. I, p. 216). If the secondary coil be a solid conductor of large dimensions, the losses in it owing to eddy currents may be large. For this reason, when the frequency is greater than 50, the secondary conductor is generally stranded.

Values of the resistances.

If l_1 and l_2 be the lengths of the wires used in the primary and secondary windings of a transformer, then in general we have $l_1/l_2 = n_1/n_2$. If also S_1 and S_2 be the areas of the cross sections of the windings and if the current densities in each of the windings be the same at full load, we have

$$S_1/S_2 = A_1/A_2 = n_2/n_1.$$

Hence if ρ be the resistivity of copper

$$R_1/r_2 = (\rho l_1/S_1)/(\rho l_2/S_2) = (l_1/l_2)(S_2/S_1) = (n_1/n_2)^2.$$

In most transformers the ratio of the resistances of the windings has approximately this value.

Constant potential transformer with no magnetic leakage.

If we make the assumption that there is no magnetic leakage, that is, that all the flux generated in the primary passes through the secondary, the equations can be written down without difficulty. When the secondary load x is non-inductive we can write

$$e_1 = R_1 i_1 + n_1 \frac{\partial \phi}{\partial t} \quad \text{.................(1)},$$

$$-n_2 \frac{\partial \phi}{\partial t} = (r_2 + x)\, i_2 \quad \text{....................(2)},$$

$$\phi \mathcal{R} = n_1 i_1 + n_2 i_2 \quad \text{....................(3)},$$

and

$$V_2 = x A_2 \quad \text{.............................(4)}.$$

In these equations the symbols have their usual meanings. It must be noticed that the flux ϕ and the reluctance $4\pi\mathcal{R}/10$ are not single valued functions of $n_1 i_1 + n_2 i_2$, as they have different values for a given value of this variable depending on whether it is increasing or diminishing.

From equation (1) we have

$$e_1 - R_1 i_1 = n_1 \frac{\partial \phi}{\partial t}.$$

If we square each side of this equation and take the mean values for a whole period, we find that

$$V_1^2 - 2R_1 W_1 + R_1^2 A_1^2 = n_1^2 V^2,$$

where W_1 is the mean value of $e_1 i_1$, that is, the mean power given to the primary, and V is the effective value of $\partial\phi/\partial t$. If we now write $V_1 A_1 \cos \psi_1$ for W_1, we see by drawing a triangle (Fig. 134), the sides of which are equal to V_1, $n_1 V$ and $R_1 A_1$ respectively, that the angle between V_1 and $R_1 A_1$ will be equal to ψ_1. In Fig. 134, OB is the applied potential difference V_1 and OA is $R_1 A_1$.

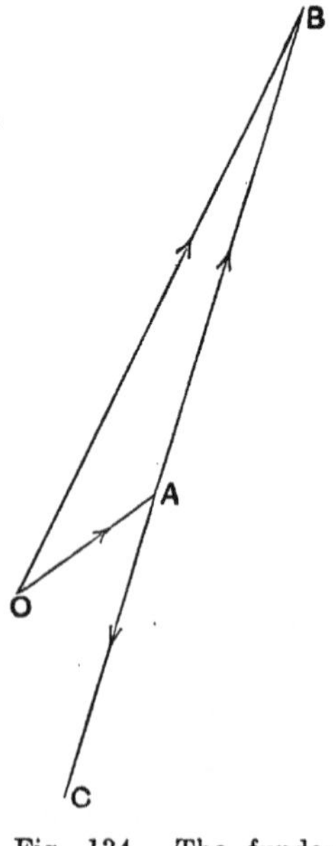

Fig. 134. The fundamental diagram of a transformer. $OB = V_1$, $OA = R_1 A_1$, $AB = n_1 V$, $AC = n_2 V = V_2 + r_2 A_2$. The angle BOA equals ψ_1.

From the diagram we see that we may suppose the applied

potential difference OB to be replaced by its two components OA and AB respectively. The component AB neutralises the back electromotive force due to the variations of the flux in the primary coil, and the component OA drives the current A_1 through the resistance R_1. Now, in commercial transformers, whether the iron circuit be open or closed, $R_1 A_1$ is rarely as great as the hundredth part of V_1, even at full load. Hence the lines OB and AB in Fig. 134 are nearly coincident, and the phase difference between the applied P.D. and the electromotive force set up by the varying flux of induction in the core is always nearly 180 degrees.

Again from (2) and (4) we get

$$n_2 V = V_2 + r_2 A_2,$$

and from (2) the phase of i_2 is in opposition to that of v, and thus, if we produce BA to C in Fig. 134 and make AC equal to $(n_2/n_1) AB$, then AC will represent $V_2 + r_2 A_2$ in magnitude and phase. We always have

$$V_1^2 - 2R_1 W_1 + R_1^2 A_1^2 = n_1^2 V^2$$
$$= (n_1/n_2)^2 (V_2 + r_2 A_2)^2.$$

When there is no secondary load, A_2 is zero, and we get

$$W_0 - R_1 A_0^2 = (1/2R_1) \{V_1^2 - R_1 A_0^2 - (n_1/n_2)^2 V_2^2\},$$

where W_0 is the power taken by the primary in this case and A_0 is the magnetising current. If we measure, therefore, R_1, V_1, A_0 and V_2, when the secondary is on open circuit, and if we know the ratio of n_1 to n_2, we can find the core losses at no load by this formula. In practice, however, the formula is of little use, as we are measuring the small difference between the large numbers V_1^2 and $(n_1/n_2)^2 V_2^2$, and so a small error made in measuring either V_1 or V_2 will introduce a large error into the calculated value of the core loss.

$B_{max.}$ is nearly constant at all loads.

In practice, $R_1 A_1$ (OA in Fig. 134) is always very small. It follows that $n_1 V$ (AB) is very nearly equal to V_1 (OB), and, therefore, when V_1 is maintained constant, $n_1 V$ will also be practically constant. Now since v equals $\partial\phi/\partial t$, we have

$$v = \partial (SB)/\partial t,$$

and thus (p. 281) $\quad V = 4kfSB_{max}\, 10^{-8}$ volts,

where k is the form factor of the applied P.D. It follows that if the shape of the applied P.D. wave and the frequency be maintained constant, $B_{max.}$ will also be constant. At full load, the difference between OB and AB, which in closed iron circuit transformers equals R_1A_1, is generally less than one per cent. Hence $B_{max.}$ varies by about one per cent. only, between no load and full load.

The hysteresis and eddy current losses in the core.

As the power lost owing to hysteresis is proportional to $B_{max.}^{1\cdot6}$, it follows that it diminishes by about 1·6 per cent. only, between no load and full load, provided that the shape of the applied wave is always the same. If the shape of the wave alters, the hysteresis loss may vary largely, although V_1 is kept constant. The eddy current losses in the core, on the other hand, have practically the same value at all loads, when V_1 is constant, although the shape of the applied wave alters considerably. If R_1A_1 at full load be one per cent. of V_1, the eddy current losses at full load would be about two per cent. less than at no load.

Resultant ampere turns.

When the maximum value of R_1i_1 is negligible compared with the maximum value of $n_1\partial\phi/\partial t$, we can write

$$n_1\partial\phi/\partial t = e_1,$$

and therefore

$$\phi = (1/n_1)\int e_1\partial t,$$

the constant term being zero because ϕ is a purely alternating function. We see that in this case ϕ depends only on the value of e_1. Now to each value of ϕ as it increases there is a definite value of the magnetising force, and therefore a definite value of the reluctance $\mathcal{R}$. Similarly to each value of ϕ as it diminishes there is a definite value of $\mathcal{R}$. We thus see that, in this case, $\phi\mathcal{R}$ depends only on the shape and magnitude of the primary voltage e_1. It is therefore independent of what is happening in the secondary circuit, and it is therefore the same function of e_1 at all loads. Hence by equation (3) the resultant magnetising turns $n_1i_1 + n_2i_2$ must be the same at all loads. We therefore have

$$n_1i_1 + n_2i_2 = n_1i_0,$$

where i_0 is the current in the primary when there is no load on the secondary. Since a linear relation connects i_1, i_2 and i_0, we can construct (see Vol. I, p. 305) a triangle (Fig. 135) the sides of which are A_1, A_0 and $(n_2/n_1) A_2$ respectively. The angles of this triangle will give the phase differences between the various currents. Again, by equations (1) and (2) given above, we always have

$$e_1 = R_1 i_1 - (n_1/n_2)(r_2 + x) i_2,$$

and thus a linear relation connects e_1, i_1 and i_2, and therefore their vectors lie in a plane. If OY (Fig. 135) be the position of the vector representing V_1, then OY, OD and OC will be in one plane and the angle YOD equals ψ_1 where $\cos\psi_1$ is the power factor of the primary.

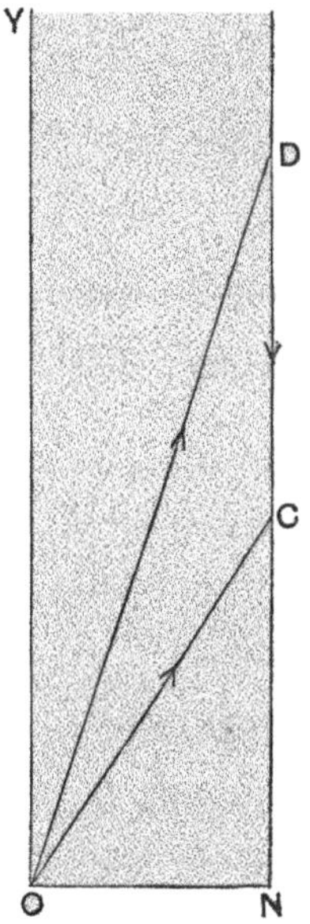

Fig. 135. The currents in the primary and secondary of a transformer, on a non-inductive load, with no magnetic leakage. $OD = A_1$, $OC = A_0$, $DC = (n_2/n_1) A_2$.

If the angle YOC be ψ_0, we have, by trigonometry,

$$\begin{aligned} A_1 \sin\psi_1 &= ON \\ &= A_0 \sin\psi_0 \;\ldots\ldots\ldots\ldots\ldots\ldots (a) \\ &= \text{a constant}, \end{aligned}$$

and

$$A_1 \cos\psi_1 - (n_2/n_1) A_2 = A_0 \cos\psi_0 \;\ldots (b),$$

and therefore

$$\tan\psi_1 = n_1 A_0 \sin\psi_0 / (n_2 A_2 + n_1 A_0 \cos\psi_0) \;\ldots\ldots\ldots (c).$$

These equations are useful in practical work, and enable us to determine accurately the primary power factor for any secondary current, when the load is non-inductive.

Example. The following are the data for a Swinburne open iron circuit transformer of the 'Hedgehog' type: $V_1 = 2400$ volts, $n_1/n_2 = 24$, $A_0 = 0{\cdot}70$ ampere and $W_0 = 84$ watts. Let us suppose that we require to find the primary current and power factor when the secondary current is 50 amperes.

Since $$W_0 = V_1 A_0 \cos\psi_0 = 84,$$

it follows that $\cos\psi_0 = 0{\cdot}05$, and therefore $\sin\psi_0 = 1{\cdot}00$.

Hence, by (c),

$$\tan\psi_1 = n_1 A_0 \sin\psi_0/(n_2 A_2 + n_1 A_0 \cos\psi_0) = 0{\cdot}33,$$

and therefore ψ_1 is 18·3 degrees and $\cos\psi_1 \doteq 0{\cdot}95$.

Finally from (a) we have

$$A_1 = A_0 \sin\psi_0/\sin\psi_1 = 2{\cdot}23 \text{ amperes.}$$

The secondary voltage.

Looking back at Fig. 134, by projecting OAB on OB, we see that

$$V_1 = n_1 V + R_1 A_1 \cos\psi_1,$$

for the cosine of the angle ABO is always practically unity.

Thus, from (b), for a non-inductive load,

$$V_1 = (n_1/n_2)(V_2 + r_2 A_2) + R_1\{(n_2/n_1) A_2 + A_0 \cos\psi_0\},$$

and hence

$$V_2 = (n_2/n_1) V_1 - \{r_2 + (n_2/n_1)^2 R_1\} A_2 - (n_2/n_1) R_1 A_0 \cos\psi_0 \quad \ldots(d).$$

This formula is a useful one. The last term $(n_2/n_1) R_1 A_0 \cos\psi_0$ is generally negligible.

Example.

The data for a 15 kilowatt Ferranti transformer are as follows. The resistances when warm of the primary and secondary coils are 2·75 and 0·0061 ohms respectively. The applied primary voltage is 2400, the ratio (n_1/n_2) of the turns is 24 and the power W_0 taken by the transformer on no load is 240 watts. We have

$$r_2 + (n_2/n_1)^2 R_1 = 0{\cdot}0061 + (1/24)^2\, 2{\cdot}75$$
$$= 0{\cdot}011.$$

We also have $V_1 A_0 \cos\psi_0 = 240$,

and therefore $A_0 \cos\psi_0 = 0{\cdot}1$.

When the secondary current is 150 amperes, we have

$$V_2 = (1/24)\,2400 - 0{\cdot}011\,.\,150 - (1/24)\,0{\cdot}275$$
$$= 100 - 1{\cdot}65 - 0{\cdot}01$$
$$= 98{\cdot}34.$$

Output.

The rating of a transformer depends on the permissible voltage drop at the secondary terminals. If we assume that a two per cent. drop is the maximum permissible, the rating of the transformer would be the power in the secondary when the voltage drop is two per cent. In this case we get by formula (*d*)

$$A_2' = \{(n_2/n_1)\,V_1 - (n_2/n_1)\,R_1 A_0 \cos\psi_0\}/[50\,\{r_2 + (n_2/n_1)^2 R_1\}],$$

where A_2' is the maximum permissible current in the secondary. Hence the rating of the transformer is

$$(49/50)\,\{(n_2/n_1)\,V_1 - (n_2/n_1)\,R_1 A_0 \cos\psi_0\}\,A_2',$$

and, since $(n_2/n_1)\,R_1 A_0 \cos\psi_0$ is always very small, this may be written

$$[49\,\{n_2 V_1/(50 n_1)\}^2]/\{r_2 + (n_2/n_1)^2 R_1\}.$$

Efficiency.

The efficiency of a transformer is the ratio of the power utilised in the external load on the secondary, to the power taken by the primary. We can obtain a formula for η by means of the formulae (*b*) and (*d*) given above. We have by (*b*)

$$A_2 = (n_1/n_2)\,(A_1 \cos\psi_1 - A_0 \cos\psi_0)$$
$$= (n_1/n_2)\,\{(W_1 - W_0)/V_1\}.$$

We have also by (*d*)

$$V_2 = (n_2/n_1)\,V_1 - \{(n_1/n_2)\,r_2 + (n_2/n_1)\,R_1\}\,(A_1 \cos\psi_1 - A_0 \cos\psi_0) - (n_2/n_1)\,R_1 A_0 \cos\psi_0$$
$$= (n_2/n_1)\,V_1 - (n_2/n_1)\,Q A_1 \cos\psi_1 + (n_1/n_2)\,r_2 A_0 \cos\psi_0,$$

where $Q = R_1 + (n_1/n_2)^2\,r_2$.

Now since $\eta = A_2 V_2/W_1$, it follows that

$$\eta = (1 - W_0/W_1)\,\{1 - QW_1/V_1^2 + (n_1/n_2)^2\,(r_2 W_0/V_1^2)\}.$$

In ordinary transformers $(n_1/n_2)^2\,(r_2 W_0/V_1^2)$ is negligible, and hence

$$\eta = (1 - W_0/W_1)\,(1 - QW_1/V_1^2) \quad \text{...............}(e).$$

We have also, since η equals W_2/W_1,

$$W_2 = (W_1 - W_0)\,(1 - QW_1/V_1^2).$$

Thus when we are given W_2 we can always find W_1, and hence the efficiency of the transformer for a given secondary load. Again,

$$W_1 - (W_0 + W_2) = (Q/V_1^2)\,W_1\,(W_1 - W_0).$$

If we plot out therefore the copper losses $W_1 - W_0 - W_2 + R_1 A_0^2$ as a function of the power W_1 taken by the primary, we get a parabola.

It easily follows from (e) that the efficiency is a maximum when

$$W_1 = V_1 \sqrt{W_0/Q}, \quad \text{or} \quad A_1 \cos \psi_1 = \sqrt{W_0/Q},$$

and we have $$\eta_{\text{max.}} = \{1 - \sqrt{QW_0}/V_1\}^2.$$

Examples.

In the Ferranti transformer considered above Q equals 6·34 when the windings are warm. By (e) the efficiency of this transformer when the primary is taking 12 kilowatts is given by

$$\eta = (1 - 240/12000)(1 - 6{\cdot}34 \times 12000/2400^2) = 0{\cdot}967.$$

Its efficiency at this load is thus 96·7 per cent.

In a three kilowatt open iron circuit transformer Q is 53·4 ohms, V_1 is 2400 and W_0 is 121 watts. The load W_1 at which the transformer has its maximum efficiency is

$$W_1 = 2400\sqrt{121/53{\cdot}4}$$
$$= 3{\cdot}614 \text{ kilowatts}.$$

The efficiency at this load

$$= \{1 - (53{\cdot}4 \times 121)^{\frac{1}{2}}/2400\}^2$$
$$= 93{\cdot}4 \text{ per cent}.$$

The following table shows the effect on the percentage efficiency of a variation in the copper and iron losses in the Ferranti transformer considered above.

Load in kilowatts	Real efficiency	No iron losses	No copper losses	Copper and iron losses each halved
0·646	64·4	99·9	64·4	82·2
16·8	96·8	98·2	98·7	98·4

Regulation on inductive load.

Let V_2 be the voltage across the secondary load and $\cos \psi$ be the power factor of this load. The component of the induced voltage AC (Fig. 134) in the direction of the current vector will be $V_2 \cos \psi + r_2 A_2$ and at right angles to this direction the component is $V_2 \sin \psi + X_2 A_2$, where X_2 is the reactance of the secondary coil. Hence we have

$$(V_2 \cos\psi + r_2 A_2)^2 + (V_2 \sin\psi + X_2 A_2)^2 = AC^2$$
$$= (n_2/n_1)^2 (V_1^2 - 2R_1 W_1 + R_1^2 A_1^2),$$

and therefore

$$\{V_2 + (r_2 \cos\psi + X_2 \sin\psi) A_2\}^2$$
$$= (n_2/n_1)^2 V_1^2 [\{1 - (R_1 A_1 \cos\psi_1)/V_1\}^2 + (R_1 A_1 \sin\psi_1)^2/V_1^2]$$
$$- (r_2 \sin\psi - X_2 \cos\psi)^2 A_2^2.$$

Taking the square root of each side and noticing that the first term on the right-hand side is much greater than the second and that $R_1^2 A_1^2/V_1^2$ is small compared with unity, we get

$$V_2 + (r_2 \cos\psi + X_2 \sin\psi) A_2$$
$$= (n_2/n_1) \{V_1 - R_1 A_1 \cos\psi_1 + (R_1 A_1 \sin\psi_1)^2/2V_1\}$$
$$- (r_2 \sin\psi - X_2 \cos\psi)^2 A_2^2/\{2n_2 V_1/n_1\} \quad \ldots(f),$$

very approximately.

Now since the resultant of the magnetising turns remains practically constant at all loads, we see, by resolving the ampere turns along and perpendicular to the vector of V_1, that

$$n_1 A_1 \cos\psi_1 = n_2 A_2 \cos\psi + n_1 A_0 \cos\psi_0,$$

and $$n_1 A_1 \sin\psi_1 = n_2 A_2 \sin\psi + n_1 A_0 \sin\psi_0.$$

Hence substituting for $n_1 A_1 \cos\psi_1$ in (f) and simplifying we get

$$V_2 = (n_2/n_1) V_1 - RA_2 \cos\psi - X_2 A_2 \sin\psi - (n_2/n_1) R_1 A_0 \cos\psi_0$$
$$- \frac{n_1 (r_2 A_2 \sin\psi - X_2 A_2 \cos\psi)^2}{2n_2 V_1} + \frac{n_2 (R_1 A_1 \sin\psi_1)^2}{2n_1 V_1} \quad \ldots(g),$$

where $$R = r_2 + (n_2/n_1)^2 R_1.$$

In commercial transformers we may write $n_2^2 R_1 = n_1^2 r_2$, and substituting for $A_1 \sin\psi_1$ in (g) and simplifying, we get

$$V_2 = (n_2/n_1) V_1 - ZA_2 \cos(\alpha - \psi) - (n_2/n_1) R_1 A_0 \cos\psi_0$$
$$- \frac{n_1}{2n_2 V_1} \left\{ (X_2 A_2 \cos\psi - r_2 A_2 \sin\psi)^2 \right.$$
$$\left. - \left(\frac{n_1}{n_2} r_2 A_0 \sin\psi_0 + r_2 A_2 \sin\psi \right)^2 \right\} \quad \ldots\ldots(h),$$

where $Z^2 = R^2 + X_2^2$, and $\tan\alpha = X_2/R$.

In many practical cases the last two terms on the right-hand side of (h) can be neglected and so

$$V_2 = (n_2/n_1)\, V_1 - ZA_2 \cos(\alpha - \psi) \quad \ldots\ldots\ldots\ldots(i).$$

For a given value of A_2, therefore, the voltage drop is a maximum when $\psi = \alpha$. It is interesting to notice that when $\psi = -(\pi/2 - \alpha)$ the secondary voltage has approximately the same value at all loads. We shall now describe a practical method of measuring X_2 and Z.

The short circuit test.

If we short circuit the secondary terminals of a transformer through an ammeter of small resistance, it will be found that we can get full load secondary current by applying to the primary terminals a voltage V_s the value of which is two or three per cent. of the normal voltage. The magnetic flux in the core in this case is only about a fiftieth of its normal value, and hence since we are supposing that the magnetic leakage is negligible the resultant of the magnetising turns acting on it must be very small. Therefore $n_1 A_1$ must be practically equal to $n_2 A_2$ and in opposition in phase to it. But at full load on ordinary working $n_1 A_1 = n_2 A_2$ very approximately and hence, in the short circuited case, when the current A_2 in the secondary has its full load value so also has the current A_1 in the primary. From Fig. 134, we get

$$r_2^2 A_2^2 + X_2^2 A_2^2 = (n_2/n_1)^2 \{V_s^2 - 2R_1 W_s + R_1 A_1^2\} \ldots(j),$$

where W_s is the reading of the wattmeter in the primary circuit.

The values of r_2 and R_1 can be readily found by using direct current. If E_2, for instance, be the voltage across the secondary terminals when the direct current through the secondary is A_2, we have $r_2 = E_2/A_2$. When the values of n_2 and n_1 are known therefore, X_2 is given by (j) and hence Z and α can be found. In commercial testing, owing to the low value of the flux in the core, the hysteresis and eddy current losses can be neglected. We may therefore write

$$W_s = R_1 A_1^2 + r_2 A_2^2.$$

Also, since we generally have $n_2^2 R_1 = n_1^2 r_2$, equation (j) may be written

$$X_2 A_2 = (n_2/n_1)\,(V_s^2 - 4R_1^2 A_1^2)^{\frac{1}{2}} \ldots\ldots\ldots\ldots\ldots(k).$$

Let us consider the case of a 500 K.W. 5000/500 transformer, that is, a 500 kilowatt transformer which transforms down from 5000 to 500 volts. In this case $n_1/n_2 = 10$, and suitable values for the primary and secondary resistances would be $R_1 = 0{\cdot}25$ ohm and $r_2 = 0{\cdot}0025$ ohm respectively, as these values would make the copper loss at full load about one per cent. of the output. We shall also suppose that $V_1 A_0 \cos\psi_0 = 5000$ and $\psi_0 = 45°$. If the short circuit voltage V_s be 150, we get by (k)

Numerical example.

$$1000X_2 = (1/10)\,(150^2 - 50^2)^{\frac{1}{2}} = 14{\cdot}14.$$

We also have

$$R = r_2 + (n_2/n_1)^2 R_1 = 0{\cdot}005,$$

and thus, $1000R = 5$, $1000Z = 15$ and $\alpha = 70°\,32'$. Substituting in (h) we get

$$V_2 = 500 - 0{\cdot}015 A_2 \cos(70°\,32' - \psi) - 0{\cdot}025$$
$$- (14{\cdot}14 \cos\psi + 25/A_2)\,(14{\cdot}14 \cos\psi - 25/A_2 - 5\sin\psi)\, A_2^2/(2\,.\,10^8).$$

The following table gives the voltage drop at various power factors when the current in the secondary has its full load value of 1000 amperes.

$\cos\psi$	Voltage drop	$\cos\psi$	Voltage drop	$\cos\psi$	Voltage drop
1	6·0	0·6	14·5	0·2	14·9
0·9	11·3	0·5	14·9	0·1	14·6
0·8	13·0	0·4	15·0	0·0	14·1
0·7	13·9	0·3	15·0		

With a leading power factor of 0·94 the voltage drop would only be about 1 at full load.

When transformers are working in parallel with one another it is necessary not only that their voltage ratios be adjusted with high accuracy but also that the angles ψ_0 and α should be the same for each, where $\cos\psi_0$ is the power factor of the primary on open circuit and $\tan\alpha = X_2/R$. This ensures that the secondary voltages are in phase and have the same values at all loads, and hence there are no 'circulating currents' between the secondaries. It ensures also that all the primary currents are in phase with one another so that their sum equals the current in the leads. If these conditions are not

Parallel operation.

satisfied the efficiency of the transformation by the transformers may be much smaller than the efficiency of the transformation by a single large transformer.

Equivalent net-work.

When the secondary load has inductance or capacity, the problem can be advantageously attacked by considering the equivalent net-work. Let us suppose that there is no magnetic leakage, and that we have a choking coil L and a condenser K in series in the secondary circuit. The equations are

$$e_1 = R_1 i_1 + n_1 \partial\phi/\partial t$$

and $$-n_2 \partial\phi/\partial t = (r_2 + x)\, i_2 + N . \partial i_2/\partial t + (1/K) \int i_2 \partial t.$$

They may also be written in the form

$$e_1 - R_1 i_1 = n_1 \partial\phi/\partial t,$$

and $$e_1 - R_1 i_1 = (n_1/n_2)^2 (r_2 + x)\, i' + (n_1/n_2)^2 N . \partial i'/\partial t + \{1/(n_2/n_1)^2 K\} \int i' \partial t,$$

where $i' = -(n_2/n_1)\, i_2$.

These equations suggest the following equivalent net-work. Let us suppose that the primary T of the transformer has zero resistance, and that an external resistance R_1 is put in series with it. We shall also suppose that the potential difference is applied across the two in series. Connect a resistance $(n_1/n_2)^2 (r_2 + x)$, a choking coil with self-induction $(n_1/n_2)^2 N$ and a condenser with capacity $(n_2/n_1)^2 K$ in series, and place this circuit as a shunt across the primary terminals. The above equations show us that the primary current will be equal to the current in R_1 in magnitude and phase and that the secondary current will be equal to n_1/n_2 times the current in the circuit shunting the transformer and will be in opposition in phase to it.

If i be the current in the imaginary primary coil T, then

$$i_1 = i + i'$$

or $$n_1 i_1 + n_2 i_2 = n_1 i.$$

If R_1 be zero, i will obviously be constant, and hence as before we find that

$$n_1 i_1 + n_2 i_2 = n_1 i_0.$$

All the formulae given above can easily be proved by means of this equivalent net-work.

Inductive and condenser loads.

Replacing a transformer by means of its equivalent net-work is also useful in practical work, as it enables us to tell at once what will happen in special cases. Suppose for example that we put an inductive coil N in series with the secondary. Replacing the transformer by its equivalent net-work we get Fig. 136. If $r_2 + x$ be very small, the net-work will act simply like a choking coil, and so the primary current will lag nearly ninety degrees behind the applied P.D. and the primary and secondary currents will be nearly in opposition in phase. If N were zero and $r_2 + x$ very small, then the primary and secondary currents would be nearly in opposition in phase and the primary current would be nearly in phase with the applied P.D.

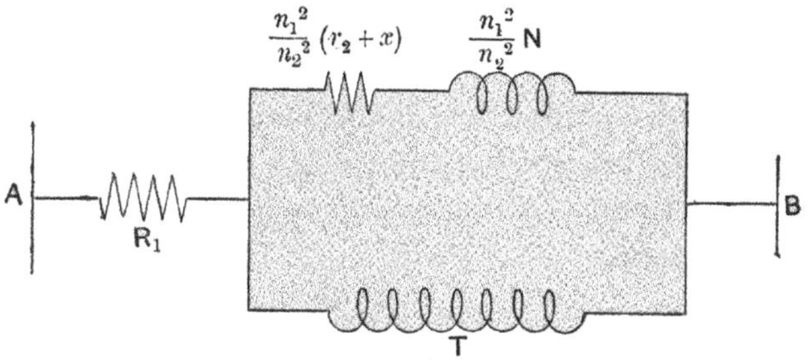

Fig. 136. Equivalent net-work of a transformer on an inductive load, when there is no magnetic leakage. T acts in the same way that the primary of the transformer would if it had no resistance and the secondary was open circuited.

Resonance with transformers.

When we put a condenser load across the secondary terminals we can see at once from the diagram that in certain cases the primary current will be in advance of the applied P.D. in phase. Hence the transformer as a whole will act like a condenser, and if there is inductance in series with it, we can have resonance and a dangerous rise of the potential difference between certain parts of the circuit. In the early days of electric lighting these resonance effects caused a great deal of trouble to electrical engineers. We can also see that in certain cases resonance of currents will take place in the net-work, a very small primary current giving rise to a very large secondary current.

It has to be remembered however that, when we have condensers in the circuit, the current wave is generally considerably distorted and alters in shape as we vary the capacity and resistance in the circuit. Hence in this case diagrams obtained on the supposition that the wave shape does not alter have only a limited use. The following experiment illustrates this.

The primary circuit of a small transformer converting from 100 to 200 volts was connected across the hundred volt mains of a supply company. Across the secondary terminals a condenser, of capacity two microfarads, was placed in series with an adjustable resistance. When this resistance was zero the current in the primary was 0·67 ampere. As the resistance was increased the primary current diminished, attaining a minimum value of 0·62 ampere when R was 35 ohms. It then increased to a maximum value of 0·915 ampere when R was 1500 ohms, and it finally diminished to 0·74 ampere, its value on open circuit when R was infinite. The secondary current however continually diminished as R increased. The alteration in the shape of the current wave was proved by the ratio of the volts at the condenser terminals to the secondary current continually increasing as the resistance in the circuit increased.

REFERENCES

J. A. Fleming, 'Experimental Researches in Alternate-Current Transformers,' *Journ. of the Inst. of El. Eng.*, Vol. 21, p. 594, 1892.

S. Beeton, C. P. Taylor and J. M. Barr, 'Experimental Tests on the Influence of the Shape of the Applied Potential Difference Wave on the Iron Losses of Transformers,' *Journ. of the Inst. of El. Eng.*, Vol. 25, p. 474, 1896.

A. Russell, 'Formulae for Transformers,' *The Electrician*, Vol. 38, p. 725, 1897.

—— 'The efficiency of the Alternating Current Transformer,' *The Electrician*, Vol. 40, p. 720, 1898.

E. B. Rosa and M. G. Lloyd, 'The Determination of the Ratio of Transformation and of the Phase Relations in Transformers,' *Bulletin of the Bureau of Standards*, Vol. 6, p. 1, 1909.

P. G. Agnew and T. T. Fitch, 'The Determination of the Constants of Instrument Transformers,' *Bulletin of the Bureau of Standards*, Vol. 6, p. 281, 1901.

CHAPTER XI

The transformation of polyphase currents by single phase transformers. Primaries connected in four wire star. Primaries connected in three wire star. Three phase transformer. Mesh to star. Mesh to mesh. Three phase to two phase. Three phase to single phase. Three single phase transformers connected in star. Boosting transformer. Reducing the pressure. Increasing the pressure. Boosting. Variable induction transformer. Compensator. Compensator for arc lamps. Current direction indicator. References.

Transformation of polyphase currents by single phase transformers.

THE transformation of polyphase currents, from high pressure to low pressure or *vice versâ*, by means of stationary transformers, is practically as simple as the corresponding problem in single phase working. To effect the transformation we use either polyphase transformers or groups of three single phase transformers. In either case they may be connected in star or in mesh. We shall first consider the case of three single phase transformers, connected in mesh (Fig. 137). The three primaries are connected in series at P_1, P_2 and P_3, and the three secondaries at S_1, S_2 and S_3. P_1, P_2 and P_3 are connected with the primary system of mains at 1, 2 and 3; S_1, S_2 and S_3 with the secondary system at 1′, 2′ and 3′.

If the three transformers have the same ratio of transformation, then, neglecting magnetic leakage, we can write

$$e_1 - e_1' = v_1 = R_1 i_1 + n_1 \frac{\partial \phi}{\partial t}; \qquad 0 = r_2 i_2 + v_2 + n_2 \frac{\partial \phi}{\partial t};$$

$$e_1' - e_1'' = v_1' = R_1' i_1' + n_1 \frac{\partial \phi'}{\partial t}; \qquad 0 = r_2' i_2' + v_2' + n_2 \frac{\partial \phi'}{\partial t};$$

$$e_1'' - e_1 = v_1'' = R_1'' i_1'' + n_1 \frac{\partial \phi''}{\partial t}; \qquad 0 = r_2'' i_2'' + v_2'' + n_2 \frac{\partial \phi''}{\partial t};$$

where e_1, e_1', e_1'' are the potentials of the primary mains, and v_1, v_1' and v_1'' are the applied potential differences. The secondary potential differences and the currents in the secondary windings are denoted by v_2, v_2' and v_2'', and i_2, i_2' and i_2'' respectively.

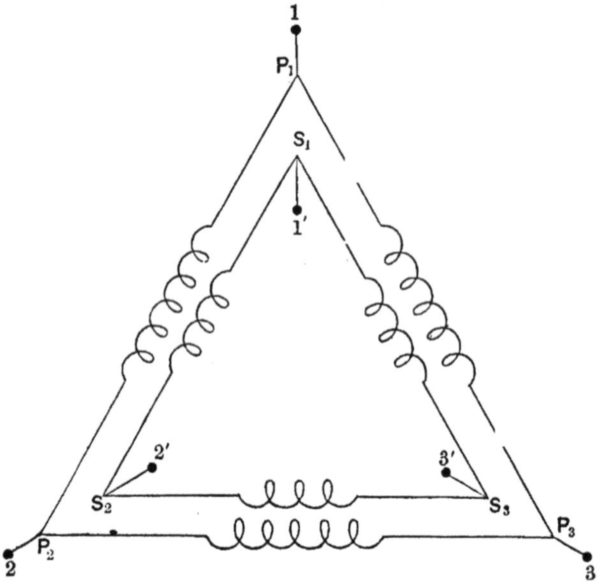

Fig. 137. Transforming three phase currents by means of three single phase transformers.

If the resistances of the primary and secondary coils of the transformers can be neglected, we have

$$v_2 = -(n_2/n_1)\, v_1, \quad v_2' = -(n_2/n_1)\, v_1', \text{ and } v_2'' = -(n_2/n_1)\, v_1''.$$

These equations show that the waves of secondary voltage are exactly similar to the primary waves, but differ from them in phase by 180°. Since $v_2 + v_2' + v_2''$ is always zero, it follows that the vectors of the secondary voltages form a triangle. The sides of this triangle (Fig. 138) are equal to V_2, V_2' and V_2'' respectively, and the supplements of its angles give the phase differences between the secondary voltages. We see that the sides of this triangle equal the sides of the primary voltage triangle multiplied by n_2/n_1, when the resistances of the transformer windings can be neglected.

When the applied potential difference follows the harmonic law, the shape of the wave of the magnetising current will be similar to that of the curve shown in Fig. 132. This curve has a large third harmonic. If the P.D. wave, therefore, between each pair of mains is sine shaped, the sum of the magnetising currents $i_1 + i_1' + i_1''$, round the mesh, will not be zero, but will equal three times the sum of the harmonic terms in the Fourier series representing the current, whose frequencies are given by $3(2n+1)f$, where n is zero or a positive integer. In the primary mesh, therefore, we have at all loads a local current component which,

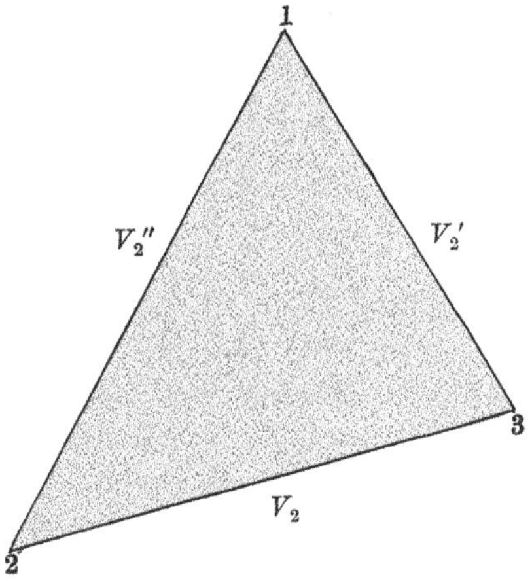

Fig. 138. Transforming three phase currents by means of three single phase transformers. The sides of this triangle equal the secondary voltages.

since the third harmonic is much the most important, is practically sine shaped and has a frequency $3f$.

If the resistances of the primary coils are negligible, the secondary potential differences will be of the same shape as the primary potential differences, provided that the magnetic leakage is negligible, the secondary coils mesh-connected, and the secondary loads balanced and non-inductive. The secondary currents, in this case, will be of the same shape as the applied P.D. waves. We see that the additional components of the primary current, necessary to prevent these secondary currents producing a flux which would upset the balance of the back and the applied electromotive forces, must have the same shape as the applied P.D. waves and will be in phase with them. As the balanced load, therefore, on the secondary increases, the shape of the primary currents becomes more like the shape of the applied potential difference waves, and the power factor consequently approaches unity. This effect also ensues when the secondaries are connected either in three wire or four wire star, provided that the load is also connected in three wire or four wire star. If, however, when the secondary coils are connected in star, the load is connected in mesh, the secondary P.D. waves will not be in opposition in phase and will not, in general, be similar to the applied P.D. waves. Thus the additional components of the primary current due to the load will differ in phase and, in general, also, in shape from the primary P.D. wave, and the power factor will therefore be low (see Vol. I, Chap. x). Hence, we see that, when the primaries are in mesh and the secondaries are in star, the secondary load must not be mesh-connected.

Primaries connected in four wire star.

Let us now consider three single phase transformers with their primaries star-connected, and let us suppose that their common junction is connected, either through the earth or through the 'neutral main,' with the common junction of the armature windings of the generator, so that we have a four wire star system. If the waves of P.D. between the mains and the common junction be sine shaped, the magnetising currents will be shaped as in Fig. 132, and the current in the neutral wire $i_1 + i_1' + i_1''$ will be practically sine shaped and have a frequency $3f$. Whatever the shape of the applied P.D. waves, the current in the neutral wire will be represented by terms the frequencies of which are given by $3(2n+1)f$. When the secondary coils and load are both star-connected, then, as the load increases, the shape of the primary current wave becomes more like that of the applied P.D. wave and the time-lag between the two waves diminishes. The primary power factor, therefore, will be high when the transformer is loaded. If, however, the secondary windings be mesh-connected and the load be star-connected, the secondary current waves and, therefore, also the corresponding components of the primary current waves will neither be in phase with, nor, as a rule, will they be similar to, the applied P.D. waves. Hence, the power factor will be low, and thus this connection must not be used.

Primaries connected in three wire star.

The case when the primary coils are connected in three wire star, that is, when their common junction is insulated, is interesting, and deserves careful consideration. The sum of the three currents flowing into the common junction must be zero at every instant. From the symmetry of the arrangement, the time-lag between any two of the three currents must be one-third of a period, and, since the sum of the three must be zero, the Fourier series for each must not contain terms whose frequencies are given by $3(2n+1)f$. The currents, therefore, cannot be shaped like the curve in Fig. 132, since this curve has a large harmonic of frequency $3f$. It follows that the P.D. waves between the mains and the centre of the star cannot be sine shaped.

Let us suppose that the instantaneous value of the P.D. between two of the mains is given by $F(t)$, and let $\psi(t)$ and $\psi(t+T/3)$ give the values of the potential differences between these two mains and the centre of a star, the arms of which are non-inductive and equal. We must have

$$F(t) = \psi(t) - \psi(t+T/3)$$
$$= v_{1.x} - v_{2.x},$$

where $v_{1.x}$ and $v_{2.x}$ are the voltages across the primary terminals of two of the transformers. Now, when we write $t+T/3$ for t in $v_{1.x}$ we must get $v_{2.x}$. It follows, therefore, that we may write

$$v_{1.x} = \psi(t) + \chi(3t),$$

where $\chi(3t)$ is a periodic alternating function of period $T/3$.

Hence, if the star wave $\psi(t)$ produce a magnetising current wave i_1, which has harmonics of frequency $3(2n+1)f$, the shape of the applied wave will assume a form $\psi(t) + \chi(3t)$, which gives a magnetising current wave that is free from these harmonics. When the common junction of the three primary windings, therefore, is disconnected from the fourth wire, the star-wave form will generally change, and this change will always increase the effective value of the star voltage, although the mesh voltages remain the same. It is worth noticing that this proves that the mesh and star voltages cannot, in general, be represented by the sides of an equilateral triangle and the three lines joining the angular points to the centre. They can be represented, however, by the edges of a tetrahedron. F. J. Dykes has shown that, in some cases, the star voltage, on open circuit, rises more than ten per cent. when the central connection is insulated.

In Fig. 139, *m* is an oscillograph record of the mesh voltage, and the curves, *S* and *S*, are oscillograph records of the corresponding star voltages on open circuit, the secondary coils being star-connected. It is interesting to notice that the general

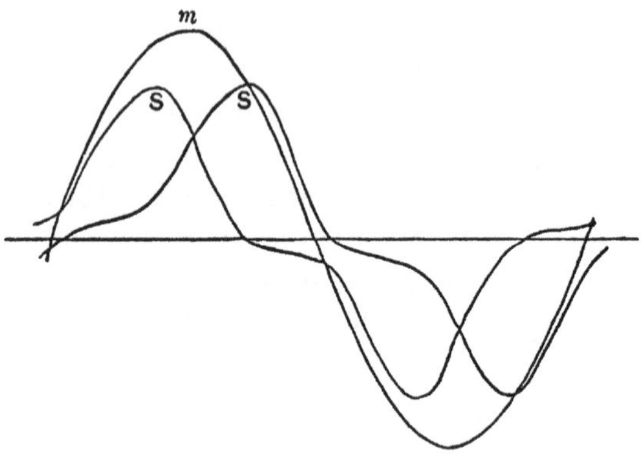

Fig. 139. Three single phase transformers connected in star. The curve *m* gives the shape of the mesh wave and the curves *S*, *S* give the shape of the star waves.

characteristics of the waves *S* and *S* are not unlike the general characteristic of the P.D. wave (Fig. 133) necessary to produce a sine wave of magnetising current. During an appreciable fraction of the first quarter of a period, the curve is approximately parallel to the zero line, and during the second quarter of a period it rises and falls rapidly.

When the mesh wave is distorted as in Fig. 140 it will be seen that the general shape of the star wave *S* still remains the same. The curves *S* and *S*, in Figs. 139 and 140, are shown with a time-lag of 60 degrees between them, one of them giving the voltage from one main to the centre of the star, and the other giving the voltage from the centre of the star to the other main. Thus, if we add the ordinates of the S, S curves together, we should get the m curve. For both figures the author is indebted to F. J. Dykes.

If the secondary coils be mesh-connected, it is found that there is practically no change in the shape of the primary star waves when the centre of the star is disconnected from the neutral wire. There is, however, a local current, the harmonics of which have frequencies $3(2n+1)f$, flowing in the secondary mesh; the magnetising effect of this current and of the primary current,

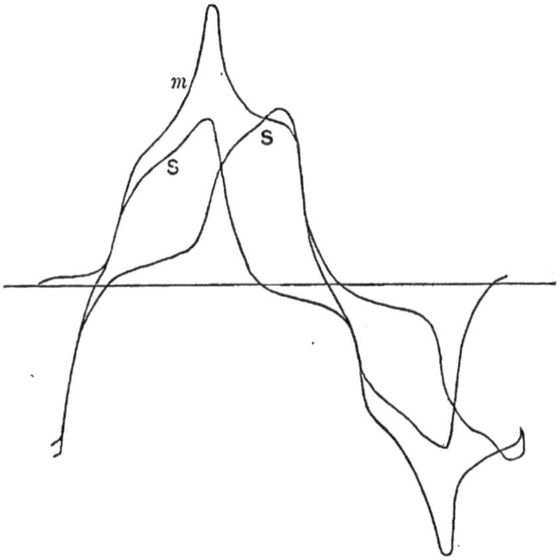

Fig. 140. Three single phase transformers connected in star; m is the mesh wave and S, S are the corresponding star waves.

which, of course, can contain no harmonics whose frequencies are $3(2n+1)f$, practically produce the four wire star wave.

Three phase transformer.

The design of a simple form of three phase transformer is indicated in Fig. 141. Thin iron plates, with two rectangular holes stamped out of them, are placed over one another and wound with six coils as in the figure. We shall first consider the case when both primary and secondary windings are connected in star and when the centre of the star is insulated.

Let e_1 be the potential difference between P (Fig. 141) and the common junction of the primary windings. If there is no magnetic leakage, our equations are

$$e_1 = R_1 i_1 + n_1 \frac{\partial \phi}{\partial t},$$

$$0 = r_2 i_2 + e_2 + n_2 \frac{\partial \phi}{\partial t},$$

where e_2 is the potential difference between S and the common

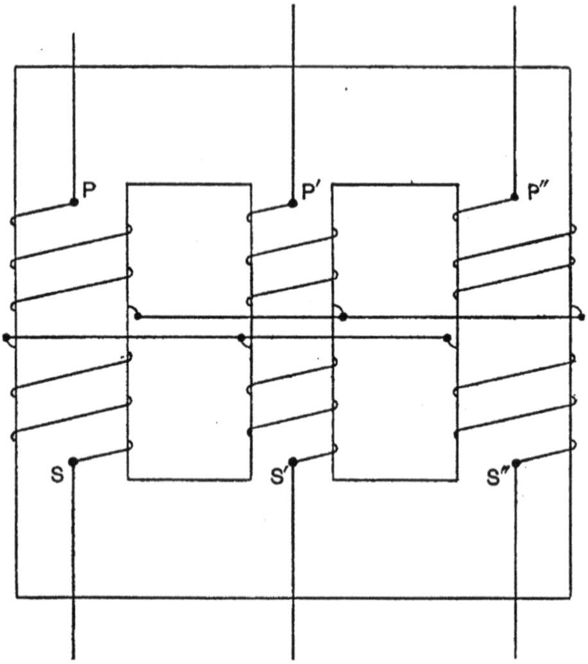

Fig. 141. Three phase transformer, primary and secondary circuits having star windings.

junction of the secondary windings. When $R_1 i_1$ is negligible, we have

$$e_2 + r_2 i_2 = -(n_2/n_1)\, e_1.$$

In a similar way we can find equations for e_2' and e_2''. These equations are identical in form with the corresponding equations for a single phase transformer, and can be discussed in the same way. When, in addition, the term $r_2 i_2$ can be neglected, we see that the secondary waves are in exact opposition in phase to the primary, and that the ratio of their effective values is n_2/n_1.

The above equations show us that if e_1, e_1' and e_1'' are similar and equal waves, then, on the above assumptions, e_2, e_2' and e_2'' will also be equal and similar. It also follows that ϕ, ϕ', ϕ'', the magnetic fluxes embraced by the three currents, will each follow the same law and be equal in magnitude; consequently, since their sum is always zero when there is no leakage, the phase difference between any two of them will be 120 degrees. Let $4\pi\mathcal{R}_{1.2}/10$ be the reluctance of each of the magnetic circuits $PSS'P'$ and $P'S'S''P''$ (Fig. 141). Then, if the reluctance of the circuit $PSS''P''$ be $4\pi\mathcal{R}_{1.3}/10$, our magnetic equations are

$$\phi = n_1(i_1 - i_1')/\mathcal{R}_{1.2} - n_1(i_1'' - i_1)/\mathcal{R}_{1.3},$$
$$\phi' = n_1(i_1' - i_1'')/\mathcal{R}_{1.2} - n_1(i_1 - i_1')/\mathcal{R}_{1.2},$$
$$\phi'' = n_1(i_1'' - i_1)/\mathcal{R}_{1.3} - n_1(i_1' - i_1'')/\mathcal{R}_{1.2}.$$

As the paths of the magnetic lines are in iron, the quantities $\mathcal{R}_{1.2}$ and $\mathcal{R}_{1.3}$ are not constant, as their values depend on the magnetising forces. If we assume that they are constant, and if the three cores (Fig. 141) have equal cross-sectional areas, $\mathcal{R}_{1.3}$ will be greater than $\mathcal{R}_{1.2}$, and so the magnetising current in the middle core will be less than in the outer cores and the arrangement will be unsymmetrical. If, however, we design the transformer so that $\mathcal{R}_{1.3}$ equals $\mathcal{R}_{1.2}$ when the applied magnetic forces are the same, we must make the section of the middle core less than that of either of the outer cores. In this case the flux density and consequently the hysteresis and eddy current losses will be greater in it, and this will again upset the balance on the primary side of the transformer. It is practically impossible, therefore, to make this type of transformer so that the currents and voltages will be absolutely symmetrical.

When the resistances of the primary and secondary coils are negligible, the secondary mesh voltages v_2, v_2' and v_2'' can be written down easily. We have

$$v_2 = e_2 - e_2' = (n_2/n_1)(e_1' - e_1) = -(n_2/n_1)v_1,$$
$$v_2' = e_2' - e_2'' = (n_2/n_1)(e_1'' - e_1') = -(n_2/n_1)v_1',$$
$$v_2'' = e_2'' - e_2 = (n_2/n_1)(e_1 - e_1'') = -(n_2/n_1)v_1''.$$

Thus the secondary mesh voltages are in opposition in phase to the primary mesh voltages, and the ratio of their effective values is n_2/n_1.

One curious effect noticed with this type of transformer is the large increase in the primary magnetising currents when the centre of the primary star is connected with the neutral main. There are now harmonics of frequency $3(2n+1)f$ in the magnetising waves, and as these harmonic currents all produce fluxes in phase with one another, the return path for the corresponding flux must be through the air, and thus the leakage is excessive. Owing to the high reluctance of the path of this leakage flux the corresponding requisite magnetising currents will be large, and the harmonic currents of frequency $3(2n+1)f$, more particularly the harmonic current of triple frequency, will have appreciable amplitudes. This is found to be the case in practice.

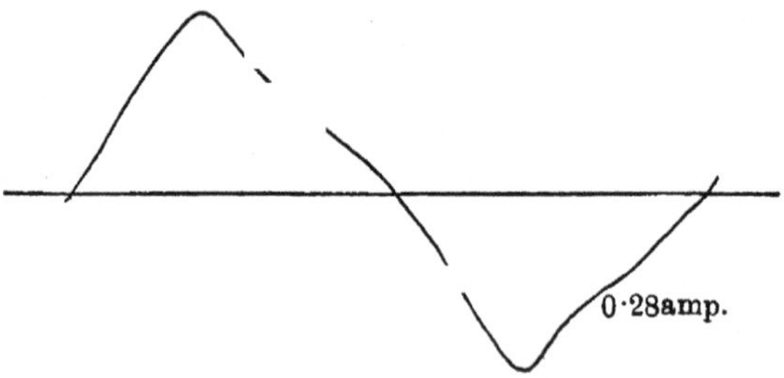

Fig. 142. Oscillograph record of the magnetising current in the winding of the middle core of a three phase transformer when the common junction of the primary windings is insulated.

The oscillograph records shown in Figs. 142 and 143 were obtained by F. J. Dykes from a transformer of this type with the three cores of equal sectional area. Fig. 142 gives the magnetising current of the middle core when the centre of the primary star is insulated. The effective value of this current was only 0·28 ampere. When the centre of the primary star was connected with the neutral main, the current assumed the shape shown in Fig. 143 and its effective value rose to 2·4 amperes. If a fourth core were provided, this rise in the value of the magnetising current would be prevented, but then the regulation afforded by a three phase transformer would be impaired seriously, as any want of balance in the secondary load would send a flux of fundamental frequency through the fourth core.

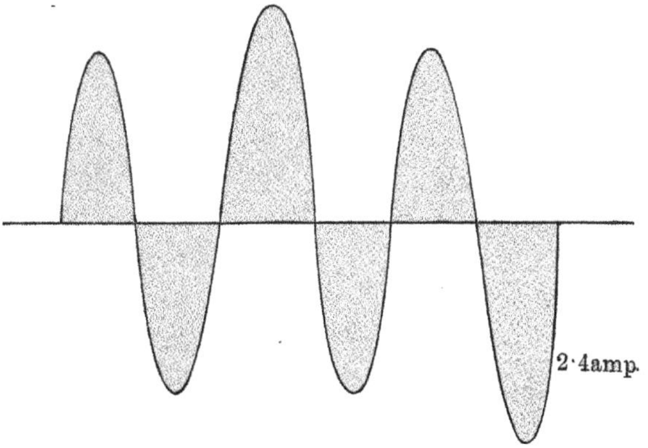

Fig. 143. Oscillograph record of the magnetising current when the common junction is connected with the neutral wire.

Mesh to star.

In the three phase transformer shown in Fig. 144, the primary windings are connected in mesh and the secondary in star. Let v_1, v_1' and v_1'' be the potential differences

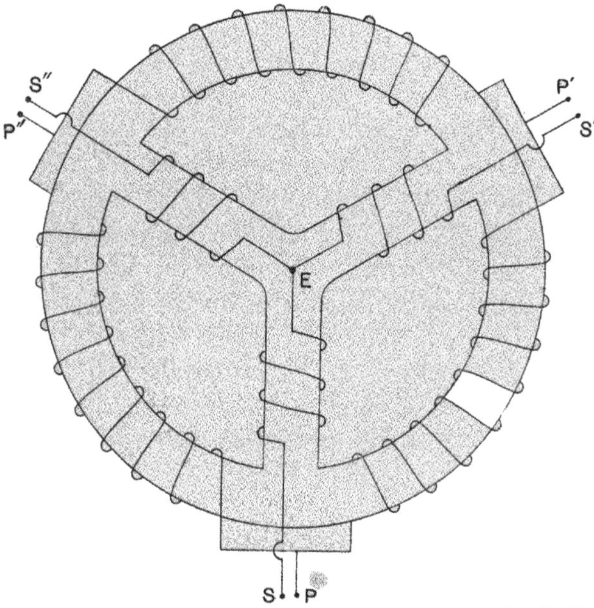

Fig. 144. Three phase transformer. Primary with mesh winding and secondary with star winding.

between the points P and P', P' and P'', and P'' and P respectively. Let also ϕ, ϕ' and ϕ'' be the magnetic fluxes embraced by the windings between the same three points. Then we have

$$v_1 = R_1 i_1 + n_1 \frac{\partial \phi}{\partial t}, \quad v_1' = R_1' i_1' + n_1 \frac{\partial \phi'}{\partial t}, \quad v_1'' = R_1'' i_1'' + n_1 \frac{\partial \phi''}{\partial t},$$

where R_1, R_1' and R_1'' are the resistances of the primary windings. Now $v_1 + v_1' + v_1''$ is always zero. Therefore, when the primary resistances are negligible,

$$\frac{\partial \phi}{\partial t} + \frac{\partial \phi'}{\partial t} + \frac{\partial \phi''}{\partial t} = 0.$$

Let v_2 be the potential difference between the secondary terminals S and S' (Fig. 144); then we have

$$- v_2 = r_2 i_2 - r_2' i_2' + n_2 \frac{\partial}{\partial t}(\phi - \phi'') - n_2 \frac{\partial}{\partial t}(\phi' - \phi).$$

When the resistances of the secondary windings are negligible, we can therefore write

$$- v_2 = n_2 \frac{\partial}{\partial t}(2\phi - \phi' - \phi'')$$

$$= 3n_2 \frac{\partial \phi}{\partial t} = 3\,(n_2/n_1)\,v_1.$$

Hence, the secondary mesh voltages are in exact opposition in phase to the primary mesh voltages, and the ratio of transformation is $n_1/3n_2$. When the effective values of the primary potential differences are all equal to one another, the secondary potential differences will also be equal. The effective value of the voltage between S and the centre point E of the secondary windings, in this case, will be $1/\sqrt{3}$ times the voltage between S and S'. It will, therefore, be equal to $\sqrt{3}\,(n_2/n_1)$ times the voltage between P and P'. The effective value of the flux in the radial limbs of the transformer is thus $\sqrt{3}$ times its value in the circumference of the stampings.

Mesh to mesh.

In Fig. 145 both the primary and the secondary windings of the three phase transformer are connected in mesh. With the same notation as in the preceding paragraph and neglecting the resistances of the primary and the secondary windings, we have

$$v_1 = n_1 \frac{\partial \phi}{\partial t}; \quad -v_2 = n_2 \left(\frac{\partial \phi'}{\partial t} - \frac{\partial \phi''}{\partial t}\right);$$

and thus
$$-v_2 = \frac{n_2}{n_1}(v_1' - v_1'').$$

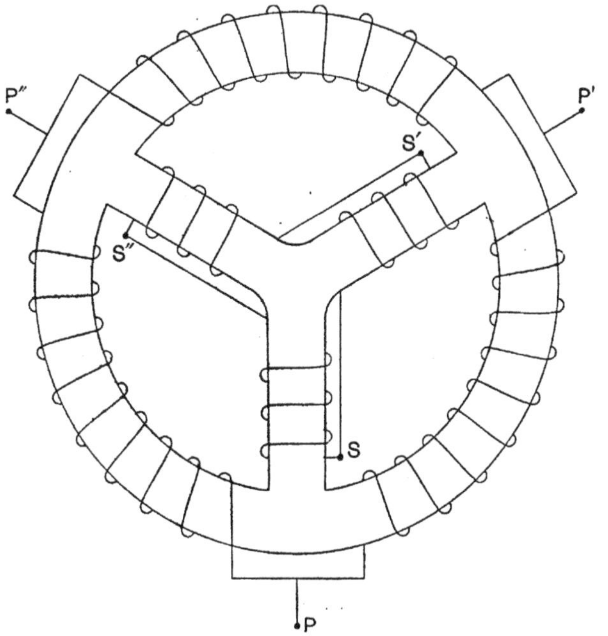

Fig. 145. Three phase transformer. The primary and secondary windings are both connected in mesh.

When the effective values of the primary potential differences are all equal, then

$$V_2 = \{n_2\sqrt{3}/n_1\} V_1.$$

Hence the ratio of transformation is $n_1/(n_2\sqrt{3})$.

The methods of transforming three phase to two phase currents or *vice versâ* by means of special transformers were explained in Vol. I, Chap. XVII. These transformations could also be made by means of three single phase transformers. If the potential differences across any

Three phase to two phase.

two of the three phase mains be V, it is necessary that two of the transformers be wound for an applied P.D. equal to $V/2$ and the third for an applied P.D. equal to $\sqrt{3}\,V/2$. The primaries of the two equal transformers are connected in series between the mains 1 and 2, and their common junction is connected to the main 3 through the primary of the other transformer. If the ratio of transformation of the two equal transformers be n_1/n_2, and that of the third transformer be $\sqrt{3}\,n_1/(2n_2)$, the secondaries of the two equal transformers connected in series will give a secondary voltage $n_2\,V/n_1$, and the secondary of the third transformer will give a secondary voltage $2n_2/(\sqrt{3}n_1)\times\sqrt{3}\,V/2$, that is, $(n_2/n_1)\,V_1$, which is in quadrature with the other secondary voltage. The same arrangement can also be used to transform two phase to three phase currents.

Three phase to single phase.

A single phase circuit may be supplied from any two of the secondary terminals of a three phase transformer. When, however, the single phase circuit is carrying the full load of the transformer, the potential drop at the secondary terminals is large and the heating of one of the secondary and one of the primary windings is excessive. One method which has been proposed of getting over this difficulty is to connect the secondary windings as follows. Let S_1S_2, $S_1'S_2'$ and $S_1''S_2''$ be the terminals of the three secondary windings, then, if we join S_2 and S_1', and S_2' and S_2'', so that we have two of the windings in series and the third in 'cross series' with them, we can get single phase currents between the terminals S_1 and S_1''.

Neglecting the primary resistances, we have, if the primaries are connected in mesh,

$$e_1 - e_1' = v_1 = n_1\frac{\partial\phi}{\partial t},$$

$$e_1' - e_1'' = v_1' = n_1\frac{\partial\phi'}{\partial t},$$

$$e_1'' - e_1 = v_1'' = n_1\frac{\partial\phi''}{\partial t}.$$

Therefore
$$\frac{\partial\phi}{\partial t}+\frac{\partial\phi'}{\partial t}+\frac{\partial\phi''}{\partial t}=0.$$

Again, if v_2 be the P.D. between S_1 and S_1'', we get

$$0 = 3r_2 i_2 + v_2 + n_2 \frac{\partial \phi}{\partial t} + n_2 \frac{\partial \phi'}{\partial t} - n_2 \frac{\partial \phi''}{\partial t},$$

and thus

$$2n_2 \frac{\partial \phi''}{\partial t} = 3r_2 i_2 + v_2,$$

and

$$2 \frac{n_2}{n_1} v_1'' = 3r_2 i_2 + v_2.$$

If we neglect $3r_2 i_2$ in comparison with v_2, we get $V_1/V_2 = n_1/(2n_2)$. Hence we see that V_2 is twice the P.D. between the terminals of one of the secondary coils. Also, if we suppose that the magnetising currents of the primary coils are negligible, we must have, at every instant, the currents in the primaries equal and opposite to the currents in the secondaries. Since the same

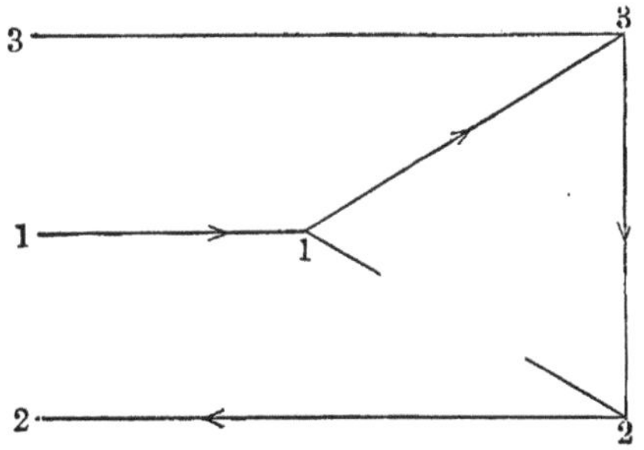

Fig. 146. Transforming three phase currents to single phase currents. The currents in the primary mains.

current flows in each of the secondaries, all the primary currents must be equal in magnitude, two of them flowing in the same direction round the mesh (Fig. 146) and the third flowing in the opposite direction. The currents will also be in step with one another, and, since the current in the branch 2—3 equals the current in the branch 3—1, there will be no current in the main 3. Hence, the currents for this transformer will all be supplied by the mains 1 and 2. Although all the primary coils are equally heated, the load is only on one pair of the supply mains. This method, therefore, does not distribute the load between the primary mains. The voltage drop also is large in practice, owing to the high resistance and leakage inductance of the secondary winding.

Three single phase transformers connected in star.

The case of three single phase transformers connected in star with their secondaries connected two in series and one in cross series (Fig. 147) is instructive. P, P' and P'', the primary terminals of the transformers, are connected with the three phase mains and their three other terminals are joined together. The single phase

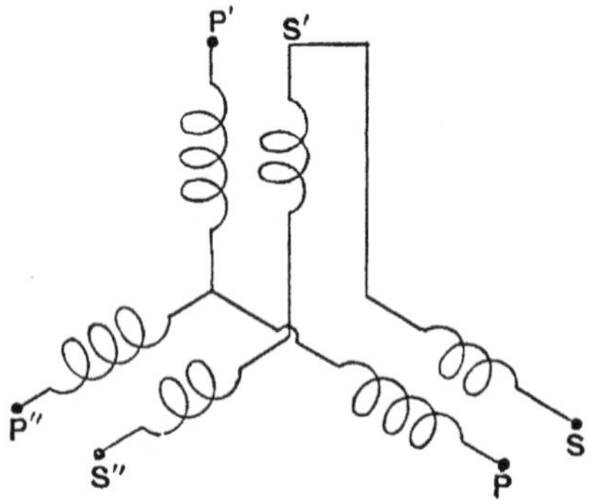

Fig. 147. Three single phase transformers. The primaries are connected in star and the secondaries are connected two in series and one in cross series. Single phase currents from the terminals, S and S''.

currents are obtained from S and S''. The problem can be solved analytically without difficulty. The solution shows that the arrangement acts like a transformer with excessive magnetic leakage. The load on the primary also is not balanced, and so the arrangement is of very little practical use.

Boosting transformer.

If the primary of a single phase transformer is connected with the supply mains, we get a certain pressure V_2 across the secondary terminals. If we now join one of the primary terminals and one of the secondary terminals together, by means of a wire, we get another pressure between the other primary and the other secondary terminal, and we can take electric ene gy from these two terminals. When a transformer is used in this fashion it is called an auto-transformer or a boosting transformer, or simply a booster, and the pressure can be boosted positively or negatively so that the pressure can be greater or less than the applied potential difference, depending on which of the primary and secondary terminals are connected by the conducting link. To a first approximation we can regard the primary as a battery having a voltage V_1 and the secondary as a battery having a voltage $-V_2$. When they are connected in series the boosted voltage is $V_1 - V_2$, and when they are connected in opposition or cross series the voltage is $V_1 + V_2$.

In Fig. 148 the connections are shown of a transformer connected as a negative booster, the pressure in the secondary circuit X being less than the applied pressure. ADB is the primary coil of the transformer which is connected between the supply mains, and AC is the secondary coil. The secondary load is placed across B and C. In this case,

Reducing the pressure.

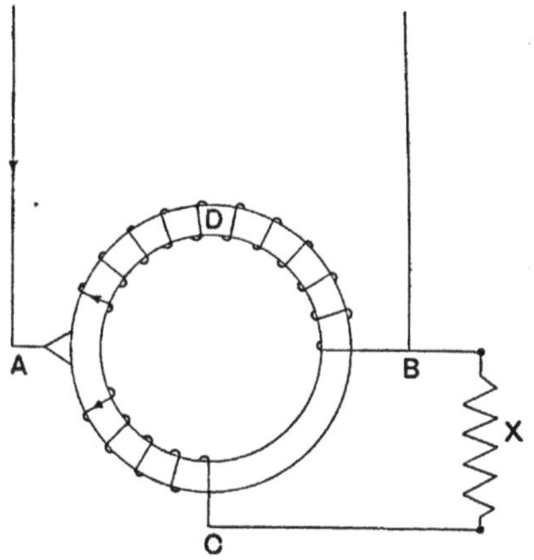

Fig. 148. Reducing the pressure by using a transformer as a negative booster.

when the resistance X is very great, the currents in both the divided circuits on the ring tend to magnetise the core in the same direction. If e_{BA} denote the P.D. between B and A, we have

$$e_{BA} = R_1 i_1 + n_1 \partial\phi/\partial t,$$

where R_1 is the resistance and n_1 the number of turns of the winding ADB. If the effective value of the P.D. between B and A be maintained constant, and if $R_1 A_1$ be negligible compared with it, we see that the effective value of $n_1 \partial\phi/\partial t$ is practically constant at all loads. To a first approximation, therefore, we may assume that the maximum value of the flux is constant at all loads, and that $n_1 i_1 + n_2 i_2$ is the same function of the time at all loads, where i_2 is the current and n_2 the number of turns in the coil AC. If i_0 be the current in ADB when CB is on open circuit, we have $n_1 i_1 + n_2 i_2 = n_1 i_0$, approximately, at all loads. At full load the effective value A_2 of i_2 is much greater than A_0. We see, therefore, that at full load i_1 and i_2 must differ in phase by nearly 180°.

We also have

$$e_{CA} = r_2 i_2 + n_2 \partial\phi/\partial t.$$

Hence

$$e_{BC} = R_1 i_1 - r_2 i_2 + (n_1 - n_2)\,\partial\phi/\partial t$$
$$= (n_1 - n_2)\,\partial\phi/\partial t, \text{ approximately.}$$

The effective value of the potential difference between B and C is therefore less than that between A and B which we suppose to be kept constant. Now, as the resistance X is diminished, the current in AC is increased, more lines of induction thread the core ADB and the induced electromotive force in the primary is in opposition to the applied potential difference, so that the current in the coil ADB continues to diminish until the electromotive force generated by the varying flux in the core becomes greater than the applied potential difference. The primary current then begins to increase again. The current in ADB is now in opposition to the applied potential difference, and the coil ADB is giving energy to the other branches of the circuit being actuated by the coil AC.

The action of a negative booster can readily be understood from a diagram (Fig. 149). Let us first consider the case when X is infinite. The triangle OB_0D is the primary voltage triangle of the transformer. OD represents the applied potential difference in magnitude and phase. B_0D represents the back electromotive force set up by the variation of the flux of induction in the core, and OB_0 is $R_1 A_0$, where A_0 is the magnetising current. Now, if there is no magnetic leakage, the E.M.F. caused by the flux in AC will equal $(n_2/n_1)\,DB_0$, where n_1 and n_2 are the number of turns in ADB and AC respectively (Fig. 148). Measure DC_0 equal to $(n_2/n_1)\,DB_0$, and join OC_0, then OC_0 gives the boosted voltage on open circuit.

Let a current now flow in the external circuit X; the resultant magnetising turns acting on the core are increased and the flux of induction is therefore increased. Let B_0B (Fig. 149) be the small increase of the back electromotive force due to the increased flux and make DC equal to $(n_2/n_1)\,DB$. Then OC is the pressure across X. Let R_1, r_2 and A_1, A_2 be the resistances and currents in the primary and secondary coils of the transformer. Then OB equals R_1A_1 and OC equals $(r_2+x)\,A_2$, where x is the resistance of the non-inductive load. OB and OC give the phases of the currents A_1 and A_2 respectively, and we see that when the load is heavy they are practically in opposition in phase. The phase of the resultant ampere-turns must always be very nearly coincident with OB_0. The magnitude of this resultant, however, does not remain constant but slightly increases.

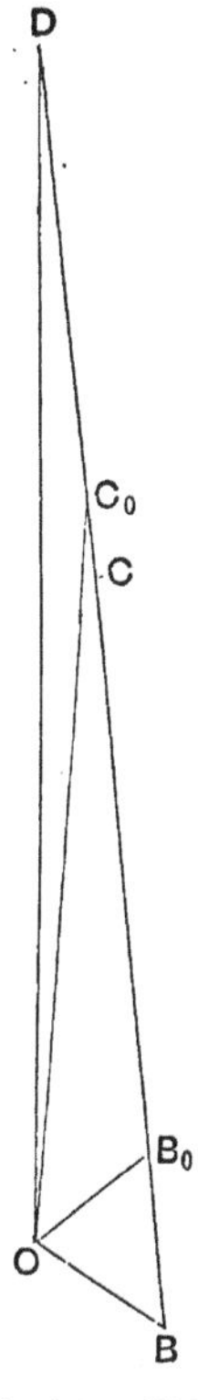

Fig. 149. Voltage diagram for a negative booster. OD is the applied P.D. and OC is the reduced pressure.

Since OB_0 gives approximately the phase of the resultant ampere-turns, we have

$$n_1A_1 \sin(\psi_1 - \psi_0) = n_2A_2 \sin(\psi_0 - \psi_2),$$

where ψ_0, ψ_1 and ψ_2 are the angles B_0OD, BOD and COD respectively. Also, since ψ_2 is small and the resultant ampere-turns are approximately n_1A_0, we have

$$n_1A_1 \cos\psi_1 + n_2A_2 = n_1A_0 \cos\psi_0.$$

We see that when A_2 is greater than

$$(n_1A_0 \cos\psi_0)/n_2,$$

ψ_1 must be greater than 90°.

Again, by construction,

$$DB = \{n_1/(n_1 - n_2)\}\, BC,$$

and therefore

$$\{(n_1 - n_2)/n_1\}\,(V_1 - R_1A_1 \cos\psi_1) = V_2 + r_2A_2 - R_1A_1 \cos\psi_1.$$

Hence, we have

$$V_2 = \{(n_1 - n_2)/n_1\}\, V_1 - \{r_2 + (n_2/n_1)^2 R_1\}\, A_2,$$

approximately.

The connections of a transformer when used as a booster are shown in Fig. 150. It will be seen that the currents tend to magnetise the core in opposite directions. In Fig. 151, OD is the applied potential difference, B_0D the back electromotive force in the primary coil, and OB_0 is $R_I A_0$. A transformer connected either as a booster or as a negative booster is also called an 'auto-transformer.'

Increasing the pressure. Boosting.

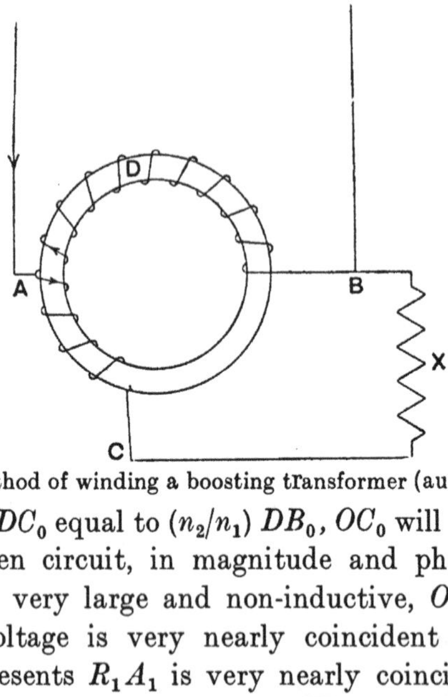

Fig. 150. Method of winding a boosting transformer (auto-transformer).

If we make DC_0 equal to $(n_2/n_1)\, DB_0$, OC_0 will give the boosted voltage, on open circuit, in magnitude and phase. When the resistance X is very large and non-inductive, OC the vector of the boosted voltage is very nearly coincident with OC_0, and OB which represents R_1A_1 is very nearly coincident with OB_0. Since there is a demagnetising effect, CB must be less than C_0B_0, but, if there is no magnetic leakage, DC is equal to $(n_2/n_1)\, DB$. In practice, the diminution of the flux in the core is very slight, and so, to a first approximation, we can suppose that the flux is constant at all loads. It has to be remembered that in the ordinary transformer OB is generally less than the hundredth part of OD even at full load.

Assuming that the resultant magnetising ampere-turns are represented in phase by OB_0 and that they are equal to n_1A_0 at all loads, we get

$$n_1 A_1 \sin(\psi_0 - \psi_1) = n_2 A_2 \sin(\psi_0 + \psi_2),$$

and since ψ_2 is very small, we also have

$$n_1 A_1 \cos\psi_1 - n_2 A_2 = n_1 A_0 \cos\psi_0, \text{ approximately.}$$

Again since

$$CB = \{(n_1 + n_2)/n_1\} DB,$$

we have

$$V_2 + r_2 A_2 - R_1 A_1 \cdot \cos \psi_1$$
$$= \{(n_1 + n_2)/n_1\} \{ V_1 - R_1 A_1 \cos \psi_1\},$$

and thus

$$V_2 = \{(n_1 + n_2)/n_1\} V_1 - \{r_2 + (n_2/n_1)^2 R_1\} A_2,$$

approximately.

The formulae for the voltage drop on the secondary of a negative booster, and for the voltage drop on the secondary of a transformer, having the same ratio of transformation

$$(n_1 - n_2)/n_1$$

and the same primary and secondary resistances R_1 and r_2 respectively, are

$$V_2 = \{(n_1 - n_2)/n_1\} V_1 - \{r_2 + (n_2/n_1)^2 R_1\} A_2$$

and

$$V_2 = \{(n_1 - n_2)/n_1\} V_1 - [r_2 + \{(n_1 - n_2)/n_1\}^2 R_1] A_2.$$

When n_2 is less than $n_1/2$, the negative booster gives the smaller drop. Hence, if we wish to reduce the pressure by less than fifty per cent., it is advisable to use a negative booster rather than a transformer. Similarly if we wish to raise the pressure by less than fifty per cent. a booster is preferable to a transformer.

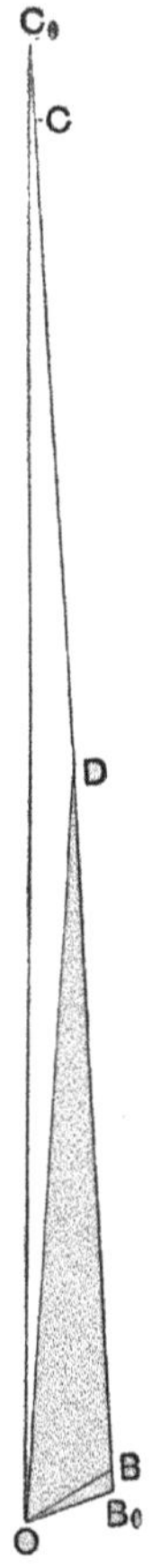

Fig. 151. Voltage diagram for a booster. *OD* is the applied P.D. and *OC* is the boosted pressure.

Variable induction transformer.

A useful method of varying the pressure on supply mains is illustrated in Figs. 152 and 153. The variable induction transformer consists of a laminated iron ring with a secondary coil wound round it, half of the coil being wound in one direction and half in the other. One end of this coil is connected with one primary main and the other is a secondary terminal, the other secondary terminal being connected directly with the primary main. The primary is wound on a bundle of iron stampings shaped as in the figure, and is capable of rotation round an axis coincident with the axis of the cylindrical ring. In the position shown in Fig. 152 both the primary and secondary currents tend to magnetise the halves of the ring

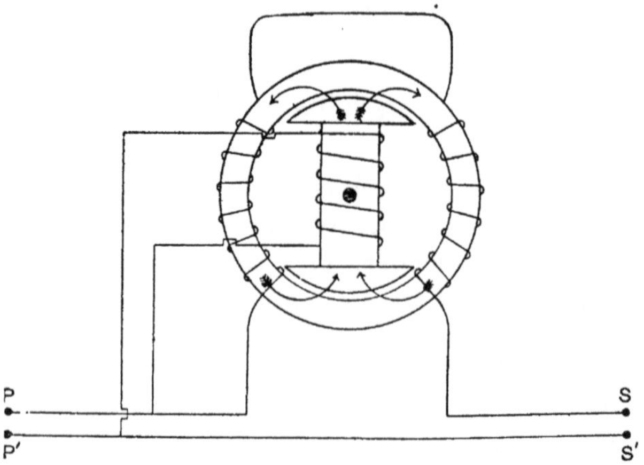

Fig. 152. Variable induction transformer. Position of rotating primary when the boosted pressure between S and S' is a maximum.

in the same direction. The induced pressure V' in the main winding between P and S is practically in opposition in phase to

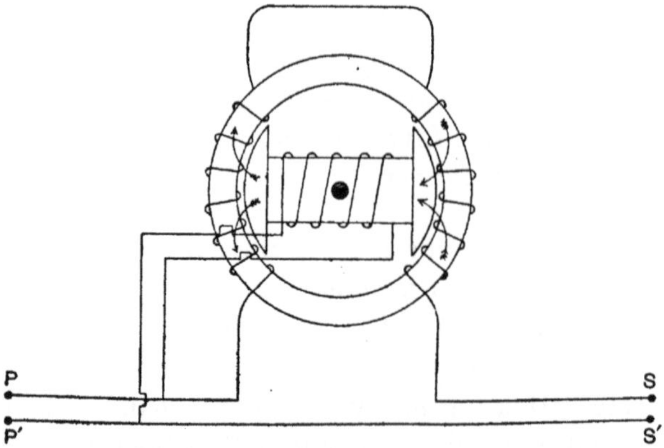

Fig. 153. Variable induction transformer. Position of rotating primary when the pressure between S and S' equals the pressure between P and P'.

the pressure V between P and P'. The pressure between S and S', therefore, must be $V + V'$. Since the applied P.D. on the primary side is maintained constant, the flux must be approximately constant, and hence we can easily write down formulae for the boosted voltage. In this case the reluctance of the magnetic circuit is considerable owing to the air-gaps, and the magnetising current of the primary is greater than for the closed iron circuit transformer.

If we rotate the primary through ninety degrees (Fig. 153) it will be seen from the figure that the induced electromotive forces neutralise one another, and so the pressure between S and S' equals the pressure between P and P'. If we rotate it through another ninety degrees, it will act as a negative booster and the pressure between S and S' will be less than that between P and P'.

Compensator.

The connections of an iron ring wound as a compensator are shown in Fig. 154. A and C are the terminals for the applied voltage, and the secondary loads are placed between various terminals connected with points on the wire coiled round the ring. Consider one of these circuits AB,

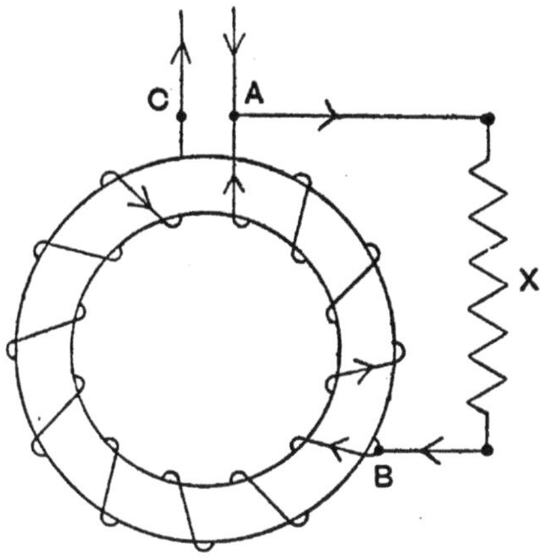

Fig. 154. Compensator. A and C are the terminals for the applied P.D. A and B are the secondary terminals (auto-transformer).

for example, in Fig. 154, and suppose that there are n_2 turns in the coil AB, and n_1 turns in the coil BC, so that there are $n_1 + n_2$ turns of wire round the ring. If V_1 be the effective value of the potential difference between A and C, $n_2 V_1/(n_1 + n_2)$ will be the potential difference, on open circuit, between A and B. We shall now find a formula to show how this voltage is maintained as the non-inductive load in the circuit X is increased, and compare the result with that obtained when we wind the ring as an ordinary transformer.

Let e_1 and e_2 be the potential differences between A and C, and between A and B respectively. Let i be the current in X, i_2 the current in AB, and i_1 the current in BC, which will also be the current in the mains. Let r be the resistance of one turn of the winding so that n_2r and n_1r are the resistances of AB and BC respectively. Our equations are

$$e_1 = xi + n_1\left(ri_1 + \frac{\partial\phi}{\partial t}\right),$$

$$e_2 = xi = n_2\left(ri_2 + \frac{\partial\phi}{\partial t}\right),$$

and $$i_1 = i_2 + i.$$

Thus $$xi = (n_2/n_1)(e_1 - xi) - n_2ri,$$

and therefore $$i = (n_2/n_1)\, e_1/[x\{(n_1 + n_2)/n_1\} + n_2r].$$

When, therefore, there is no magnetic leakage, i is always in phase with the applied potential difference. Hence

$$V_2\{(n_1 + n_2)/n_1\} = (n_2/n_1)\, V_1 - n_2rA,$$

and thus $$V_2 = \{n_2/(n_1 + n_2)\}\, V_1 - \{n_1n_2/(n_1 + n_2)\}\, rA$$
$$= \{n_2/(n_1 + n_2)\}\, V_1 - \{n_1/(n_1 + n_2)\}\, r_2A \ldots (a),$$

where r_2 is the resistance of the coil AB.

If we had wound the same ring as a transformer, having $n_1 + n_2$ turns on the primary and n_2 turns on the secondary, with wire of the same size as for the compensator so that more copper would be required, then for the secondary voltage of the transformer we should have

$$V_2 = \{n_2/(n_1 + n_2)\}\, V_1 - [r_2 + \{n_2/(n_1 + n_2)\}^2 R_1]\, A$$
$$= \{n_2/(n_1 + n_2)\}\, V_1 - \{(n_1 + 2n_2)/(n_1 + n_2)\}\, r_2A \ldots\ldots (b).$$

Comparing (a) with (b), we see that the compensator regulates better than the transformer.

In Fig. 155, let the phases of V_1, A_1 and A_2 be represented by OV_1, OK and OM respectively, and let OK and OM represent also the magnitudes of A_1 and A_2. Now, since

$$\dot{i}_1 = \dot{i}_2 + \dot{i},$$

it follows, by the triangle of vectors, that KM must represent A in magnitude and phase. Hence KM must be parallel to OV_1. Since the applied potential difference is maintained constant, the flux of induction in the ring will also be constant if the resistances of the windings are negligible. Therefore, $n_1\dot{i}_1 + n_2\dot{i}_2$, the resultant magnetising ampere-turns, must be equal to $n_1\dot{i}_0$. The vector value of this resultant must also be constant in magnitude and direction. To find this resultant we have to find the resultant of $n_1 . OK$ and $n_2 . OM$ in Fig. 155. Divide KM in L so that $n_1 . KL$ equals $n_2 . ML$. By the triangle of vectors we can replace $n_1 . OK$ by $n_1 . OL$ and $n_1 . LK$. We can also replace $n_2 . OM$ by $n_2 . OL$ and $n_2 . LM$. But by construction $n_1 . LK$ equals $n_2 . ML$, and since they are acting in opposite directions they balance. Hence OL represents the resultant magnetising ampere-turns in phase, and when there is no load on the secondary we see that OL equals A_0.

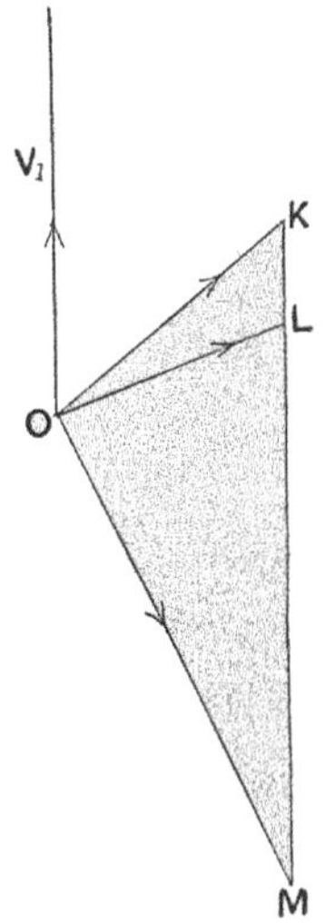

Fig. 155. Diagram of the currents in a compensator.

Let the angles V_1OK and V_1OM be ψ_1 and ψ_2 respectively, and let the angle V_1OL equal ψ_0, then, resolving along OV_1, we have

$$n_1 A_1 \cos\psi_1 + n_2 A_2 \cos\psi_2 = (n_1 + n_2) A_0 \cos\psi_0.$$

Also $\quad LM = \{n_1/(n_1 + n_2)\} A$, and $LK = \{n_2/(n_1 + n_2)\} A$.

We see from Fig. 155 that as A increases, A_1 (OK in the figure) continually increases. A_2 (OM) on the other hand at first diminishes to a minimum value. It is then in quadrature with V_1. It now begins to increase and is ultimately nearly in opposition to V_1, showing that the coil AB is acting like the secondary of a transformer which has BC for its primary.

Let us consider next the case of a compensator with n equal secondary circuits, and suppose that there are non-inductive resistances, $x_1, x_2, \ldots x_n$, placed across the terminals. If e be the value of the applied potential difference, and $e_1, e_2, \ldots$ be the values of the P.D.s across the secondary circuits, we have

$$e = e_1 + e_2 + \ldots + e_n,$$
$$e_1 = n_1 r i_1 + n_1 \partial\phi/\partial t = x_1 i_1',$$
$$e_2 = n_1 r i_2 + n_1 \partial\phi/\partial t = x_2 i_2',$$
$$\ldots\ldots\ldots\ldots\ldots\ldots\ldots\ldots\ldots\ldots$$
$$e_n = n_1 r i_n + n_1 \partial\phi/\partial t = x_n i_n',$$

where r is the n_1th part of the resistance of one of the coils, and $i_1, i_2, \ldots$ are the currents in the various coils.

We have $$i_1 + i_1' = i_2 + i_2' = \ldots = i_n + i_n'.$$
Therefore $$i_1 - i_2 = i_2' - i_1', \text{ etc.}$$

We also have
$$x_1 i_1' - x_2 i_2' = n_1 r (i_1 - i_2) = n_1 r (i_2' - i_1'),$$
and thus
$$(x_1 + R)\, i_1' = (x_2 + R)\, i_2' = (x_3 + R)\, i_3' = \ldots = k,$$
where R equals $n_1 r$.

Now $$e_1 + e_2 + \ldots + e_n = x_1 i_1' + x_2 i_2' + \ldots = e,$$
and therefore $$k\Sigma\{x/(x + R)\} = e.$$
Hence $$V_1 = \{x_1/(x_1 + R)\}\, V/\{\Sigma x/(x + R)\},$$
and the values of $V_2, V_3, \ldots V_n$ can be written down by symmetry. We see that when there is no magnetic leakage $V_1, V_2, \ldots$ are all in phase with V.

If there are only two sections, $V_1 + V_2$ will be equal to V, and V_1 will be greater than V_2, when x_1 is greater than x_2. The above equations show that the smaller the value of R the better will be the regulation. Compensators are very useful in practice for subdividing high pressure service.

Compensator for arc lamps. To illustrate the regulating power of compensators, the following experiment was made on a small compensator for subdividing a one-hundred volt service into two fifty volt circuits for use with arc lamps. The weight of the compensator complete was just under 30 pounds. When the pressure of the supply was 104 volts, the frequency being 84, the magnetising current was 0·33 of an ampere. The effective value of the potential difference across either *AB* or *BC* (Fig. 156) was 52 volts in this case. When the load between *B* and *C* was taking 30 amperes more than that between *A* and *B*, the P.D. across *BC* was 50, and across *BA* it was 54 volts. Therefore for small differences in the load this compensator regulated extremely well. When a current of 40 amperes was flowing in one secondary circuit and the other was open, the current in the main was 20 amperes, and this was practically its value in each half of the

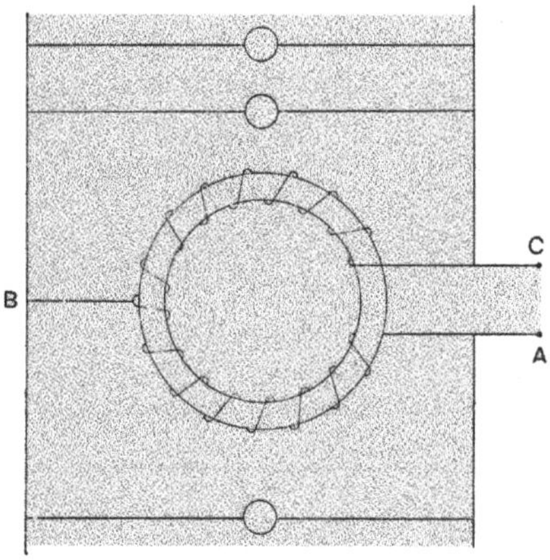

Fig. 156. Compensator for subdividing a 100 volt service.

ring, the currents in the ring windings being practically in opposition in phase. We see that in the secondary circuit we have a circulating local current of 20 amperes superposed on the main current of 20 amperes.

Current direction indicator.

In Central Stations where alternators are running in parallel, a transformer of special design is sometimes used to indicate the direction in which a particular current is flowing. One form of current direction indicator is shown in Fig. 157. The winding round the two outside limbs of the transformer is connected to the bus bars, and the winding on the inside limb is in series with the alternator. When the alternator is supplying current to the mains, the currents in the left and middle windings magnetise the core embraced by the windings connected to the green lamp and so it lights up. In this case the core embraced by the windings connected to the

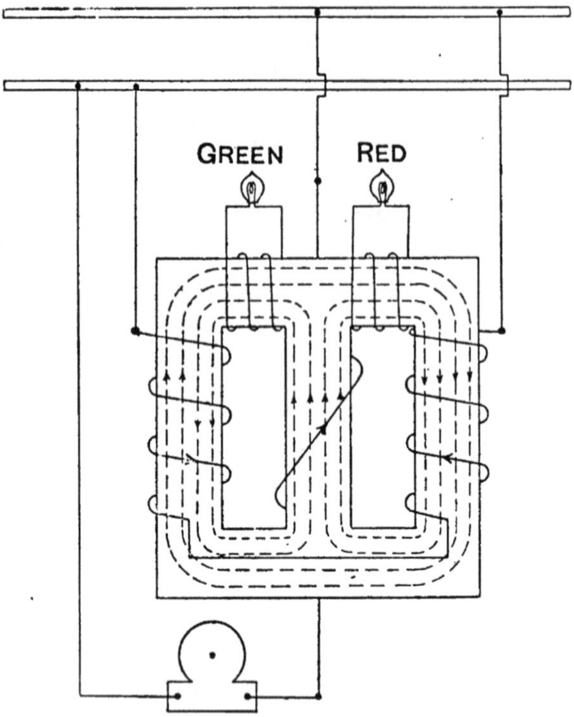

Fig. 157. Current direction indicator or discriminating transformer. When the red lamp lights the generator is receiving current from the bus bars instead of supplying current to the bus bars.

red lamp is only feebly magnetised. If, however the current in the middle winding reverses in direction, the red 'amp will light up and the green lamp go out. The arrow heads indicate that this is the case illustrated in the figure.

REFERENCES

A. RUSSELL, 'Formulae for Transformers,' *The Electrician*, Vol. 38, p. 725. March, 1897.

E. W. COWAN and L. ANDREWS, 'The Arrangement and Control of Long-distance Transmission Lines,' *Journ. of the I.E.E.*, Vol. 32, p. 901, 1903.

CHAPTER XII

Magnetic leakage. Secondary on open circuit. Loaded transformer. Magnetic equations. Fundamental equations. The value of k_1. Transformer diagram. Leakage lag in the secondary. Method of finding n. Equations connecting the currents. Formula for the secondary voltage. Efficiency of a leaky transformer. Effects of wave shape. Experimental tests on the effects of wave shape. Core losses. Induction density in the core. Experiment with Ganz machine (peaky waves). Method of calculation. Experiment with Wechsler machine (rounded waves). Efficiency formulae. Mathematical example. Equivalent net-work. Transformer diagram. Leakage reactance. Secondary P.D. drop on an inductive load. High voltage transformers. The heating of transformers. References.

Magnetic leakage.

We shall now investigate the effect of magnetic leakage on the working of the alternating current transformer. Let us consider the case of a constant pressure single phase transformer having n_1 turns in the primary coil and n_2 turns in the secondary. In practice, the magnetic lines due to the primary and secondary currents are not necessarily linked with all the primary and all the secondary turns. When we are dealing with the constant pressure transformer the error introduced by this assumption is not large, and, to a first approximation, the theory is in agreement with experiment. In practical work, however, a second approximation is necessary. In calculating, for instance, the difference of the pressures between the secondary terminals at no load and full load, an error of only one per cent. in the determination of V_2 may introduce an error of fifty per cent. into the calculated value of this pressure drop. We also need to know the various causes which produce this drop.

If ϕ_m denote the flux linked with the primary and secondary coils, we can write

$$\phi_m = (n_1 i_1 + n_2 i_2)/\mathcal{R},$$

where $\mathcal{R}$ is a variable quantity which, when we neglect the effects of eddy currents, can be determined from the hysteresis loop of the iron in the core taken between the maximum and minimum magnetising forces to which it is subjected. On page 279 we wrote $n_1 \partial\phi_a/\partial t$ for the back electromotive force due to the leakage flux from the primary, where ϕ_a is the mean value per turn of the primary leakage flux. If we suppose that the copper of the primary has infinite conductivity, so that there are no lines of force in the copper itself and that none of the leakage lines pass into the iron or cause eddy currents, this expression is strictly correct. In practice it is only approximately true. If the primary consist of a thick solid copper conductor, so that many of the lines of force embrace only a fraction of the current and the eddy current losses are appreciable, the error due to our assumption may be large. In most practical cases however the error is small, and we shall write $n_1 i_1/\mathcal{R}_a$ for ϕ_a, where $\mathcal{R}_a$ is a constant.

By Ohm's law the equation to determine the primary current is

$$i_1 = \left(e_1 - n_1 \frac{\partial \phi_m}{\partial t} - n_1 \frac{\partial \phi_a}{\partial t}\right) \bigg/ R_1,$$

or

$$e_1 = R_1 i_1 + n_1 \frac{\partial \phi_m}{\partial t} + n_1 \frac{\partial \phi_a}{\partial t} \quad \text{.................}(1).$$

In a similar way we can show that the equation for the secondary current is

$$-e_2 = r_2 i_2 + n_2 \frac{\partial \phi_m}{\partial t} + n_2 \frac{\partial \phi_b}{\partial t} \quad \text{..............}(2),$$

where e_2 is the secondary P.D. and r_2 is the resistance of the secondary winding.

Secondary on open circuit.

When the secondary is on open circuit, i_2 and ϕ_b are both zero, and therefore

$$-e_2 = n_2 \frac{\partial \phi_m}{\partial t}.$$

Hence by (1)

$$e_1 + \frac{n_1}{n_2} e_2 = R_1 i_1 + n_1 \frac{\partial \phi_a}{\partial t} \quad \text{.................}(3).$$

If e_1 and e_2 were in opposition in phase, the ratio of e_1 to e_2 would be constant (see Vol. I, Chapter x). The equation given above shows that this can only be rigorously true when R_1 is zero and either there is no primary leakage or the ratio of ϕ_m to ϕ_a is constant.

When the resistance of the primary can be neglected, we have

$$e_1 = n_1 \frac{\partial}{\partial t}(\phi_m + \phi_a),$$

and therefore $\phi_m + \phi_a$ is independent of the secondary load and depends only on the shape and the magnitude of the wave of the applied potential difference. Hence also, by the differential calculus, $\phi_m + \phi_a$ has a turning value, that is, a maximum or a minimum value, when e_1 vanishes. If we neglect the effects of eddy currents in the core, then, on open circuit, ϕ_m has a maximum value when i_1 has a maximum value. If we make the assumption that ϕ_a is in a constant ratio to i_1, ϕ_a will have a maximum value when i_1 has its maximum value. We see, therefore, that at the instant when the magnetising current of the transformer has a maximum value, the applied potential difference is zero. Similarly the applied potential difference vanishes when the magnetising current has a minimum value.

It also follows that when R_1 is negligible, e_2 vanishes with e_1, for at this instant the flux in the iron has a turning value. The time-lag between the primary and secondary voltages when the secondary is on open circuit is therefore 180 degrees. The phase difference between them, however, is not 180 degrees unless the ratio of ϕ_m to ϕ_a is constant, for the shapes of their waves are different in other cases.

Loaded transformer.

When the secondary of the transformer has a non-inductive load, we get from (1) and (2)

$$e_1 + \frac{n_1}{n_2} e_2 = R_1 i_1 - \frac{n_1}{n_2} r_2 i_2 + n_1 \frac{\partial \phi_a}{\partial t} - n_1 \frac{\partial \phi_b}{\partial t}.$$

At the instant when e_1 is zero we have, when R_1 is negligible,

$$e_2 = n_2 \frac{\partial \phi_a}{\partial t} - n_2 \frac{\partial \phi_b}{\partial t} - r_2 i_2.$$

In a constant pressure transformer the maximum values of the quantities on the right-hand side of this equation are small compared with the maximum value of e_2. Hence, when e_1 is zero, e_2 and consequently also i_2, since we are considering a non-inductive load, is small. But, when e_1 is zero, $\phi_m + \phi_a$ has a turning value and hence also the resultant magnetising turns must have a turning value. The positive turning value must be equal to $n_1 I_0$, since the maximum value of $\phi_m + \phi_a$ is the same at all loads. At this instant therefore i_1 must be practically equal to its maximum value I_0 on open circuit, since i_2 is small. The

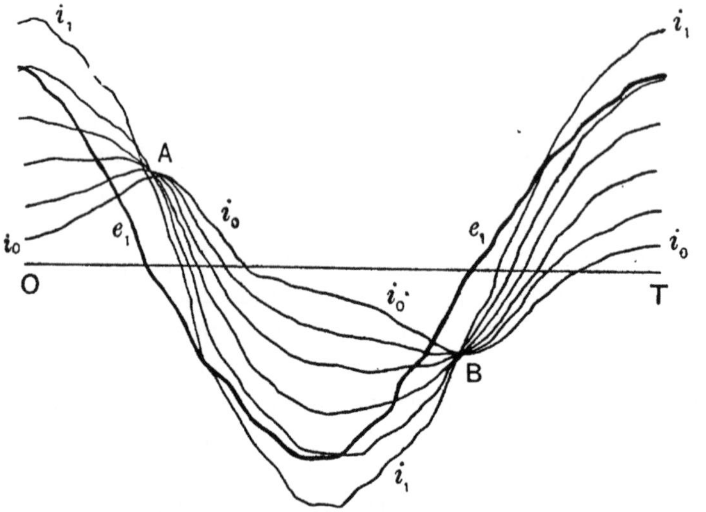

Fig. 158. The primary current waves of a Gaulard transformer at various non-inductive loads when e_1 is the shape of the applied potential difference wave. A and B are approximately the maximum and minimum heights of the magnetising current wave i_0.

curves in Fig. 158 show the applied potential difference wave and the waves of the primary current for various non-inductive loads on the secondary of a transformer. The curves were drawn by the ondograph of Hospitalier. It will be seen that all the primary current curves pass approximately through the turning points of the no-load primary current curve. These points are on the ordinates through the points where the P.D. wave cuts the time axis.

With the notation of this chapter we have

Magnetic equations.

$$\phi_m = (n_1 i_1 + n_2 i_2)/\mathcal{R},\ \phi_a = n_1 i_1/\mathcal{R}_a \text{ and } \phi_b = n_2 i_2/\mathcal{R}_b.$$

Hence

$$\phi_m + \phi_a = (n_1 i_1 + n_2 i_2)/\mathcal{R} + n_1 i_1/\mathcal{R}_a = (n_1 i_1 + k_1 n_2 i_2)/k_1 \mathcal{R} \ldots (4),$$

where

$$k_1 = \mathcal{R}_a/(\mathcal{R}_a + \mathcal{R}).$$

We also have

$$\phi_m + \phi_a = (n_1 i_1 + n_2 i_2)(1/\mathcal{R} + 1/\mathcal{R}_a) - n_2 i_2/\mathcal{R}_a,$$

and therefore $\phi_m = k_1(\phi_m + \phi_a) + k_1 n_2 i_2/\mathcal{R}_a.$

The equation (1) may be written in the form

Fundamental equations.

$$e_1 = R_1 i_1 + n_1 \frac{\partial}{\partial t}(\phi_m + \phi_a) \text{.................}(a),$$

and equation (2) is

$$-n_2 \frac{\partial \phi_m}{\partial t} = R_2 i_2 + n_2 \frac{\partial \phi_b}{\partial t},$$

where $R_2 = x + r_2$, x being the non-inductive load.

Hence by the preceding paragraph

$$-n_2 \frac{\partial}{\partial t}\{k_1(\phi_m + \phi_a)\} = R_2 i_2 + \frac{n_2^2}{\mathcal{R}_a}\frac{\partial}{\partial t}(k_1 i_2) + \frac{n_2^2}{\mathcal{R}_b}\frac{\partial i_2}{\partial t} \text{...}(b).$$

In Chapter VIII we showed that the equations to the air core transformer are

$$e_1 = R_1 i_1 + L_1 \frac{\partial}{\partial t}\{i_1 + (M/L_1) i_2\}$$

and

$$-M \frac{\partial}{\partial t}\{i_1 + (M/L_1) i_2\} = R_2 i_2 + L_2 \sigma \frac{\partial i_2}{\partial t}.$$

Comparing these equations with the equations (a) and (b) given above it will be seen that, when there is no iron in the core,

$$n_1(\phi_m + \phi_a) = L_1 i_1 + M i_2,$$

and $k_1 n_2(\phi_m + \phi_a) = M i_1 + (M^2/L_1) i_2 = (M/L_1) n_1(\phi_m + \phi_a).$

Therefore $k_1 n_2/n_1 = M/L_1.$

We see also, by comparing the equations, that

$$L_1 = n_1^2(1/\mathcal{R} + 1/\mathcal{R}_a), \quad L_2 = n_2^2(1/\mathcal{R} + 1/\mathcal{R}_b), \quad M = n_1 n_2/\mathcal{R},$$

and $\sigma = 1 - M^2/L_1 L_2 = 1 - \mathcal{R}_a \mathcal{R}_b/(\mathcal{R}_a + \mathcal{R})(\mathcal{R}_b + \mathcal{R}).$

If the primary applied potential difference be maintained constant and the primary resistance be negligible, we have shown that

$$L_1 i_1 + M i_2 = L_1 i_0.$$

Substituting $k_1 n_2/n_1$ for M/L_1 in this equation we find that

$$n_1 i_1 + k_1 n_2 i_2 = n_1 i_0,$$

an equation which holds for the iron core transformer (see formula (6), p. 340).

It follows from the preceding section that the equations for the iron core and the air core transformer are identical when we assume that k_1 is constant. We have already shown that when the effective value of the wave of the applied P.D. is maintained constant but its shape varied, the maximum value of the flux in the iron core varies. Hence the reluctance of the core and therefore also the value of k_1, which equals $\mathcal{R}_a/(\mathcal{R}_a + \mathcal{R})$, will vary with the wave shape. We shall make the assumption, when finding the approximate formulae required in practical work, that for a given effective value and for a given wave shape of the applied P.D. we can find the equations connecting the effective values of the currents and volts, and the mean value of the power, as if k_1 had a constant value.

The value of k_1.

In practice $\mathcal{R}_a$ is much greater than $\mathcal{R}$ except for two brief intervals every period, and so k_1 during nearly the whole period is approximately equal to unity. When $n_1 i_1 + n_2 i_2$ is zero, ϕ_m is finite owing to remanence, $\mathcal{R}$ must therefore be zero and so k_1 is unity. When $n_1 i_1 + n_2 i_2$ lies between zero and $n_1 i_c$, where $n_1 i_c$ is the value of the magnetising turns required to produce the coercive force, $\mathcal{R}$ is negative and varies from zero to $-\infty$. Hence k_1 must vary from 1 to ∞ and from $-\infty$ to 0 in the time that $n_1 i_1 + n_2 i_2$ takes to increase to its maximum value and diminish to zero again. When $n_1 i_1 + n_2 i_2$ is a little greater than $n_1 i_c$, k_1 attains a value k which is nearly equal to unity, and it retains this value during the time that $n_1 i_1 + n_2 i_2$ takes to increase to its maximum value and diminish to zero again.

Since k_1 equals $\mathcal{R}_a/(\mathcal{R}_a + \mathcal{R})$ we see that when $\mathcal{R}$ has the value $-\mathcal{R}_a$, k_1 is infinite. It is obvious therefore that if we only consider the instantaneous values of the variables the assumption that k_1 is constant is inadmissible. We can see from equation (6), p. 340, namely,

$$n_1 i_1 + k_1 n_2 i_2 = n_1 i_0,$$

that when k_1 is infinite i_2 is zero. Since, on our assumptions, ϕ_b vanishes with i_2, it follows that the flux in the core is the same as when the secondary is on open circuit, and so $i_1 = i_0$. Hence $k_1 n_2 i_2$ is zero when k_1 is infinite, and when k_1 is large its value is small and hence the effective value of $k_1 n_2 i_2$ may be written $k n_2 A_2$, where k is a fraction nearly equal to unity, without appreciable error, provided that the time taken by $n_1 i_1 + n_2 i_2$ to increase from zero to $n_1 i_c$ be short compared with the time it takes to reach its maximum value. It is more difficult to see the magnitude of the error introduced into our equations by the assumption that k_1 is constant in (*b*). It amounts to assuming that the hysteresis of the core is negligible, and that its permeability is constant. These assumptions are not admissible when we are considering instantaneous values; but when we are considering effective values, especially when the coercive force is small compared with the maximum magnetising force, as in the case of an open iron circuit transformer or a closed iron transformer working at a high flux density in the core, the equations deduced on this assumption are sufficiently accurate for all practical purposes. The formulae often express what happens with an accuracy which is within the limits of experimental error over the whole range of the permissible loads.

Transformer diagram.

The equations (*a*) and (*b*) given above can, when we may regard k_1 as constant, be studied readily by means of the diagram given in Fig. 159, which is almost identical with the fundamental diagram of the transformer when there is no magnetic leakage (Fig. 134, p. 290). In Fig. 159 OC represents the effective value of V_1 of the applied potential difference, OB represents $R_I A_I$, the electromotive force required to drive the current A_1 through the resistance R_1, and BC represents the effective value of $n_1 \partial (\phi_m + \phi_a)/\partial t$, that is, the effective value of the back electromotive force due to the varying flux linked with the primary circuit. These three electromotive forces balance one another, and therefore their vectors always form a triangle whatever may be the load on the transformer. In practice, OB is about one per cent. of OC at full load, and therefore, when the applied potential difference OC is maintained constant, BC will be approximately constant at all loads. Hence, the maximum value Φ of the flux of induction linked with the primary is approximately constant at all loads. If the shape of the wave of the applied P.D. does not alter, this flux will be about one per cent. less at full load than at no load.

Let us now consider the equation (*b*) for the current in the secondary; assuming that k_1 is constant it may be written in the form

$$-k_1 n_2 \frac{\partial}{\partial t}(\phi_m + \phi_a) = R_2 i_2 + (n_2^2/\mathcal{R}_b + k_1 n_2^2/\mathcal{R}_a)\frac{\partial i_2}{\partial t}.$$

If we produce CB (Fig. 159) to D and make

$$BD/BC = k_1 n_2/n_1 = n,$$

BD will represent the effective value of $-k_1 n_2 \frac{\partial}{\partial t}(\phi_m + \phi_a)$ in magnitude and phase. If there were no magnetic leakage, k_1 would be equal to unity. In actual transformers the value of k_1 is nearly equal to unity, but it varies with the shape of the applied wave.

When the resistance of the primary is negligible and the secondary is on open circuit, the electromotive force at its terminals is in phase with BD, but when there is a non-inductive load x on the secondary, the terminal potential difference is in phase with the current and is not in phase with BD. The resistance R_2 of the secondary circuit equals $r_2 + x$, and we may consider that its self-inductance is $n_2^2/\mathcal{R}_b + k_1 n_2^2/\mathcal{R}_a$, and that it is acted on by an electromotive force the vector of which is represented by BD. If BF represent

$$V_2 + r_2 A_2,$$

DF will represent the electromotive force due to the inductance of our hypothetical secondary circuit. Since this inductance is constant, BF and DF are at right angles to one another. We shall call DF the leakage E.M.F. of the transformer. It vanishes only when the primary and secondary leakages are zero.

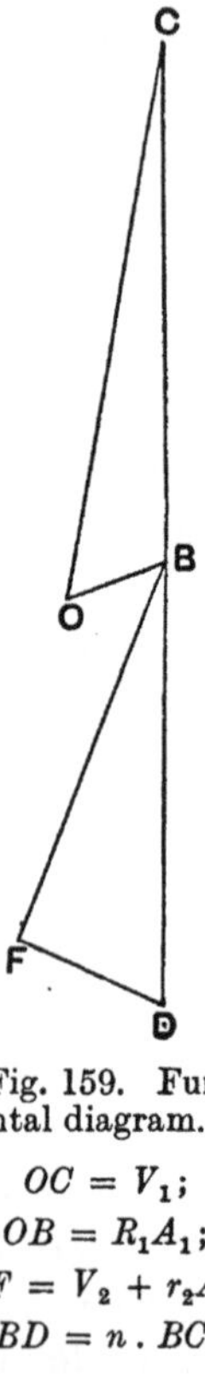

Fig. 159. Fundamental diagram.

$OC = V_1$;
$OB = R_1 A_1$;
$BF = V_2 + r_2 A_2$;
$BD = n \, . \, BC$.

To a first approximation BC is constant at all loads, and therefore also BD is nearly constant. If the wave of the applied P.D. were sine shaped, we should have DF equal to

$$\omega\,(n_2^2/\mathcal{R}_b + k_1 n_2^2/\mathcal{R}_a)\,A_2$$

approximately. We can therefore suppose that the leakage E.M.F. is approximately proportional to the secondary current.

Leakage lag in the secondary.

We shall call the angle DBF in Fig. 159 the angle of leakage lag in the secondary, and we shall denote it by θ. When the secondary current is zero, DF is zero, and therefore V_2 is in phase with BD. According to this diagram, if R_1 were zero, V_2 and V_1 would be in opposition in phase. We have seen earlier in this chapter that, although e_2 and e_1 vanish at the same instant, yet V_2 and V_1 are not in opposition in phase as the wave shapes of e_2 and e_1 are different. In practice it is almost impossible to detect any difference in shape between oscillograph records of e_2 and e_1, hence it will be seen from the numerical examples worked out in Vol. I, Chap. X, that the phase difference must be very nearly 180 degrees. Hence this error, which is due to the assumption we are making that k_1 is a constant, is, from the graphical point of view, a negligible one.

If the wave of E.M.F., the vector of which is BD, be sine shaped we have

$$\sin\theta = \omega\,(n_2^2/\mathfrak{R}_b + k_1 n_2^2/\mathfrak{R}_a)\,A_2/BD.$$

Now even at full load on ordinary transformers θ is very rarely as great as ten degrees. Hence, since the sine of a small angle is approximately equal to its circular measure, no great error is introduced, provided that the shape of the secondary E.M.F. wave does not alter, by the assumption that the angle of leakage lag in the secondary is proportional to the secondary current.

Method of finding n.

In Fig. 159 the angle BOC is the phase difference ψ_1 between the primary applied P.D. and the primary current. The cosine of this angle is the power factor of the primary circuit. Now, when the secondary is on open circuit, OB is $R_1 . A_0$, where A_0 is the magnetising current and the angle BOC is ψ_0. Hence we have

$$n = k_1 n_2/n_1 = BD/BC = E_2/(V_1 - R_1 A_0 \cos\psi_0) \;\ldots\ldots(5),$$

where E_2 is the effective value of the secondary E.M.F. on open circuit. This equation enables us to determine n easily. As a rule $R_1 A_0 \cos\psi_0$ is negligible compared with V_1.

From (4) we see that

$$k_1(\phi_m + \phi_a) = (n_1 i_1 + k_1 n_2 i_2)/\mathcal{R}.$$

Equations connecting the currents.

Also, from (1), we have

$$\phi_m + \phi_a = \frac{1}{n_1}\int_0^t (e_1 - R_1 i_1)\,\partial t,$$

if t be reckoned from the instant when $\phi_m + \phi_a$ is zero. If, therefore, we neglect $R_1 i_1$ in comparison with e_1, the wave of $\phi_m + \phi_a$ will practically be constant in magnitude and shape at all loads, provided that the effective value and the shape of the applied P.D. wave be maintained constant. The magnetising turns that produce $\phi_m + \phi_a$ must therefore be represented by the same function of the time at all loads, and hence

$$(n_1 i_1 + n_2 i_2)/\mathcal{R} + n_1 i_1/\mathcal{R}_a = n_1 i_0 (1/\mathcal{R} + 1/\mathcal{R}_a),$$

or

$$n_1 i_1 + k_1 n_2 i_2 = n_1 i_0 \quad \text{.......................(6)},$$

where k_1 equals $\mathcal{R}_a/(\mathcal{R}_a + \mathcal{R})$.

At full load, the maximum value of $\phi_m + \phi_a$ is slightly less than at no load owing to the term $R_1 i_1$ becoming appreciable. The magnetising turns $n_1 i_0$ are therefore also slightly less. In practice, however, the maximum value of $n_1 i_1$ at full load is much greater than the maximum value of $n_1 i_0$, and equation (6) shows that the difference between these two large quantities is always equal to a small quantity. Hence it is unnecessary to make the one or two per cent. correction to the small term $n_1 i_0$ as this correction is considerably within the possible errors of observation.

Making now the supposition that k_1 is constant, the equation (6) shows that the vectors A_1, A_2 and A_0 are in one plane. We see that the resultant of the magnetising turns $n_1 A_1$ and $k_1 n_2 A_2$ equals $n_1 A_0$. Hence resolving the vector values of the ampere-turns along and perpendicular to CD, we have

$$n_1 A_1 \cos\psi_1 - k_1 n_2 A_2 \cos\theta = n_1 A_0 \cos\psi_0,$$

and

$$n_1 A_1 \sin\psi_1 - k_1 n_2 A_2 \sin\theta = n_1 A_0 \sin\psi_0.$$

We may write

$$\left.\begin{aligned} nA_2\cos\theta &= A_1\cos\psi_1 - A_0\cos\psi_0 \\ \text{and}\quad nA_2\sin\theta &= A_1\sin\psi_1 - A_0\sin\psi_0 \end{aligned}\right\} \quad \text{............(7)},$$

where $n = k_1 n_2/n_1$

Hence we may use any of the three following equations to find θ:

$$\sin\theta = (A_1 \sin\psi_1 - A_0 \sin\psi_0)/nA_2 \quad\text{.............(8)},$$

$$\cos\theta = (A_1 \cos\psi_1 - A_0 \cos\psi_0)/nA_2 \quad\text{.............(9)},$$

and

$$\tan\theta = (A_1 \sin\psi_1 - A_0 \sin\psi_0)/(A_1 \cos\psi_1 - A_0 \cos\psi_0)\text{...(10)}.$$

If W_0 and W_1 express the power given to the primary when the secondary is on open and closed circuit respectively, we have

$$W_0 = V_1 A_0 \cos\psi_0 \quad\text{and}\quad W_1 = V_1 A_1 \cos\psi_1.$$

Equation (9) may therefore be written in the form

$$\cos\theta = (W_1 - W_0)/nV_1A_2 \quad\text{...............(11)}.$$

It will be seen that equation (10) is independent of n, and θ is calculated from the readings taken on the primary ammeter and wattmeter only. Equation (8) is also useful in finding θ, but equations (9) and (11) can only be used when all the quantities involved have been determined with the greatest accuracy. This is due to the fact that $\cos\theta$ only differs from unity by about 1·5 per cent. at full load, and so an error of one per cent. made in measuring $W_1 - W_0$ will make a large error in the value of θ deduced from (11).

It follows from Fig. 159 that

Formula for the secondary voltage.

$$\cos\theta = BF/BD, \quad\text{or}\quad BF = BD\cos\theta.$$

We also have

$$V_2 + r_2A_2 = n\cos\theta \,.\, BC = n\cos\theta\,(V_1 - R_1A_1\cos\psi_1),$$

approximately, and therefore by (9) we get

$$V_2 = n\cos\theta \,.\, V_1 - \{r_2 + (n\cos\theta)^2 R_1\}\,A_2 - n\cos\theta \,.\, R_1A_0\cos\psi_0$$
$$= n\cos\theta \,.\, V_1 - (n\cos\theta)^2\,QA_2 - n\cos\theta \,.\, R_1A_0\cos\psi_0,$$

where $$Q = R_1 + r_2/(n\cos\theta)^2.$$

If E_2 be the value of V_2 on open circuit, then, since $\cos\theta$ equals unity and A_2 is zero, we have

$$E_2 = nV_1 - nR_1A_0\cos\psi_0,$$

which agrees with (5).

We may therefore write

$$V_2 = E_2 \cos\theta - (n\cos\theta)^2 QA_2 \quad \text{............(12)}.$$

This equation gives us the following approximate equation to find the value of θ:

$$\cos\theta = \{V_2 + (r_2 + n^2R_1)A_2\}/E_2 \quad \text{............(13)}.$$

When an accurate electrostatic voltmeter is available and the alternator gives a steady effective E.M.F., this is a good method of finding θ.

Efficiency of a leaky transformer.

By equations (11) and (12) we have

$$A_2 = (W_1 - W_0)/nV_1\cos\theta,$$

and $V_2 = n\cos\theta\,(V_1 - R_1A_0\cos\psi_0) - (n\cos\theta)^2 QA_2$.

The efficiency η, therefore, is given by

$$\begin{aligned}\eta &= V_2A_2/W_1\\ &= (1 - W_0/W_1)(1 - R_1A_0\cos\psi_0/V_1 - n\cos\theta\,.\,QA_2/V_1)\\ &= (1 - W_0/W_1)(1 - QW_1/V_1^2 + W_0r_2/V_1^2n^2\cos^2\theta) \quad \text{...(14)}.\end{aligned}$$

Hence in ordinary transformers we can use the equation

$$\eta = (1 - W_0/W_1)(1 - QW_1/V_1^2) \text{............(15)}$$

to determine the efficiency.

Since Q equals $R_1 + (r_2/n^2)\sec^2\theta$, and even at full load the value of $\sec^2\theta$ is only about three per cent. greater than unity, it follows that no practical error is introduced by finding Q from the equation

$$Q = R_1 + r_2/n^2.$$

Slight magnetic leakage therefore makes very little difference in the efficiency of a transformer at a given load. If the transformer be rated by its output when the potential drop at the secondary terminals is 'x' per cent., we see from (12) that the greater the leakage for a given quantity of copper and iron the smaller will be the permissible maximum output. Hence designers of constant pressure transformers endeavour to reduce the magnetic leakage to a minimum.

Effects of wave shape.

The shape of the applied wave of P.D. has an important effect on the working of transformers, especially at light loads. If we neglect the primary resistance and reckon the time from the instant when e_1 is zero and increasing we have from (1)

$$\phi_m + \phi_a = \frac{1}{n_1}\int_0^t e_1 \partial t - \Phi = A_t/n_1 - \Phi \ldots\ldots\ldots(16),$$

where Φ is the maximum value of $\phi_m + \phi_a$. Hence if A denote the area of the positive half of the applied wave we have

$$\Phi = A \,.\, 10^8/2n_1.$$

Let v_m and $\bar{v}$ denote the mean height and the height of the centre of gravity of the applied wave respectively, then by Vol. I, p. 127, $V_1^2 = 2v_m\bar{v}$, and thus since $A = v_m\,(T/2) = v_m/2f$, we have

$$\Phi = V_1^2 \,.\, 10^8/(8fn_1\bar{v}) \quad \ldots\ldots\ldots\ldots\ldots\ldots(17).$$

It follows that for a given value of V_1 and a given frequency, the maximum value of $\phi_m + \phi_a$ varies inversely as $\bar{v}$. Since on no load ϕ_m and ϕ_a have their maximum values at the same instant, namely, when the current is a maximum, it follows that $\Phi_m + \Phi_a = \Phi$. In this case Φ_a is generally negligible compared with Φ_m, and hence Φ_m varies inversely as $\bar{v}$. In approximate work this assumption is generally permissible.

If we have a family of equivolt curves (see Vol. I, Chap. IV) those with a high centre of gravity, generally called 'pointed' or 'peaky' curves, produce a smaller induction than those with a low centre of gravity, called 'rounded' or 'square-shouldered' curves. Hence the induction density, for a given effective value of the applied P.D., varies with the shape of the wave, and therefore the magnetising current and the core losses vary considerably with it. It is therefore quite impossible to predict the efficiency and the load for a given secondary drop, that is, the power of the transformer when connected with a given circuit, unless we know the shape of the P.D. that will be applied to it.

Experimental tests on the effects of wave shape.

In 1895, G. Roessler made careful tests to find out how the working of a small transformer varied with the shape of the wave of the applied P.D. We shall apply our formulae to his experimental results to see how closely our theory agrees with experiment. The test is a severe one as the transformer experimented on had a capacity of about half a kilowatt only, and the effects of the resistance of the primary are appreciable at heavy loads. The transformer is of the closed iron circuit type. The author has shown in *The Electrician*, Vol. 42, p. 567, that his formulae apply even more closely to the open iron circuit type.

The following are the constants of the transformer on which the experiments were carried out. The iron plates used in the core were half a millimetre thick, and were well insulated. The total weight of the iron was 8·168 kilogrammes (18·01 lbs.). The numbers, n_1 and n_2, of turns in the primary and secondary respectively were 132 and 265. The primary and secondary resistances, R_1 and r_2, were 0·179 and 0·775 ohm when cold, and 0·214 and 0·943 ohm when hot, respectively, the resistances being heated by running the transformer for five hours at full load. As only a Cardew voltmeter was used, which took 35 watts at 120 volts, the open circuit volts could not be measured directly. Roessler gives 117·5 as the open circuit voltage, and this value has been adhered to in our calculations.

Core losses.

The transformer was tested first with a P.D. obtained from a 5·5 kilowatt four pole machine by Ganz and Co., in which the field magnets rotated, and afterwards

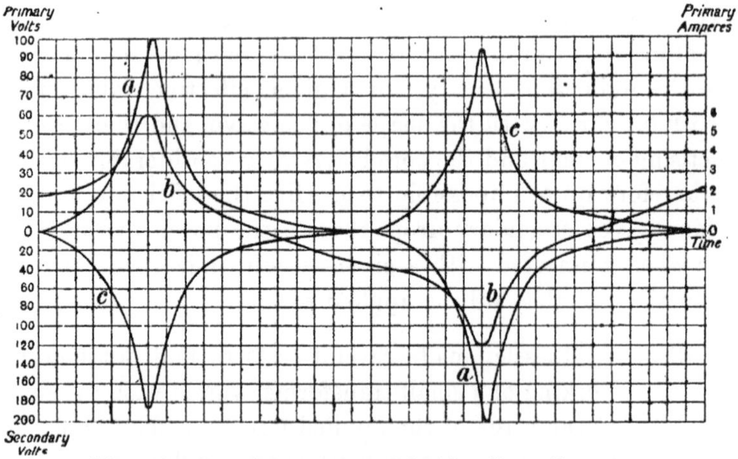

Fig. 160. Transformer connected with a Ganz Alternator.

(*a*) Primary potential difference wave.

(*b*) Primary current wave at half full load.

(*c*) Secondary potential difference wave at half full load. The scale of the secondary volts is half that of the primary.

with a P.D. from a small 0·5 kilowatt machine by Wechsler and Co., with four field poles and a rotating ring armature which had four coils.

The transformer was used as a step-up transformer, the effective voltage of the applied potential difference wave being 60 and of the secondary potential difference wave about twice as much. The shape of the potential difference wave of the Ganz machine is shown by the curve '*a*' in Fig. 160. The potential difference wave '*c*' of the secondary at half load is also shown. It will be seen that the two curves are approximately similar, and that they vanish at the same instants. The secondary load is non-inductive, and so the maximum value of the secondary current occurs at the instant when the secondary voltage is a maximum. Hence at this instant the primary current has also its maximum value. When e_1, e_2 and i_2 are all zero, i_1 must be equal to the maximum value of i_0, where i_0 is the magnetising current wave. The effective value V_1 of the primary voltage wave is 60, and the height of the centre of gravity of the area of the wave is 50·8.

In Fig. 161 the corresponding curves of the transformer when

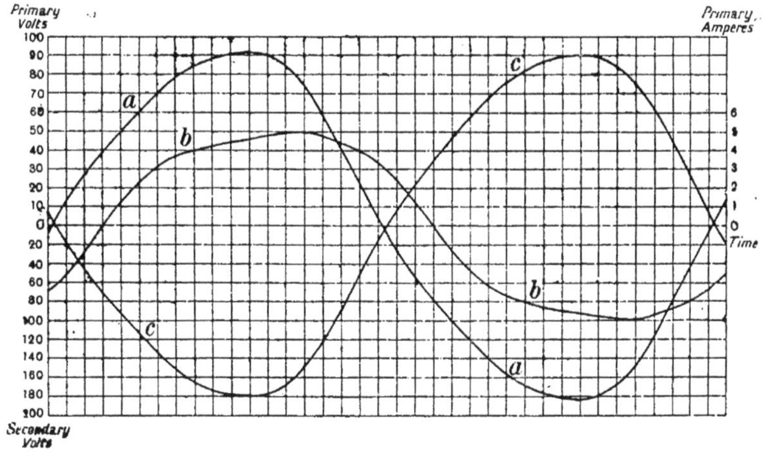

Fig. 161. Transformer connected with a Wechsler Alternator.

(*a*) Primary potential difference wave.

(*b*) Primary current wave at half full load.

(*c*) Secondary potential difference wave at half full load. The scale of the secondary volts is half that of the primary.

connected to the Wechsler alternator are shown. The effective primary voltage V_1 is 60, and the height of the centre of gravity of the applied P.D. wave is 35·7. We see from (17) that the ratio of the maximum value Φ of the flux, when the transformer is connected with the Wechsler machine to its maximum value when connected with the Ganz machine, is 50·8/35·7, that is, 1·42. Again, by Steinmetz's law, the hysteresis loss varies as the 1·6th power of the maximum induction density, and hence, if we make the assumption that the ratio of the maximum induction densities in the two cases is 1·42, the hysteresis loss will be 1·76 times greater with the Wechsler machine.

We saw in Chap. X that the eddy current loss in the core is practically constant if V_1 remain constant. We shall assume that its value is x in both tests. When connected to the Ganz machine W_0 is 34·5, and deducting the copper loss $0{\cdot}2\,(1{\cdot}46)^2$, that is, 0·4 watt (see Table I given below), we get 34·1 watts as the core loss. Similarly, we find the core loss with the rounded waves to be equal to 51·9. We have, therefore,

$$51{\cdot}9 - x = 1{\cdot}76\,(34{\cdot}1 - x),$$

and thus $x = 11$ watts approximately.

The eddy current loss cannot be calculated accurately by this method as the empirical law for the hysteresis loss is only approximately true. Assuming, however, that this value of the eddy current loss is exact, we easily find that the hysteresis loss in the iron is 2·3 watts per pound with the rounded waves, and 1·3 watts per pound with the peaky waves.

Induction density in the core.

In the experiments the mean value of the frequency f was 40·6, and the cross section of the core was 20·19 square centimetres.

Hence we find by means of (17) that with the pointed waves

$$\Phi/20{\cdot}19 = 8200,$$

and with the rounded waves

$$\Phi/20{\cdot}19 = 11700.$$

If we neglect Φ_a in comparison with Φ_m, we see that the maximum induction density in the core is 8200 with the peaky waves, and 11700 with the rounded waves.

In order that the figures may be compared readily, we have altered them so as to make the primary voltage 60 in each experiment. The values of A_1, A_2 and V_2 have been altered in the ratio of 60 to V_1, and the values of W_1 and W_2 in the ratio of 60^2 to V_1^2. The values of ψ_1 in the fifth column are calculated from the formula

$$\cos \psi_1 = W_1/V_1 A_1.$$

Experiment with Ganz machine (peaky waves).

In Table I, Roessler's results are given. The frequency of the alternating current is 40·6. The maximum induction density in the iron core is 8200. V_1 denotes the primary P.D., A_1 the primary current, W_1 the primary power, ψ_1 the phase difference between the primary current and P.D., V_2 the secondary P.D., A_2 the secondary current and η the efficiency.

The efficiency is calculated by the formula

$$\eta = W_2/W_1,$$

where W_2, the power in the external secondary circuit, equals $V_2 A_2$.

TABLE I. EXPERIMENTAL RESULTS.

Number of Experiment	V_1	A_1	W_1	ψ_1	V_2	A_2	W_2	η
1	60	1·462	34·5	66·9	—	0	0	0
2	60	1·809	71·6	48·7	117·5	0·302	35·4	0·494
3	60	2·144	99·4	39·5	117·0	0·511	59·7	0·607
4	60	2·982	156·9	28·8	116·7	1·004	117·2	0·747
5	60	3·761	205·3	24·5	115·0	1·427	164·2	0·800
6	60	4·512	252·9	20·9	114·3	1·817	207·7	0·822
7	60	5·554	317·9	17·4	114·0	2·344	267·3	0·841
8	60	6·787	390·9	16·5	112·5	2·972	334·7	0·856
9	60	7·575	433·7	17·3	111·3	3·365	374·7	0·864
10	60	8·825	509·9	15·6	110·3	3·992	440·3	0·864
11	60	—	576·9	—	108·7	4·580	498·0	0·863

In Table II the values of the various quantities, when the

applied P.D. is 60, are given. In calculating this table a knowledge of the following data only has been assumed:—

1. The magnetising power W_0, which equals 34·5 watts.
2. The magnetising current A_0, which equals 1·462 amperes.
3. The resistance constant Q, which equals 0·404 ohm.
4. The secondary voltage E_2, on open circuit, which is taken to be 117·5.

The angle θ (see the foot of the page) has been taken equal to (6·5)/3 degrees per ampere of secondary current.

The values of A_2 in this table are taken directly from Table I.

TABLE II. CALCULATED VALUES.

Number of Experiment	A_2 (from I)	ψ_1	A_1	W_1	V_2	W_2	η
1	0·000	66·9	1·462	34·5	117·5	0	0
2	0·302	49·2	1·786	70·0	117·0	35·3	0·505
3	0·511	40·9	2·084	94·5	116·7	59·6	0·631
4	1·004	29·2	2·912	153·0	115·9	116·0	0·763
5	1·427	24·0	3·678	202·0	115·3	164·0	0·815
6	1·817	21·1	4·428	248·0	114·4	208·0	0·838
7	2·344	18·7	5·439	309·0	113·5	266·0	0·861
8	2·972	17·5	6·673	382·0	112·3	334·0	0·874
9	3·365	17·0	7·446	427·0	111·5	375·0	0·878
10	3·992	16·9	8·714	500·0	110·2	440·0	0·880
11	4·580	17·1	9·946	570·0	108·8	498·0	0·880

The value of n is first found. We have, by (5),

$$n = E_2/(V_1 - R_1 A_0 \cos\psi_0) = 1{\cdot}96.$$

Method of calculation.

The values of the angle θ, in the first row of the following table, are calculated by formula (8), namely,

$$\sin\theta = (A_1 \sin\psi_1 - A_0 \sin\psi_0)/nA_2.$$

Number of Experiment	2	3	4	5	6	7	8	9	10	11
θ by (8)	1·0	1·0	2·6	4·4	4·2	3·9	5·7	7·8	7·5	—
θ by (10)	1·0	1·0	2·6	4·3	4·1	3·8	5·6	7·8	7·4	—
Probable θ ...	0·7	1·1	2·1	3·1	4·0	5·0	6·5	7·2	8·7	10·0

In the second row the figures are found by formula (10). The probable values of θ were found by plotting out the values of θ and A_2 given in the first two rows of the above table and drawing, through the origin, the straight line which makes the average deviation of these points from it a minimum.

If we calculate θ by the cosine formula (9), we find that in each experiment $\cos\theta$ comes out greater than unity, showing that there is probably a small error (one or two per cent.) in the determination of W_0. The leakage may be expressed by saying that the lag due to leakage when connected with the Ganz machine is 2·2 degrees per ampere of secondary current.

In Table II the column headed A_2 is taken from Table I. The column headed ψ_1 is calculated by the formula

$$\tan\psi_1 = (A_0 \sin\psi_0 + nA_2 \sin\theta)/(A_0 \cos\psi_0 + nA_2 \cos\theta),$$

which follows at once from (7). We see from the column headed ψ_1 in Table II that this angle attains its minimum value before the secondary current increases to 4·58 amperes.

The column headed A_1 is calculated by the formula

$$A_1 = A_0 \sin(\psi_0 - \theta)/\sin(\psi_1 - \theta) \quad \text{...........(18)}.$$

This formula follows readily from (7). W_1 is now got by evaluating $V_1 A_1 \cos\psi_1$. V_2 is calculated by (12), and hence, in our case, since $(n\cos\theta)^2 Q = 1{\cdot}5$ approximately, we have

$$V_2 = 117{\cdot}5 \cos\theta - 1{\cdot}5 A_2.$$

W_2 equals $V_2 A_2$, since the load is non-inductive, and the last column η is the ratio of W_2 to W_1. This method of calculating the efficiency at various loads is however not to be commended as it is affected by the errors made in calculating A_1, ψ_1 and V_2. A much simpler and more accurate method is given below.

Experiment with Wechsler machine (rounded waves).

In Table III the results of Roessler's experiments on this transformer when connected with a Wechsler machine are given. When necessary we have, as before, reduced his readings so as to make the effective primary voltage 60 in all the tests. The frequency was the same as in the preceding test. The flux density in the core, however, is now much higher, being approximately equal to 11700 (p. 346). The same notation as in Table I is employed, and the formulae used in calculating Table IV are the same as those used in calculating Table II.

TABLE III. EXPERIMENTAL RESULTS.

Number of Experiment	V_1	A_1	W_1	ψ_1	V_2	A_2	W_2	η
1	60	2·100	53·0	65·1	—	0	0	0
2	60	2·405	88·9	52·0	117·4	0·301	35·4	0·398
3	60	2·700	115·4	44·6	117·4	0·522	61·4	0·532
4	60	3·457	171·0	34·5	116·2	0·998	116·0	0·677
5	60	4·221	223·9	27·9	115·4	1·431	165·0	0·737
6	60	5·009	274·1	24·2	115·5	1·846	213·2	0·777
7	60	5·960	334·7	20·6	114·6	2·352	269·6	0·806
8	60	7·203	411·4	17·8	113·5	3·005	339·8	0·829
9	60	7·986	457·5	17·3	112·6	3·393	382·1	0·835
10	60	9·223	532·3	15·8	111·5	4·021	448·3	0·842
11	60	—	602·6	—	110·2	4·644	511·8	0·849

Comparing Tables III and I, we see that with the rounded waves a larger magnetising current and more power on no load are required. The voltage drop at the secondary terminals is more rapid with the peaky waves, but the efficiencies are higher. For example, with the peaky waves the efficiency is 49·4 per cent., and with the rounded waves it is 39·8 per cent., when the load is 35·4 watts.

TABLE IV. CALCULATED VALUES.

Number of Experiment	A_2 (from III)	ψ_1	A_1	W_1	V_2	W_2	η
1	0	65·1	2·100	53·0	117·5	0	0
2	0·301	52·1	2·401	88·5	117·0	35·2	0·398
3	0·522	45·0	2·683	114·0	116·7	60·9	0·535
4	0·998	34·2	3·432	170·0	116·0	116·0	0·680
5	1·431	28·1	4·178	221·0	115·3	165·0	0·746
6	1·846	24·3	4·937	270·0	114·6	212·0	0·784
7	2·352	20·9	5·875	329·0	113·3	266·0	0·809
8	3·005	17·9	7·106	406·0	112·8	339·0	0·835
9	3·393	17·2	7·875	451·0	112·2	381·0	0·843
10	4·021	15·8	9·080	524·0	111·2	447·0	0·853
11	4·644	15·3	10·30	596·0	110·1	511·0	0·858

In Table IV the values of the various quantities are calculated from the following data:—

1. W_0 equals 53·0 watts.
2. A_0 equals 2·10 amperes.
3. Q equals 0·404 ohm.
4. E_2 equals 117·5 volts.

The angle θ has been taken equal to 1·1 degree per ampere of secondary current.

Tables III and IV show a very satisfactory agreement. We find n from the formula

$$n = E_2/(V_1 - R_1 A_0 \cos \psi_0) = 1{\cdot}998.$$

The angles of leakage lag calculated by the same formulae as before are given in the following table.

Number of Experiment	2	3	4	5	6	7	8	9	10	11
θ by (8)	0·0	0·0	1·6	1·6	2·3	2·4	2·9	4·0	4·4	—
θ by (10)	0·0	0·0	1·5	1·4	2·3	2·3	2·8	4·0	4·3	—
Probable θ ...	0·3	0·6	1·1	1·6	2·0	2·6	3·3	3·8	4·4	5·2

Efficiency formulae.

The most direct method of calculating the efficiency is by means of the approximate formula (15),

$$\eta = (1 - W_0/W_1)(1 - QW_1/V_1^2).$$

With the peaky waves W_0 equals 34·5 watts and Q equals 0·404 ohm. Hence

$$\eta = (1 - 34{\cdot}5/W_1)(1 - 0{\cdot}404\, W_1/3600).$$

The following calculated efficiencies were obtained by this formula.

Number of Experiment	2	3	4	5	6	7	8	9	10	11
Observed η ...	0·494	0·607	0·747	0·800	0·822	0·841	0·856	0·864	0·864	0·863
Calculated η	0·514	0·645	0·767	0·813	0·839	0·860	0·871	0·875	0·879	0·879

The discrepancy between the observed and calculated values is probably due to an error in the measurement of W_0. With the rounded waves the formula for the efficiency is

$$\eta = (1 - 53/W_1)(1 - 0{\cdot}404\,W_1/3600).$$

Number of Experiment	2	3	4	5	6	7	8	9	10	11
Observed η ...	0·398	0·532	0·677	0·737	0·777	0·806	0·829	0·835	0·842	0·849
Calculated η	0·400	0·534	0·677	0·744	0·782	0·806	0·831	0·839	0·847	0·850

It will be seen that the observed and calculated values agree very closely.

In such a small transformer if we wish a three figure accuracy we cannot neglect the last term in formula (14). This slightly increases all the calculated values of η in the tables given above. The maximum correction is for the rounded waves at full load, and equals + 0·003.

In these calculations no attempt has been made to take into account the differences in the core loss at no load and at full load. In the actual transformer, however, the core loss is appreciably less at full load than at no load. The back E.M.F. in the primary at full load (BC in Fig. 159) is $V_1 - R_1A_1 \cos \psi_1$ nearly, and this equals 60 – 2, that is, 58 volts. Hence the induction density at full load is 3·3 per cent. less than at no load, and therefore the hysteresis loss is 5·3 per cent. less. The eddy current loss will be 6·6 per cent. less, owing to the diminished value of the effective E.M.F. causing the eddy currents.

Mathematical example.

The height of the centre of gravity of the E.M.F. wave produced by an alternator A is twice as great as that produced by an alternator B, and the effective value of the volts in each case is 2500. A ten kilowatt transformer tested with the former alternator gives a core loss of 100 watts on open circuit, and a tenth of this loss is due to eddy currents. If Q equals 10 ohms, let us find the efficiency of this transformer at one-twentieth, one-tenth, one-half, and at full load, (1) when tested with A, and (2) when tested with B.

Since the height of the centre of gravity of the wave produced by B is only one-half the height of that of A, the maximum induction density in the core is twice as great when tested on B. Hence, by Steinmetz's law, the core loss when tested on B will be

$$2^{1 \cdot 6}(100 - 10) + 10; \text{ that is, } 282{\cdot}8 \text{ watts.}$$

Using the formulae

$$\eta_A = (1 - 100/W_1)(1 - 10\, W_1/2500^2)$$

and

$$\eta_B = (1 - 283/W_1)(1 - 10\, W_1/2500^2),$$

we get the following table:

Load ...	1/20th	1/10th	1/2	Full load
Efficiency with A...	80·0	89·9	97·2	97·4
Efficiency with B...	43·4	71·5	93·5	95·6

This example illustrates the necessity of giving the height of the centre of gravity of the area of the applied P.D. wave used during the test, as well as its effective value.

The constructions given in the preceding chapter for polyphase transformers, boosters and compensators may be extended in a similar way so as to take the leakage of the magnetic lines into account. From these constructions formulae may be deduced which are useful in practical work. They show, for instance, the appreciable effects produced in ordinary working by alterations in the shape of the applied P.D.

Equivalent net-work.

When the secondary load has capacity or inductance, one of the best ways of considering the problem is to use an equivalent net-work as in Chapter X. Suppose, for example, that the load is an inductive coil (x, N). Assuming that k_1 is constant, equations (a) and (b), given on p. 335, may be written

$$e_1 = R_1 i_1 + n_1 \frac{\partial}{\partial t}(\phi_m + \phi_a),$$

$$-k_1 n_2 \frac{\partial}{\partial t}(\phi_m + \phi_a) = (r_2 + x)\, i_2 + N \frac{\partial i_2}{\partial t} + (n_2^2/\mathcal{R}_b + k_1 n_2^2/\mathcal{R}_a)\frac{\partial i_2}{\partial t}.$$

They may also be written

$$e_1 - R_1 i_1 = n_1 \frac{\partial}{\partial t}(\phi_m + \phi_a)$$

$$= -\{(r_2 + x)/n\} i_2 - (N/n)\frac{\partial i_2}{\partial t} - (1/n)(n_2^2/\mathcal{R}_b + k_1 n_2^2/\mathcal{R}_a)\frac{\partial i_2}{\partial t}$$

$$= \{(r_2 + x)/n^2\} i' + (N/n^2)\frac{\partial i'}{\partial t} + (1/n^2)(n_2^2/\mathcal{R}_b + k_1 n_2^2/\mathcal{R}_a)\frac{\partial i'}{\partial t},$$

where $i' = -ni_2$, n being equal to $k_1 n_2/n_1$. We can also write

$$I = i_1 - i'.$$

These equations suggest the equivalent net-work shown in Fig. 162. A non-inductive resistance R_1 is placed in series with an imaginary choking coil T which acts in exactly the same way

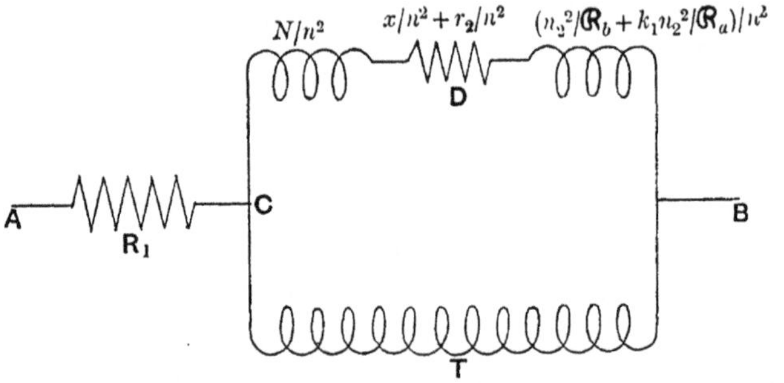

Fig. 162. Equivalent net-work of a leaky transformer on an inductive load (x, N). T acts in the same way as the primary of the transformer would, if it had no resistance and the secondary was on open circuit. k_1 is the leakage constant and n equals $k_1 n_2/n_1$.

as the primary coil of the transformer would, if it had zero resistance and the secondary was on open circuit. Across the terminals of this choking coil are placed in series two choking coils $(0, N/n^2)$ and $\{0, (1/n^2)(n_2^2/\mathcal{R}_b + k_1 n_2^2/\mathcal{R}_a)\}$ and also a non-inductive resistance $(r_2 + x)/n^2$. If the primary P.D. be applied across A and B, the current in R_1 will be equal to the current in the primary coil of the actual transformer when the load is (x, N), and the current in the secondary will be equal to $1/n$ times the current in CDB and will be in opposition to it in phase. The secondary potential difference V_2 will be equal to $1/n$ times the P.D. across the inductive coil $(x/n^2, N/n^2)$ in the circuit CDB and will be in opposition to it in phase.

In the ordinary transformer R_1 is negligible, hence the applied P.D. at the terminals of T (Fig. 162) is constant at all loads. I is therefore constant at all loads and we have

$$\dot{i}_1 + n\dot{i}_2 = \dot{i}_0$$

as before (p. 340).

Transformer diagram.

In Fig. 163 let OY give the phase of the applied potential difference and OA represent the magnetising current. Then if OP represents the primary current vector A_1 for a particular load, AP will represent nA_2 when the primary resistance R_1 is negligible, for in this case we always have

$$\dot{i}_1 + n\dot{i}_2 = \dot{i}_0.$$

If we now suppose that the applied P.D. wave is sine shaped

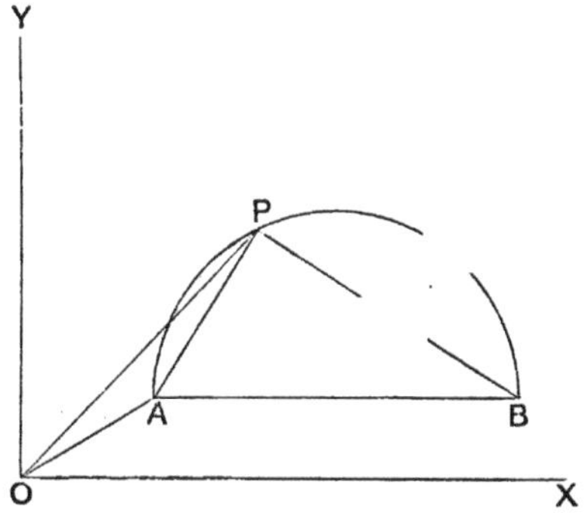

Fig. 163. Transformer diagram. Locus of P is a circle. OP is the primary current vector and OY gives the phase of the applied P.D.

and that θ is the phase difference between the P.D. applied to CDB (Fig. 162) and i', so that θ is the angle we have defined as the angle of magnetic leakage, then, when the secondary load is non-inductive, we have

$$\sin\theta = (\omega/n^2)\,(n_2^2/\mathcal{R}_b + k_1 n_2^2/\mathcal{R}_a)\,A'/V_1$$
$$= (\omega A_2/nV_1)\,(n_2^2/\mathcal{R}_b + k_1 n_2^2/\mathcal{R}_a).$$

Draw AB (Fig. 163) parallel to OX and PB perpendicular to AP. The angle ABP equals the angle between AP and OY and is therefore equal to θ. We also have

$$AB = AP/\sin\theta = nA_2/\sin\theta = n^2 V_1/\{\omega\,(n_2^2/\mathcal{R}_b + k_1 n_2^2/\mathcal{R}_a)\}.$$

Thus, when the load is non-inductive and the applied potential difference V_1 is maintained constant, the locus of P is a circle. It has to be remembered that in proving this theorem we have assumed that the applied wave of P.D. is sine shaped and that R_1 is negligible.

We shall now make the further assumption that r_2 is negligible. In this case P coincides with B so that AB equals nA_s, where A_s represents the short circuit current in the secondary. Therefore

$$nV_1/A_s = (n_2^2/\mathcal{R}_b + k_1 n_2^2/\mathcal{R}_a)\,\omega.$$

Leakage reactance.

The quantity $(n_2^2/\mathcal{R}_b + k_1 n_2^2/\mathcal{R}_a)\,\omega$ is called the 'leakage reactance.' It can be measured very easily. If we short-circuit the secondary terminals of a transformer through an ammeter of negligible resistance and inductance, and gradually increase the P.D. applied to the primary terminals until the secondary current equals the full load current A_2 of the transformer, we have

$$nV_1'/A_2 = (n_2^2/\mathcal{R}_b + k_1 n_2^2/\mathcal{R}_a)\,\omega,$$

where V_1' is the applied P.D. when A_2 is the short-circuit current in the secondary. This expression may be taken as a measure of the leakage of the transformer, and can be employed usefully in conjunction with the diagram given in Fig. 163. If the applied P.D. wave be not a sine curve, then the leakage reactance can still be measured by the ratio of nV_1' to A_2, but the equation given above has to be modified.

High voltage transformers.

For the economical transmission of power very high voltages are necessary. In some cases pressures greater than 50000 volts are used. In these cases the transformers are of large size, and, even when the losses are only one per cent. of the full load output, this may represent twenty or thirty kilowatts. Special arrangements have then to be employed to keep the transformers cool. There are three methods in general use. In the first method the transformers are cooled by currents of air produced by electric fans. In the second method they are immersed in oil, contained generally in an iron case which is corrugated so as to increase the cooling surface. In the third method we have large spirals of brass tubing, through whcih water is kept circulating, immersed in the oil so as to keep it cool. Both core and shell type transformers are employed, the magnetic material usually consisting of plates of a steel alloy. Numerous ventilating ducts are made, through which the oil or air circulates when the transformer is working. The coils are arranged in layers, so that wires at great differences of potential are kept well apart from one another. This construction also admits of sandwiching the primary and secondary coils, and so making the magnetic leakage a minimum.

In Fig. 164 is shown the efficiency curve of one of the 2340 kilovolt-ampere transformers made by the Oerlikon Company for the power transmission plant at Caffaro (see p. 113). The transformer is in a cast-iron case containing oil, and water cooling

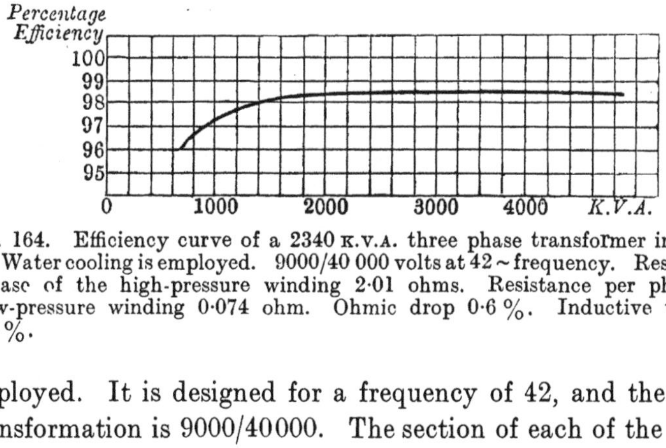

Fig. 164. Efficiency curve of a 2340 K.V.A. three phase transformer in oil at 60° C. Water cooling is employed. 9000/40 000 volts at 42 ~ frequency. Resistance per phase of the high-pressure winding 2·01 ohms. Resistance per phase of the low-pressure winding 0·074 ohm. Ohmic drop 0·6 %. Inductive voltage drop 5 %.

is employed. It is designed for a frequency of 42, and the ratio of transformation is 9000/40000. The section of each of the three cores is rectangular, and the cores are arranged side by side. The windings are of copper strip insulated by presspahn. The high-pressure coils are each divided into 36 sections, and are wound outside the low-pressure coils. The resistance per phase of the high-pressure winding when warm is 2·01 ohms, and the resistance per phase of the low-pressure winding is 0·074 ohm. The temperature of the oil during the test was 60° C. The iron losses at all loads are approximately 20·5 kilowatts. The drop in volts due to the resistances of the primary and secondary coils at full load is 0·6 per cent. and the reactive drop, obtained by finding the primary voltage required to get full load secondary current in the short-circuited secondary coil, is five per cent. of the open circuit secondary voltage. A pressure of 60000 volts was maintained between the high tension and low tension coils, in parallel, and the iron case for half-an-hour in order to test the electric strength of the insulating material between them. The transformer and its oil case are shown in Fig. 165.

Fig. 165. 2340 K.V.A. transformer for a 40 000 volt transmission line.

The oil used for transformers is generally a mineral oil, and great care is taken to secure that it is free from acid, alkali, or water. The electric strength of the oil is much greater than that of air, and, unlike solid dielectrics, in the event of a spark passing it is at once extinguished and the electric strength and insulating properties of the oil are not weakened. When the transformer is connected with the supply mains, the oil in the ventilating ducts and in contact with the coils is warmed and rises to the surface. The oil in contact with the brass spira tubes, surrounding the transformer, through which water is kept circulating, sinks, and thus a continual circulation of the oil is maintained. The heat developed in the transformer is carried away by the convection currents in the oil and conducted into the water or radiated from the case. A 2400 kilovolt-ampere transformer requires approximately three gallons (13·6 kilogrammes) of water per minute for cooling.

Large high voltage transformers usually have a thermometric alarum fitted to them, so that in the event of the temperature of the oil getting too high, owing to a diminished flow of the water or for any other reason, a bell is rung or a gong is sounded.

The heating of transformers.

In practical work it is found that the larger the size of the transformer the greater the difficulty experienced in keeping it cool. This is due to the fact that the area of the cooling surface increases only as the square of the linear dimensions whilst the weight of the copper and iron used, and consequently the heat generated in them for given current and flux densities, increases as the cube of the linear dimensions. As a rule, therefore, not only are larger transformers made heavier in proportion, but more attention is paid to the methods adopted for keeping them cool. It is customary also so to design transformers of large capacity that their efficiencies are higher than those of small transformers.

REFERENCES

J. A. Fleming, 'Experimental Researches in Alternate-current Transformers, *Journ. of the Inst. of El. Engin.*, Vol. 21, p. 594, 1892.

G. Roessler, 'Das Verhalten von Transformatoren unter dem Einflusse von Wechselströmen verschiedenen periodischen Verlaufs.' Paper read to the Verband Deutscher Electrotechniker, Munich, July 6, 1895.

A. Russell, 'Magnetic Leakage in the Alternating Current Transformer,' *The Electrician*, Vol. 42, pp. 567 et seq., 1899.

CHAPTER XIII

Alternating current motors. Asynchronous motors. Faraday's cube. Foucault's disc. Induction motors. Stator. Slip. The torque when the rotor is a copper cylinder. The efficiency of the rotor. The magnetic field in the air-gap. Three phase motor. The rotor current. The leakage factor. Formulae for the torque. The efficiency of the motor. The circle diagram. Stator resistance neglected. Numerical example. Reversing the direction of rotation. Transformer analogy. Equivalent net-work. Speed greater than synchronism. Testing induction motors. High speed and low speed motors. References.

Alternating current motors.

In the utilisation of alternating currents to supply motive power for industrial purposes, many types of motor are employed. Some of these motors run at the same speed at all loads, whilst the speed of others varies with the load. They may conveniently be divided, therefore, into synchronous and asynchronous machines. The theory of synchronous motors has already been discussed in Chapters v and vi. We saw that an ordinary single phase or polyphase alternator will run as a synchronous motor when connected with the supply mains in the proper manner. If $2p$ be the number of poles of the field magnets, n the number of revolutions per second and f the frequency, then, since the angular velocity of the rotor $2\pi n$ equals ω/p where ω is $2\pi f$, we must have n always equal to f/p. The only method of altering the speed in this type of machine, therefore, is to alter either the frequency f of the supply current or the number of poles of the field magnets.

Asynchronous motors. Asynchronous motors may be divided roughly into induction motors and commutator motors. The operation of induction motors depends on the torque produced on a suitable rotor when placed in a rotating magnetic field. The fundamental methods of producing rotating magnetic fields are described in Vol. I, Chapter XVIII, and an investigation is also made of some of their properties. In practice, the speed of the ordinary type of induction motor only varies by about five per cent. from no load to full load. For most practical purposes, therefore, we may regard the induction motor as a constant speed motor. Most forms of commutator motor are variable speed motors. In this chapter we shall only consider the elementary theory of the induction motor, delaying the consideration of the theory of commutator motors until Chapter XVI.

Faraday's cube. Faraday showed that when a metallic cube was placed in a rotating magnetic field the cube revolved in the same direction as the field. This rotation is due to the reaction of the currents induced in the mass of metal on the magnetic field. The mechanical forces produced tend to make the induced currents a minimum, and thus act so as to rotate the cube in the same direction as the field. If the cube were perfectly free to move, it would rotate with the same angular velocity as the field, and no induced currents would be generated after it had attained synchronism. If the cube had to perform work in overcoming friction, it would rotate at a less speed than synchronism, and the induced currents acting on the field would produce the couple required to do the necessary work.

Foucault's disc. In Foucault's classical experiment of a copper disc rotating in a strong magnetic field, the energy expended in making the disc rotate at constant speed, when the magnetic field has become steady, is converted mainly into heat generated by the eddy currents induced in it. The rest of the energy is expended in overcoming mechanical and air friction. Assuming that all the energy is expended in heating the disc, we can write

$$g\omega = 4{\cdot}2H \quad \text{..........................(i).}$$

In this formula g is the torque in 10^7 dyne-centimetre units acting on the disc, ω is its angular velocity and H is the heat in calories generated per second in the disc. Hence g can be found, when H and ω are known.

Induction motors.

In the experiment of Faraday's cube the rotation takes place in exactly the same manner whether the rotating magnetic field is produced by rotating direct current electromagnets or by means of alternating currents. The torque produced by the induced currents is small, and hence little power could be obtained from a motor constructed on this principle. The earliest induction motor, which was invented by Ferraris and constructed in 1885, consisted simply of a copper cylinder placed

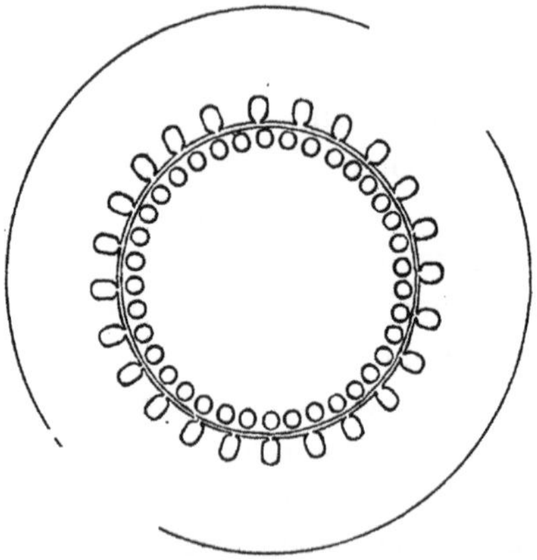

Fig. 166. Form of stator and rotor stampings for an induction motor. The slots which receive the stator windings are open. The rotor is of the squirrel-cage type, the holes round the circumference receiving the copper conductors.

in a rotating magnetic field. The principle of its action is therefore identical with that of Faraday's cube. To obtain an appreciable torque from this kind of motor, we must have large induced currents in a strong magnetic field. To get a strong magnetic field, it is necessary to have that part of the path of the flux which is in non-magnetic media as short as possible. One way of doing this is to construct the rotor of circular iron stampings so that it forms a cylinder, the diameter of which is only slightly less than the inner diameter of the stator. In the slots of the stator are wound the coils which produce the rotating magnetic field. The torque produced in a rotor of this type, due to the hysteresis and to the eddy currents in the iron of the rotor, is small. A motor constructed in this manner is called a 'hysteresis motor.' If holes are made (Fig. 166) near the circumference of the rotor, parallel to its axis, the holes being evenly distributed round the rotor, and, if copper conductors are placed in them, the ends of the conductors being all short-circuited at each face of the cylinder, a very powerful torque is obtained. If the iron is supposed to be removed from this rotor, the copper bars with the copper short-circuiting plates at the ends will be similar to a squirrel-cage. Hence this type of rotor is generally called a 'squirrel-cage rotor.' It was described and patented by Dolivo-Dobrowolsky in 1889. Instead of having plates or rings of copper to short-circuit the rotor windings, platinoid or other high resistance metal is sometimes used. The effect of this is, as we shall see presently, to increase the starting torque, but it lowers the efficiency of the motor. In small motors the advantage of an increased starting torque more than counterbalances the small decrease in the efficiency.

Stator. The stator of an induction motor consists of centre hole circular iron stampings about 20 mils (0·51 mm.) in thickness. These are usually built up inside a cast-iron case, so that they form a hollow cylinder with slots (Fig. 166) along the inside parallel to the axis of the cylinder. The winding of the stator of a polyphase motor is simple. It may be made up of rectangular former-wound coils, that is, coils which are wound into shape on a rectangular wooden block before being fixed on the stator, or it may have a regular bar winding as in the case of a polyphase alternator (see Chapter II). When rectangular coils are used they are connected in star, except in the case of large low tension motors, in which case they are sometimes connected in mesh. It is difficult to arrange the crossings at the ends of the conductors neatly in three phase stators. For this reason it is customary to place all the coils which belong to one phase at a distance apart equal to twice the polar pitch, so that the current at any instant goes round all the coils belonging to one phase in the same direction. This is sometimes called a 'hemitropic winding.'

The form of the section of the slot (Fig. 167) has a considerable effect on the working of the motor. If the slot be closed so that it forms a tunnel through the iron of the stator, there is appreciable magnetic leakage round the bottom of the slot, some of the lines of force due to the stator current encircling the primary coils only.

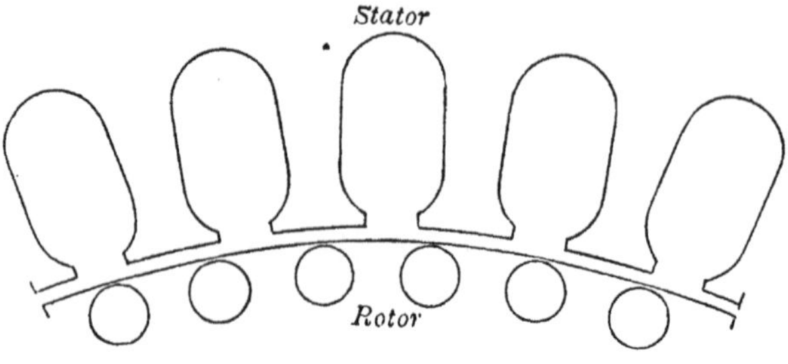

Fig. 167. Shape of the slots in the stator and rotor of an induction motor. The rotor slots are now usually half-open like the stator slots.

In this case, however, the current that the machine takes when running at approximately synchronous speed at no load, that is, the magnetising current, is very small. When former-wound coils are used in the stator circuits, the slots are simply rectangular in shape. The magnetising current of this type of motor is high compared with that of motors which have nearly closed slots or which have tunnels for the stator windings.

Slip. Let us suppose that the magnetic field due to the stator currents rotates with a constant angular velocity ω_1. Let us also suppose that the angular velocity of the rotor is ω_2. The slip s of the rotor is the ratio of the excess of the angular velocity of the magnetic field over the angular velocity of the rotor, to the angular velocity of the field. In symbols, we have $s = (\omega_1 - \omega_2)/\omega_1$, or $\omega_2 = \omega_1(1 - s)$. The percentage slip is $100\,s$. When the rotor is at rest, ω_2 is zero, and hence the slip is unity and the percentage slip is 100. If the rotor were rotating synchronously with the field, both the slip and the percentage slip would be zero. We shall denote the relative angular velocity $\omega_1 - \omega_2$ of the stator field and the rotor by ω, so that $\omega = s\omega_1$.

The torque when the rotor is a copper cylinder.

Let us suppose that the rotor is a copper cylinder and that the stator produces a pure rotating magnetic field. Impress on both the stator and the rotor an angular velocity $-\omega_1$ equal and opposite to that of the stator field. The magnetic field due to the stator currents is now fixed in space and the rotor is revolving with an angular velocity $-(\omega_1 - \omega_2)$. The magnetic field produced by the currents induced in the rotor will rotate round it in the positive direction with the angular velocity $\omega_1 - \omega_2$. This follows since the induced currents flow in such a direction that they tend to prevent the magnetic flux due to the stator currents from entering the cylinder. Hence the magnetic field due to the rotor currents, and therefore also the resultant magnetic field in the air-gap, will be fixed in magnitude and direction. We have thus a copper cylinder rotating in a fixed magnetic field and the torque g acting on it must obviously be constant. The power given to it by the field is $g(\omega_1 - \omega_2)$, that is $g\omega$, and this must equal the heat generated in the rotor per second. The frequency of the currents induced in the cylinder is $\omega/2\pi$. It depends only on the relative angular velocity ω of the field and the cylinder, and hence, if ω_1 be constant, the frequency of the induced currents varies as the slip s. When ω_1 is constant, we also see that the torque multiplied by the slip is proportional to the power expended in heating the rotor.

The efficiency of the rotor.

Let us still suppose that the rotor of the motor is a copper cylinder. The magnetic field due to the stator currents produces a torque g on the rotor, and as the field rotates with an angular velocity ω_1 the power given to the rotor is $g\omega_1$. We have seen that the power expended in heating the rotor is $g(\omega_1 - \omega_2)$. Hence $g\omega_2$ is the power available for producing rotation in the cylinder and overcoming the resisting torque due to the load, the friction of the bearings, etc. We shall define the efficiency η_r of the rotor as the ratio of the mechanical power $g\omega_2$ developed in it, to the total power $g\omega_1$ received. Thus we have

$$\eta_r = \omega_2/\omega_1 = 1 - s, \text{ and } s = 1 - \eta_r,$$

where s is the slip.

In general, whatever form the rotor may have, $G\omega$ is the average value of the power expended in heating it, and $G\omega_1$ is the average value of the power it receives, where G is the average torque. Hence the formulae given above still hold when η_r denotes the ratio of the average mechanical power to the average total power given to the rotor. It is to be noticed that η_r fixes a superior limit to the commercial efficiency of the motor.

The magnetic field in the air-gap.

Let us now consider a three phase induction motor the currents in the stator windings of which follow the harmonic law. When adjacent coils belonging to one phase of the stator winding are wound in opposite directions, let $2p$ be the number of coils per phase, and let a be the step from the centre of one coil to the centre of the next coil of the same phase winding, so that $2pa$ is the circumference of the stator. If the winding be hemitropic we suppose that p is the number of coils in a phase winding, and that $2a$ is the distance between adjacent coils. In either case, owing to the very minute air-gap used in practice, we can assume that the circumference of the rotor is also $2pa$. Let $\omega_1/2\pi$ be the frequency of the polyphase currents which supply the stator so that the angular velocity of the rotating field will be ω_1/p, whichever winding be used. If the angular velocity of the rotor be ω_2/p, the relative angular velocity of the rotor and the field due to the stator will be $(\omega_1 - \omega_2)/p$ or ω/p, and this is the rate at which the stator flux will cut the windings of the rotor. The flux due to the induced currents in the rotor windings will rotate relatively to the rotor at the same speed ω/p as the stator flux, and its angular velocity in space will therefore be $\omega/p + \omega_2/p$, that is ω_1/p.

Now impress on both the rotor and the stator an angular velocity $-\omega_1/p$. Both the stator and the rotor fields will be brought to rest, and we shall have the rotor revolving in a fixed magnetic field with an angular velocity $-(\omega_1 - \omega_2)/p$. In order to calculate the instantaneous value g of the torque we need to know the value of the currents generated in the rotor. We need to make some assumption, therefore, as to the distribution of the flux in the air-gap. In practice jutting-out (salient) poles are never used and the windings are well distributed, we may therefore suppose that the distributions of the magnetic flux due to the currents in the stator and rotor respectively can be represented by sine curves. The ordinates of these curves represent the distribution of the magnetic flux in the air-gap of the motor.

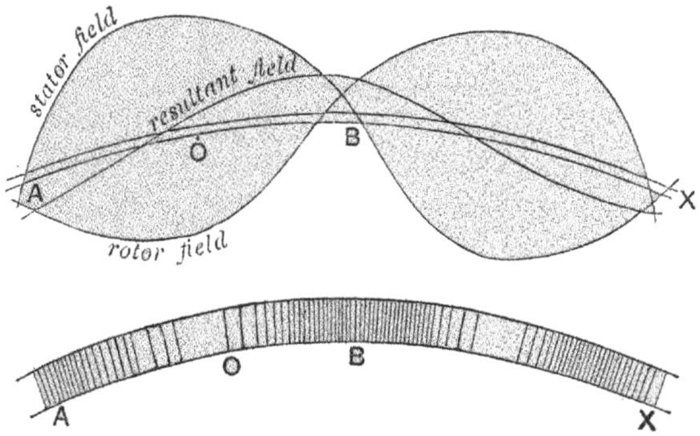

Fig. 168. Sine distribution of the magnetic field in the air-gap of an induction motor. *AB* is a magnified image of the air-gap, the flux density being represented by the number of lines per unit length. There are $2p$ bunches of lines round the air-gap, neighbouring bunches pointing in opposite directions.

Let us assume that the origin O from which the abscissae are measured rotates round the air-gap with the angular velocity ω_1/p of the gliding magnetic field, and that the intensity h_1 of the field due to the stator currents is a maximum at O. The intensity of the field at a point P may be written

$$h_1 \cos(\pi x/a) - h_2 \cos(\pi x/a - \alpha),$$

where $-h_2 \cos(\pi x/a - \alpha)$ is the intensity of the field at P due to the rotor currents, and x is the length of OP measured along the circumference of the rotor. We have prefixed the negative sign to h_2 as the induced currents in the rotor tend to prevent the magnetic flux from entering it.

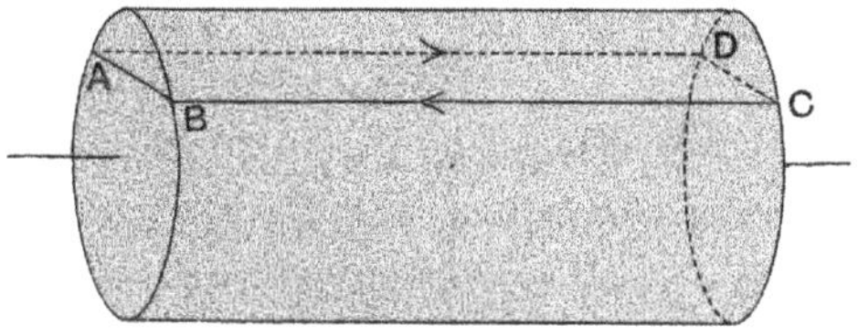

Fig. 169. *AD* and *BC* are two of the rotor conductors. *AB* and *CD* are connecting pieces.

Let us now consider a complete turn of the rotor winding (Fig. 169) formed by two conductors AD and BC and their end connections. The conductors are placed in slots, and are parallel to the axis of the rotor. Let b denote the breadth of the coil, that is, the distance, measured along the air-gap, between the axes of the two conductors. At any instant let x be the abscissa of a

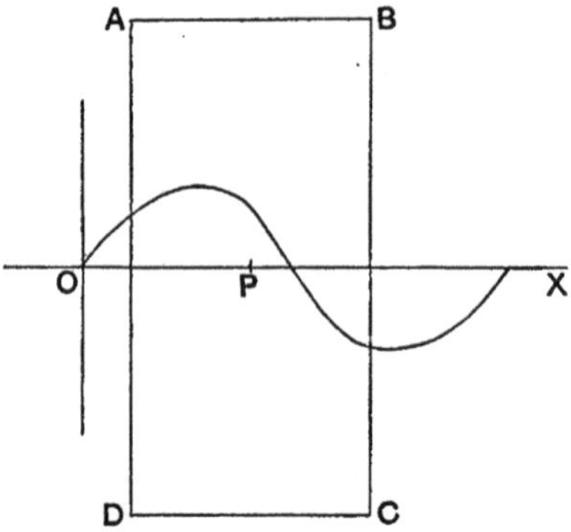

Fig. 170. $ABCD$ is one of the rotor circuits (Fig. 169). The magnetic flux is perpendicular to the plane of the paper, and its density is given by the ordinates of the sine curve drawn in a plane through OX at right angles to the plane of the paper.

point P (Fig. 170), at the centre of this coil, from the moving origin. If l be the length of the parallel conductors and ϕ be the total flux embraced by the coil at this instant, we have

$$\phi = \int_{x-b/2}^{x+b/2} hl\,dx'$$

$$= \int_{x-b/2}^{x+b/2} \{h_1 \cos(\pi x'/a) - h_2 \cos(\pi x'/a - \alpha)\}\, l\,dx'$$

$$= (al/\pi)\,[h_1 \sin(\pi x'/a) - h_2 \sin(\pi x'/a - \alpha)]_{x-b/2}^{x+b/2}$$

$$= (alh_1/\pi)\{\sin(\pi x/a + \pi b/2a) - \sin(\pi x/a - \pi b/2a)\}$$
$$- (alh_2/\pi)\{\sin(\pi x/a + \pi b/2a - \alpha) - \sin(\pi x/a - \pi b/2a - \alpha)\}$$

$$= (2alh_1/\pi)\sin(\pi b/2a)\cos(\pi x/a)$$
$$- (2alh_2/\pi)\sin(\pi b/2a)\cos(\pi x/a - \alpha)$$

$$= \Phi_1 \cos(\pi x/a) - \Phi_2 \cos(\pi x/a - \alpha) \quad \text{.........................}(a),$$

where $\Phi_1 = (2al/\pi)\sin(\pi b/2a)\,.\,h_1,$

and $\Phi_2 = (2al/\pi)\sin(\pi b/2a)\,.\,h_2.$

Φ_1 is the maximum value of the primary flux embraced by the given coil of the rotor, and Φ_2 is the maximum value of the secondary flux embraced by the same coil. To a first approximation Φ_1 will be proportional to I_1, where I_1 is the maximum value of the current in a stator conductor. We may therefore write $\Phi_1 = MI_1$, where M is a constant.

Similarly, we can write $\Phi_2 = L_2 I_2$, where L_2 is a constant.

Three phase motor.

The problem of the working of an ideal three phase induction motor on a balanced load admits of an exact solution. In this case we find that the assumption of simple harmonic currents in both the stator and rotor windings is permissible. If the load on the three phase windings be unbalanced this assumption would not be permissible, but as this could only

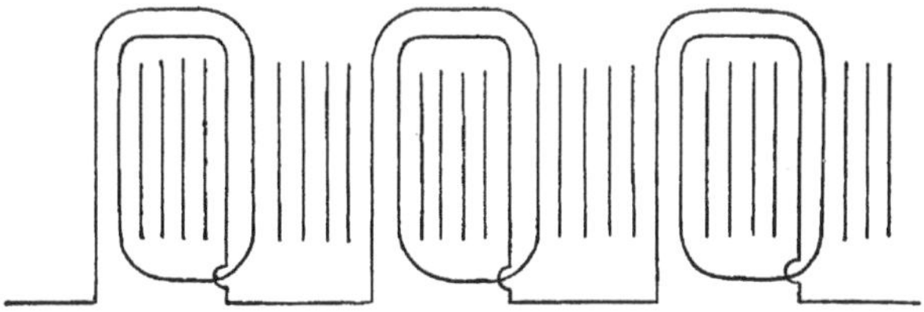

Fig. 171. One phase of the winding of the stator of a three phase induction motor.

arise if the design or construction of the machine is faulty we need not consider it. In single phase motors we shall see that the currents cannot be sine shaped at all speeds. Hence the problem presents special difficulties. We shall therefore first discuss the problem of the three phase motor.

Let us suppose that the three stator circuits are symmetrically situated at angular distances apart of 120°. Let i_1, R_1, $L_{1.1}$, $L_{1.2}$ and $L_{1.3}$ be the current, the resistance, the self-inductance, the mutual inductance between the first and the second circuit, and between the first and the third circuit respectively. We shall suppose that the three rotor circuits are also symmetrically situated. Let i_1', R_1', $L'_{1.1}$, $L'_{1.2}$ and $L'_{1.3}$ be the corresponding quantities for the first of the rotor circuits. We shall denote the mutual inductance between the first coil of the stator winding and the first coil of the rotor winding by $m_{1.1'}$, and similarly the mutual inductance between the pth winding of the stator and the qth winding of the rotor will be denoted by $m_{p.q'}$. The value of $m_{p.q'}$ at any instant depends on the relative positions of the pth circuit of the stator and the qth circuit of the rotor. If ω_2 be the angular velocity of the rotor, $m_{p.q'}$ is a periodic function of the time, the frequency of which is $\omega_2/(2\pi)$.

Let e_1, e_2, e_3 be the potential differences applied to the first, second and third windings of the stator respectively. Then, we have

$$e_1 = R_1 i_1 + L_{1.1}\frac{\partial i_1}{\partial t} + L_{1.2}\frac{\partial i_2}{\partial t} + L_{1.3}\frac{\partial i_3}{\partial t} + \frac{\partial}{\partial t}\{m_{1.1'}\,i_1' + m_{1.2'}\,i_2' + m_{1.3'}\,i_3'\} \quad \ldots(1)$$

and two similar equations.

For the rotor circuits we also have three equations, the first of which is

$$0 = R_1' i_1' + L'_{1.1}\frac{\partial i_1'}{\partial t} + L'_{1.2}\frac{\partial i_2'}{\partial t} + L'_{1.3}\frac{\partial i_3'}{\partial t} + \frac{\partial}{\partial t}\{m_{1'.1} i_1 + m_{1'.2} i_2 + m_{1'.3} i_3\} \quad \ldots(2).$$

Since the machine is symmetrical, we have

$$R_1 = R_2 = R_3 = R, \quad R_1' = R_2' = R_3' = R',$$
$$L_{1.2} = L_{1.3} = L_{2.3}, \text{ and } L'_{1.2} = L'_{1.3} = L'_{2.3}.$$

As a first simplification let us suppose that

$$m_{1.1'} = M_1 \cos \omega_2 t, \quad m_{1.2'} = M_1 \cos(\omega_2 t + 2\pi/3)$$

and $$m_{1.3'} = M_1 \cos(\omega_2 t + 4\pi/3).$$

From the definition of mutual inductance, $m_{1.1'} = m_{1'.1}$. Hence

$$m_{1'.1} = M_1 \cos \omega_2 t, \quad m_{1'.2} = M_1 \cos(\omega_2 t - 2\pi/3)$$

and $$m_{1'.3} = M_1 \cos(\omega_2 t - 4\pi/3).$$

We also have $i_1 + i_2 + i_3 = 0$, and $i_1' + i_2' + i_3' = 0$. Hence since the load is balanced we shall suppose that

$$i_1 = I\cos(\omega_1 t - \beta), \quad i_2 = I\cos(\omega_1 t - \beta - 2\pi/3)$$

and $$i_3 = I\cos(\omega_1 t - \beta - 4\pi/3)$$

satisfy the equations (1) and (2). Since the frequency of the currents in the rotor windings is $\omega/(2\pi)$, we shall suppose that

$$i_1' = I'\cos(\omega t - \beta'), \quad i_2' = I'\cos(\omega t - \beta' - 2\pi/3)$$

and $$i_3' = I'\cos(\omega t - \beta' - 4\pi/3)$$

satisfy the same equations. Making these assumptions we easily find that

$$m_{1.1'}i_1' + m_{1.2'}i_2' + m_{1.3'}i_3' = (3/2)\,M_1 I' \cos(\omega_1 t - \beta')$$

and

$$m_{1'.1}i_1 + m_{1'.2}i_2 + m_{1'.3}i_3 = (3/2)\,M_1 I \cos(\omega t - \beta).$$

Hence the equations (1) and (2) become

$$e = RI\cos(\omega_1 t - \beta) - L\omega_1 I \sin(\omega_1 t - \beta) - M\omega_1 I' \sin(\omega_1 t - \beta') \ldots (3)$$

and

$$0 = R'I'\cos(\omega t - \beta') - L'\omega I' \sin(\omega t - \beta') - M\omega I \sin(\omega t - \beta) \ldots (4),$$

where we have written e for e_1, and M, L and L′ for $3M_1/2$, $L_{1.1} - L_{1.2}$ and $L'_{1.1} - L'_{1.2}$ respectively.

From equation (4) we can readily deduce useful equations connecting the stator and rotor currents. As equation (4) is true for all values of t it is true when $\omega t = 0$ and when $\omega t = \pi/2$. Thus we get

The rotor current.

$$-R'I'\cos\beta' = L'\omega I' \sin\beta' + M\omega I \sin\beta$$

and

$$R'I'\sin\beta' = L'\omega I' \cos\beta' + M\omega I \cos\beta.$$

Hence we find that

$$(M\omega I)^2 = (R'I')^2 + (L'\omega I')^2,$$

$$MI\cos(\beta' - \beta) = -L'I'$$

and

$$M\omega I \sin(\beta' - \beta) = R'I'.$$

Writing $A_1\sqrt{2}$, $A_2\sqrt{2}$ and α for I, I' and $\beta' - \beta$ respectively, we get

$$M\omega A_1 = \{R'^2 + (L'\omega)^2\}^{\frac{1}{2}} A_2 \ldots\ldots (5),$$

$$MA_1 \cos\alpha = -L'A_2 \ldots\ldots (6),$$

$$M\omega A_1 \sin\alpha = R'A_2 \ldots\ldots (7),$$

and

$$\tan\alpha = -R'/(L'\omega) \ldots\ldots (8).$$

In these equations α is the phase difference between i_1 and i_1'.

We see from (8) that if R' were zero then α would be π, and so the stator and rotor currents would always be in opposition in phase. If α_s be the value of α at the start, *i.e.* when $\omega_2 = 0$ and ω is therefore ω_1, we have

$$\tan\alpha_s = s\tan\alpha \ldots\ldots (9),$$

where s is the slip. Equation (5) may be written

$$M\omega_1 A_1 = \{(R'/s)^2 + (L'\omega_1)^2\}^{\frac{1}{2}} A_2 \ldots\ldots (10).$$

Hence as s diminishes from 1 to 0, A_2/A_1 diminishes from $M\omega_1/\{R'^2 + (L'\omega_1)^2\}^{\frac{1}{2}}$ to 0.

The relations between the stator and rotor currents are shown graphically in Fig. 172.

The leakage factor.

It is convenient to define the leakage factor σ of the ideal three phase induction motor by the equation (see Vol. I, p. 338)

$$\sigma = 1 - M^2/(LL') = 1 - (9/4)\, M_1^2/\{(L_{1.1} - L_{1.2})\,(L'_{1.1} - L'_{1.2})\} \quad \text{.........(11).}$$

Noticing that $M^2/L' = \mathrm{L}\,(1 - \sigma)$, we see that (6) may be written

$$-\mathrm{L}\,(1 - \sigma)\, A_1 \cos\alpha = MA_2 \text{ ...(12).}$$

When there is no magnetic leakage σ is zero, and when there are no magnetic linkages σ is unity. In good machines, σ generally lies between 0·03 and 0·05. Its value depends mainly on the number and depth of the slots and on the size of the air-gap. The greater the air-gap and the deeper the slots, the greater will be the magnetic leakage, and hence the greater the value of σ. The calculation of σ directly from the dimensions of the machine, and the data of its magnetic circuit, would be very difficult. Designers of induction motors use empirical formulae, but, as a rule, these formulae are obtained by studying the results of tests on different types of motors and not from theoretical considerations.

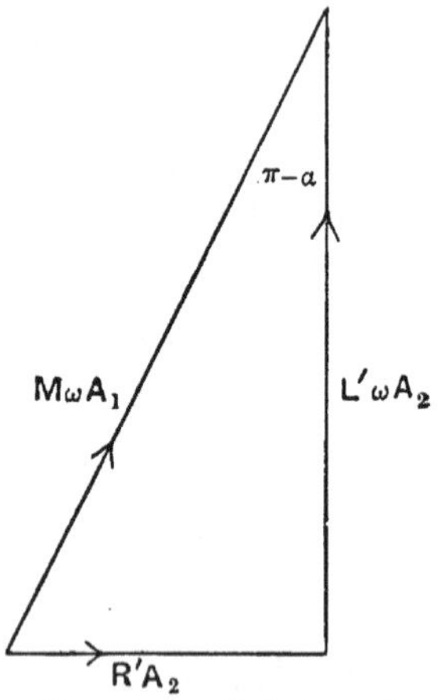

Fig. 172. Relations between the rotor and stator currents.

Formulae for the torque.

If ψ be the angle of lag of the current $I\cos(\omega_1 t - \beta)$ behind the applied potential difference, equation (3) may be written

$$V\cos(\omega_1 t - \beta + \psi) = RA_1\cos(\omega_1 t - \beta) - L\omega_1 A_1 \sin(\omega_1 t - \beta)$$
$$- M\omega_1 A_2 \sin(\omega_1 t - \beta') \text{ ....(13),}$$

where V is the effective value of the P.D. applied to one phase of the stator winding and A_1 and A_2 are the effective values of the stator and rotor currents. As (13) must be true for all values of t, we find, on putting $\omega_1 t - \beta = 0$ and $= \pi/2$ respectively, that

$$V \cos\psi = RA_1 \quad + M\omega_1 A_2 \sin\alpha \qquad (14)$$

and

$$V \sin\psi = L\omega_1 A_1 + M\omega_1 A_2 \cos\alpha \qquad (15),$$

where $\alpha = \beta' - \beta$. Multiplying both sides of (14) by A_1, we get

$$VA_1 \cos\psi = RA_1^2 + M\omega_1 A_1 A_2 \sin\alpha.$$

Now $VA_1 \cos\psi$ is the power given to one phase of the primary winding. This is expended in heating the primary winding and giving power to the rotor. Since RA_1^2 is the heat generated in the winding, the power given to the rotor is $M\omega_1 A_1 A_2 \sin\alpha$. If G be the torque due to the current in this winding, we have therefore

$$G\omega_1 = M\omega_1 A_1 A_2 \sin\alpha,$$

since the angular velocity of the uniformly rotating field is ω_1.

Hence

$$G = MA_1 A_2 \sin\alpha \qquad (16).$$

Substituting for $\sin\alpha$ from (7), this becomes

$$G = (R'/\omega) A_2^2 = (R'/\omega_1) A_2^2/s \qquad (17).$$

By squaring (14) and (15) and adding, we get

$$(V/A_2)^2 = (Z_1 Z_2/M\omega)^2 + (M\omega_1)^2 + 2\,(Z_2/s)\,(R \sin\alpha + L\omega_1 \cos\alpha),$$

where $Z_1^2 = R^2 + L^2\omega_1^2$ and $Z_2^2 = R'^2 + L'^2\omega^2$, noticing that by (5), we have $M\omega A_1 = Z_2 A_2$.

Hence using (6) and (7) we get

$$A_2^2 = \frac{V^2}{(Z_1 R'/Ms\omega_1)^2 + 2RR'/s + (L'/M)^2\,(R^2 + L^2\omega_1^2\sigma^2)} \qquad (18).$$

Therefore, by (17)

$$G = \frac{(R'/\omega_1)\,V^2}{(Z_1 R'/M\omega_1)^2/s + (L'/M)^2\,(R^2 + L^2\omega_1^2\sigma^2)\,s + 2RR'} \qquad (19).$$

As the slip s varies, the numerator of this fraction remains constant. The denominator, however, has a minimum value when

$$s = (R'/L'\omega_1)\,\{Z_1/(R^2 + L^2\omega_1^2\sigma^2)^{\frac{1}{2}}\} \qquad (20).$$

Now s cannot have values greater than unity or less than zero. If the value of s, therefore, given by (20) is less than unity, G is a maximum for this value of s and

$$G_{\text{max.}} = \frac{MV^2}{2RM\omega_1 + 2\,(L'/M)\,\{Z_1\,(R^2 + L^2\omega_1^2\sigma^2)^{\frac{1}{2}}\}} \qquad (21).$$

Hence if R' be not greater than $(L'\omega_1/Z_1)(R^2 + L^2\omega_1{}^2\sigma^2)^{\frac{1}{2}}$, so that the value of s found from (20) is not greater than unity, $G_{\text{max.}}$ is independent of R', that is, of the resistance of the rotor winding. If however R' be greater than this value, then the torque G has its greatest value when the slip s is unity, that is, at the moment of starting. In this case also the greater the value of R' the less will be the starting torque. We see, therefore, that the resistance

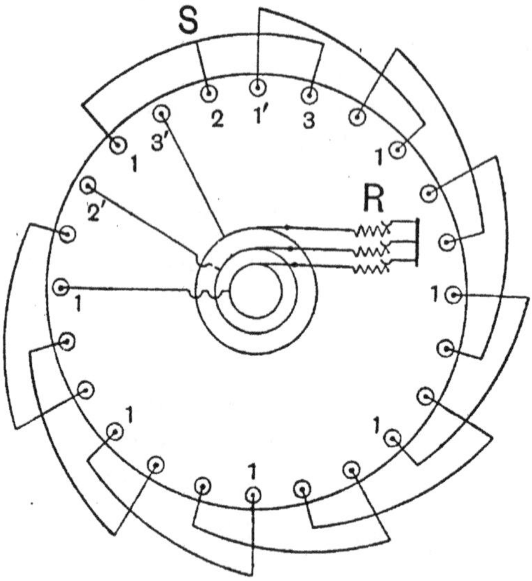

Fig. 173. Rotor of a three phase induction motor with slip rings by means of which resistances can be interpolated in the rotor circuits.

R_s' of one phase of the secondary winding in order to get the maximum torque at the instant of starting is given by

$$R_s' = L'\omega_1\left(\frac{R^2 + L^2\omega_1{}^2\sigma^2}{R^2 + L^2\omega_1{}^2}\right)^{\frac{1}{2}} \quad \ldots\ldots\ldots\ldots\ldots(22)$$

$$= \sigma L'\omega_1 \text{ approximately} \quad \ldots\ldots\ldots\ldots\ldots(23),$$

since, in practice, R is small compared with $L\omega_1$.

In Fig. 173 a method of winding the rotor of a three phase induction motor is indicated which enables resistance to be inserted in each of the three rotor circuits during the start. When the rotor has got up speed the resistances of the rheostat R are short circuited. It will be seen that the three phase windings 1, 2 and 3 are connected in star at S and at the rheostat.

Since the three stator windings are equal, the electric power given to the stator is $3RA_1^2 + 3G\omega_1$ and the useful mechanical power obtained from the rotor is $3G\omega_2$.

The efficiency of the motor.

Hence if η be the efficiency of the motor

$$\eta = \frac{3G\omega_2}{3RA_1^2 + 3G\omega_1}$$

$$= \frac{1 - s}{1 + RA_1^2/G\omega_1}$$

$$= \frac{1 - s}{1 + s\,(R/R')\,(A_1/A_2)^2}$$

and thus

$$\eta = \frac{1 - s}{1 + (R/R')\,\{(R'/M\omega_1)^2/s + (L'/M)^2 s\}} \quad \text{......(24).}$$

We see that, except when R is zero, η is always less than $1 - s$. We also see that the greater the value of M and the higher the frequency, the higher will be the efficiency. For a given value of s, if R' vary, then η has a maximum value when $R' = L'\omega_1 s$.

The problem of the three phase induction motor is usually discussed graphically in practice. As this method is most instructive, we shall now prove the 'circle diagram' by means of the equations given above and point out some of its properties.

By means of equations (12), (14) and (15) we get

The circle diagram.

$$VA_1 \cos\psi = RA_1^2 - L\omega_1 A_1^2 (1 - \sigma) \sin\alpha \cos\alpha \text{...(25)}$$

and

$$VA_1 \sin\psi = L\omega_1 A_1^2 - L\omega_1 A_1^2 (1 - \sigma) \cos^2\alpha \text{ ...(26).}$$

If we choose the axes of coordinates so that the axis of Y gives the phase of V_1, the coordinates x and y of the extremity of the current vector are given by

$$x = A_1 \sin\psi \quad \text{and} \quad y = A_1 \cos\psi.$$

Hence we have

$$Vy = R\,(x^2 + y^2) - L\omega_1\,(x^2 + y^2)\,(1 - \sigma) \sin\alpha \cos\alpha$$

and

$$Vx = L\omega_1\,(x^2 + y^2) - L\omega_1\,(x^2 + y^2)\,(1 - \sigma) \cos^2\alpha.$$

Thus, eliminating α, we get

$$\left(\frac{Vx}{x^2 + y^2} - L\omega_1\right)^2 + \left(\frac{Vy}{x^2 + y^2} - R\right)^2 = \{L\omega_1\,(1 - \sigma)\cos\alpha\}^2$$

$$= -L\omega_1\,(1 - \sigma)\left\{\frac{Vx}{x^2 + y^2} - L\omega_1\right\},$$

and simplifying, we have finally

$$(x - \bar{x})^2 + (y - \bar{y})^2 = r^2 \qquad \text{.................(27)},$$

where $$\bar{x} = VL\omega_1 (1 + \sigma)/\{2 (R^2 + L^2\omega_1{}^2\sigma)\} \qquad \text{.........(28)},$$

$$\bar{y} = VR/(R^2 + L^2\omega_1{}^2\sigma) \qquad \text{.........................(29)},$$

and $$r = VL\omega_1 (1 - \sigma)/\{2 (R^2 + L^2\omega_1{}^2\sigma)\} \qquad \text{.........(30)}.$$

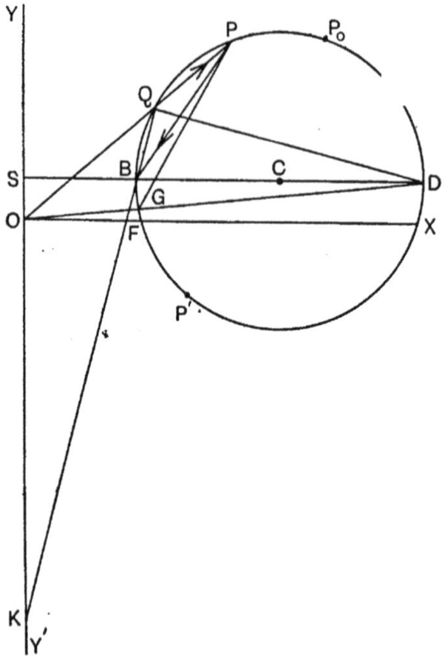

Fig. 174. The circle diagram of a three phase induction motor. $OP = A_1$. $PG = (M\omega_1/Z_1) A_2$. PB gives the phase of A_2. SK/SK_0 = the slip s, where K_0 (not shown in the figure) is the position of K when P is at P_0.

Thus the extremity of the line representing the vector of the current in a phase winding of the stator always lies on the circumference of the circle the coordinates of the centre of which are $\bar{x}$ and $\bar{y}$, and the radius of which is r.

From the equations for $\bar{x}$, $\bar{y}$ and r given above it is easy to see that as the magnetic leakage increases, and therefore as σ increases, $\bar{y}$ and r both diminish. Provided also that $L\omega_1/R$ is greater than unity, which it is in practice, $\bar{x}$ also diminishes as the magnetic leakage increases. It is interesting to notice that when $R = L\omega_1$, we have

$$\bar{x} = V/2R; \quad \bar{y} = V/\{R (1 + \sigma)\} \quad \text{and} \quad r = V(1 - \sigma)/\{2R (1 + \sigma)\}.$$

Hence in this case as σ increases from 0 to 1, $\bar{x}$ remains constant, $\bar{y}$ diminishes from V/R to $V/(2R)$ and R diminishes from $V/(2R)$ to zero.

By eliminating σ between the equations (28) and (29) we get

$$\frac{\bar{x}}{(L^2\omega_1{}^2 - R^2)/(2L\omega_1)} - \frac{\bar{y}}{R} = \frac{V}{L^2\omega_1{}^2 - R^2}.$$

Hence, as the magnetic leakage increases, the centre of the circle moves along the straight line represented by this equation, $\bar{y}$ diminishing from V/R to $VR/(R^2 + L^2\omega^2)$ and r diminishing from $VL\omega_1/(2R^2)$ to zero.

It is also easy to see that

$$\sigma = \frac{\bar{x} - r}{\bar{x} + r} \quad \text{........................(31)}.$$

It follows therefore that the ratio of the abscissa of the near end of the diameter of the circle which is parallel to OX to the abscissa of the far end of this diameter equals the leakage coefficient. Hence, when σ is zero, $\bar{x} = r$.

Let (x_0, y_0) and (x_s, y_s) be the coordinates of the extremities G and P_0 of the primary current vector when the rotor is running at synchronous speed and when it is at rest respectively. In the former case A_2 is zero and so, from (14) and (15), $\tan\psi_0 = L\omega_1/R$. If BCD (Fig. 174) be the diameter of the circle which is drawn parallel to OX, we have

$$\tan YOD = \frac{\bar{x} + r}{\bar{y}} = \frac{L\omega_1}{R}.$$

Hence G must lie on OD. Since $y_0 = VR/(R^2 + L^2\omega_1{}^2)$ it is less than $\bar{y}$, and G is therefore the point where OD cuts the circle.

If A_0 be the value of A_1 at synchronous speed we have $V = Z_1 A_0$, where $Z_1 = (R^2 + L^2\omega^2)^{\frac{1}{2}}$. Hence substituting in (14) and (15) we get

$$(M\omega_1/Z_1)\, A_2 \sin\alpha = A_0 \cos\psi - A_1 \cos\psi_0 \quad \text{......(32)}$$

and

$$(M\omega_1/Z_1)\, A_2 \cos\alpha = A_0 \sin\psi - A_1 \sin\psi_0 \quad \text{......(33)},$$

where $\tan\psi_0 = L\omega_1/R$. Squaring these equations and adding we get

$$(M\omega_1/Z_1)^2 A_2{}^2 = A_0{}^2 + A_1{}^2 - 2A_0A_1 \cos(\psi_0 - \psi).$$

In Fig. 174, $OG = A_0$, $OP = A_1$ and the angle $POG = \psi_0 - \psi$. It follows, therefore, that

$$PG = (M\omega_1/Z_1)\, A_2 \quad \text{.....................(34)}.$$

We also deduce from (32) and (33) that

$$A_0^2 = A_1^2 + (M\omega_1/Z_1)^2 A_2^2 - 2\,(M\omega_1/Z_1)\, A_1 A_2 \cos(\pi - \alpha + \pi/2 - \psi_0).$$

Hence the angle $OPG = \pi - \alpha + \pi/2 - \psi_0$. But the angle BPG = the angle BDO = the angle $DOX = \pi/2 - \psi_0$, and therefore the angle $OPB = \pi - \alpha$. The equation (34) shows that PG is proportional to the current in a phase winding of the rotor. We also see that the phase difference α between the currents in the stator and the rotor equals the angle between PB and OP produced.

If we join QB and produce it to meet OY' in K, the angle SBK = the angle $QBD = \pi/2 -$ the angle $QDB = \alpha - \pi/2$. Hence by (8)

$$\tan SBK = L'\omega/R' = s \tan SBK_0 = SK/SB,$$

where $\tan SBK_0 = L'\omega_1/R'$, so that OK_0 is the value of OK when P is at P_0 and OP_0 represents the starting current. Hence, since SB is constant for all positions of P, we see that the slip $s = SK/SK_0$.

We saw in Vol. I, p. 345, that OBP_0 is a straight line when R' is zero. In practice, R' is finite and therefore C is below OP_0. Let us now consider how the currents vary as the angular velocity of the rotor increases from 0 to ω_1. We see at once that the current A_1 diminishes continually from its starting value OP_0 to its minimum value OM (see Vol. I, p. 345) where OMC is a straight line. It then increases to OG. The current A_2 continually diminishes until at synchronous speed its value is zero. The phase difference α between the currents is nearly 180° at the start and continually diminishes towards its minimum value of 90°. The power factor $\cos\psi$ of the stator circuit has its maximum value (see Vol. I, p. 345) when OP is a tangent to the circle, it then diminishes, having the value R_1/Z_1 at synchronous speed. This value is obviously less than the starting value $\cos P_0OX$. As P moves from P_0 to G, Q moves from Q_0 to D, the slip SK varying from unity to zero.

If we draw OL (not shown in Fig. 174) at right angles to PB and meeting PB produced in L, then by (6)

$$PL = -A_1 \cos\alpha = (L'/M)\, A_2.$$

The area of the triangle POL

$$= (1/2)\, PL.PO \sin\alpha = (1/2)\,(L'/M)\, A_1 A_2 \sin\alpha.$$

Hence by (16) the torque on the rotor equals $(2M^2/L')$ times the area of the triangle POL. If PN be drawn at right angles to OX, $V.PN$ is the power given to a stator winding. This is obviously a maximum when PN passes through C.

It will be seen that the graphical method is both instructive and useful. We shall now consider the case when the resistance of the stator windings can be neglected, as in this case the diagram is simpler.

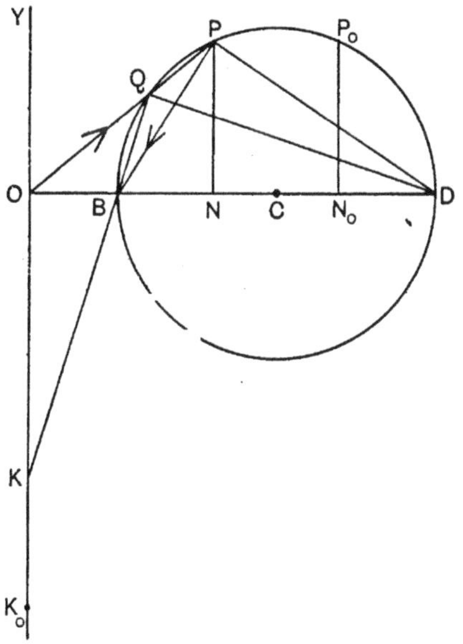

Fig. 175. Simplified circle diagram ($R=0$). $OP=A_1$, $OB=A_0$, $OD=A_0/\sigma$, $PB=(M/L)\,A_2$, $O\hat{P}B=O\hat{K}B=\pi-\alpha$, $s=OK/OK_0$, $\eta_r=\omega_2/\omega_1=1-s=KK_0/OK_0$, $G=G_s\,(PN/P_0N_0)$, $OP_0=$ the starting point current.

Stator resistance neglected.

When $R=0$, we have $\bar{y}=0$ and therefore the centre of the circle which is the locus of P (Fig. 175) lies on OD. If K be at K_0 when OP has its initial value OP_0, then since the angle OKB equals $\pi-\alpha$, we have $s=OK/OK_0$. When R is zero, the efficiency η_r of the rotor is given by

$$\eta_r=\omega_2/\omega_1=1-s=KK_0/OK_0.$$

If G denote the torque per phase on the rotor, we have

$$G\omega_1=VA_1\cos\psi=V\,.\,PN,$$

and hence $$G = G_s\,(PN/P_0N_0),$$
where G_s is the starting torque.

The power factor of the stator circuit has its maximum value when OP is a tangent to the circle. In this case

$$OP = A_0/\sqrt{\sigma} \text{ and } (\cos\psi)_{\text{max.}} = r/OC = (1-\sigma)/(1+\sigma).$$

We also have $PB = (M/L)A_2$. Hence since $PB^2 = BN \,.\, BD$, we see that the power expended in heating the phase winding of the rotor is proportional to BN. We shall now find formulae for A_1, $\cos\psi$ and G in terms of the slip s. Noticing that $R = 0$ and $V = L\omega_1 A_0$, we get by (25) and (26)

$$A_0\cos\psi = -A_1(1-\sigma)\sin\alpha\cos\alpha \quad \text{............(35)}$$

and $$A_0\sin\psi = A_1 - A_1(1-\sigma)\cos^2\alpha \quad \text{............(36).}$$

Hence squaring and adding we get

$$A_0^2 = A_1^2 - 2A_1^2(1-\sigma)\cos^2\alpha + A_1^2(1-\sigma)^2\cos^2\alpha,$$

and therefore

$$A_1 = A_0(1+\tan^2\alpha)^{\frac{1}{2}}/(\sigma^2+\tan^2\alpha)^{\frac{1}{2}}.$$

If α_s be the initial value of α, so that $\tan\alpha_s = -R'/L'\omega_1$, we get by (9)

$$A_1 = A_0(s^2+\tan^2\alpha_s)^{\frac{1}{2}}/(\sigma^2 s^2+\tan^2\alpha_s)^{\frac{1}{2}} \text{.........(37).}$$

From (35) and (36) we get

$$-\tan\psi = \frac{\sigma+\tan^2\alpha}{(1-\sigma)\tan\alpha}$$

and therefore

$$\cos\psi = \frac{-(1-\sigma)\tan\alpha}{(\sigma^2+\tan^2\alpha)^{\frac{1}{2}}(1+\tan^2\alpha)^{\frac{1}{2}}}$$
$$= \frac{-s(1-\sigma)\tan\alpha_s}{(\sigma^2 s^2+\tan^2\alpha_s)^{\frac{1}{2}}(s^2+\tan^2\alpha_s)^{\frac{1}{2}}} \quad \text{......(38).}$$

We also have

$$G\omega_1 = VA_1\cos\psi = L\omega_1 A_0 A_1\cos\psi,$$

and thus by (35)

$$G = -L(1-\sigma)A_1^2\sin\alpha\cos\alpha$$
$$= -LA_0^2(1-\sigma)\tan\alpha/(\sigma^2+\tan^2\alpha)$$
$$= -LA_0^2(1-\sigma)\,s\tan\alpha_s/(\sigma^2 s^2+\tan^2\alpha_s) \quad \text{......(39).}$$

It is easy to see that G has its maximum value when $s = \tan\alpha/\sigma$.

Let us suppose that the leakage factor σ for a given induction motor is 0·1 and that $\tan \alpha_s = 9/400$. Let us also suppose that we wish to know how the primary current, primary power factor and the torque vary with the slip. Substituting for σ and $\tan \alpha_s$ in equations (37), (38) and (39) we get the equations to the required curves which can then easily be plotted (Fig. 176). It will be seen that the starting torque is less than half the value of the maximum torque and that the maximum power factor is not obtained until the slip is about 0·07.

Numerical example.

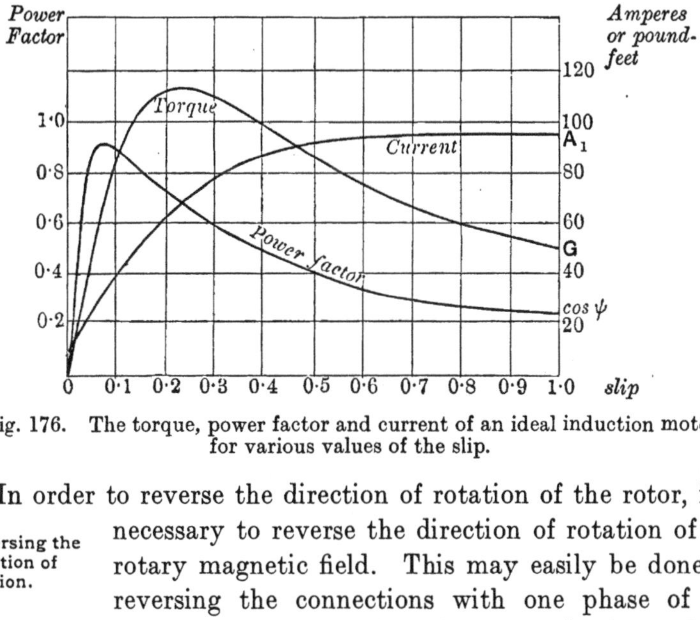

Fig. 176. The torque, power factor and current of an ideal induction motor for various values of the slip.

Reversing the direction of rotation.

In order to reverse the direction of rotation of the rotor, it is necessary to reverse the direction of rotation of the rotary magnetic field. This may easily be done by reversing the connections with one phase of the stator winding. Let us suppose that the motor is three phase. Before the alteration is made we may express the gliding magnetic field in the air-gap (see Vol. I, p. 433) by

$$H \sin \omega t \cos (\pi x/a) + H \sin (\omega t + 2\pi/3) \cos (\pi x/a + 2\pi/3)$$
$$+ H \sin (\omega t + 4\pi/3) \cos (\pi x/a + 4\pi/3),$$

which is equal to $(3H/2) \sin (\omega t - \pi x/a)$.

After reversing the leads across the first phase the field is represented by

$$- H \sin \omega t \cos (\pi x/a) - H \sin (\omega t + 4\pi/3) \cos (\pi x/a + 2\pi/3)$$
$$- H \sin (\omega t + 2\pi/3) \cos (\pi x/a + 4\pi/3),$$

that is, by $- (3H/2) \sin (\omega t + \pi x/a)$.

The fields in the two cases are therefore rotating in opposite directions.

Transformer analogy.

The circle diagram (Fig. 174) giving the locus of the extremity of the vector of the current in a phase winding of the stator of an ideal three phase induction motor is identical with that giving the locus of the extremity of the current vector for the primary circuit of a certain ideal single phase transformer (Vol. I, p. 344). The resistance of the primary winding of this transformer is R, its inductance is L, the mutual inductance between the windings is M and the self-inductance of the secondary is L′. The no-load current of the transformer equals the current per phase of the stator when the rotor is running at synchronous speed. The starting current per stator phase of the motor equals the primary current of the transformer when the secondary is short circuited.

When the rotor is at rest the varying magnetic flux, due to the stator coils, which is linked with its windings induces an E.M.F. equal to $M\omega_1 A_1$ per rotor phase. At this instant we have

$$M\omega_1 A_1 = \{R'^2 + (L'\omega_1)^2\}^{\frac{1}{2}} A_2.$$

When the rotor is running there will still be an electromotive force induced in its windings, but as the relative speed of the rotor and the rotating magnetic field has been reduced in the ratio of s to 1, the E.M.F. developed will now be $sM\omega_1 A_1$, *i.e.* $M\omega A_1$.

Hence $$M\omega A_1 = \{R'^2 + (L'\omega)^2\}^{\frac{1}{2}} A_2,$$

and thus $$M\omega_1 A_1 = \{(R'/s)^2 + (L'\omega_1)^2\}^{\frac{1}{2}} A_2.$$

It will be seen that this equation is the same as (10). We see therefore that the resistance of the secondary winding of the equivalent transformer is R'/s when s is the slip of the rotor. From the mathematical point of view therefore the two problems are identical. We can thus readily discuss the effects produced by putting condensers in the stator and rotor circuits.

In the theory of the motor given above, however, it has to be remembered that we have neglected the effects of hysteresis and eddy currents in the magnetic circuits. We have also neglected the effects of the eddy currents induced in the other parts of the machine. We must not expect therefore to obtain a very close agreement between experimental results and our theory even when the applied P.D. is approximately sine shaped.

Equivalent net-work.

We have shown in Chapter XII, p. 354, that, for purposes of calculation, we may replace a leaky transformer by a simple equivalent net-work. We can also use the same net-work for calculations in connection with induction motors. If s be the slip produced by a given mechanical load, the electrical load in the secondary circuit of the auxiliary transformer is a non-inductive resistance the value of which is R'/s. Thus if we place a non-inductive resistance $R'/(sn^2)$ in the secondary branch of the net-work which is equivalent to this transformer, the primary current, and therefore the current in the stator of the motor, can be found.

Since the useful mechanical power that we can get from a motor equals $3(1/s - 1) R'A_2^2$, it will be seen that it is immaterial whether we use a low ratio of transformation M/L, and therefore thick wire coils with few turns on the rotor, or a high ratio and fine wire coils of many turns, provided that the total heat developed in the rotor is the same in both cases at the same slip. The value of the leakage factor σ, however, which depends on the values of M, L and L′, has a great influence on the working of the motor. In practical work it has been noticed that the number of bars used in constructing squirrel-cage rotors is not of great importance. Manufacturers generally use squirrel-cage rotors, containing the same number of bars, for stators wound for various voltages. They also use the same rotors for machines with two phase and three phase stator windings.

Speed greater than synchronism.

Let us now consider what happens when the rotor of a three phase motor is driven mechanically at a speed higher than that corresponding to synchronism. The equations (3) and (4) still hold, but ω_2 is now greater than ω_1 so that the slip s is negative. The extremity P of the vector representing the current in a phase winding of the stator still lies on the circle (Fig. 174). When P coincides with F the slip is negative and the power taken from the mains is zero. At higher speeds of the rotor P lies on the arc $FP'X$, and power is given to the mains, the maximum possible amount of power being given when P' is vertically below C. At still higher speeds when P' lies above OX the motor takes power from the mains. It is to be noticed that however high the speed of the rotor the current per phase taken from the mains cannot be greater than OP_m where OCP_m is a straight line.

Testing induction motors.

If the power W_1 supplied to the stator of an induction motor be read by means of a suitable wattmeter, and a mechanical load $G_2\omega_2$ be applied to the rotor pulley by means of a Prony brake or other simple form of absorption dynamometer, the efficiency η is given by the formula

$$\eta = G_2\omega_2 / W_1.$$

In this formula G_2 is in 10^7 dyne-centimetre units. This method is simple and can be applied easily when the motor is small. In the case of large motors the method is troublesome and expensive, and therefore approximate electrical methods are employed.

It is useful first of all to construct a circle diagram similar to Fig. 174. In order to do this we need to know the power and the current taken by the stator at no load, and also the power taken by the stator at stand-still. With our usual notation,

$$W_0 = 3 V_1 A_0 \cos\psi_0,$$

and thus

$$\cos\psi_0 = W_0 / (3 V_1 A_0),$$

if the motor be three phase and V_1 is the voltage applied at the terminals of a stator winding.

Similarly we have

$$\cos\psi_s = W_s / (3 V_1 A_s),$$

where the suffix s gives the values of the quantities at stand-still. To construct the diagram we draw a line OG (Fig. 174) equal to A_0 and make the angle YOG equal to ψ_0.

We draw a line OP_0 equal to A_s and inclined to OY at an angle ψ_s. We then join P_0G, and draw a line bisecting P_0G at right angles and cutting the line OG at C, which is very approximately the centre of the circle required.

For example in a three phase induction motor, star-connected,

W_0 was equal to 1824 watts, A_0 to 8·8 amperes, and $\sqrt{3}V_1$ was 460 volts. Thus

$$\cos\psi_0 = 1824/(\sqrt{3} \times 8{\cdot}8 \times 460) = 0{\cdot}260,$$

and therefore $$\psi_0 = 74^\circ{\cdot}9.$$

In determining the current at stand-still, precautions have to be taken, otherwise the large currents generated may damage the stator or rotor windings. It is customary to reduce the applied potential difference, preferably by diminishing the excitation of the generator. The amperes and watts per phase are then read for various values of the applied voltage, and from the results the values of the corresponding quantities at the normal voltage are deduced by drawing curves. For the motor considered above the following results were obtained:

Applied voltage	Amperes per phase	Watts per phase
139	45	774
186	64	1587
234	86	2842

Plotting these results on sectional paper and drawing smooth curves through the points, we deduce that the current would be 184·5 amperes per phase and the power 13000 watts per phase when the applied mesh voltage is 460. We have therefore

$$3V_1A_s\cos\psi_s = 13000 \times 3,$$

and thus $$A_s\cos\psi_s = 13000\sqrt{3}/460 = 48{\cdot}9.$$

Therefore also $$\cos\psi_s = 48{\cdot}9/184{\cdot}5,$$

and $$\psi_s = 74^\circ{\cdot}6.$$

We have thus sufficient data to construct the circle diagram and hence we can find the slip and the efficiency at various loads. When the applied waves are sine shaped, this method gives satisfactory results, but even in this case it is advisable to check our results by some more direct electrical method. For instance, suppose that W_1, W_0, A_1, A_0, R_1 and s can be measured.

If the motor is three phase, η is given approximately by the formula

$$\eta = (1 - s)\{W_1 - W_0 - 3R_1(A_1^2 - A_0^2)\}/W_1.$$

Thus values of η can be found for various loads.

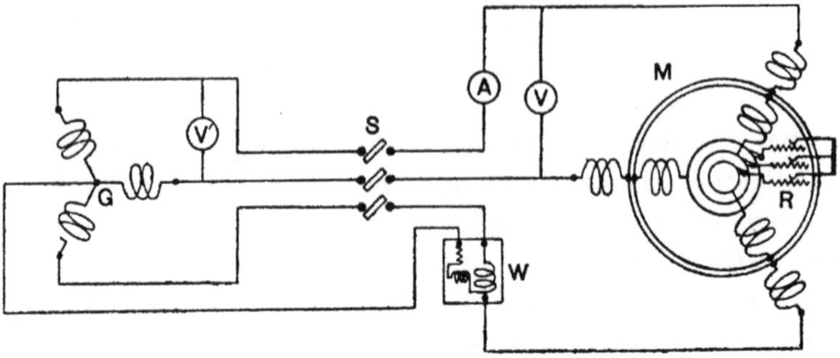

Fig. 177. Connections for testing a three phase induction motor. G is the generator and M is the motor. The rotor starting resistances are at R; S is a triple pole switch and W a wattmeter.

The following are the results of a test on a 32 H.P. three phase induction motor with eight poles. The connections for

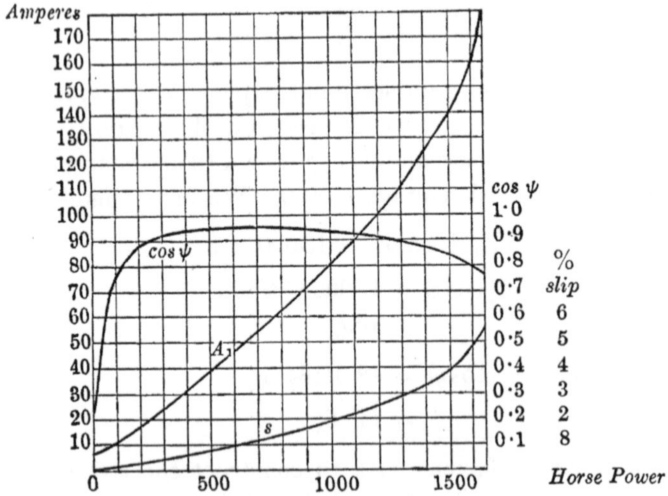

Fig. 178. Three phase induction motor (Oerlikon).

A_1 = the current in one phase of the stator windings.

$\cos\psi$ = the power factor.

s = the percentage slip.

The abscissae give the mechanical output in horse power. One kilowatt is 1·34 horse power.

the test are shown in Fig. 177. To start the motor, resistance is introduced into the rotor windings by means of brushes pressing on three slip rings. When the rotor is running at its usual speed, these resistances are short circuited. The voltage between the supply mains was 460. The frequency was 50, so that at synchronous speed the rotor revolved $60f/p$, that is,

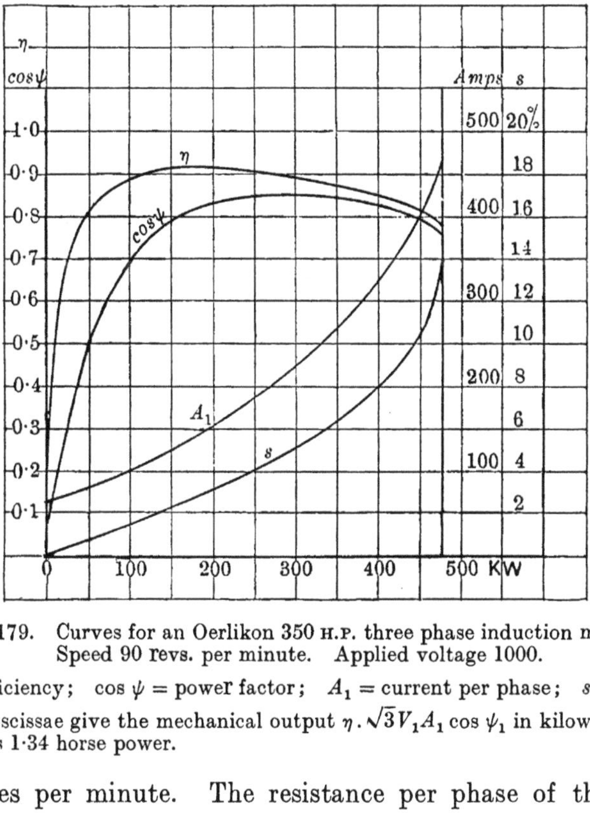

Fig. 179. Curves for an Oerlikon 350 H.P. three phase induction motor. Speed 90 revs. per minute. Applied voltage 1000.

η = efficiency; $\cos\psi$ = power factor; A_1 = current per phase; s = slip.

The abscissae give the mechanical output $\eta . \sqrt{3}V_1A_1 \cos\psi_1$ in kilowatts. One kilowatt is 1·34 horse power.

750 times per minute. The resistance per phase of the stator windings when warm was 0·19 ohm and the no-load current was 8·8 amperes. Thus

$$3A_0^2R_1 = 3 \times (8{\cdot}8)^2 \times 0{\cdot}19 = 44.$$

The wattmeter reading W_0 at no load was 1824. At the normal load A_1 was 36·6 amperes, W_1 was 26800 watts, and the number of revolutions per minute was 730. Thus

$$s = (750 - 730)/750 = 0{\cdot}0267$$

and $$1 - s = 0{\cdot}9733.$$

Therefore, by the formula on p. 386,

$$\eta = 0{\cdot}9733\,(26800 - 1824 - 0{\cdot}57 \times 36{\cdot}6^2 + 44)/26800 = 0{\cdot}88.$$

Thus the efficiency is 88 per cent.

The results of tests on a 720 horse power three phase induction motor having eight poles are shown in Fig. 178. The effective

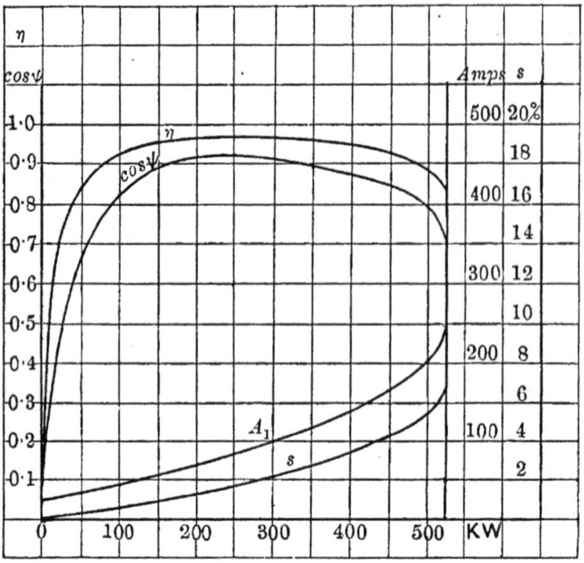

Fig. 180. Curves for an Oerlikon 350 H.P. three phase induction motor. Speed 980 revs. per minute. Applied voltage 2000.
η = efficiency; $\cos\psi$ = power factor; A_1 = current per phase; s = slip.

value of the applied potential difference was 6000 volts per phase and the frequency was 25. At synchronous speed the current was 7·5 amperes and the power expended was 15000 watts. In starting the motor the potential difference applied to its terminals was gradually increased. At 1000 volts it began to turn slowly. The stator resistance per phase was 0·67 ohm and the rotor resistance was 0·0035 ohm. It was found that the leakage factor σ was 0·0268. The slip at full load was 1·3 per cent., that is, s was 0·013, the maximum value of the efficiency 95·0 per cent. and of the power factor 0·952.

High speed and low speed motors. The curves shown in Figs. 179 and 180 were obtained by testing two induction motors of the same power built by the Oerlikon Company. They were both intended to work pumps, but whilst one had to run at the high speed of 980 revolutions per minute the other had to run at only 90 revs. per minute. The applied voltage for the high speed motor was 2000 per phase and for the low speed motor 1000 per phase, the frequency being 50. At synchronous speed the low speed machine took 63·5 amperes per phase and the power factor was 0·089. The corresponding numbers for the high speed machine were 17·1 and 0·14 respectively.

The following table gives the most important of the results obtained. The data for an ordinary three phase induction motor of the same power made by the Oerlikon Company are also added for purposes of comparison.

Revs. per minute	Weight in Kgs.		Percentage efficiency	Power factor	Percentage slip
	Total	Active			
980	5500	2200	94	0·93	1·9
370	—	3200	94	0·90	2·0
90	19000	6100	90	0·87	4·0

REFERENCES

D. C. and J. P. JACKSON, *Alternating Currents and Alternating Current Machinery.*

A. S. MCALLISTER, *Alternating Current Motors.*

B. A. BEHREND, *The Induction Motor.*

ALFRED HAY, *Alternating Currents.*

A. POTIER, 'Sur les Courants Polyphasés,' *International Physical Congress at Paris,* Report 3, p. 197, 1900.

LARMOYER, 'Mesure du Rendement des Moteurs Asynchrones,' *Assoc. Ing. Él., Liége,* Bull. 3, p. 29, 1903.

T. LEHMANN, 'Diagramme rigoreux du Moteur Asynchrone Polyphasé,' *L'Éclairage Électrique,* Vol. 36, p. 281, 1903.

CHAPTER XIV

The equivalent ampere-turns of the inducing windings of asynchronous motors. The equivalent ampere-turns of separate phase windings. The equivalent ampere-turns of superposed phase windings. Varying the speed of induction motors. Several sources of supply available. Motors in cascade. Test of a combination induction motor. The Heyland motor. Polyphase motors in single phase circuits. Two phase motors worked by a single phase machine. Single phase motors. Numerical example. Theory of single phase motor. Starting devices for single phase motors. The starting of induction motors having a rotating field. Three phase induction motor running at half speed. Effect of halving the applied pressure. Starting devices for polyphase induction motors. Induction generators. Frequency transformers. Starting a frequency transformer. References.

The equivalent ampere-turns of the inducing windings of asynchronous motors.

WE shall now show how to calculate the ampere-turns which are equivalent to the mean value of the magnetising forces acting on the armature. Let us suppose that the winding of one phase of the stator of the asynchronons motor is similar to that of the field magnets of a direct current machine, so that when a current is flowing in it we get p segments of South, and p segments of North polarity with spaces between the adjacent segments which are subjected to no magnetising force. We shall also suppose that there are two slots per pole and per phase (Fig. 181). We have therefore, in the first place, to find the equivalent ampere-turns of an elementary rectangular coil, two of the sides of which are placed in two parallel slots on the inner circumference of the stator. Let the coil have N turns, and let the distance measured along the air-gap between the axes of the two slots be denoted by b'. Let the distance also measured along the air-gap, between the middle points of two adjacent coils of the same phase, be a. The polar step of the flux in the rotor will also be practically equal to a, as the air-gap is always very narrow.

Let O' be the point on the stator midway between the axes of the two slots containing the elementary coil considered, and let O be the point on the rotor opposite to O' when the current i in the elementary coil has its maximum value I. If x be the distance, measured along the air-gap, of O from O' at the time t, then, if the current follow the harmonic law and we assume that the slip is zero, we may write

$$i = I \cos (2\pi/T)\, t = I \cos (\pi x/a),$$

where T is the period of the applied P.D.

Let us now find the mean magnetising force in ampere-turns acting during the time T on a point Y (Fig. 181) on the circumference of the rotor. From symmetry this will be equal to

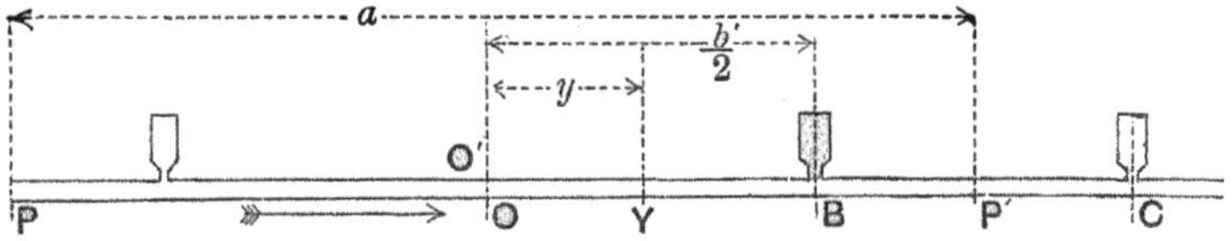

Fig. 181.   Equivalent ampere-turns of the stator windings.

the mean value over the half-period. Let y be the distance of Y from O. Then, during the passage from Y to B the ampere-turns acting on Y are Ni; from B to C, they are zero, and for the rest of the half-period they are $-Ni$. The mean value, therefore, of the ampere-turns $\mathscr{F}'$ acting on Y is given by

$$\mathscr{F}' = (1/a) \left\{ \int_0^{b'/2-y} NI \cos (\pi x/a)\, \partial x - \int_{a-y-b'/2}^{a} NI \cos (\pi x/a)\, \partial x \right\}$$

$$= (NI/a) \left\{ \int_0^{b'/2-y} \cos (\pi x/a)\, \partial x + \int_0^{b'/2+y} \cos (\pi x/a)\, \partial x \right\}$$

$$= (2/\pi)\, NI \sin (\pi b'/2a) \cos (\pi y/a).$$

The mean magnetising force $\mathscr{F}$, therefore, from P to P' (Fig. 181) is given by

$$\mathscr{F} = (2/\pi)\, NI \sin (\pi b'/2a) \cdot (1/a) \int_{-a/2}^{+a/2} \cos (\pi y/a)\, \partial y$$

$$= (4/\pi^2)\, NI \sin (\pi b'/2a).$$

This formula could also be obtained by putting $b = a$ in the formula given on p. 43. For a simple wave winding $b' = a$, and thus

$$\mathscr{F} = (4/\pi^2)\, NI = 0\cdot 4 NI \text{ nearly.}$$

We shall next find formulae for the magnetising forces due to the currents in polyphase windings. We shall first consider a separate phase winding (Fig. 182) with an even number of slots per pole and per phase. In separate phase windings the conductors belonging to one phase are placed in $2pn_1$ slots, which form $2p$ groups each containing n_1 slots, and no conductors belonging to other phases are placed in these slots. Since we are merely concerned with finding the equivalent ampere-turns acting on the magnetic circuits linking the coils of the stator and rotor, we may suppose

The equivalent ampere-turns of separate phase windings.

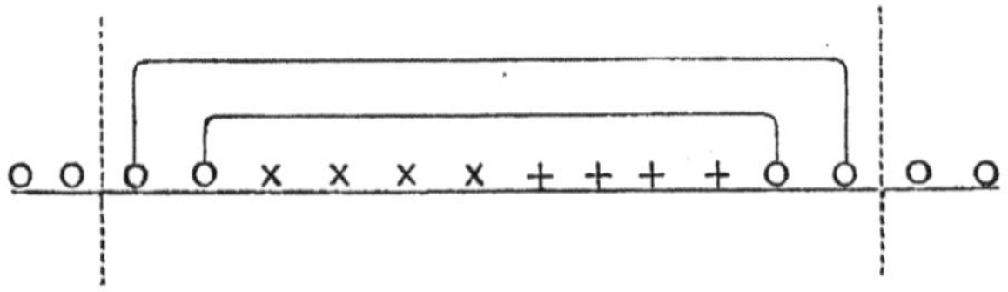

Fig. 182. Separate phase winding with an even number of slots per pole and per phase. The slots marked O, X or + contain conductors belonging to one phase only.

that the ends of the conductors are joined by connecting wires so arranged that the conductors and connecting wires form a number of elementary coils. We can then find the equivalent ampere-turns for each coil by the formula given above, and hence the resultant ampere-turns can be found by adding the results together, since the integral of the sum of n quantities is equal to the sum of the integrals of the n quantities.

Since n_1 is the number of slots per phase in the polar step, qn_1 is the total number of slots in it, and $2pqn_1$ is the total number of slots on the stator. Let N_1 be the total number of conductors per pole and per phase, so that $2pN_1$ is the total number of conductors in a phase winding. Let us first consider the coil belonging to the windings of one phase which has the greatest breadth b_1. If a be the polar step, we have $b_1/a = (qn_1 - 1)/qn_1$, and thus $\pi b_1/2a = (\pi/2)(1 - 1/qn_1)$. Similarly if $b_2, b_3, \ldots$ be the breadths of the other coils we have

$$\pi b_2/2a = (\pi/2)(1 - 3/qn_1); \quad \pi b_3/2a = (\pi/2)(1 - 5/qn_1); \ldots.$$

Hence, noticing that the number of coils is $n_1/2$ and that the number of turns in a coil is N_1/n_1, we get for the resultant equivalent ampere-turns $\mathcal{I}$ of the q phase windings per pole

$$\begin{aligned}\mathcal{I} &= (4q/\pi^2)(N_1/n_1) I \{\sin(\pi/2 - \pi/2qn_1) + \sin(\pi/2 - 3\pi/2qn_1) + \ldots\} \\ &= (4q/\pi^2)(N_1/n_1) I [\cos(\pi/2qn_1) + \cos(3\pi/2qn_1) + \ldots \\ &\qquad + \cos\{(n_1 - 1)\pi/2qn_1\}] \\ &= (2q/\pi^2) N_1 I \sin(\pi/2q)/\{n_1 \sin(\pi/2qn_1)\}.\end{aligned}$$

Now let n be the total number of slots and N the total number of conductors belonging to a phase winding. We have $n_1 = n/2pq$ and $N_1 = N/2p$. Substituting these values of n_1 and N_1 in the above formula we get

$$\mathcal{I} = (2q^2/\pi^2) NA\sqrt{2} \sin(\pi/2q)/\{n \sin(p\pi/n)\} \quad \ldots\ldots(1),$$

where A is the effective value of the current in a winding.

Formula (1) still holds when there is an odd number n_1 of slots per pole and per phase. To prove this, let us consider the arrangement of the conductors indicated in Fig. 183. When there is an

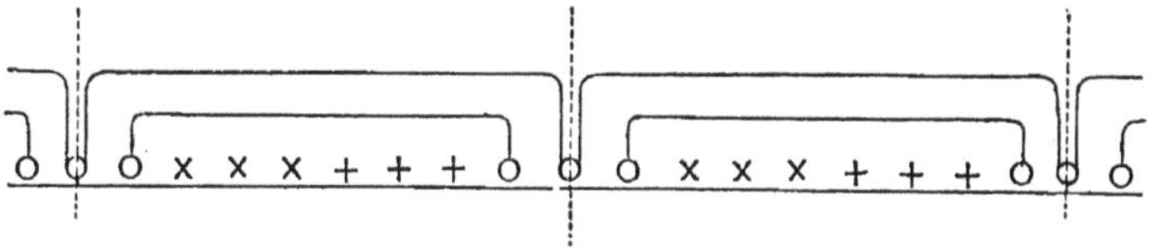

Fig. 183. Separate phase winding with an odd number of slots per pole and per phase.

even number of conductors in each slot we can suppose that the conductors in the middle slot of a phase group are divided into two equal sets, $N_1/2n_1$ of the conductors forming a coil with one set of conductors on the right, and the other half of them forming a coil with the corresponding conductors on the left. When there is an odd number of conductors in each slot, we may suppose that one of the conductors in the middle slot is split into two, each half carrying an equal current, and thus we can still make the assumption that the conductors are divided into two equal sets. In either case, we can write

$$\begin{aligned}\mathcal{I} &= (2q/\pi^2)(N_1/n_1) I [\cos 0 + 2\cos(\pi/qn_1) + 2\cos(2\pi/qn_1) + \ldots \\ &\qquad + 2\cos\{(n_1 - 1)/2\}(\pi/qn_1)] \\ &= (2q^2/\pi^2) NA\sqrt{2} \sin(\pi/2q)/\{n \sin(p\pi/n)\},\end{aligned}$$

which is the same as formula (1).

Let us finally consider the equivalent ampere-turns of a superposed phase winding. In one form of this winding, the slots only contain conductors belonging to one phase; in which case alternate slots contain conductors of different phases. In another form (Fig. 184) each slot contains an equal number of conductors of different phases. These windings are sometimes employed to obtain a flux distribution which will be approximately sine shaped.

The equivalent ampere-turns of superposed phase windings.

Let n_1 denote the number of slots per pole containing conductors in one phase. Then whether n_1 be odd or even we may suppose the ends of the conductors joined as in Fig. 184. In this case

$$\pi b_1/2a = (\pi/2)(1 - 2/qn_1); \quad \pi b_2/2a = (\pi/2)(1 - 6/qn_1); \ldots.$$

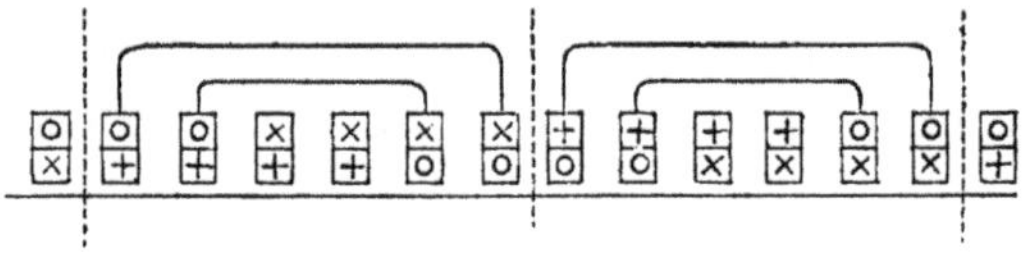

Fig. 184. Superposed phase winding in which each slot contains an equal number of conductors of different phases. Connections of O phase shown.

Therefore

$$\mathscr{X}' = (4q/\pi^2)(N_1/n_1) I \left[\cos(\pi/qn_1) + \cos(3\pi/qn_1) + \ldots + \cos\{(n_1 - 1)\pi/qn_1\}\right]$$

$$= (q^2/\pi^2) NA\sqrt{2} \sin(\pi/q)/\{n \sin(p\pi/n)\} \quad \ldots\ldots\ldots\ldots(2),$$

noticing that n is equal to pqn_1 in this case.

By comparing formulae (1) and (2) we see that when N, p and n have the same values for a separate phase winding and a superposed phase winding,

$$\mathscr{X}' = \mathscr{X} \cos(\pi/2q).$$

Thus the magnetising force, and therefore the flux, is always greater with the separate phase winding.

In proving the formulae given above we have made the assumption that the total number of slots is a multiple of the number of poles ($2p$), so that n_1 is an integer. In alternators and induction motors as usually constructed the number of slots is frequently not a multiple of the number of poles. This considerably increases the difficulty of calculating the equivalent ampere-turns. Approximate values, however, can be obtained from formulae (1) and (2), and these formulae, which are due to C. F. Guilbert, are used in practice for this purpose.

Varying the speed of induction motors.

Ordinary forms of induction motors are practically constant speed machines as the slip is very small. Even at full load it is sometimes only two or three per cent. In most machines the torque on the rotor is a maximum when the slip is less than twenty-five per cent., and if we increase the load so that the slip is greater than this, the rotor rapidly slows down to rest. One method of getting a good output from a motor at different speeds is to have several sources of alternating current supply each at a different frequency. This method is limited in its application, but when motors are directly coupled to heavy rotating apparatus in which a considerable amount of kinetic energy is stored the method is useful, as economies can be effected by first starting the motors from the low frequency mains and then switching them on to higher frequency mains as their speed increases.

Several sources of supply available.

At the Sugar Refinery of Cambrai, induction motors are employed to turn sugar turbines. When rotating at their normal speed the kinetic energy stored in each turbine is 125 000 kilogramme-metres (904 000 foot-pounds). In starting a motor coupled to one of these turbines the expenditure of energy is about 300 000 kilogramme-metres, owing to the unavoidable losses due to the resistances of the motor itself during the start and to friction. To minimise these losses three sources of polyphase currents having frequencies of 21, 35 and 50 are employed. The current which has a frequency of 50 is supplied by a Boucherot alternator. The currents having the two other frequencies are obtained by means of a smaller alternator running at 420 revolutions per minute, and driven by means of a belt from the flywheel of the larger machine. This machine is of the inductor type with two fixed armatures, the rotor having three polar projections on one side and five on the other, so that the frequencies of the induced currents are in the ratio of 3 to 5.

In starting the motor at successively increasing frequencies, the losses during the start are considerably diminished. In addition, instead of all the kinetic energy stored in the turbines being lost during the stopping of a turbine, a considerable proportion of it may be recovered by switching the motor in turn on to circuits of diminishing frequency. The reason of this will be understood from the circle diagram of the induction motor (Fig. 174) described in the preceding chapter.

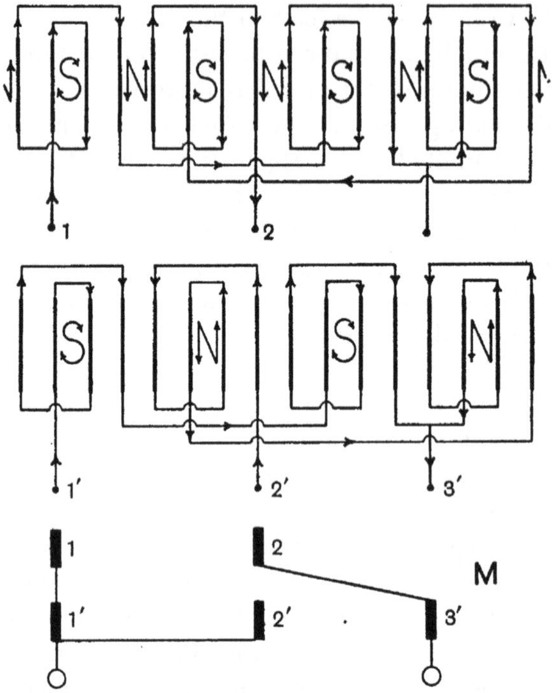

Fig. 185. Varying the number of poles of the stator winding of an induction motor.

When the supply frequency is fixed, it is necessary to modify the design of an induction motor if a variable speed be desired. Three methods of doing this are used in practice. In the first method the polar pitch of one set of poles equals the winding pitch, and the polar pitch of the other set of poles equals a multiple of the winding pitch. It is easy to arrange, for instance, by means of a special switch (Fig. 185), that the number of poles is either $2n$ or n. In the first position of the switch M (Fig. 185) contact is made between the parts 1 and 2 of the switch and the terminals 1 and 2 of the windings as shown in the top diagram. The currents will flow as shown in the top part of the figure, dividing the air-gap into eight portions of North and South polarity respectively. In the second position of the switch contacts are made between the parts $1'$, $2'$ and $3'$ of the switch and the terminals $1'$, $2'$ and $3'$ of the windings. It will be seen that there are now only four segments of North and South polarity. In the latter case the rotor will run at double the speed it does in the former. In this motor the rotor is generally of the squirrel-cage type, and hence we are confined to two speeds only, as it is difficult in practice to interpolate resistance in the windings of this kind of rotor.

The second method is to vary the applied potential difference by means of a compensator (p. 325) and have a suitable resistance in the rotor circuit. This method is not recommended. The efficiency of this type of motor is very low at the slow speed, and so a large motor must be employed for a comparatively small load. The third method is simply to vary the resistance of the rotor circuit. This procedure lowers the efficiency of the motor so much that it is only permissible when variable speeds are very seldom required.

It will be seen that, of the methods of varying the speed of induction motors, the most satisfactory is to design the motor so that the number of poles of the stator windings can be altered by some simple commutating device We thus get two speeds at which the motor will run, and its speed does not appreciably vary from the set speed as the load increases. In the second and third methods, if the speed is to be maintained constant, every change of the load makes it necessary to readjust either the applied potential difference or the rotor resistance.

Motors in cascade.

When we have two induction motors, one of which has slip rings on the axis of its rotor, it is possible to connect them so that they run either in parallel or 'in cascade.' To connect them in cascade the stator terminals of the motor, which has slip rings on the axis of its rotor, are connected with the supply mains. The stator terminals of the second motor are connected with brushes pressing on the slip rings of the rotor of the first machine. The two rotors are also directly connected together by a suitable mechanical coupling so that they run at the same speed. In electric traction both rotors are mounted on an axle of the locomotive.

In Fig. 186 two induction motors are shown connected in cascade. The first motor is connected with the three phase mains M through a compensating transformer, so that the voltage applied

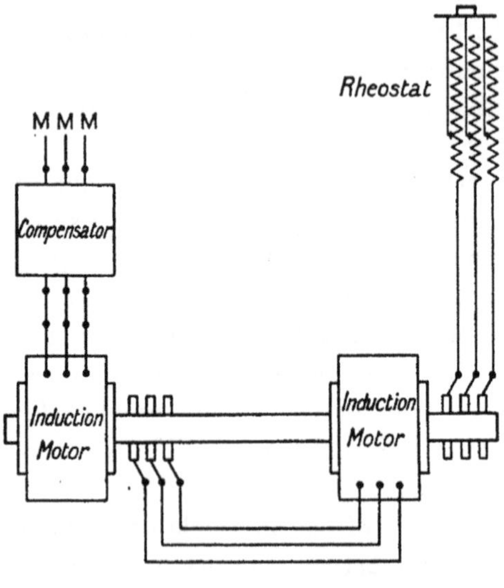

Fig. 186. Motors in cascade.

to it can be varied. The rotor of the second motor has slip rings by means of which its windings are put in series with a rheostat, so that a good starting torque can be obtained.

Let s_1 be the slip of the first rotor, and n the number of revolutions per second corresponding to synchronism, then the actual speed of the rotor is $n(1 - s_1)$. Let $2p$ be the number of poles in the stator winding, then the frequency of the alternating currents induced in the rotor conductors is pns_1. The frequency of the alternating currents supplied to the second machine is pns_1 and therefore, if $2p$ be the number of poles in its stator winding, the synchronous speed of its rotor is ns_1. Thus, if s_2 be the slip of the second machine, $ns_1(1 - s_2)$ will be the number of revolutions made per second by its rotor. Since the two rotors are direct coupled they must run with the same angular velocity, and so we have

$$n(1 - s_1) = ns_1(1 - s_2)$$

and hence

$$s_1 = 1/(2 - s_2).$$

Now the second machine is running under normal conditions, and its slip s_2 must consequently be small. The following table gives the value of $1 - s_1$ for various values of s_2. Since the rotor

s_2 ...	0·01	0·02	0·03	0·04	0·05	0·06	0·07	0·08	0·09	0·10
$1 - s_1$	0·498	0·495	0·492	0·490	0·487	0·484	0·482	0·479	0·476	0·474

makes $n(1 - s_1)$ revolutions per second, we see that, even when the second machine is heavily loaded, the speed only differs from $n/2$ by about five per cent.

Let ω_1 be the angular velocity of the rotating field of the first machine, and let ω_2 be the angular velocity of the rotor. By definition $\omega_1 = \omega_2 + s_1\omega_1$, and thus, if G be the torque acting on the rotor of the first machine, we have

$$G\omega_1 = G\omega_2 + Gs_1\omega_1.$$

Now $G\omega_1$ is the total power given to the first rotor, and $G\omega_2$ is the mechanical power given directly to it, and thus $Gs_1\omega_1$ is the electric power generated in its conductors together with the power expended in hysteresis and eddy current losses. We may therefore write

$$Gs_1\omega_1 = G'\omega_2 + W_0,$$

where $G'\omega_2$ is the mechanical power given to the rotor of the second machine and W_0 represents the losses due to the heating of the two rotors and of the stator of the second machine. We thus deduce the following expressions for the efficiency η_r, p. 365,

$$\begin{aligned}\eta_r &= (1 + G'/G)(\omega_2/\omega_1)\\ &= (1 + G'/G)(1 - s_1)\\ &= 1 - s_1 W_0/(W_0 + G'\omega_2).\end{aligned}$$

Hence the greater the value of the torque G' acting on the rotor of the second machine, and the smaller the value of W_0, the greater will be the efficiency of the motors working in cascade.

Since it is immaterial whether we supply the stator or the rotor of an induction machine with the alternating currents from the generator, it is sometimes more convenient in practice to supply them to the rotor of the second machine. In this case the starting resistances can be connected across the stator terminals of the second machine.

Let us now consider the case of m motors working in cascade. We have as before

$$s_{m-1} = 1/(2 - s_m),$$
$$s_{m-2} = 1/(2 - s_{m-1}) = (2 - s_m)/(3 - 2s_m),$$
$$\cdots\cdots\cdots\cdots\cdots\cdots\cdots\cdots\cdots\cdots$$

and thus finally

$$s_1 = \{(m-1) - (m-2)\, s_m\}/\{m - (m-1)\, s_m\}.$$

This formula shows us that the slip of the first machine is greater than $(m-1)/m$. The angular velocity, therefore, of the rotors of m similar induction motors connected in cascade will be slightly less than the mth part of the angular velocity of the rotating magnetic field of the first motor.

Another interesting case arises when the stator windings of two machines have different numbers of poles. Let us suppose, for example, that the first machine has $2p$ poles and that the second machine has $2q$ poles. If the rotating magnetic field due to the first machine make n turns per second, the frequency of the currents in its rotor is pns_1. Thus the magnetic field due to the stator currents of the second machine will make $(p/q)\, ns_1$ revolutions per second, and hence we have

$$n(1 - s_1) = (p/q)\, ns_1 (1 - s_2),$$

and therefore $$s_1 = 1/\{1 + (p/q)(1 - s_2)\}.$$

If s_2 be very small, $n(1 - s_1)$ is equal to $\{p/(p+q)\}\, n$, that is, $f/(p+q)$, where f is the frequency of the applied potential difference. Thus, if we have two motors each of which has slip rings on its rotor, the rotors will run at the same speed whichever stator be connected with the supply mains, provided that the motors be connected in cascade in each case.

Test of a combination induction motor.

The following figures give the results of a test, made by Danielson, on a combination three phase induction motor which can be run at three speeds. The motor consists practically of two induction motors mounted on the same bed-plate and having a common shaft with three bearings. The main motor has 14 poles and its maximum output is 200 horse power. The auxiliary motor can be connected either as a two pole or a four pole motor, and its stator can be connected with the rotor circuits of the main motor. When the main motor is run alone, its speed is 428 revolutions per minute. When it is connected in cascade with the auxiliary motor, the speed is either $428\,\{7/(7+1)\}$, that is, 375, or $428\,\{7/(7+2)\}$, that is, 333 revolutions per minute, depending on whether the auxiliary motor is arranged to have two or four poles. The output of the machine is practically the same at the three speeds. In the following table, η gives the percentage efficiency of the main motor and of the main motor in cascade with the auxiliary motor, and $\cos\psi$ gives the power factor in the various cases.

Output ...	50 horse power		100 horse power		150 horse power		200 horse power	
Revolutions per minute	η	$\cos\psi$	η	$\cos\psi$	η	$\cos\psi$	η	$\cos\psi$
428	80	0·77	87	0·89	90	0·92	90	0·92
375	76	0·60	86	0·78	88	0·82	87	0·83
333	80	0·58	86	0·76	88	0·81	84	0·80

The number of revolutions per minute is $60f/(p+q)$ and thus it can be found when the frequency f is known. For instance, in the above test f is 50, p is 7, and q is 1 in one case and 2 in the other. Hence the revolutions per minute can be found at once. Since the slip is 1/8 in one case and 2/9 in the other, the frequency of the alternating currents supplied to the auxiliary machine is 6·25 in the first case and 11·1 in the second.

The Heyland motor.

In practice, the normal full load current in the stator of an induction motor is only four or five times that of the no-load current; hence at small loads the power factor of the stator circuit is low. This is objectionable in practical work, and hence several attempts have been made to raise the power factor of an induction motor. The Heyland motor, the principle of which is illustrated in Fig. 187, is one of the most successful of these attempts. Since the current in the stator of an induction motor is the same as that in the primary of a certain transformer, it follows that if we can put the equivalent of a condenser load on the rotor we can diminish the magnetising current required for the stator and so increase the power factor. This is effected in the Heyland motor as follows. The rotor is furnished with a commutator similar to those used with direct current machines. Three brushes press on this commutator, their angular distances from each other being 120 degrees. The windings of the rotor are connected with slip rings S in the usual manner, the brushes pressing on these slip

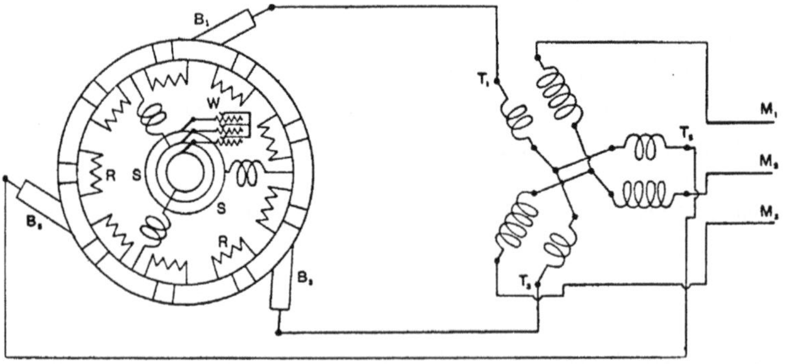

Fig. 187. Principle of Heyland compensated induction motor. Connections of rotor only are shown. S, slip rings for inserting resistances W into the circuit of the rotor windings. R, resistances connecting the segments of the commutator. T_1, T_2 and T_3 are the terminals of the step-down transformer from the mains.

rings being short circuited when full speed is attained. The segments of the commutator are all connected together by strips of high resistance metal, so that the rotor circuits are completed by means of these resistances. The three brushes on the commutator are connected with the secondary terminals of a three phase transformer, the primary being connected with the mains.

The ratio of transformation is so chosen that the pressure between the commutator brushes is only about eight volts. If the rotor were at rest, the frequency of the currents induced in its windings would be the same as the frequency of the stator currents. When it is running at its normal load, the frequency is very small and thus the impedance offered by the rotor circuits is small. The low voltage, therefore, is quite sufficient in certain positions of the brushes to provide the necessary magnetising current for the stator flux. The current in the stator is a minimum for a certain position of the brushes, and in this position the step-down transformer furnishes the leading currents in the rotor circuits which are the equivalent of the condenser load in the transformer analogy. It is found in practice that it is possible to get a power factor which is nearly equal to unity by this method. The stator windings are not shown in Fig. 187 as they are the same as those of ordinary three phase machines.

Polyphase motors in single phase circuits.

In Fig. 188 one method of connecting the stator of a three phase induction motor for use in single phase circuits is shown. The single phase supply mains M_1 and M_2 are connected with the terminals T_1 and T_2 of the stator and T_2 and T_3 are connected with the terminals of an alternating current booster (see Chap. XI) having a suitable

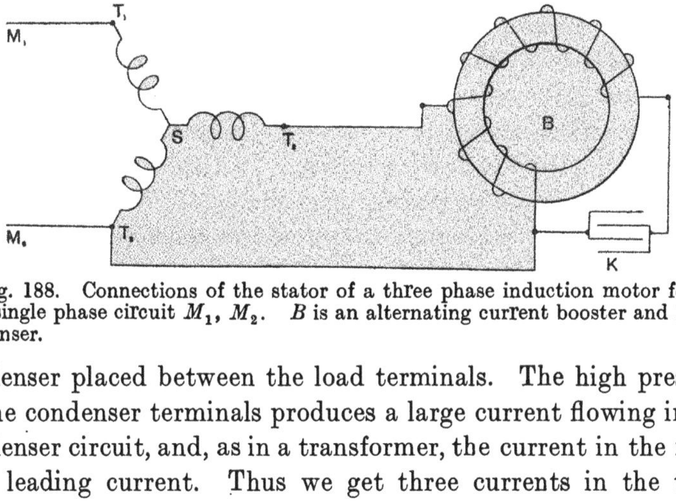

Fig. 188. Connections of the stator of a three phase induction motor for use on a single phase circuit M_1, M_2. B is an alternating current booster and K is a condenser.

condenser placed between the load terminals. The high pressure at the condenser terminals produces a large current flowing in the condenser circuit, and, as in a transformer, the current in the main is a leading current. Thus we get three currents in the three stator windings in different phases and a rotary field is produced.

Two phase motors worked by a single phase machine.

If we have a number of two phase motors each of which has two separate windings A and B, and if we connect all the A windings in parallel with a single phase alternator and also connect all the B windings of the motors in parallel, then it is found that, provided that at least one of the motors is always running, the others can be stopped and started and will work satisfactorily. This is due to the currents induced in the rotors of the running machines producing a rotating flux, which develops an electromotive force in the B circuits of the stators, and thus they will operate almost as well as when they are connected with a two phase machine. It can be shown experimentally that the electromotive force developed in the B circuit of a machine when the rotor is revolving at synchronous speed is about four-fifths of the potential difference applied to the A circuit. The machine therefore acts like a single phase transformer so far as altering the pressure is concerned. The phase of the secondary electromotive force, in this case, however, differs by a quarter of a period from that of the primary.

Single phase motors.

We saw in Vol. I, Chap. XVIII, that a rotating magnetic field can easily be produced when polyphase currents are available, and hence the design of polyphase induction motors is, comparatively speaking, simple. If two of the phases of a three phase induction motor are out of action when running, it will continue to run, but the stator current in the phase still in circuit will be three times as large as it was before the other circuits were broken. In this case we have the rotor revolving in an oscillating magnetic field. Now if Φ be the flux produced by the alternating current in the active phase of the stator winding, we can, by the principles developed in Vol. I, replace the oscillating field Φ by two rotating fields, the magnitudes of which are $\Phi/2$ revolving in opposite directions. The field rotating in the same direction as the rotor will act like the original rotating field on the rotor constraining it to revolve; its magnitude however is only one-third that of the original field for the same current in the phases. If the new rotating field therefore is to be equal to the old, the new current, assuming that the flux and current are proportional, must increase three times. The mean value of the torque produced by the field rotating in the opposite direction to the rotor is very small. We see that if in an oscillating magnetic field the rotor be brought up to speed it will operate in much the same way as it would in a rotary field. We shall now discuss the theory of single phase motors in greater detail.

Let us suppose that we have an alternating magnetic field fixed in space and that its intensity is given by $B \sin \omega_1 t$. Consider a coil of wire having n turns, placed so that it can revolve about an axis through its centre perpendicular to the lines of force of the field and perpendicular also to the axis of the coil. Let S be the mean area of the turns of wire, then if ω_2 is the angular velocity with which the coil revolves, and ϕ is the flux embraced by the coil at the time t, we have

$$\begin{aligned}\phi &= BS \sin \omega_1 t \,.\, \sin(\omega_2 t - a)\\ &= (BS/2)\left[\cos\{(\omega_1 - \omega_2)\, t + a\} - \cos\{(\omega_1 + \omega_2)\, t - a\}\right],\end{aligned}$$

where a is a constant. Thus the torque produced is the same as if the coil were fixed and we had two magnetic fields rotating with angular velocities $(\omega_1 - \omega_2)$ and $(\omega_1 + \omega_2)$ respectively in opposite directions, the intensity of each of the rotating fields being $B/2$.

Let r be the resistance, and l the self-inductance of the coil. Let e_1 and i_1 be the electromotive force and the current due to it induced in the coil by the field which is rotating with an angular velocity $\omega_1 - \omega_2$. We have

$$e_1 = -n\partial\phi/\partial t = (nBS/2)(\omega_1 - \omega_2)\sin\{(\omega_1 - \omega_2)\, t + a\},$$

and thus

$$i_1 = (nBS/2Z_1)(\omega_1 - \omega_2)\sin\{(\omega_1 - \omega_2)\, t + a - \beta_1\},$$

where

$$Z_1 = \{r^2 + (\omega_1 - \omega_2)^2 l^2\}^{1/2}, \text{ and } \cos\beta_1 = r/Z_1.$$

If G_1 be the average value of the torque due to this rotating field, we have

$$G_1(\omega_1 - \omega_2) = (n^2B^2S^2/8Z_1)(\omega_1 - \omega_2)^2\cos\beta_1,$$

and thus $G_1 = (n^2B^2S^2r/8)(\omega_1 - \omega_2)/\{r^2 + (\omega_1 - \omega_2)^2 l^2\}$.

Similarly we can show that if G_2 be the torque due to the other rotating field, we have

$$G_2 = (n^2B^2S^2r/8)(\omega_1 + \omega_2)/\{r^2 + (\omega_1 + \omega_2)^2 l^2\}.$$

Hence if G be the resultant torque on the rotor, we have

$$\begin{aligned}G &= G_1 - G_2\\ &= (n^2B^2S^2/4)\, r\omega_2(\omega_1^2 l^2 - r^2 - \omega_2^2 l^2)/\{(\omega_1^2 l^2 - r^2 - \omega_2^2 l^2)^2 + 4\omega_1^2 l^2 r^2\}.\end{aligned}$$

This formula shows us at once that the torque vanishes when ω_2 is zero and that it vanishes again when ω_2 equals $\sqrt{\omega_1^2 - r^2/l^2}$. If r is greater than $l\omega_1$, the motor will not work at all, as the torque will always act so as to prevent rotation. If r is less than $\omega_1 l$, then, as ω_2 increases, the torque G increases to a maximum value; it vanishes when ω_2 equals $\sqrt{\omega_1^2 - r^2/l^2}$, and becomes negative for values of ω_2 greater than this. The angular velocity of the rotor, therefore, can never attain synchronism with either of the rotary components of the oscillating field.

Let us now suppose that ω_2 is constant, and that we vary r. We find by differentiating G with respect to r and equating to zero that the torque has a maximum value $G_{\text{max.}}$ when r is given by

$$r = \sqrt{2\omega_1^2 l^2 - \omega_2^2 l^2} - \omega_1 l,$$

and since

$$\omega_2 = \omega_1 (1 - s),$$

this becomes

$$r = \omega_1 l \{\sqrt{1 + 2s - s^2} - 1\}.$$

When r has this value we get

$$G_{\text{max.}} = (n^2 B^2 S^2/16l)(1 - s),$$

and

$$P_{\text{max.}} = (n^2 B^2 S^2 \omega_1/16l)(1 - s)^2,$$

where $P_{\text{max.}}$ is the greatest possible value of the mechanical power given to the rotor when the slip is s. When s is small the value of r which makes the mechanical power a maximum is nearly equal to $s\omega_1 l$. If r were less than this, then, adding resistance to the rotor would increase the speed and thus increase the efficiency of the machine.

Numerical example.

We can easily draw curves to illustrate how the torque varies with the slip. For instance, let the ratio of r to $\omega_1 l$ be as 1 to 10. Then if s be the slip for which the torque is a maximum for this value of r, we have

$$\sqrt{1 + 2s - s^2} - 1 = 1/10,$$

and thus

$$1 + 2s - s^2 = 1{\cdot}21,$$

and

$$s = 0{\cdot}11 \text{ approximately}.$$

In Fig. 189 the curve A represents the torque due to the field rotating in the same direction as the rotor and the curve B represents the negative torque due to the field rotating in the opposite direction. The curve C represents the resulting torque, and is constructed by subtracting the ordinate of the curve B from that of A, and making this length the ordinate of C.

The equation to the curve A is

$$y_1 = 100\,(10 - x)/\{1 + (10 - x)^2\},$$

and the equation to the curve B is

$$y_2 = 100\,(10 + x)/\{1 + (10 + x)^2\}.$$

The equation to the curve C is

$$y = y_1 - y_2.$$

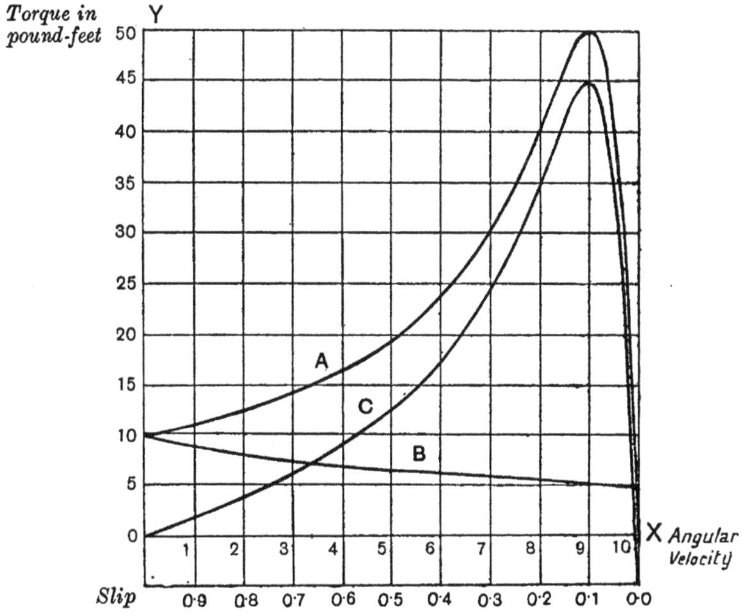

Fig. 189. The curve C gives the torque on the rotor of a single phase machine. The curves A and B give the torques due to the two rotary components of the oscillating magnetic field. These torques act in opposite directions and C is their resultant.

The maximum value of the torque occurs when the slip is about ten per cent., and the torque vanishes when the slip is the half of one per cent. and when it is unity, that is, when the rotor is at rest.

Theory of the single phase motor.

In the discussion of the torque on the rotor of a single phase motor given above we have supposed that the amplitude of the magnetic field remains constant at all loads. Owing to the resistance of the primary circuit this assumption is only permissible when obtaining a roughly approximate solution to the working of the motor. It is advisable, therefore, to discuss the problem more rigorously. Let us suppose that the single phase motor has a stator winding the constants of which are (R, L) and a single rotor winding (r, l). Let the value of the potential difference applied to the stator terminals be e and let the angular velocity of the rotor be ω_2. Let us also suppose that the mutual inductance m between the stator and rotor windings is $M \sin(\omega_2 t + \gamma)$. Then if i and i' denote the respective values of the stator and rotor currents and if we neglect hysteresis and eddy current effects, we have

$$e = Ri + L\frac{\partial i}{\partial t} + \frac{\partial}{\partial t}(mi') \quad \text{.....................(3)}$$

and

$$0 = ri' + l\frac{\partial i'}{\partial t} + \frac{\partial}{\partial t}(mi) \quad \text{.....................(4).}$$

Multiplying (3) by i and (4) by i', and adding the resulting equations, we get

$$ei = Ri^2 + ri'^2 + \frac{\partial}{\partial t}(\tfrac{1}{2}Li^2 + \tfrac{1}{2}li'^2 + mii') + ii'\frac{\partial m}{\partial t}.$$

The first two terms on the right-hand side of this equation give the power expended in heating the conductors, and the next term gives the rate at which energy is being stored in the magnetic field. The term $ii'(\partial m/\partial t)$, therefore, must represent the mechanical power expended by the rotor. If g denote the instantaneous value of the retarding torque on the rotor, $g\omega_2$ is the rate at which it is doing work, and hence

$$\begin{aligned} g\omega_2 &= ii'(\partial m/\partial t) \\ &= M\omega_2 ii' \cos(\omega_2 t + \gamma) \end{aligned}$$

and

$$g = Mii' \cos(\omega_2 t + \gamma) \quad \text{.....................(5).}$$

We see therefore that the instantaneous value of the torque vanishes when m has a maximum or a minimum value. It also vanishes whenever i or i' vanishes.

If we suppose that e has the value $E \cos \omega_1 t$ and assume that

$$i = \{E/(R^2 + L^2\omega_1^2)^{1/2}\} \cos(\omega_1 t - \alpha_1) + i_1,$$

where $\tan \alpha_1 = L\omega_1/R$, we can, proceeding as in the corresponding problem for armature reaction (p. 61), obtain series formulae for i and i'. In order to lighten the analytical labour, however, we shall discuss the much simpler problem of finding the value of e when the current i can be represented by the sine function $I\cos(\omega_1 t + \alpha)$.

Substituting $I\cos(\omega_1 t + \alpha)$ for i and $M\sin(\omega_2 t + \gamma)$ for m in (4) we easily find that

$$\begin{aligned} i' &= \{M(\omega_1 - \omega_2) I/2Z_1\}\cos\{(\omega_1 - \omega_2)t + \alpha - \gamma - \beta_1\} \\ &\quad - \{M(\omega_1 + \omega_2) I/2Z_2\}\cos\{(\omega_1 + \omega_2)t + \alpha + \gamma - \beta_2\} \quad \ldots(6), \end{aligned}$$

where the values of Z_1 and β_1 are given on p. 405 and

$$Z_2^2 = r^2 + (\omega_1 + \omega_2)^2 l^2, \text{ and } \cos\beta_2 = r/Z_2.$$

We see from (6) that, for any speed ω_2 of the rotor other than zero or ω_1, the current wave in the rotor is the sum of two sine waves the frequencies of which are $(\omega_1 - \omega_2)/(2\pi)$ and $(\omega_1 + \omega_2)/(2\pi)$ respectively. As ω_2 varies, therefore, the shape of the resultant wave continually alters. At synchronous speed, namely when $\omega_2 = \omega_1$, one of the components vanishes. At this speed the current in the rotor is sine shaped and its frequency is double that of the supply. If A' be the effective value of i' and A be the effective value of i, we see from (6) that

$$A'^2 = \left(\frac{MA}{2}\right)^2 \left\{\frac{(\omega_1 - \omega_2)^2}{Z_1^2} + \frac{(\omega_1 + \omega_2)^2}{Z_2^2}\right\} \quad \ldots\ldots\ldots(7).$$

Substituting $I\cos(\omega_1 t + \alpha)$ for i, and also the value of i' given in (6), in equation (5), we find that the mean value of the torque G acting on the rotor is given by

$$\begin{aligned} G &= \left(\frac{MA}{2}\right)^2 \left\{\frac{(\omega_1 - \omega_2)\cos\beta_1}{Z_1} - \frac{(\omega_1 + \omega_2)\cos\beta_2}{Z_2}\right\} \\ &= \left(\frac{MA}{2}\right)^2 r\left\{\frac{\omega_1 - \omega_2}{Z_1^2} - \frac{\omega_1 + \omega_2}{Z_2^2}\right\} \quad \ldots\ldots\ldots\ldots\ldots\ldots\ldots\ldots(8) \end{aligned}$$

which, provided we write $nBS/\sqrt{2}$ for MA, is the same equation as that given on p. 405. The discussion we have previously given, therefore, applies to (8).

The mechanical power given to the rotor is $G\omega_2$ and the power expended in heating it is rA'^2. Hence the rotor efficiency η_r is given by

$$\begin{aligned} \eta_r &= \frac{G\omega_2}{G\omega_2 + rA'^2} \\ &= \left(\frac{\omega_2}{\omega_1}\right)^2 \cdot \frac{l^2(\omega_1^2 - \omega_2^2) - r^2}{l^2(\omega_1^2 - \omega_2^2) + r^2}, \end{aligned}$$

and this is obviously less than ω_2/ω_1. For a given value of ω_2, also, η_r diminishes as r increases.

On substituting their values for m, i and i' in (3) we get

$$e = ZI\cos(\omega_1 t + a + \beta)$$
$$+ \frac{M^2(\omega_1 - \omega_2) I}{4Z_1}[\omega_1 \cos(\omega_1 t + a - \beta_1)$$
$$- (\omega_1 - 2\omega_2)\cos\{(\omega_1 - 2\omega_2)t + a - 2\gamma - \beta_1\}]$$
$$+ \frac{M^2(\omega_1 + \omega_2) I}{4Z_2}[\omega_1 \cos(\omega_1 t + a - \beta_2)$$
$$- (\omega_1 + 2\omega_2)\cos\{(\omega_1 + 2\omega_2)t + a + 2\gamma - \beta_2\}] \quad \ldots(9)$$

where $Z^2 = R^2 + \omega_1^2 L^2$, and $\tan\beta = \omega_1 L/R$.

Hence e is the resultant of three waves the frequencies of which are $\omega_1/2\pi$, $(\omega_1 - 2\omega_2)/2\pi$ and $(\omega_1 + 2\omega_2)/2\pi$. If the stator current wave, therefore, remain sine shaped at all speeds the shape of the applied potential difference wave will continually vary. In no case however will it be sine shaped.

If we multiply both sides of equation (9) by $I\cos(\omega_1 t + a)$ and take mean values we get

$$VA\cos\psi = RA^2 + \left(\frac{MA}{2}\right)^2 r\omega_1\left\{\frac{\omega_1 - \omega_2}{Z_1^2} + \frac{\omega_1 + \omega_2}{Z_2^2}\right\} \quad \ldots(10)$$

where $\cos\psi$ is the power factor of the stator current. Similarly we have

$$VA\sin\psi = L\omega_1 A^2 - \left(\frac{MA}{2}\right)^2 l\omega_1\left\{\frac{(\omega_1 - \omega_2)^2}{Z_1^2} + \frac{(\omega_1 + \omega_2)^2}{Z_2^2}\right\} \quad \ldots(11).$$

From (8) and (10)

$$VA\cos\psi = RA^2 + G\omega_1 + r(M^2A^2/2Z_2^2)(\omega_1 + \omega_2)\,\omega_1,$$

and hence if η be the efficiency of the motor

$$\eta = \frac{\omega_2}{\omega_1}\cdot\frac{G}{G + A^2\{R/\omega_1 + rM^2(\omega_1 + \omega_2)/2Z_2^2\}} \quad \ldots(12).$$

From (11) and (7) we have

$$VA\sin\psi = \omega_1(LA^2 - lA'^2) \quad \ldots\ldots\ldots\ldots\ldots(13).$$

We see that if the power factor be unity $LA^2 = lA'^2$.

If we drive the rotor of a single phase induction motor at a speed greater than synchronism, then in certain cases power is restored to the mains and the machine acts as a single phase induction generator. Generators of this type would not be as efficient as three phase induction generators (see p. 416) and owing also to the variable wave shape of the generated electromotive force they would run badly in parallel.

Starting devices for single phase motors.

For single phase motors special starting devices are generally necessary, since the torque is zero when the angular velocity of the rotor is zero. Small motors can be started by hand, a pull at the belt being all that is required. They may be started in either direction as they work equally well in the two cases. Motors of large size generally have a secondary winding on the stator. During the start a current which is out of phase with the current in the primary winding is sent through this winding, so that a rotating field is produced in addition to the alternating field.

There are many methods employed in practice for obtaining currents which will differ in phase. One method is to employ static or electrolytic condensers in the starting circuit. Another method employed by de Kando is to produce an unsymmetrical field by cutting out during the start some of the windings in the circuit of one pair of quadrants. Heyland has suggested having two stator windings in parallel. One of these windings is divided into two, which are put in series during the process of starting and then put in parallel for the normal working. By this means we get during the start two branched circuits with different inductances and resistances, and hence the currents are out of phase and a rotary component is superposed on the oscillating field.

We have seen that when the resistance r of the rotor circuit is given by

$$r = \omega_1 l\,(\sqrt{1 + 2s - s^2} - 1),$$

the torque is a maximum. If the angular velocity of the rotor is very small, s is practically unity, and r is $0{\cdot}414\omega_1 l$. Thus, in order to get the maximum starting torque, r should have this value. Arnò inserts resistance in the rotor circuit by means of slip rings on the rotor shaft, and adjusts the external resistance so that the starting torque is a maximum. The resistance is then gradually diminished as the speed increases. As the torque is zero when the rotor is at rest, an initial impulse has to be given to the rotor.

It is found possible to start single phase motors of large size by this method.

Steinmetz employs a second winding with its axis inclined at 60° to the axis of the first winding and with its terminals in series with a condenser. This enables the motor to start when there is no load on it, and it increases the power factor of the stator circuit.

It will be seen that in all these methods the motor has to start on a loose pulley.

In the special case when the motor drives a dynamo for charging accumulators, it can easily be brought up to speed by driving the dynamo as a motor from the accumulators. When the speed of the single phase motor coupled to the shunt wound dynamo is sufficiently high the switch for the alternating current is closed. The armature of the dynamo is now driven at increased speed, the current in the battery circuit reverses and the accumulators are charged.

The starting of induction motors having a rotating field.

When we connect the windings of the stator of an induction motor with the supply mains, we get momentarily a very large current, owing to the low power factor of the motor when the rotor is at rest or moving slowly (see Fig. 174). This current rapidly diminishes as the rotor speeds up. It is however objectionable as, unless the mains are very heavy, it causes an appreciable blink in the light given out by electric lamps in adjacent circuits. In order to diminish this starting current, the stator windings are often so arranged that they are connected in star at the start and then can be changed over to the mesh connection when the rotor has got up speed. This can readily be effected by means of a special switch. With the star connection the voltage across the windings is only $1/\sqrt{3}$, that is, 0·577... of what it is with the mesh connection. Hence by this procedure the starting current is diminished by nearly 50 per cent. To diminish the starting current still further and at the same time to increase the starting torque, a wound rotor is often employed, the windings being connected with a variable resistance. At the start the maximum resistance is inserted in the rotor windings, and this resistance is gradually cut out as the rotor gets up speed. The resistances are so chosen that the starting torque is nearly the same as the full load torque. In this case the starting current is roughly equal to the full load current.

Three phase induction motor running at half speed.

Let us consider the case of a three phase induction motor having a star-connected rotor provided with slip rings for inserting the star resistances. Suppose that the rotor is running with angular velocity ω_2 and that we raise one of the brushes pressing on the slip rings. There is now only a single circuit for the current in the rotor windings and the frequency of this current is

$$p(\omega_1 - \omega_2)/2\pi.$$

The flux produced by it will therefore be an oscillating flux. We may resolve it into two components rotating in opposite directions with angular velocities $\pm(\omega_1 - \omega_2)$ relatively to the rotor. But the rotor is rotating with an angular velocity ω_2, and thus the actual angular velocities of the rotor fields in space are $\omega_2 \pm (\omega_1 - \omega_2)$, that is, ω_1 and $2\omega_2 - \omega_1$ respectively. If the rotor start from rest with one brush raised, it will run in stable equilibrium as a synchronous motor (p. 435) when ω_2 equals $\omega_1/2$. If this brush be now put in contact with the slip ring, the rotor will speed up until its angular velocity is nearly ω_1. If we again raise the brush, it will in general continue to run with the angular velocity ω_1, but if the retarding torque be great it may slow down to half speed.

Effect of halving the applied pressure.

If V be the pressure per phase applied to the stator terminals of an induction motor, then the maximum power it can exert varies as V^2. If we halve the value of V, therefore, the 'breakdown power' will only be one-quarter as great. The starting torque also varies as V^2. Hence the starting torque also will only be a quarter as great. If the mechanical friction is great the torque may be insufficient to start the motor. It is also easy to see that the efficiency of the motor will be reduced owing to the relatively greater importance of the frictional forces. Notwithstanding the diminution in the core losses at the lower pressure, the efficiency is from 5 to 10 per cent. less.

In practice it is customary to design induction motors so that the maximum power they can exert is from 2 to $2\frac{1}{2}$ times their normal rating.

Starting devices for polyphase induction motors.

The device illustrated in Fig. 190 which is due to Fischer-Hinnen is found effective in starting motors. The starting resistances X (Fig. 190) are shunted by the inductive coils R. When the rotor is at rest, the frequency of the currents in it equals the frequency of the applied potential difference, and thus the impedance of

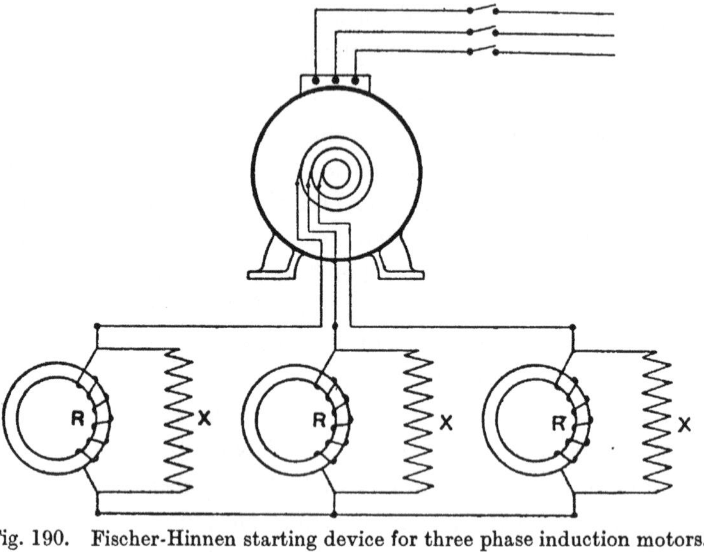

Fig. 190. Fischer-Hinnen starting device for three phase induction motors.

a coil R is high compared with the resistance X. When however the rotor is running nearly at synchronous speed, the frequency of the rotor currents is very low, and thus the impedance of a coil R is low compared with the resistance X. The resistance of the rotor circuits is thus high during the start and is very small when full speed is attained, the resistances X being practically short circuited by the inductive coils R.

The starting device due to A. P. Zani is similar to the above. During the start the external non-inductive resistances are shunted by the inductive coils shown in Fig. 191. Instead, however, of relying on the diminished impedance at low frequencies, the reluctance of the magnetic circuits is automatically considerably increased when the speed attains a given value less than the lowest working speed. The reactance of the coils is thus made negligibly small, and the starting resistances are practically short circuited. This is effected by means of centrifugal force (Fig. 192), the pole pieces flying apart and so increasing the reluctance very considerably. High efficiencies have been obtained with this type of motor.

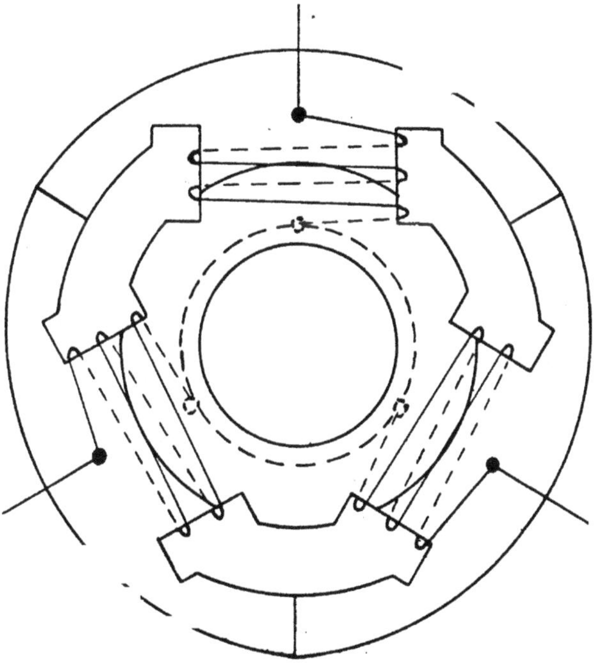

Fig. 191. A. P. Zani's starting device when revolving slowly.

In several types of starting device, Steinmetz makes use of the electrical properties of magnetite. Magnetite and materials similar to it have a high resistance at ordinary temperatures but become good conductors at high temperatures. Hence, if magnetite resistances be inserted in the rotor circuits, they offer a gradually diminishing resistance to the currents during the start, and thus they act in a similar manner to starting devices which automatically switch out resistance. Steinmetz has also used magnetite in the construction of squirrel-cage rotors. The rotor conductors are in electrical contact with the short circuiting rings only through magnetite washers, the fastening bolt being insulated both from the washer and the ring. The resistance of the rotor circuits is thus large at the start owing to the resistance of the washers, but when the washers get hot their resistance is negligible. When the rotor stops the washers cool rapidly, being in contact with a metal ring, and so the motor can safely be restarted almost immediately.

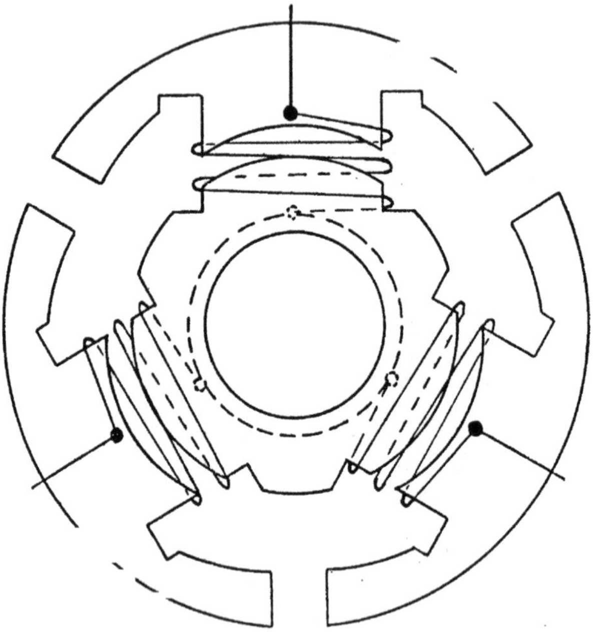

Fig. 192. A. P. Zani's starting device when running at full speed.

Induction generators.

We have seen that when the rotor of an induction motor is driven at a speed greater than that corresponding to synchronism it gives power to the stator. A motor of this kind, driven by an engine, can therefore be used to aid the polyphase alternator in supplying power. As the speed of the rotor has no effect on the frequency of the supply current, such a machine is called an 'asynchronous generator,' or sometimes, an 'induction generator.'

In the Leblanc system of distribution an ordinary polyphase alternator is used in conjunction with a number of induction machines which can all be put in parallel with the bus bars. The frequency of the alternating currents in the stator circuits is the same whatever may be the speed of the rotors. It depends merely on the speed of the polyphase alternator which is always running. Leblanc compares the *rôle* of this machine to that of a *chef d'orchestre,* as it controls the frequency of the currents in the stators of all the other machines. Each of the induction machines works practically at constant power, the regulating alternator governing the pressure between the bus bars as well as the frequency. These machines can be put in and taken out of circuit as readily as ordinary direct current dynamos, and irregularities in the speed of the engines driving them have very little effect on the supply of power.

Frequency transformers.

We have seen that when we are using an induction motor, the stator currents of which produce a pure rotating magnetic field, the frequency of the induced alternating currents in the rotor is $p(\omega_1 - \omega_2)/2\pi$, where $2p$ is the number of poles of the machine. If we have a wound rotor, and the ends of the winding are connected with slip rings mounted on the shaft, we can collect alternating currents of lower frequency from these rings. We can use the motor, therefore, to alter the frequency of the supply. In this case it acts as a frequency transformer. The theory of the induction motor shows that as we alter the load taken from the slip rings the angular velocity of the rotor, and therefore the frequency of the supply, would vary. In most practical cases it is essential that the frequency of the supply remain constant, and hence more elaborate devices for getting a supply at a new frequency are employed.

Let us suppose, for instance, that a Railway Company uses alternating current at 25 frequency for traction purposes, and that there are towns and villages in the neighbourhood of the line where there is a demand for the electric light. In many cases it will be profitable for the Railway Company to supply this demand. As a frequency of 25, however, is not suitable for lighting, it is advisable to use transformers to raise the frequency of the lighting supply. A transformer of this nature is indicated in Fig. 193.

In the diagram the synchronous motor S.M. is mechanically

coupled to the induction motor I.M. The rotors of the two machines therefore rotate with the same angular velocity ω_2. The synchronous motor and the stator of the induction motor are both connected with the low frequency mains. Let us first suppose that the latter machine is connected in such a way that the rotating magnetic field due to the currents in the stator windings rotates in the opposite direction to the rotor. If the angular velocity of this field be ω_1 and that of the rotor be $-\omega_2$, the frequency of the supply at the slip rings of the rotor will be $p(\omega_1+\omega_2)/2\pi$. The frequency has thus been increased in the ratio $(\omega_1+\omega_2)/\omega_1$. If $2p$ be the number of poles of the synchronous motor, $\omega_2=\omega_1$, and thus if the frequency of the power

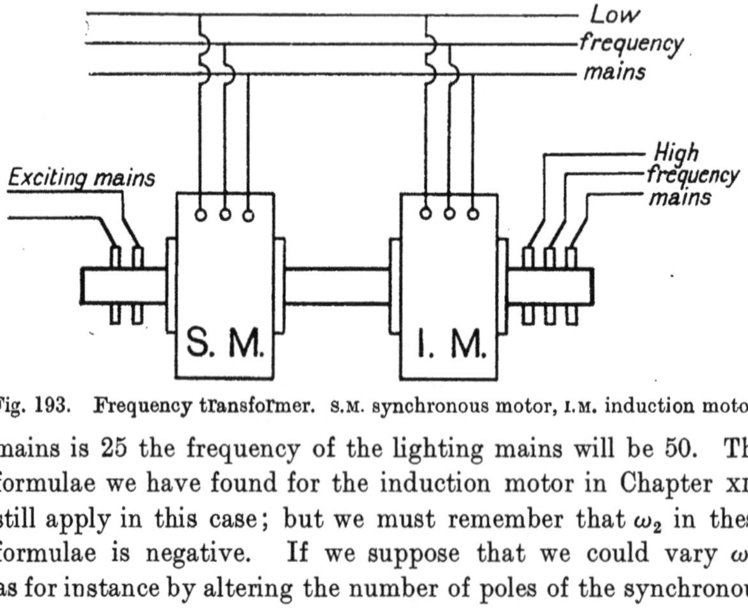

Fig. 193. Frequency transformer. S.M. synchronous motor, I.M. induction motor.

mains is 25 the frequency of the lighting mains will be 50. The formulae we have found for the induction motor in Chapter XIII still apply in this case; but we must remember that ω_2 in these formulae is negative. If we suppose that we could vary ω_2, as for instance by altering the number of poles of the synchronous motor, then the circle diagrams (Figs. 174 and 175) would still hold. For an infinite value of ω_2, α would be 180° and so P (Fig. 174) would be at M where OBM is a straight line. In general, however, we have to remember that the equivalent value of L′, and therefore of σ, alters with the load. Hence the diameter and position of the circle also change.

Let us suppose that G is the average value of the magnetic torque on the rotor of the induction motor, then neglecting frictional and eddy current losses we see that $G(\omega_1 + \omega_2)$ is the electrical power developed in the rotor. To this power the synchronous motor contributes $G\omega_2$ and the induction motor $G\omega_1$. The values of these quantities for the ideal frequency transformer can readily be computed from (19) (p. 373), it being noticed that $s = (\omega_1 + \omega_2)/\omega_1$ and that R' and L′ vary with the resistance and inductance of the load.

Let us suppose that two of the stator connections of the induction motor shown in Fig. 193 are interchanged so that the stator field now rotates in the same direction as the rotor. If $\omega_2 = \omega_1$, P will be at G (Fig. 174) and the currents in the rotor will be zero. There will be no load on the synchronous motor, and the power taken from the mains by the induction motor will only be that due to the heating of the stator windings. If the frequency of the currents developed in the rotor has to be n times that of the supply mains, we must have $(\omega_1 - \omega_2)/\omega_1 = -n$, and therefore ω_2 must equal $(n+1)\,\omega_1$. If we want to double the frequency, therefore, ω_2 must be equal to $3\omega_1$. In this case the power given by the synchronous motor through the shaft to the induction motor is $3G\omega_1$. Of this power $2G\omega_1$ is converted into electrical power in the rotor and the remainder $G\omega_1$ is in general given back as electrical power into the supply mains, the induction motor acting as a generator. We thus see that one-third of the total power generated by the synchronous motor circulates between it and the induction generator, leading to appreciable heating losses in the mains. The torque, therefore, when the slip is -2 is greater than when the slip is $+2$. This also follows at once from formula (19). As the power of the synchronous motor in the former case has to be three times that in the latter case, it will be seen that notwithstanding the higher speed the frequency transformer in the former case would be a much less efficient machine. In both cases it is to be noticed that the synchronous motor maintains the ratio of ω_1 to ω_2 absolutely constant. The ratio of the frequency of the supply current to that of the current generated $\pm\,\omega_1/(\omega_1 - \omega_2)$ is therefore also absolutely constant. A serious drawback to machines of this type is that the voltage regulation of the currents developed is very unsatisfactory.

Another use to which frequency transformers can be put is to link together two power supply systems having different frequencies. If the peak loads of the systems occur at different times, appreciable economies both in plant and working expenses can be effected in this way. The commonest type of frequency transformer employed is one consisting of a synchronous motor direct coupled to a synchronous generator. If the motor have $2p$ and the generator $2q$ poles, the ratio of the frequencies in the two circuits will be p/q. In this type of apparatus it is immaterial which machine acts as a motor and which as a generator.

Starting a frequency transformer.

Difficulty is generally experienced in starting a frequency transformer consisting of two synchronous machines. In the first place it is necessary to get the ratio of the frequencies of the two circuits as nearly as possible equal to the ratio of the number of poles of the two machines. This is done by adjusting the frequencies at each station. We synchronise the motor side in the ordinary way and then synchronise on the generator side. This latter operation can be accelerated if the stator framework is so constructed that it can be rocked round through part of a turn. This is generally effected by means of a small motor geared to it.

If the synchronous generator only supply a small fraction of the load of the supply system to which it is connected, then it may run badly in parallel and readily fall out of circuit. But if it supply a large fraction of the load it will in general run satisfactorily in stable equilibrium. Hence its capacity should be large.

These transformers can be used to couple single to polyphase circuits no matter what the frequencies or the voltages of the circuits may be. By over-exciting the field of either machine the power factor of the system to which it is connected will be improved. This procedure is often advantageous.

REFERENCES

H. M. HOBART, *Electric Motors.*

JACQUES GUILLAUME, 'Dispositif Fischer-Hinnen pour le Démarrage des Moteurs Asynchrones Triphasées,' *L'Éclairage Électrique*, Vol. 24, p. 131, July, 1900.

P. BOUCHEROT, 'Quelques Applications d'Alternateurs Compounds,' *Bulletin de la Société Internationale des Électriciens*, Vol. 2, p. 446, 1902.

E. DANIELSON, 'A Novel Combination of Polyphase Motors for Traction Purposes,' *Journ. of the Am. Inst. of El. Engin.*, Vol. 19, p. 527, 1902.

—— 'Kaskadenschaltung bei Motoren für Walzwerke,' *Elektrotechnische Zeitschrift*, Vol. 25, p. 43, 1904.

A. C. EBORALL, 'On Induction Machines and a New Type of Polyphase Generator,' *The Electrician*, Vol. 51, p. 442, 1903.

C. F. GUILBERT, 'Calculation of the Equivalent Ampere Turns of Windings for Single and Polyphase Currents,' *Electrical World and Engineer*, Vol. 43, p. 516, 1904.

W. L. WATERS, 'The Non-Synchronous Generator in Central Station and Other Work,' *Journ. of the Am. Inst. of El. Engin.*, Vol. 27, p. 156, 1908.

H. G. STOTT and R. J. S. PIGOTT, 'Test of a 15000-K.W. Steam-Engine-Turbine Unit,' *Journ. of the Am. Inst. of El. Engin.*, Vol. 29, p. 183, 1910.

P. BUNET, 'Le Problème de la Transformation de la Fréquence,' *La Lumière Électrique*, Vol. 16, (4) p. 241, 1911.

C. TURNBULL, 'Frequency Changers,' *The Electrician*, Vol. 69, p. 1021, 1912.

—— 'Les Transformateurs Rotatifs de Fréquence,' *La Revue Électrique*, Vol. 20, p. 125, 1913.

L. J. HUNT, 'The Cascade Induction Motor,' *Journ. of the Inst. of El. Engin.*, Vol. 52, p. 406, 1914.

CHAPTER XV

Gliding magnetic fields. Field produced by a wave-winding. Distributed windings. Formula for the flux in terms of the ampere-turns. The induced electromotive force in the stator winding. Two phase field. Effect of the rotor currents on the distribution of the magnetic lines in the air-gap. Influence of the harmonics of the magnetic field on the working of induction motors. Stator connected in four wire star. Current waves not sine shaped. The effect of raising a rotor brush. References.

IN discussing the theory of the induction motor we have supposed that the distribution of the magnetic flux in the air-gap follows the harmonic law. In many practical cases formulae obtained on this assumption are found to be very approximately true, and they are helpful to the electrician. In some cases, however, phenomena arise which are due to the distribution of the magnetic flux not following the harmonic law. It is therefore necessary to consider the effect of the presence of harmonics in the magnetic flux on the working of the machine.

Gliding magnetic fields.

When adjacent coils of a phase winding of the stator are wound in opposite directions, we shall assume that the number of coils is $2p$ and that the distance between their centres is a. When, however, the coils are all wound in the same direction, so that the winding is hemitropic, we shall assume that p is the number of coils and that $2a$ is the distance between the centres of adjacent coils. We shall also assume that the minimum distance between consecutive coils which are wound the same way equals the breadth of the narrowest turn of a coil (see Fig. 47, p. 87). If the field be sine shaped we can write

$$h_1 = H \cos \omega t \cos (\pi x/a).$$

In this formula $H \cos \omega t$ is the induction density in the air-gap, due to the current in No. 1 phase winding, at points on a fixed line parallel to the axis of the rotor through the centre of the face of a coil. The intensity of the field due to the current in this phase, at all points whose distance from the fixed line, measured along the air-gap, equals x, is given by h_1. If r be the inner radius of the stator, we have $2pa = 2\pi r$, and therefore $\pi/a = p/r$.

Let h be the resultant intensity of the field at a point at a distance x from the fixed line, and let the motor be three phase. Then, since at every instant we have $h = h_1 + h_2 + h_3$, we may write

$$h = H \cos \omega t \cos (px/r) + H \cos (\omega t - 2\pi/3) \cos (px/r - 2\pi/3)$$
$$+ H \cos (\omega t - 4\pi/3) \cos (px/r - 4\pi/3)$$
$$= (3/2) H \cos (\omega t - px/r).$$

Thus h will always have its maximum value $(3/2) H$ at points whose abscissae equal $(\omega r/p) t$. In general, we see that at all points on any line which moves round the air-gap with a constant linear velocity $\omega r/p$, that is, $2a/T$, the flux density is constant.

Let us now consider the magnetic field produced by a simple wave-winding having one slot per pole and per phase (see Fig. 44, p. 82). The field due to one phase will be a curve *fff*... which, with the axis $O'X$, makes up a series of rectangles (Fig. 194). By Fourier's theorem (p. 118) this curve may be represented by

Field produced by a wave-winding.

$$y = (4/\pi) h \{\sin (\pi x/a) + (1/3) \sin (3\pi x/a) + \dots\} \text{......(1)},$$

when the origin is at O'. Changing the origin to O, the middle point of the base of one of the rectangles, we get

$$y = (4/\pi) h \{\cos (\pi x/a) - (1/3) \cos (3\pi x/a) + \dots\} \text{......(2)}.$$

In this equation y is the value of the field due to the current in No. 1 phase winding at points which have x for abscissa, and $h = H \cos \omega t$. We have seen above that the resultant field Y_1, due to the first harmonic terms, is given by

$$Y_1 = (4/\pi)(3/2) H \cos (\omega t - px/r) = (6/\pi) H \cos (\omega t - \pi x/a).$$

Hence the amplitude of this field is $(6/\pi) H$, and it rotates in the positive direction with an angular velocity ω/p.

The field Y_3, due to the third harmonic terms, at points distant x from the axis is given by

$$Y_3 = -(4H/3\pi)[\cos\omega t \cos(3\pi x/a) + \cos(\omega t - 2\pi/3)\cos\{(3\pi/a)(x - 2a/3)\} + \cos(\omega t - 4\pi/3)\cos\{(3\pi/a)(x - 4a/3)\}] = 0.$$

Thus the resultant field due to the third harmonic terms is zero. In three phase motors, therefore, which have a wave-winding, the

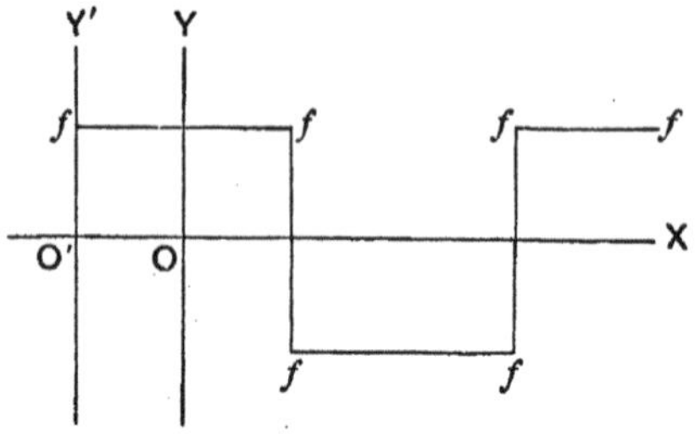

Fig. 194. The magnetic field in the air-gap produced by a single slot-winding.

third harmonic term in the resultant flux in the air-gap is always absent. Similarly this flux will contain no harmonic term the order of which is $3n$.

The field Y_5, due to the fifth harmonic, at points distant x from the axis is given by

$$Y_5 = (4H/5\pi)[\cos\omega t \cos(5\pi x/a) + \cos(\omega t - 2\pi/3)\cos\{(5\pi/a)(x - 2a/3)\} + \cos(\omega t - 4\pi/3)\cos\{(5\pi/a)(x - 4a/3)\}]$$
$$= (6H/5\pi)\cos(\omega t + 5\pi x/a).$$

Hence the maximum value of the field due to the fifth harmonic is $6H/5\pi$, and it rotates backwards with an angular velocity $\omega/5p$. We also see that the breadth of the bands of flux of opposite polarity in the magnetic distribution due to the fifth harmonic is $a/5$. Similarly we can show that the field due to the seventh harmonic rotates forwards with angular velocity $\omega/7p$ and that its amplitude is $6H/7\pi$.

The following table gives the amplitudes, the angular velocities, and the polar breadths of the harmonics of the resultant magnetic field in the air-gap of a three phase induction motor which has a stator with a wave-winding:

Order of the harmonic	$6r-1$	$6r+1$
Amplitude	$6H/\{(6r-1)\pi\}$	$6H/\{(6r+1)\pi\}$
Angular velocity	$-\omega/\{(6r-1)p\}$	$\omega/\{(6r+1)p\}$
Polar breadth	$a/(6r-1)$	$a/(6r+1)$

It will be seen that all harmonics of the orders 5, 11, 17, ... $6r-1$, ... produce fields which rotate backwards.

Distributed windings.

When we have a distributed winding (Fig. 20, p. 36) with $2m$ slots per coil, we may assume in approximate work that the curve representing the intensity of the flux is 'stepped.' For instance, if we have four slots per coil (Fig. 182, p. 392), the stepped curve has the form *fff* shown in Fig. 195.

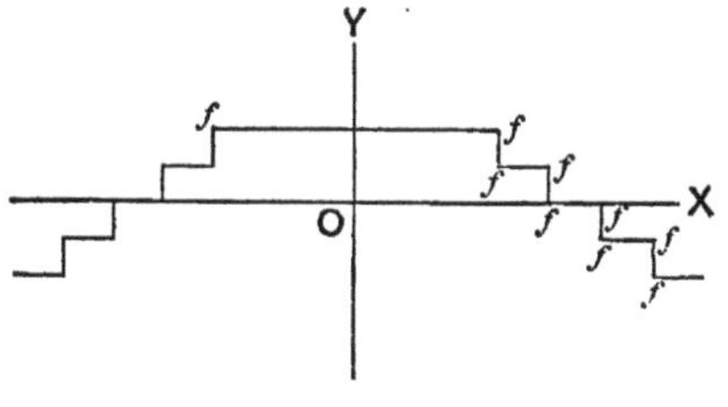

Fig. 195. The shape of the magnetic flux due to the currents in one phase of a three phase machine which has four slots per coil.

The Fourier series for the general case can be found as follows. Writing $x+b$ for x in (1), we get

$$y = (4/\pi)\, h\, [\sin\{(\pi/a)(x+b)\} + (1/3)\sin\{(3\pi/a)(x+b)\} + \dots],$$

and writing $-x$ for x in this equation, we have

$$y = -(4/\pi)\, h\, [\sin\{(\pi/a)(x-b)\} + (1/3)\sin\{(3\pi/a)(x-b)\} + \dots].$$

The equation to the resultant curve (Fig. 196) obtained by adding these two curves together is

$$y = (8/\pi)\, h\,\{\sin(\pi b/a)\cos(\pi x/a) + (1/3)\sin(3\pi b/a)\cos(3\pi x/a) + \dots\}.$$

The maximum height of this curve is $2h$; reducing it to h, the equation becomes

$$y = (4/\pi)\, h\,\{\sin(\pi b/a)\cos(\pi x/a) + (1/3)\sin(3\pi b/a)\cos(3\pi x/a) + \dots\} \quad \text{.........(3).}$$

The breadth of the rectangular waves (Fig. 196) represented by (3) is $2b$, and the minimum distance between them is $a - 2b$.

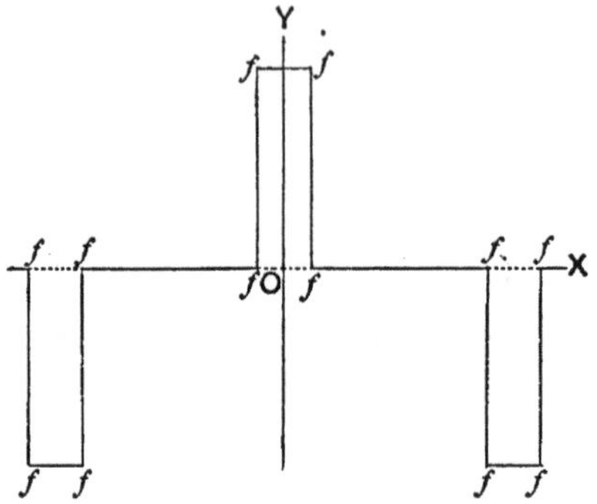

Fig. 196. Rectangular waves.

Let us first suppose that the winding has four slots per coil. In this case b will equal $5a/12$ for the inner winding and $7a/12$ for the outer winding. If $h/2$ be the value of the flux density in the air-gap produced by each winding, the resultant flux density is given by

$$y = (2/\pi)\, h\,[\{\sin(5\pi/12) + \sin(7\pi/12)\}\cos(\pi x/a) + (1/3)\{\sin 3(5\pi/12) + \sin 3(7\pi/12)\}\cos(3\pi x/a) + \dots]$$

$$= (4/\pi)\, h\,[\sin(5\pi/12)\cos(\pi x/a) - (1/3)\sin(\pi/4)\cos(3\pi x/a) + \dots] \quad \text{...........(4).}$$

When there are six slots per coil, the values of b for the windings are $7a/18$, $9a/18$, and $11a/18$. In general if there are $2m$ slots per coil, the values of b for the windings are

$$\{(2m+1)/(6m)\}\, a,\ \{(2m+3)/(6m)\}\, a,\ \dots\ \{(4m-1)/(6m)\}\, a.$$

In this case the coefficient of $\cos(\pi x/a)$ in the expansion of y equals

$$(4/\pi)\, h\, (1/m)\, [\sin\{(2m+1)/(6m)\}\,\pi + \sin\{(2m+3)/(6m)\}\,\pi + \dots],$$

and this equals $(4/\pi)\, h/\{2m \sin(\pi/6m)\}$.

Calculating the coefficients of the other terms of the expansion in the same way, we find that the flux density y at any point in the air-gap, due to the current in a phase-winding, when there are $2m$ slots per coil, is given by

$$\begin{aligned} y = (4/\pi)\, h\, [&\{1/2m \sin(\pi/6m)\} \cos(\pi x/a) \\ &- \{1/3m \sin(\pi/2m)\} \cos(3\pi x/a) \\ &+ \{1/10m \sin(5\pi/6m)\} \cos(5\pi x/a) \\ &+ \{1/14m \sin(7\pi/6m)\} \cos(7\pi x/a) - \dots] \quad \dots\dots\dots(5). \end{aligned}$$

When m is large we may write

$$1/\{2m \sin(\pi/6m)\} = 3/\pi,$$

since the sine of a small angle is approximately equal to its circular measure. Similarly

$$1/\{3m \sin(\pi/2m)\} = 2/3\pi, \text{ etc.}$$

and thus we find that, in this case,

$$\begin{aligned} y = (24/\pi^2)\, h\, [&(1/2)\cos(\pi x/a) - (1/3^2)\cos(3\pi x/a) \\ &+ \{1/(2\,.\,5^2)\}\cos(5\pi x/a) + \dots] \quad \dots\dots\dots\dots\dots\dots(6), \end{aligned}$$

approximately.

If we integrate equation (3) we find that

$$\begin{aligned} y_1 &= \int_0^x y\, dx \\ &= (4/\pi^2)\, ah\, \{\sin(\pi b/a)\sin(\pi x/a) \\ &\qquad + (1/3^2)\sin(3\pi b/a)\sin(3\pi x/a) + \dots\}. \end{aligned}$$

The shape of this curve is shown in Fig. 197. Changing the origin O' (Fig. 197) to the point O $(a/2, 0)$, we write $x + a/2$ for x in this equation, and thus

$$\begin{aligned} y_1 = (4/\pi^2)\, ah\, \{&\sin(\pi b/a)\cos(\pi x/a) \\ &- (1/3^2)\sin(3\pi b/a)\cos(3\pi x/a) + \dots\}. \end{aligned}$$

Finally writing yb for y_1, so that the maximum value of y is h, we get

$$\begin{aligned} y = (4/\pi^2)\,(a/b)\, h\, \{&\sin(\pi b/a)\cos(\pi x/a) \\ &- (1/3^2)\sin(3\pi b/a)\cos(3\pi x/a) + \dots\} \quad \dots\dots\dots(7). \end{aligned}$$

The lengths of the two parallel sides of the trapezium (Fig. 197), bisected by OY, are a and $a - 2b$ respectively, and the height of the trapezium is h.

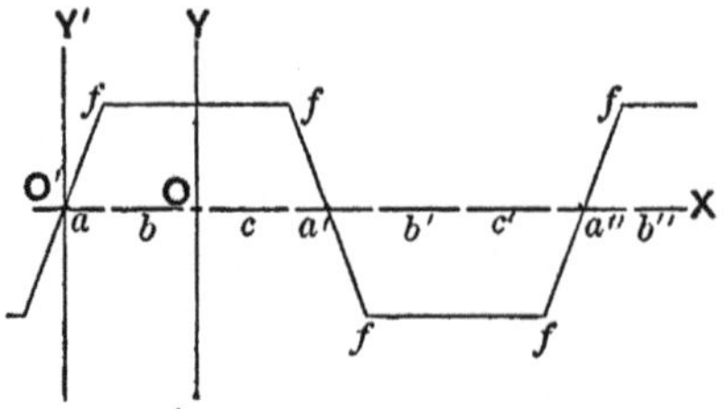

Fig. 197. Shape of the flux wave in the air-gap of a three phase machine due to the current in one phase when there is an infinite number of slots.

The curve fff... shown in Fig. 197 is a useful one. By giving various values to b we get trapeziums of all shapes. For instance when b is zero we get the series of rectangles shown in Fig. 194, and the equation (7) simplifies to (2). When $b = a/6$, the equation (7) is the same as (6), and when $b = a/2$ we get the series of triangles the equation to which is

$$y = (8h/\pi^2)\{\cos(\pi x/a) + (1/3^2)\cos(3\pi x/a) + \ldots\}.$$

Formula for the flux in terms of the ampere-turns.

Since the permeability of iron is very large compared with the permeability of air, we may, in getting approximate formulae, suppose that it is infinite. Let us suppose that $n/2$ is the number of turns in a stator coil when the adjacent coils of one phase are wound in opposite directions, and that n is the number of turns when they are wound in the same direction. In either case pn is the number of turns and $2pn$ the number of conductors per phase, since we suppose that the number $2p$ of coils in the former case is double that in the latter.

Let the current in No. 1 phase-winding be $I \cos \omega t$, and let d be the radial depth of the air-gap. Let also S be the mean cross-sectional area of the path of the flux ϕ linked either with two adjacent coils belonging to one phase or, when the winding is hemitropic, with one side of a coil. Since, on our assumption, $2d/S$ is the mean reluctance of the magnetic circuit of ϕ, we have (Vol. I, p. 72)

$$\phi = (4\pi/10)\, nI \cos \omega t \,.\, (S/2d).$$

Hence, if B_1 denote the maximum value at any instant of the flux density due to the current in No. 1 phase, we have

$$B_1 = \pi n I/5d.$$

If the wave-shape of the flux be rectangular (Fig. 194), and if the rotor be running at synchronous speed, the instantaneous value b_1 of the flux density due to the current in No. 1 phase, at all points in the air-gap which have x for their abscissa, is given by

The induced electromotive force in the stator winding.

$$b_1 = (4/\pi)(\pi n I \cos \omega t/5d)\{\cos(\pi x/a) - (1/3)\cos(3\pi x/a) + \ldots\}.$$

If b therefore be the value of the flux density at the given points, due to the currents in the three phases, we get

$$b = (1{\cdot}2nI/d)\{\cos(\omega t - \pi x/a) + (1/5)\cos(\omega t + 5\pi x/a) - (1/7)\cos(\omega t - 7\pi x/a) - \ldots\}.$$

Hence we see that the flux can be resolved into a series of waves gliding with velocities $\omega a/\pi$, $-\omega a/5\pi$, $\omega a/7\pi$, ...

The value e_1 of the back electromotive force developed in a conductor at the point where x equals $a/2$, by the bands of flux gliding with velocity $\omega a/\pi$, is given (p. 17) by

$$e_1 = (1{\cdot}2nI/d)\sin\omega t \,.\, l \,.\, (\omega a/\pi) \,.\, 10^{-8} \text{ volts},$$

where l is the length of the conductor. Similarly we have

$$e_5 = (1{\cdot}2nI/5d)\sin\omega t \,.\, l \,.\, (\omega a/5\pi) \,.\, 10^{-8} \text{ volts},$$

where e_5 is the back electromotive force developed in the conductor by the bands of flux of breadth $a/5$, which glide backwards round the air-gap with velocity $\omega a/5\pi$. If e denote the total back electromotive force developed in this conductor, we have

$$e = (1{\cdot}2nIla\omega/\pi d)(1 + 1/5^2 + 1/7^2 + 1/11^2 + 1/13^2 + \ldots)\sin\omega t \,.\, 10^{-8}.$$

Now $\pi^2/8 = \sum_1^\infty 1/(2m-1)^2$, and thus dividing each side of this equation by 3^2 and subtracting the result from the original equation, we see that $\pi^2/9$ is the sum of the series within the brackets. Hence,

$$e = (1{\cdot}2\pi/9d)\, nIla\omega \sin\omega t \,.\, 10^{-8}$$
$$= 0{\cdot}419\,(nI/d)\, la\omega \sin\omega t \,.\, 10^{-8}.$$

Since there are $2pn$ conductors per phase, the effective value

V of the back electromotive force developed per phase winding is given by

$$V = \sqrt{2} \,.\, 0{\cdot}419\,(pn^2I/d)\, la\omega \,.\, 10^{-8}$$
$$= 3{\cdot}72\,(pn^2I/d)\, laf \,.\, 10^{-8},$$

where f is the frequency.

Let us now suppose that there are $2m$ slots per coil. If B_1 denote the flux density at points inside the narrowest turn of a coil, we have, as before,

$$B_1 = \pi nI/5d.$$

Thus the equation to the flux, which will be shaped like the stepped curve shown in Fig. 195, is, by (5),

$$b_1 = (4/\pi)\,(\pi nI \cos \omega t/5d)\,[\{1/2m \sin(\pi/6m)\} \cos(\pi x/a) - \{1/3m \sin(\pi/2m)\} \cos(3\pi x/a) + \ldots].$$

Hence we find that

$$b = (0{\cdot}6nI/dm)\,[\{1/\sin(\pi/6m)\} \cos(\omega t - \pi x/a) + \{1/5 \sin(5\pi/6m)\} \cos(\omega t + 5\pi x/a) + \{1/7 \sin(7\pi/6m)\} \cos(\omega t - 7\pi x/a) + \ldots].$$

The amplitude, therefore, of the field gliding round the air-gap with velocity $\omega a/\pi$ is $k_1 nI/d$, where k_1 equals $0{\cdot}6/m \sin(\pi/6m)$.

The following table shows how k_1 varies for different values of m:

m	1	2	3	4	5	6	∞
k_1	1·200	1·159	1·152	1·149	1·148	1·148	1·146

If $k_5 nI/d$ denote the amplitude of the fifth harmonic field, we have

$$k_5 = (k_1/5)\,\{\sin(\pi/6m)/\sin(5\pi/6m)\}.$$

Similarly we may write

$$k_7 = (k_1/7)\,\{\sin(\pi/6m)/\sin(7\pi/6m)\}, \text{ etc.}$$

When m equals unity, k_5 is the fifth part of k_1. When m equals 2, k_5 equals $0{\cdot}054k_1$, and, when m is infinite, k_5 equals $0{\cdot}04k_1$. Similarly we can show that when m is greater than unity, the amplitudes of the seventh and higher harmonic fields are small compared with k_1. The higher harmonic fields, also, glide much more slowly than the fundamental field. The electromotive forces, therefore, which they develop in the various wires will be much smaller, and so in approximate work they can be neglected. Hence b is given approximately by the equation

$$b = \{0\cdot6/m \sin(\pi/6m)\}(nI/d)\cos(\omega t - \pi x/a).$$

The abscissae of the axes of the slots in which are embedded the conductors of the coil which has the origin at the centre of its polar face are

$$\pm\{(2m+1)/6m\}a, \quad \pm\{(2m+3)/6m\}a, \ldots \pm\{(4m-1)/6m\}a,$$

and the number of conductors in a slot is n/m. Hence since the band of flux is gliding with a velocity $\omega a/\pi$, the instantaneous value e of the back electromotive force developed in No. 1 phase winding is given by

$$\begin{aligned} e &= p(n/m)\{0\cdot6/m \sin(\pi/6m)\}(nI/d)[\cos\{\omega t - (2m+1)\pi/6m\} \\ &\qquad - \cos\{\omega t + (2m+1)\pi/6m\} + \ldots]\,l(\omega a/\pi)\,.\,10^{-8} \\ &= 1\cdot2p(n^2/m^2)(l\omega a/\pi)\{1/d\sin(\pi/6m)\}I\sin\omega t[\sin\{(2m+1)\pi/6m\} \\ &\qquad + \ldots + \sin\{(4m-1)\pi/6m\}]\,.\,10^{-8} \\ &= 2\cdot4plaf(n^2/m^2)\{1/d\sin(\pi/6m)\}I\sin\omega t \\ &\qquad \times[\{\sin(\pi/2)\sin(\pi/6)\}/\sin(\pi/6m)]\,.\,10^{-8}. \end{aligned}$$

Hence the effective value V of the back E.M.F. per phase is given by

$$V = \beta\,.\,(pn^2I/d)\,laf\,.\,10^{-8},$$

where

$$\begin{aligned} \beta &= 0\cdot6\sqrt{2}/\{m^2\sin^2(\pi/6m)\} \\ &= 0\cdot849/\{m^2\sin^2(\pi/6m)\}. \end{aligned}$$

When m is unity, β equals 3·39. We have shown on p. 430 that its true value in this case is 3·72. Hence the error introduced in this case by neglecting the higher harmonics is less than ten per cent. When there are four slots per coil, β is approximately 3·17, and when m is large β is approximately 3·10.

If A_0 be the magnetising current per phase at synchronous speed, we have $I = A_0\sqrt{2}$, and thus A_0 is given by the formula

$$A_0 = (d/1\cdot2)\{m^2\sin^2(\pi/6m)\}V\cdot10^8/(pn^2laf),$$

and since RA_0 is generally negligible compared with V, we may assume in this formula that V is the applied P.D. Hence, on our assumptions, we see that the magnetising current per phase at synchronous speed varies directly as the radial depth of the air-gap and the applied P.D., and inversely as the frequency, the mean cross-sectional area of the coils, the number of coils and the square of the number of turns per coil.

Two phase field.

We shall now consider the magnetic field in the air-gap of a two phase induction motor which has two separate windings. Let us suppose that the winding of one phase is similar to that shown in Fig. 63, p. 104, but let us suppose that there are $2m$ slots per coil. We shall also suppose that there are n turns per coil. There will be m steps in the flux curve (Fig. 195), due to the current in one phase. If the distance between the centres of consecutive coils of one phase be $2a$, and if b_1 and b_2 be the values of the flux densities at the points whose abscissae equal x, due to the currents in the two phases, we have, by (5),

$$b_1 = (4/\pi)(\pi nI \cos \omega t/5d)[\{1/2m \sin(\pi/6m)\} \cos(\pi x/a) - \{1/3m \sin(\pi/2m)\} \cos(3\pi x/a) + \ldots]$$

and

$$b_2 = (4/\pi)\{\pi nI \cos(\omega t - \pi/2)/5d\}[\{1/2m \sin(\pi/6m)\} \cos(\pi x/a - \pi/2) - \{1/3m \sin(\pi/2m)\} \cos(3\pi x/a - 3\pi/2) + \ldots].$$

Thus, if b is the value of the resultant flux at points the abscissae of which equal x, we get

$$\begin{aligned} b &= b_1 + b_2 \\ &= (4/\pi)(\pi nI/5d)[\{1/2m \sin(\pi/6m)\} \cos(\omega t - \pi x/a) \\ &\qquad - \{1/3m \sin(\pi/2m)\} \cos(\omega t + 3\pi x/a) \\ &\qquad + \{1/10m \sin(5\pi/6m)\} \cos(\omega t - 5\pi x/a) + \ldots]. \end{aligned}$$

We may therefore suppose that the flux is the resultant of a series of waves. If the order of the wave be $4r - 1$, it glides backwards with angular velocity $\omega/(4r - 1)$, and the breadth of the band of flux is $a/(4r - 1)$. If the order of the wave be $4r + 1$, it glides forwards with angular velocity $\omega/(4r + 1)$, and its breadth is $a/(4r + 1)$.

It is to be noticed that this case is unlike the corresponding problem for three phase fields, for here none of the harmonics cancel out. In three phase fields we only have harmonics of the order $6r \pm 1$. In two phase fields we may have all the harmonics of the order $2r + 1$, and, in general, the third harmonic cannot be neglected. The calculation of the back E.M.F. in the stator windings when the rotor is running at synchronous speed can be made in the same way as for three phase motors.

Effect of the rotor currents on the distribution of the magnetic lines in the air-gap.

In this chapter we have assumed that the currents in the stator windings follow the harmonic law, and that there are no rotor currents. We have also assumed that the hysteresis and eddy current losses are negligible. Our results are approximately correct when the rotor is running at synchronous speed, or when the brushes are lifted from the slip rings so that the rotor windings are on open circuit. We shall now consider how the rotor currents modify our results.

Let ϕ be the total flux linked with a stator winding at a particular instant, and let e be the applied potential difference, then, when the resistance of the stator windings can be neglected, we have

$$e = \partial\phi/\partial t \quad \text{..........................(8).}$$

Now when the slip is appreciable we may regard ϕ as the resultant of two component fluxes ϕ_1 and ϕ_2, where ϕ_1 is the flux due to the currents in the stator windings and ϕ_2 is the flux due to the rotor currents. Equation (8) shows that we always have

$$\phi_1 + \phi_2 = \phi_s,$$

where ϕ_s is the flux due to the currents in the stator windings at synchronous speed. Hence the time flux wave of the stator windings of an induction motor, when their resistance is negligible, will have the same shape and the same magnitude at all loads. We can therefore apply the theorems concerning the harmonics of the field and the back E.M.F. induced in the rotor windings, which we have proved above, to machines under load.

Influence of the harmonics of the magnetic field on the working of induction motors.

When the stator has a simple rectangular winding and there is no overlapping, the shape of the resultant field in the air-gap is roughly similar to the rectangles shown in Fig. 194. If the resistance of the stator windings be negligible, this will also be the shape at all loads. If the stator be supplied with sine shaped currents, there is no third harmonic, and the amplitude of the fifth is only equal to the fifth part of that of the first, and it rotates backwards with an angular velocity $\omega_1/5p$. Now, when the motor is loaded, the slip is only two or three per cent., and so the angular velocity of the rotor is only slightly less than ω_1/p. Hence the slip of the rotor relative to the fifth harmonic of the flux will be nearly equal to $(\omega_1/5p + \omega_1/p)/(\omega_1/5p)$, that is, 6. We see from Fig. 174, p. 376, that the backward torque produced at this slip will be very small, and, since the amplitude of the fifth harmonic is only one-fifth that of the first harmonic, its effect on the working of the machine will be negligible. The slip of the rotor relative to the seventh harmonic which rotates in the forward direction is $(\omega_1/7p - \omega_1/p)/(\omega_1/7p)$, that is, -6. As its amplitude is only one-seventh that of the first, the effect produced will be less than that produced by the fifth harmonic. The effects of the 11th, 13th, 17th,... harmonics will be still more minute. In this case, therefore, when finding approximate formulae, the sine curve assumption is permissible.

Stator connected in four wire star.

When the stator windings of a three phase induction motor are connected in star and the neutral point is insulated, there can be no third harmonics in the current waves, for the sum of the instantaneous values of the three currents to the neutral point must always be zero. When, however, the neutral point is connected with the fourth wire of a four wire three phase system, and the magnetic circuit of the motor is highly saturated, the third harmonics in the current waves may be considerable. These harmonics will tend to magnetise the stator ring in such a way that an alternating magnetic field of $6p$ poles will be produced. We should expect, therefore, that, as in the case of a star-connected three phase transformer, connected in four wire star (see p. 306), there would be an appreciable difference in the working of the motor when the neutral point is insulated and when it is connected with the fourth wire.

Current waves not sine shaped.

When the current waves in the stator windings are not sine shaped, the amplitudes of the third and higher harmonic magnetic fields in the air-gap may be large. In this case, therefore, when the rotor is speeding up, we should expect that it would sometimes run in stable equilibrium at speeds which are submultiples of its synchronous speed. This sometimes happens in practice.

The effect of raising a rotor brush.

When the rotor has a three phase winding and slip rings for inserting resistance into the rotor windings, we can make it run at half-speed by preventing one of the rotor brushes making contact before and after the start. This phenomenon, however, is not due to the presence of harmonics in the magnetic field of the air-gap, and can be explained simply by the properties of rotating and oscillating magnetic fields. At half-speed the frequency of the alternating currents induced in the closed phase winding of the rotor is only half that of the stator currents. The oscillating magnetic field due to it can be resolved into two rotary fields, one of which will have an angular velocity in space equal to the angular velocity of the stator field, and the other will be fixed in direction (p. 413). At this speed ($\omega_1/2$) the frequency of the rotor currents is $(p\omega_1/2)/(2\pi)$. Hence the rotor is rotating synchronously with the pulsations of the current in its windings and acts as a synchronous motor.

REFERENCES

M. B. FIELD, 'On Oscillatory and Rotatory Magnetic Fields and the Theory of the Single Phase Motor,' *Electrical Review*, Vol. 44, p. 194, Feb. 1899.

E. NOAILLON, 'Quelques considérations sur les champs tournants,' *Bulletin de l'Association des Ingénieurs-Électriciens*, Liége, Vol. 3, p. 42, Feb. 1903.

F. T. CHAPMAN, 'The Magnetic Field of the Three Phase Induction Motor' *The Electrician*, Vol. 74, p. 318, 1914.

CHAPTER XVI

Commutator motors. Alternating current series motor. Compensating devices. Theory of the series motor. Formulae for the alternating current series motor. Circle diagram. Equations for the direct current series motor. The formula for the torque. Alternating current shunt motor. Repulsion motor. Induction commutator motor. Other forms of motor. References.

Commutator motors.

In an ordinary direct current self-exciting motor, if we reverse the connections of its terminals with the mains, the direction of the magnetomotive force in the field coils and the direction of the current in the armature windings are both reversed. If, therefore, the new magnetic force acting on the field magnets be greater than the coercive force, the field flux will be reversed, and hence the torque will still be in the same direction. It is well known, in practice, that in order to make the motor run in the opposite direction, it is necessary to reverse the direction of the current either in the field magnet windings or in the armature, but not in both. Reversing both, by altering the polarity of the motor terminals, has no practical effect either on the direction of rotation or on the efficiency of the machine. It follows, therefore, that at very low frequencies, every direct current self-exciting motor when supplied with alternating currents will tend to act as a motor, as the torque always acts in the same sense in whichever direction the current is flowing. With high frequencies it is easy to see that the effects of eddy currents in screening the magnetising force due to the currents in the field windings from the interior of the field magnets may considerably modify the action of the motor. With ordinary direct current magnets the loss due to eddy currents would be excessive, and so it is essential that the field magnets be built up of thin iron plates. When this is done, both series and shunt motors will work when supplied from alternating current mains, and if the necessary modifications in their design be made and suitable devices employed to prevent excessive sparking, etc., their efficiency will be high. Both of these are types of 'commutator motors' and we shall now give an elementary theory of their action.

Alternating current series motor.

The motor shown in Fig. 198 is similar to an ordinary direct current series motor with a ring armature. The field magnets are laminated and M_1 and M_2 are connected through a suitable starting resistance with constant potential supply mains. In order to simplify the theory we shall assume for the present that the permeability of

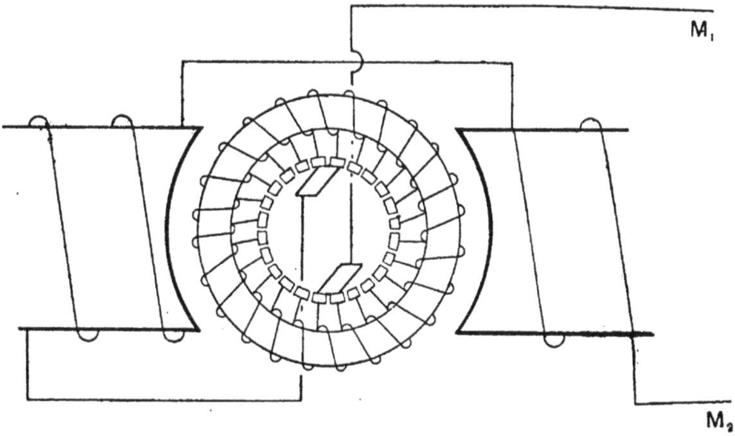

Fig. 198. Alternating current series motor.

the iron is constant; the field flux will then be in phase with the supply current. The flux will therefore vanish twice during the period of the alternating current, and the torque, also, will vanish twice. Unless special devices are employed, however, there will be excessive sparking at the brushes and so various devices are employed in practice to prevent this sparking.

Compensating devices.

In a direct current motor the armature reaction produces a transverse magnetisation of the field. A similar distortion of the field is produced in an alternating current series motor. The transverse magnetisation induces large currents in the coils short circuited by the commutator brushes and hence there is violent sparking when the current in one of these short circuited coils is broken owing to the moving commutator segment ceasing to make contact with the brush. This sparking can be reduced by using carbon brushes. One effect of using these brushes is to increase the resistance of the path for the current in a short circuited coil, thus diminishing this current. As the energy stored up in the field round the current depends on the square of the current there is much less energy available to produce sparking when the current is diminished, and so in some cases by the use of carbon brushes the effects produced by the sparking can be made negligible. The use of carbon brushes, however, lowers the efficiency of the machine, as they get very hot and absorb an appreciable amount of power.

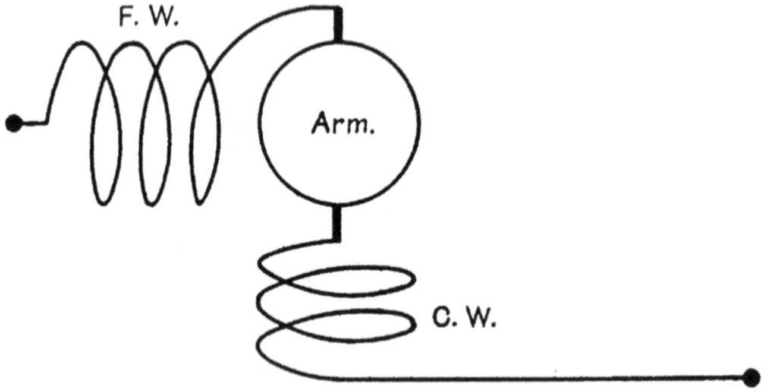

Fig. 199. Series motor with compensating winding C.W. (Conductive compensation.)

A method frequently employed in practice for neutralising the transverse field is to use a special series compensating winding C.W. (Fig. 199) as in direct current machines. This winding produces a field in direct opposition to the field produced by the currents in the armature and at right angles, therefore, to the main field. By properly choosing the number of turns in the compensating coil the transverse field can be almost neutralised. This method is generally referred to as the method of 'conductive' or 'forced' compensation.

Another method is indicated in Fig. 200. The resistance of the permanently short circuited coil C.W. is very small and the mutual inductance between it and the currents in the armature circuits is large. Hence the currents in this coil are practically in opposition in phase to the currents in the armature. The magnetomotive force acting at right angles to the main field is therefore very small and the transverse field is therefore appreciably reduced. The presence of C.W. reduces the impedance of the armature and hence a larger current flows through the field windings F.W. The main field is consequently strengthened, and as the current in the armature is greater, the torque is very appreciably increased. This method of neutralising the transverse field is known as the method of 'inductive' compensation. Whether conductive or inductive compensation is used, it is best to distribute the windings round the armature so as to neutralise its reactance as much as possible.

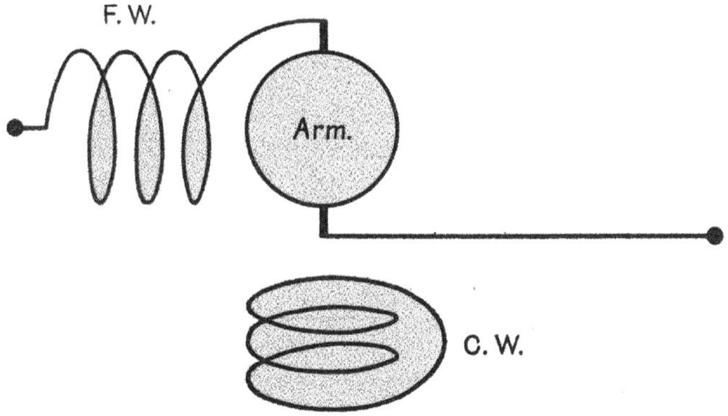

Fig. 200. Series motor with compensating winding C.W. (Inductive compensation.)

Let us now consider the two pole machine shown in Fig. 201.

Theory of the series motor. Let us suppose that the brushes B_1, B_2 are rubbing on a commutator and that the line joining them passes through the axis of the armature and makes an angle a with OY which is perpendicular to the line OX joining the centres of the polar faces. We shall assume that the radial distribution of the flux round the ring armature follows the cosine law, having its maximum density at the points opposite the middle points of the polar faces and being zero at points on the vertical line through O. Owing to the high permeability of iron we need only consider the radial component of the flux. Let θ be the angle which the plane of one of the armature windings makes with OX. Then, if ϕ_1 be the flux embraced by this winding, and $\Phi \sin \Omega t$ be the total flux entering the armature, we have

$$\phi_1 = (\Phi/2) \sin \Omega t \sin \theta \quad \text{.....................(1)},$$

since at any instant the flux through any winding varies as $\int_0^\theta \cos \theta \, \partial\theta$, that is, as $\sin \theta$, and we suppose that the flux divides equally along each half of the armature. It is to be noticed that $\Omega/2\pi$ is the frequency of the supply current.

We shall suppose that the angular velocity ω of the armature is constant. If e_1 denote the electromotive force developed in the winding, we have

$$\begin{aligned} e_1 &= -\, \partial\phi_1/\partial t \\ &= -\, (\Phi/2) \{\Omega \cos \Omega t \,.\, \sin \theta + \omega \sin \Omega t \,.\, \cos \theta\}, \end{aligned}$$

since $\partial\theta/\partial t = \omega$.

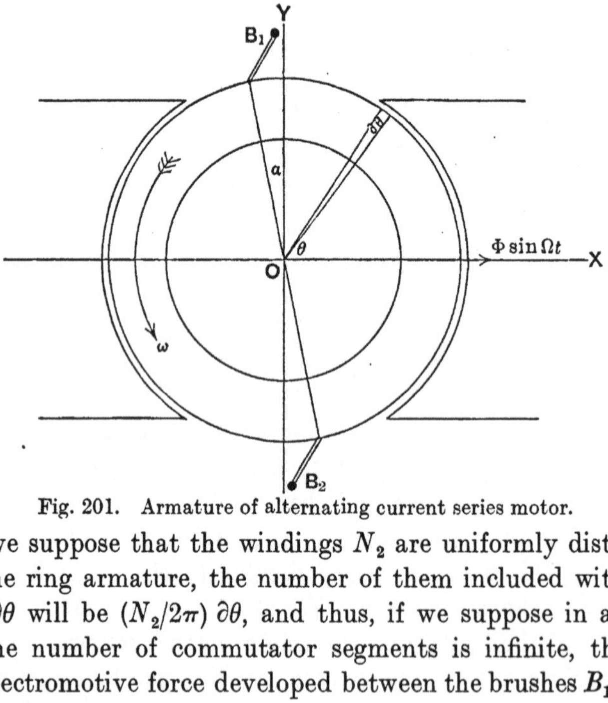

Fig. 201. Armature of alternating current series motor.

If we suppose that the windings N_2 are uniformly distributed over the ring armature, the number of them included within the angle $\partial\theta$ will be $(N_2/2\pi)\, \partial\theta$, and thus, if we suppose in addition that the number of commutator segments is infinite, the total back electromotive force developed between the brushes B_1 and B_2 is given by

$$\begin{aligned} e &= -\, (N_2/2\pi) \int_{-\pi/2+\alpha}^{\pi/2+\alpha} e_1 \partial\theta \\ &= (N_2\Phi/2\pi) \{\Omega \sin \alpha \,.\, \cos \Omega t + \omega \cos \alpha \,.\, \sin \Omega t\} \quad \text{...(2)}. \end{aligned}$$

Putting $\tan \delta = (\Omega/\omega) \tan \alpha,$

we get

$$e = \{\omega\Omega\,(N_2\Phi/2\pi) \sin (\Omega t + \delta)\}/(\Omega^2 \cos^2 \delta + \omega^2 \sin^2 \delta)^{\frac{1}{2}} \ldots(3).$$

This formula shows that the angle of time lag between e and the flux, and therefore, since we are assuming that the permeability is constant, between e and the current, is $-\delta$. Also since the frequency of e is $\Omega/2\pi$, it equals the frequency of the supply current.

It is easy to see from first principles that the frequency of the back electromotive force e will, on our assumptions, be equal to the frequency of the flux or of the applied alternating potential difference. We have assumed that the angular velocity of the commutator is constant, that the armature is perfectly symmetrical, and that the number of segments of the commutator is infinite. It follows that at the times $t, t + T, t + 2T, \ldots$ the back electromotive force between the brushes must be the same, for the fluxes and the currents are identical at these instants. The frequency, therefore, of the back electromotive force is the same as that of the applied potential difference. In practice, the number of commutator segments is finite, and there are generally slots on the armature, so that the reluctance of the magnetic circuit varies periodically with the angular velocity of the rotor. For both these reasons, harmonics, the periods of which depend on ω, will be introduced into the expression for e and this is found to be the case in actual working. In making a first rough calculation, however, we can neglect these higher harmonics as their amplitudes are rarely large. In addition, a harmonic, the period of which equals the period of rotation of the rotor, is sometimes introduced owing to the axis of the rotor being slightly out of truth.

In the particular case when α is zero, formula (3) becomes

$$e = (\omega N_2/2\pi)\,\Phi \sin \Omega t.$$

The back electromotive force, therefore, is simply proportional to the product of the angular velocity and the flux. It is also in exact opposition in phase to the current. This formula may be used for machines furnished with compensating windings.

In the general case, when α is not zero, the phase difference between e and the current i is δ, where $\tan\delta = (\Omega/\omega)\tan\alpha$. In Fig. 201 we have made α positive, the brushes being displaced in the direction of rotation. In this position the magnetomotive force of the armature currents tends to strengthen the field. Similarly, when α is negative, that is, when the brushes are moved backwards, the armature reaction tends to demagnetise the field, and the phase difference between e and i is less than π.

If we denote the current i by $I \sin \Omega t$, the formula (3) for e may be written in the form

$$e = (N_2\Phi/2\pi I)(\omega \cos\alpha \,.\, i + \sin\alpha \,.\, \partial i/\partial t),$$

and therefore $$e = M\omega \cos\alpha \,.\, i + M \sin\alpha \,.\, \partial i/\partial t \quad\text{.............(4)},$$

where $$M = N_2\Phi/2\pi I.$$

When the rotor is at rest, that is, when ω is zero, the only E.M.F. induced in each half of the rotor winding is due to the effects of the mutual induction between the field magnet windings and the armature coils. Since we have supposed that the permeability of the iron is constant, and we are neglecting hysteresis and eddy currents, there will be a constant mutual inductance coefficient between the field magnet coils and either half of the armature windings between the brushes. Putting ω equal to zero in equation (4) we see that the mutual inductance is $M \sin\alpha$. It vanishes with α, and is positive or negative according as α is positive or negative, that is, according as the brushes are moved forwards or backwards from their normal position.

Formulae for the alternating current series motor.

By means of formula (4) we can easily find formulae for the working of the alternating current series motor. We shall assume that the applied potential difference wave can be represented by $E \sin(\Omega t + \beta)$ and that the resistance of the electric circuit between the main terminals of the machine is R, so that R includes the resistance of the field coils, the resistances of the halves of the rotor winding in parallel and the resistance introduced by the brushes and connecting leads. Let L_1 and L_2 be the self-inductances of the field magnet coils and of the halves of the rotor circuit which are in parallel between the brushes. The equation to find the current i is

$$E \sin(\Omega t + \beta) = Ri + L_1 \frac{\partial i}{\partial t} + M \sin\alpha \frac{\partial i}{\partial t} + L_2 \frac{\partial i}{\partial t} + e,$$

and therefore by (4)

$$E \sin(\Omega t + \beta) = (R + M\omega \cos\alpha)\, i + (L_1 + L_2 + 2M \sin\alpha) \frac{\partial i}{\partial t}$$

$$= \rho i + \lambda \frac{\partial i}{\partial t},$$

where $\rho = R + M\omega \cos\alpha$, and $\lambda = L_1 + L_2 + 2M \sin\alpha$.

Solving this equation we find that

$$i = \{E \sin(\Omega t + \beta - \gamma)\}/(\rho^2 + \lambda^2\Omega^2)^{\frac{1}{2}} \quad \text{...........(5)},$$

where $\tan\gamma = \lambda\Omega/\rho$, or $\cos\gamma = \rho/(\rho^2 + \lambda^2\Omega^2)^{\frac{1}{2}}$.

If V and A denote the effective values of $E \sin(\Omega t + \beta)$ and i, we have

$$V = A\,(\rho^2 + \lambda^2\Omega^2)^{\frac{1}{2}},$$

and $$W = VA \cos\gamma = A^2\rho = A^2R + A^2M\omega \cos\alpha,$$

where W is the total power given to the motor.

On our assumptions, the total power given to the motor is expended in heating the conductors (A^2R) and in turning round the rotor. If G denote the average value of the torque produced by the electrical forces acting on the rotor, $G\omega$ will be the measure of the average power given to it. Thus we have

$$A^2R + A^2M\omega \cos\alpha = A^2R + G\omega,$$

and hence $$G = M \cos\alpha \,.\, A^2 \quad \text{......................(6)}.$$

The torque is therefore proportional to the square of the current and to the cosine of the angle α. Since the cosine of a small angle differs little from unity, we see that moving the brushes through a small angle on either side of the central position does not appreciably alter the value of the torque for a given current. This torque, however, is a maximum when α is zero, that is, in the central position.

The power factor $\cos\gamma$ of the motor circuit is given by

$$\cos\gamma = \rho/(\rho^2 + \lambda^2\,\Omega^2)^{\frac{1}{2}}$$

$$= (R + M\omega \cos\alpha)/\{(R + M\omega \cos\alpha)^2 + (L_1 + L_2 + 2M \sin\alpha)^2\Omega^2\}^{\frac{1}{2}} \quad \text{...(7)}.$$

It continually increases, therefore, as the angular velocity ω of the armature increases. For a given value of ω, however, any increase in the frequency of the supply current will diminish the power factor.

The efficiency η of the motor is given by the formula

$$\eta = G\omega/(A^2 R + G\omega)$$
$$= M\omega \cos \alpha/(R + M\omega \cos \alpha) \quad \text{..............(8)}.$$

Hence the efficiency, also, increases as the angular velocity increases.

The power taken from the mains is $VA \cos \gamma$ and this may be written in the form

$$V^2 (R + M\omega \cos \alpha)/\{(R + M\omega \cos \alpha)^2 + (L_1 + L_2 + 2M \sin \alpha)^2 \Omega^2\}.$$

If V remain constant and ω vary, this expression has its maximum value when

$$R + M\omega \cos \alpha = (L_1 + L_2 + 2M \sin \alpha)\, \Omega.$$

The maximum value of the power taken from the mains is, therefore,

$$V^2/\{2\,(L_1 + L_2 + 2M \sin \alpha)\, \Omega\},$$

and the power factor in this case is $1/\sqrt{2}$, that is, 0·71 nearly. The efficiency η, also, at this load is given by

$$\eta = 1 - R/\{(L_1 + L_2 + 2M \sin \alpha)\, \Omega\}.$$

The following equations give expressions for the useful power $G\omega$ given to the rotor. We have

$$G\omega = M\omega \cos \alpha \,.\, A^2$$
$$= V^2 M\omega \cos \alpha/\{(R + M\omega \cos \alpha)^2 + (L_1 + L_2 + 2M \sin \alpha)^2 \Omega^2\}.$$

Thus, if V remain constant and ω vary, the useful power has its maximum value when

$$R^2 + (L_1 + L_2 + 2M \sin \alpha)^2 \Omega^2 = M^2 \omega^2 \cos^2 \alpha \quad \text{......(9)},$$

that is, when the impedance between the terminals of the machine with the rotor at rest equals the apparent increase in the resistance of the circuit due to the action of mutual induction. The maximum value of the useful power equals $V^2/2\,(R + M\omega \cos \alpha)$, where $M\omega \cos \alpha$ is found from (9). The power factor $\cos \gamma$, in this case, is given by

$$\cos \gamma = \{(R + M\omega \cos \alpha)/(2M\omega \cos \alpha)\}^{\frac{1}{2}},$$

where $M\omega \cos \alpha$ is determined by (9).

In the particular case when R is zero, the efficiency is 100 per cent. at all loads. The power, in this case, has its maximum value $V^2/2M\omega \cos \alpha$, where ω is given by

$$\omega = (L_1 + L_2 + 2M \sin \alpha)\, \Omega/M \cos \alpha,$$

and the power factor is $1/\sqrt{2}$. In general, when R is not zero, the useful power attains its maximum value at a higher speed than that at which the power taken from the mains is a maximum. The efficiency is, therefore, higher in the former case, and the power factor is greater than $1/\sqrt{2}$.

Since the current continually increases as the speed diminishes, the torque, which is proportional to the square of the current, continually increases also, attaining its maximum value when the rotor is at rest. There is little fear, therefore, of the machine being pulled up by a temporary increase of the load, and, owing to the inductance of the circuit, there is much less risk of it being damaged by a temporary overload than in the case of the direct current machine.

Circle diagram. An instructive method of studying the working of the series motor is by means of the circle diagram shown in Fig. 202. Using our previous notation, we have

$$V^2 = (\rho A)^2 + (\lambda \Omega A)^2.$$

Hence, if Ox represent V, Op represent ρA and xp represent $\lambda \Omega A$, Opx will be a right-angled triangle and the locus of p will be the circle described on Ox as diameter. Make $Op' = M\omega \cos \alpha \,.\, A$, then $pp' = R.A$, for $\rho = R + M\omega \cos \alpha$. The tangent of the angle $xp'p$ is $\lambda\Omega/R$, the angle is therefore constant, and thus the angle $xp'O$ is constant and the locus of p' is also a circle. The angle xOp is the phase difference between the applied P.D. and the current. When the rotor is at rest, p is at p_0, Op_0 is a tangent to the circle $Op'x$, and the power factor is $R/(R^2 + \lambda^2\Omega^2)^{\frac{1}{2}}$. As the angular velocity increases, p moves along the circle, and when the angular velocity is very great, p will be close to x.

Now, it is easy to find lines on this diagram the lengths of which are proportional to the magnitudes of the variable quantities we have to consider. Draw pn and $p'n'$ perpendicular to Ox, and make the angle xOs equal to the angle $xp'O$. Produce xp' to meet Os in s, and draw $p'k$ at right angles to Op to meet Ox in k. Then, it is easy to prove the following relations.

The power factor $= Op\,(1/V)$. The current $= xp\,(1/\lambda\Omega)$.

The torque $= xn\,(VM\cos a/\lambda^2\Omega^2)$. The input $= pn\,(V/\lambda\Omega)$.

The output $= p'n'\,(V/\lambda\Omega)$. The efficiency $= Ok\,(1/V)$.

The angular velocity $= Os\,\{(R^2+\lambda^2\Omega^2)^{\frac{1}{2}}/MV\cos a\}$.

The variables are therefore proportional to the lengths of the lines given outside the brackets on the right-hand side of these equations.

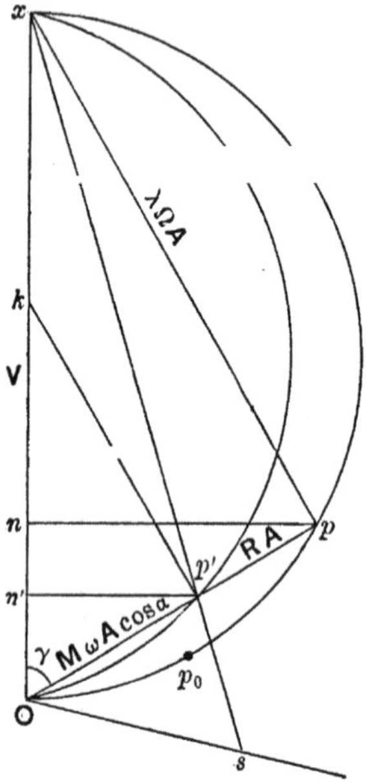

Fig. 202. Diagram of the theoretical series motor. $Ox = V =$ the applied P.D. Op represents $(R + M\omega\cos a)\,A$. As the speed of the armature increases p moves up the semicircle Opx from its initial position p_0. If the angle $xOs =$ the angle $xp'O$, Os is proportional to the angular velocity.

We see at once from the diagram that the torque continually diminishes as the angular velocity increases. We also see that the input is a maximum when n coincides with the centre of the circle. In this case the power factor is obviously $1/\sqrt{2}$. The output which is proportional to $p'n'$ is a maximum when n' is at the centre of the circle, and the power factor and the angular velocity are both greater than in the preceding case. The efficiency Ok also continually increases as the angular velocity increases.

Equations for the direct current series motor.

In order to get the corresponding approximate equations for the direct current series motor, we only need to put Ω equal to zero in the above formulae. Since $\Phi = 4\pi N_1 I/10\mathcal{R}$, where $\mathcal{R}$ is the reluctance and I the current in amperes, we find that M equals $N_1 N_2/5\mathcal{R}$. Hence, if E be the applied P.D. we have

$$E = RI + MI\omega \cos a$$
$$= RI + (N_1 N_2/5\mathcal{R})\, I\omega \cos a.$$

Therefore $\omega = (5\mathcal{R}/N_1 N_2 \cos a)\{(E - RI)/I\}$.

Again, since $G = M \cos a \,.\, A^2$,

we have $G = (N_1 N_2 \cos a/5\mathcal{R})\, I^2 = (N_2/2\pi)\,(\Phi \cos a)\, I$.

When $N_1 I$ is so large that the iron is saturated, Φ may be considered constant, and hence we see that the torque is approximately proportional to I. It must be noticed that no account has been taken of armature reaction, which can only be wholly neglected in a few cases. The value of Φ, and therefore also of M, depends on the angle a. When a forward lead is given to the brushes, the armature reaction magnetises the field, and when the brushes are moved backwards, that is, in the opposite direction to the rotation, from the normal position, the field is partially demagnetised. Hence, as we move the brushes forward, Φ and M increase, and as we move them backwards Φ and M diminish. The assumption that these quantities are constant is therefore sometimes inadmissible. If the motors, however, whether for direct or alternating current work, have suitable compensating windings, the armature reaction is negligible and the above formulae are approximately correct.

The formula for the torque.

In finding the formula for the torque of an alternating current motor, we made the assumption that the field flux was in phase with the current. Let us now consider the problem a little more closely. In actual machines the field flux does not vanish when the current vanishes, owing to the remanent magnetism, and the current attains an appreciable value before the polarity is reversed. Hence the current and the flux are not in phase with one another. We must therefore consider what effect this has on the torque. There are also appreciable backward torques due to eddy currents, friction of the bearings and brushes, and air friction. It is permissible however to neglect these when obtaining approximate formulae.

If we assume that the current and the flux follow the harmonic law, we may write $I \sin \Omega t$ for the current, and express the flux ϕ by the equation (p. 277)

$$\phi = -\Phi_r \cos \Omega t + \Phi \sin \Omega t.$$

Hence, proceeding as on p. 440, we find that the back electromotive force developed between the brushes is given by

$$e = (N_2/2\pi)\left\{\omega \cos a\,(-\Phi_r \cos \Omega t + \Phi \sin \Omega t) + \sin a \frac{\partial}{\partial t}(-\Phi_r \cos \Omega t + \Phi \sin \Omega t)\right\}.$$

Now, if we suppose that the power ei given to the rotor is wholly expended in producing the useful torque $g\omega$, we have

$$g\omega = ei.$$

But ω is practically constant, and thus, taking mean values over a whole period, we obtain

$$G\omega = (N_2 \Phi I/4\pi)\,\omega \cos a + (N_2 \Phi_r I/4\pi)\,\Omega \sin a,$$

and therefore

$$G = (N_2 A^2/2\pi I)\,(1/\omega)\cos(a-\beta)\,\{\omega^2 \Phi^2_{\text{max.}} + (\Omega^2 - \omega^2)\,\Phi_r^2\}^{1/2},$$

where $\Phi_{\text{max.}} = (\Phi^2 + \Phi_r^2)^{1/2}$ and $\tan \beta = \Omega\Phi_r/(\omega\Phi)$.

If a be zero, the torque equals $(N_2 A^2/2\pi I)\,(\Phi^2_{\text{max.}} - \Phi_r^2)^{1/2}$, and thus the greater the value of the remanent flux Φ_r, for given values of $\Phi_{\text{max.}}$ and A, the smaller will be the torque. On the other hand, if a be positive, that is, if the brushes be displaced in the direction of the rotation, the remanence increases the torque at the speed ω provided that $\tan \beta$ be less than

$$2 \tan a/(\omega^2/\Omega^2 - \tan^2 a).$$

Alternating current shunt motor.

In Fig. 203 are shown the connections of a simple shunt wound motor. In practice, M_1 and M_2 are connected with constant potential supply mains, and so, if we neglect the armature reaction, the magnetic field is practically constant at all loads. As the shunt circuit acts like a choking coil, the current in it will lag by nearly a quarter of a period behind the applied P.D. Since the armature circuit is in parallel with the windings of the field magnets, the current in it at the start will be approximately in phase with the field flux, provided that the armature acts like a choking coil. The initial torque, therefore, in this case will be high. If, on the other hand, the current in it be approximately in phase with the applied P.D., the initial torque will be very small.

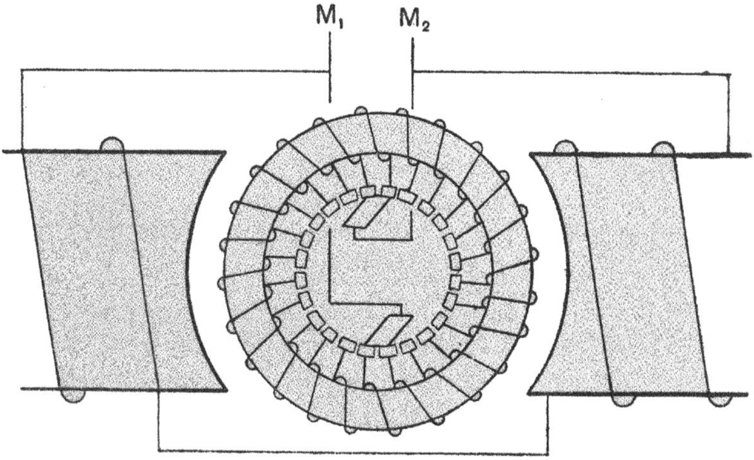

Fig. 203. Alternating current shunt motor.

Hence if the power factor of the machine be high, the torque will be small and *vice versâ*. We can also see from Fig. 203 that there will be excessive sparking at the brushes, as the coils short circuited by them are in a rapidly varying magnetic field. For commercial working, therefore, we must modify the machine so as to raise its power factor. We must also devise means to prevent sparking at the brushes.

One method of raising the power factor of a shunt motor is to put a condenser of suitable capacity in series with the shunt windings so that resonance ensues and the current in the field windings is approximately in phase with the applied P.D. The difficulty of this method, which was used by Stanley and Kelly, is that small changes in the frequency upset the relation between the capacity K and the self-inductance L which is required for resonance, namely, $\omega^2 LK = 1$. Small changes in the shape of the wave of the applied P.D. also produce considerable effects in the working of this type of motor.

An ordinary shunt motor will work satisfactorily when the current for the armature is supplied from a pair of the mains of a two phase system of supply and the current for the shunt from another pair, the mains being chosen so that the applied potential differences differ in phase by ninety degrees.

Repulsion motor.

The principle of the repulsion motor is illustrated in Fig. 204. The poles of the stator are made of iron stampings and are excited by an alternating current obtained from the supply mains. The rotor is practically an ordinary direct

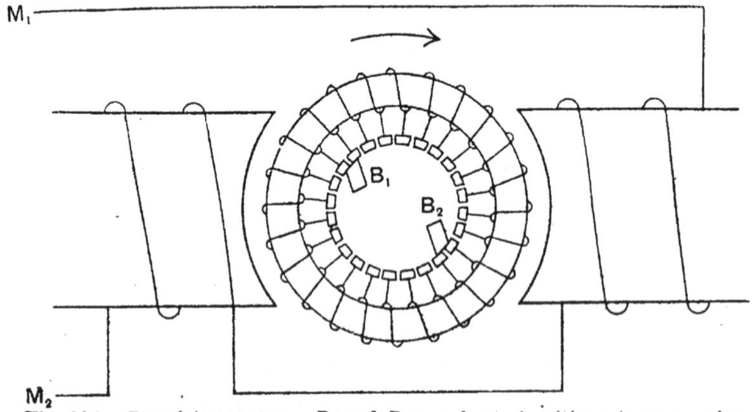

Fig. 204. Repulsion motor. B_1 and B_2 are short circuiting pieces pressing on the commutator. If B_1 and B_2 be joined by a wire, the machine will act as an induction commutator motor and will rotate in the opposite direction.

current armature with a ring or drum winding, and a commutator on which slide two short circuiting brushes B_1 and B_2. Let us suppose that the planes of the short circuited coils make angles of 45° with the direction of the magnetic field. In this position the induced currents will produce forces which tend to move the coil from a stronger to a weaker region of the magnetic field. The torque produced will thus be in the direction of the arrow head.

An objection is sometimes urged against this type of motor on the ground that only part of the armature windings is utilised. Since, however, the coils are only traversed by intermittent currents, the average heating is much less than if they were always in the circuit. The permissible intensity of the current in the conductors is therefore higher. The mere fact that all the conductors are not carrying current at the same time is thus, from the point of view of the manufacturer, not a serious matter.

If we place a copper ring in an oscillatory magnetic field due to an alternating current in a coil of wire, and if some of the lines of force are linked with the ring, the phase difference between the current induced in it and the magnetic field will, in general, be nearly 180°. The induced currents will be practically in opposition in phase to the inducing currents, just as the primary and secondary currents in a transformer are, when the secondary is on short circuit. By imagining the coils replaced by their equivalent magnetic shells, we see that there will be repulsion between them, the opposing faces being practically always of the same polarity. It is also easy to see that there is, in general, a couple tending to turn the ring so that its plane is parallel to the direction of the field. When it is in this position, there will be no induced currents and, therefore, no electromagnetic forces acting on it. If we displace it slightly so that it embraces part of the oscillatory field, the induced currents always produce a torque tending to make the flux embraced by the ring a minimum.

Let us suppose that we have a circular loop of wire placed so that its plane makes an angle θ with the lines of force in an oscillatory magnetic field. If $H \cos \Omega t$ represent the strength of the field, R and L the resistance and the inductance of the loop of wire and S its area, and if it be prevented from turning, we have

$$\frac{\partial}{\partial t}(HS \sin \theta \cos \Omega t) = Ri + L\frac{\partial i}{\partial t},$$

where i is the current in the wire. Thus, solving the equation, we get

$$i = -HS\Omega \sin \theta \sin (\Omega t - a)/(R^2 + L^2\Omega^2)^{\frac{1}{2}},$$

where $\tan a = L\Omega/R$.

Replacing the loop of wire by its equivalent magnetic shell, we see that the mean value G of the torque on the loop is proportional to the mean value of $H \cos \omega t \,.\, i \cos \theta$, and thus

$$G = kHS\Omega \sin \theta \cos \theta \sin \alpha/(R^2 + L^2\Omega^2)^{\frac{1}{2}},$$

where k is a constant. This may be written in the form

$$G = (k/2)\, HSL\Omega^2 \sin 2\theta/(R^2 + L^2\Omega^2).$$

The torque is therefore a maximum when θ is 45°.

When the amplitude of the alternating magnetic field is not the same at all points, we have, in addition to the torque on the coil, acting so as to make the flux embraced by it a minimum, forces acting which tend to move the coil from places where the field is strong to places where it is weak.

This can easily be seen by imagining the coil replaced by its equivalent magnetic shell. The electromagnetic repulsion will obviously be greater on the side on which the field is stronger.

Induction commutator motor.

Let us now consider the effect produced by joining the brushes B_1 and B_2 in Fig. 204 by a conductor. It will be seen that electromotive forces will be generated in each half of the ring and the resultant voltage between the brushes will only be zero when the rotor is at rest and the line joining the brushes is at right angles to the line joining the middle points of the polar faces. In general, therefore, there will be a current in each half of the ring and in the conductor. The induced polarities in each half of the ring will be pointing in the same direction and will be opposite to the polarity of the inducing magnet. Hence there will be attraction, and the ring will rotate in the opposite direction to that indicated by the arrow in the figure. The induced currents in the coils short circuited by the brushes will, as in the last form of motor, produce a torque opposing the motion, and this will lower the efficiency of the machine. Special precautions, therefore, have to be taken to prevent the currents in the coils short circuited by the brushes from attaining large values. For this reason, in some cases the windings are connected with the commutator by means of strips of high resistance metal, or the brushes are laminated in such a way that they offer a great resistance to the transverse flow of current across them.

Making the usual assumptions, it is easy to obtain approximate equations for the working of this type of motor. The E.M.F. developed in the rotor by induction and rotation, as in the case of the series motor, may be written in the form

$$M\omega \cos \alpha \,.\, i_1 + M \sin \alpha \frac{\partial i_1}{\partial t},$$

where the symbols have their usual meaning (p. 442). The equation for the current in the rotor or secondary circuit is therefore of the form

$$0 = M\omega \cos \alpha \,.\, i_1 + M \sin \alpha \frac{\partial i_1}{\partial t} + R_2 i_2 + l_2 \frac{\partial i_2}{\partial t} \;\ldots\ldots(a),$$

where R_2 is the resistance of the conductor connecting the brushes, in series with the windings of each half of the armature in parallel, and l_2 is the self-inductance of this circuit. The equation for the current in the magnetising circuit is

$$e = R_1 i_1 + l_1 \frac{\partial i_1}{\partial t} + M \sin \alpha \frac{\partial i_2}{\partial t} \;\ldots\ldots\ldots\ldots\ldots(b),$$

where e denotes the applied P.D. Equations (a) and (b) may be taken as the required approximate equations.

Multiplying (a) by i_2 and (b) by i_1, adding the equations together and taking mean values, we get

$$VA_1 \cos \gamma = R_1 A_1^2 + R_2 A_2^2 + M\omega \cos \alpha \,.\, A_1 A_2 \cos \alpha_{1.2},$$

where $\alpha_{1.2}$ is the phase difference between A_1 and A_2. If we neglect hysteresis and eddy current losses, the power $VA_1 \cos \gamma$ given to the motor will be expended in heating $R_1 A_1^2 + R_2 A_2^2$, and in giving mechanical power $G\omega$ to the rotor. We have, therefore,

$$G\omega = M\omega \cos \alpha \,.\, A_1 A_2 \cos \alpha_{1.2},$$

and thus $$G = M \cos \alpha \,.\, A_1 A_2 \cos \alpha_{1.2}.$$

If we make the assumption that e obeys the harmonic law, complete solutions of the linear equations (a) and (b) can easily be obtained. To a first approximation they represent the working of the motor for a given position of the brushes. Owing to armature reaction, however, if we vary α, we also vary M, and this effect is very noticeable in practice. It may be reduced by means of special windings which neutralise, to a considerable extent, the field produced by the armature currents. Slits, parallel to the axis of the rotor, are sometimes made in the poles so as to increase the transverse reluctance, and thus diminish the intensity of the transverse field produced by armature reaction. In most motors of this class the power factor is low at all loads, and for a given output they are much heavier than induction motors. The starting torque, however, is large.

Other forms of motor.

The Arnold motor is partly a repulsion motor and partly an induction motor. It starts as a repulsion motor and, once it has attained a suitable speed, the commutator is automatically short circuited and the machine runs as an ordinary single phase induction motor. A good starting torque is thus secured, and the efficiency when loaded is satisfactory. The power factor is also much higher than for the repulsion motor.

The Latour motor is virtually a combination of a repulsion motor and a series motor. The separate field magnet coils of the ordinary repulsion motor are replaced by a distributed wave winding embedded in slots. After passing through the wave windings the current passes through the rotor by means of brushes, the line joining which is at right angles to the line joining the brushes B_1 and B_2 in Fig. 204. This motor has a high efficiency and a high power factor.

The principle of the Winter-Eichberg motor is practically the same as that of the Latour motor. Instead, however, of letting the main current pass through the rotor, an auxiliary current is obtained from the secondary terminals of a transformer, the primary of which is connected with the mains, and this current is led into the rotor by means of brushes pressing on a commutator, in a similar manner to the way the main current is led into the rotor of the Latour motor. The speed of the rotor can easily be varied within wide limits by regulating the pressure applied to the rotor brushes.

REFERENCES

C. P. Steinmetz, *Alternating Current Phenomena.*

S. Sheldon and E. Hausmann, *Electric Traction and Transmission Engineering.*

E. Arnold and J. L. La Cour, *Die Wechselstromtechnik,* Vol. 5.

F. B. Crocker and M. Arendt, *Electric Motors.*

J. A. Fleming, 'Electromagnetic Repulsion,' *Electrician,* Vol. 26, p. 567, 1891.

G. T. Walker, 'Repulsion and Rotation produced by Alternating Electric Currents,' *Phil. Trans.,* Vol. 183, p. 179, 1891.

Paul Girault, 'Sur les Moteurs Série à Courants Alternatifs Simples,' *L'Industrie Électrique,* Vol. 12, p. 51, 1903.

A. C. Eborall, 'Electric Traction with Alternating Currents,' *Electrician,* Vol. 52, p. 327, 1903.

A. Blondel, 'Notes sur les Moteurs Monophasés à Collecteurs,' *L'Éclairage Électrique,* Vol. 37, p. 321, 1903.

J. Bethenod, 'Sur la Théorie du Moteur Série Compensé Monophasé,' *L'Éclairage Électrique,* Vol. 41, p. 5, 1904.

—— 'Sur le Dimensionnement des Moteurs Monophasés à Collecteur,' *L'Éclairage Électrique,* Vol. 45, p. 201, Nov. 1905.

F. Creedy, 'The Alternating Current Series Motor,' *Journ. of the Inst. of El. Engin.,* Vol. 35, p. 45, 1905.

V. A. Fynn, 'A New Single-Phase Commutator Motor,' *Journ. of the Inst. of El. Engin.,* Vol. 36, p. 324, 1906.

J. Fischer-Hinnen, 'Single Phase Commutator Motors,' *Electrician,* Vol. 63, 1909.

N. Shuttleworth, 'Polyphase Commutator Machines and their Application,' *Journ. of the Inst. of El. Engin.,* Vol. 53, p. 439, 1915.

CHAPTER XVII

The transformation of alternating to direct currents. Single phase rotary converter. The heat developed in the armature coils. Armature reaction. The alternating component of the direct voltage. Two pole polyphase converter. Heating of the armature of a polyphase converter. Six phase converter. The voltage ratio. Armature reaction in polyphase converters. The alternating component of the direct voltage. Finding the armature reaction from the characteristic curve. Compounding a rotary converter. Starting converters. Parallel running. Split pole converter. Inverted rotary converters. Regulators. Synchronous booster. Data of a 200-kilowatt rotary converter. Double current generators. The La Cour motor converter. Three wire direct current system. References.

The transformation of alternating to direct currents.

IT is found in practice that electricity can be generated very economically when the generating units are large and when the ratio of the average output of the station to the maximum output is high. This ratio is called the 'load factor' of the station. The maximum value of the load factor is unity, and it has this value when the load is continuous and has always its maximum value. As a rule, the more diversified the nature of the load supplied by a station the higher will be the load factor. It is now the customary practice to build large generating stations in places where coal is cheap, where abundant water can be had for the boilers and condensers, and where rents are low. As these stations are generally at a considerable distance from the distributing substations, it is necessary to transmit the electric power at high pressures in order to avoid either excessive losses due to the heating of the mains or a very heavy initial outlay on copper for them. Steam turbines are usually employed to drive high voltage three phase generators, and the electricity is transmitted by three core underground cables or by overhead wires. At the substations the power is generally transformed to lower pressures and converted into direct current for transmitting power to electric tramways or for lighting. One great advantage of converting the alternating into direct current is that we can use accumulators for storing the electric energy and thus easily diminish the fluctuations of the load and so raise the load factor. The conversion may be done by means of motor generator sets (see p. 206) either with or without intermediate transformers. The motor part of the set may consist of a three phase synchronous or induction motor, and the generator is simply an ordinary direct current dynamo. The efficiency of the combination is $\eta_1 \eta_2$, where η_1 is the efficiency of the motor and η_2 is the efficiency of the dynamo.

Instead of having two separate machines to convert the alternating into direct current, we can place the alternating and direct windings on one armature. The alternating current necessary to drive the machine as a motor can be supplied through slip rings on the shaft of the armature, and the direct current collected from a commutator on the same shaft. A still further simplification can be made by combining the two windings into one so that the alternating and direct currents flow in the same conductors. The winding of the armature is practically the same as that of a direct current machine and the commutator bars are connected with it in the same way. The slip rings are connected with armature conductors whose angular distances from one another equal $360°/pq$, where $2p$ is the number of poles, and q the number of phases. A machine of this type is called a 'rotary converter.'

Single phase rotary converter.

We shall first consider the single phase rotary converter. Let us suppose that we have a direct current bipolar dynamo with a ring-wound armature (Fig. 205) and that two commutator bars at an angular distance apart of 180° are connected with two slip rings on the shaft. This machine will be a simple form of rotary converter. Let us suppose that the brushes pressing on S_1 and S_2 are connected with the alternating current mains. Let us also suppose that the field magnets are separately excited by direct current. If the frequency of the supply be $\omega/2\pi$, the machine will run as a synchronous motor at the speed ω (Chapter V). For instance, if the frequency were 50, the armature would make $(\omega/2\pi)$ 60, that is, 3000 revolutions per minute when the machine is running as a synchronous motor. In this case, if the armature reaction and the resistance of the armature windings be negligible, the applied P.D. has its maximum value when *ab* (Fig. 205) is vertical and is zero when *ab* is horizontal. As the armature rotates, an electromotive force is developed between the brushes B_1 and B_2 in

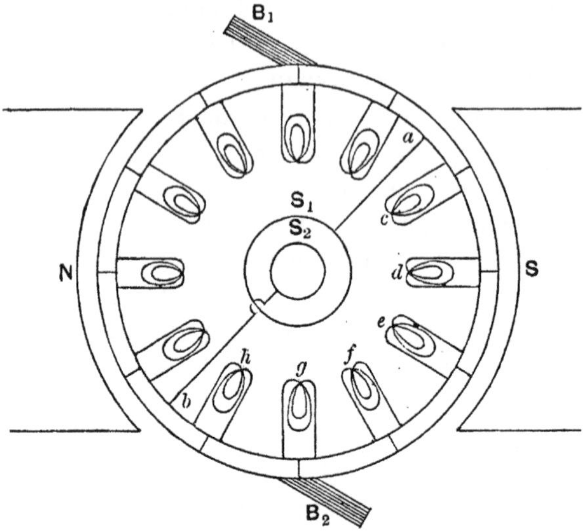

Fig. 205. Simple form of single phase rotary converter. Alternating currents are supplied by the slip rings S_1 and S_2 and direct currents are taken from B_1 and B_2.

exactly the same way as in a direct current machine. If the number of the commutator segments be infinite, this electromotive force E will be constant, and will equal the maximum value of the applied potential difference. Hence, if the applied potential difference e follow the harmonic law, we may write

$$e = E \cos \omega t,$$

when we reckon t from the instant when *ab* is vertical. If V denote the effective value of e, we have

$$V/E = 1/\sqrt{2} = 0{\cdot}707, \text{ nearly} \quad \ldots\ldots\ldots\ldots\ldots(1).$$

We see that, when the applied wave is sine shaped, the voltage on the direct current side obtained from a rotary converter is about forty per cent. greater than the effective value of the applied alternating voltage.

In the proof given above we have neglected the effects of armature reaction. In getting approximate formulae we may neglect the effects of armature reaction, but we shall see later on that in the case of the single phase rotary converter, the armature reaction is appreciable. It not only distorts the wave of the P.D. between the slip rings but it also introduces an alternating current component into the potential difference on the direct current side.

Let us suppose that the brushes B_1 and B_2 (Fig. 205) are joined through a resistance. A direct current will now flow in the armature windings, in addition to the alternating current, and will magnetise it in such a way that the magnetic forces produced by the direct current component of the field will tend to stop its motion. The direct currents therefore flow in the same direction as they do when the machine is acting as a dynamo. When the alternating current is in phase with the applied electromotive force, we see that the alternating current component in the windings will always be flowing in opposition to the direct current component, as the former component produces the rotation.

Let A be the effective value of the alternating current, C the direct current, $\cos\psi$ the power factor and η the efficiency of the converter. The power expended on the machine is $VA\cos\psi$ and its output is EC. Hence

$$\eta = EC/(VA\cos\psi) \text{ and } A = EC/(\eta V\cos\psi).$$

If the current and the potential difference waves follow the harmonic law and if the voltage drop in the armature windings be negligible when compared with V, we have

$$I = \{2/(\eta\cos\psi)\}\, C \quad \text{......................(2)},$$

where I is the maximum value of the current.

We see therefore that I will be greater than $2C$. In practice it is possible for η to be as great as 0·95 and for $\cos\psi$ to be within about one per cent. of unity.

We shall suppose that there are $2m$ coils evenly spaced round the circumference of the ring, so that the planes of consecutive coils make an angle π/m with each other. Let us first consider the power expended in heating the various coils. When ab is in the position shown in Fig. 205, the current i in the coil c will be given by

The heat developed in the armature coils.

$$i = (I/2) \cos (\omega t - \psi) - C/2,$$

where $\cos \psi$ is the power factor of the load and the coil c is to the right of the brush B_2. When it is to the left of this brush we have

$$i = (I/2) \cos (\omega t - \psi) + C/2.$$

When t is zero the line ab is vertical. Let us suppose that the plane of the coil c makes an angle $\pi/2m$ with ab at this instant. The current i in the coil will be given by

$$i = (I/2) \cos (\omega t - \psi) - C/2,$$

from $$t = 0 \text{ to } t = T/2 - T/4m,$$

and $$i = (I/2) \cos (\omega t - \psi) + C/2,$$

from $$t = T/2 - T/4m \text{ to } T/2.$$

Hence if r be the resistance of this coil, the mean value of ri^2 for the half of a period, and therefore the mean value W_1 of the power expended in heating the coil is given by

$$\begin{aligned} W_1 &= (2r/T) \int_0^{T/2-T/4m} i^2 \partial t + (2r/T) \int_{T/2-T/4m}^{T/2} i^2 \partial t \\ &= (2r/T) \int_0^{T/2} \{(I^2/4) \cos^2 (\omega t - \psi) + C^2/4\} \partial t \\ &\quad - (ICr/T) \left\{ \int_0^{T/2-T/4m} \cos (\omega t - \psi) \partial t - \int_{T/2-T/4m}^{T/2} \cos (\omega t - \psi) \partial t \right\} \\ &= r (I^2/8 + C^2/4) - (r/\pi) \sin (\pi/2m + \psi) \,.\, IC \\ &= r (C/2)^2 \{1 - (4/\pi) \sin (\pi/2m + \psi) \,.\, (I/C) + (1/2) (I/C)^2\}. \end{aligned}$$

Similarly if W_2 denote the mean power expended in the coil d (Fig. 205), we get

$$W_2 = r (C/2)^2 \{1 - (4/\pi) \sin (3\pi/2m + \psi) \,.\, (I/C) + (1/2) (I/C)^2\}.$$

When ψ is less than $\pi/2 - \pi/m$, that is, in the normal condition of working, we see that W_2 is less than W_1 and therefore the heating of the coil d will be less than that of c. When ψ is zero, the nearer the coil is to a segment connected with a slip ring the hotter will be the coil. This result is well known in practical work.

The average value W_m of the power expended in heating the mth coil is given by

$$W_m = r(C/2)^2 [1 - (4/\pi) \sin\{(2m-1)\pi/2m + \psi\} . (I/C) + (1/2)(I/C)^2]$$

$$= W_1 + (2/\pi) rIC \cos(\pi/2m) \sin\psi.$$

Hence W_m is only equal to W_1 when ψ or C is zero.

In Fig. 205, the heating of the coils e and f will be a minimum and the heating of the coils c and h a maximum, when ψ is very small. In general, the coils for which $\sin\{(2p-1)(\pi/2m) + \psi\}$ is smallest are the coils which heat most, and the coils for which $(2p-1)(\pi/2m) + \psi$ is nearest to $\pi/2$ are the coils which heat least. A maximum value to the power expended in heating a coil is $r(C^2/4 + I^2/8)$ and a minimum value is $r(C^2/4 + I^2/8 - CI/\pi)$.

If W denote the mean value of the total power expended in heating the armature, we have, since the heating of the two sides of the armature is obviously symmetrical,

$$W = 2(W_1 + W_2 + \dots + W_m)$$

$$= 2mr(C/2)^2 [1 - (4/\pi)\cos\psi/\{m \sin(\pi/2m)\} . (I/C) + (1/2)(I/C)^2] \quad \dots\dots\dots\dots(3).$$

Now $m \sin(\pi/2m)$ increases as m increases, and thus the total heating is greater the more we distribute the windings on the armature. The increase of the heating, however, due to this cause is very small as $m \sin(\pi/2m)$ only increases by about three per cent. as m increases from 4 to infinity.

If the voltage drop in the armature windings be negligible compared with V, we have by (2) and (3)

$$W = 2mr(C/2)^2 [1 - 8/\{\pi\eta m \sin(\pi/2m)\} + 2/\eta^2 \cos^2\psi].$$

Hence, when m is large, we have

$$W = 2mr(C/2)^2 [1 - 16/\pi^2\eta + 2/\eta^2 \cos^2\psi] \quad \dots\dots\dots(4).$$

The power expended in heating the armature, if the machine

were acting merely as a dynamo giving the same output, would be $2mr\,(C/2)^2$, and since $\eta \cos\psi$ is less than unity we see that the heating of the armature of the converter will be greater than this. If η and $\cos\psi$ are each equal to unity, we have

$$W = 1{\cdot}38 \,.\, 2mr\,(C/2)^2\,;$$

and if they are both equal to 0·9, we have

$$W = 2{\cdot}25 \,.\, 2mr\,(C/2)^2.$$

Notice that $2mr$ is the resistance of all the armature windings in series and mr is the resistance of each half of the armature windings.

The formulae given above could also be proved as follows. We can regard the current in every coil as the resultant of two alternating currents, one of which $(I/2)\cos(\omega t - \psi)$ follows the harmonic law and the other $\pm\, C/2$ is rectangular in shape. Now by Volume I, p. 280, the phase difference ψ' between a rectangular shaped wave and a sine shaped wave, when the time lag between them is

$$-\,\pi/2 + (2p-1)\,(\pi/2m) + \psi,$$

is given by

$$\begin{aligned}\cos\psi' &= -\,(2\sqrt{2}/\pi)\cos\{-\,\pi/2 + (2p-1)\,\pi/2m + \psi\}\\ &= -\,0{\cdot}9003\sin\{(2p-1)\,\pi/2m + \psi\}.\end{aligned}$$

Hence we find at once that

$$\begin{aligned}W_p = r\,[I^2/8 + C^2/4 - 2\,(I/2\sqrt{2})\,(C/2)\,(2\sqrt{2}/\pi)\\ \times \sin\{(2p-1)\,\pi/2m + \psi\}],\end{aligned}$$

and proceeding as before we get the same formula for W.

Armature reaction.

Since, on the alternating current side, a rotary converter acts like a synchronous motor, it runs at exactly the same speed at all loads. If $\omega/2\pi$ be the frequency of the supply, the angular velocity of a bipolar machine will be ω. We shall now calculate the magnetising forces due to the currents in the armature windings of the machine shown in Fig. 205. We shall assume that the currents in each half of the armature windings are equal. Let us first consider the effect of the alternating current $(I/2)\cos(\omega t - \psi)$. At the time t, ab in Fig. 205 makes an angle ωt with the vertical, and thus, if a current i flowing in each half of the ring produces a magnetising force $2ki$ in the direction ab, where k is a constant, we see that the vertical component of the magnetising force at the time t will be $2ki \cos \omega t$ and the horizontal component will be $2ki \sin \omega t$. Hence, substituting $(I/2) \cos (\omega t - \psi)$ for i and taking the mean value of $2ki \cos \omega t$ from t equal to zero to t equal to $T/2$, we get for the mean value of the vertical, that is, the transverse component $\mathscr{H}_t'$ of the magnetising force, due to the alternating current, the equation

$$\mathscr{H}_t' = (k/2)\, I \cos \psi,$$

and similarly we have

$$\mathscr{H}' = (k/2)\, I \sin \psi,$$

where $\mathscr{H}'$ is the mean value of the direct component of the magnetising force.

The magnetising force due to the direct current in the armature windings will act in the vertical direction, and its magnitude will be $-kC$, since it must act in the direction opposite to that of the magnetising force due to the alternating currents. If, therefore, $\mathscr{H}_t$ and $\mathscr{H}$ are the mean values of the transverse and direct components of the magnetising forces due to both the direct and alternating currents, we have

$$\left.\begin{aligned} \mathscr{H}_t &= k\{(I/2)\cos\psi - C\} \\ \text{and} \qquad \mathscr{H} &= k\,(I/2)\sin\psi \end{aligned}\right\} \quad \ldots\ldots\ldots\ldots\ldots\ldots(5).$$

By (2) we see that $\mathscr{H}_t$ is also given by

$$\mathscr{H}_t = \{(1-\eta)/\eta\}\, kC \quad \ldots\ldots\ldots\ldots\ldots\ldots(6).$$

If η were unity the transverse component would vanish, and if it were 0·8, we should have $\mathscr{H}_t = kC/4$. If η were 0·5, $\mathscr{H}_t$ would be equal to kC, and for smaller values of the efficiency it would be greater than kC. Unless therefore the efficiency of a rotary converter be very low the transverse magnetisation of the field is much smaller than when the machine is acting as a direct current dynamo having the same output.

In Fig. 206 a diagram is shown of the magnetising forces acting on the field. If ψ is lagging $\mathscr{H}$ acts in the same direction as the magnetising force due to the field magnet ampere-turns (see p. 167). Now by diminishing the value of the direct current exciting the field we increase ψ, and thus increase the magnetising effect of the armature reaction. Some machines in which the armature reaction is large will run with no direct current excitation at all. In this case, however, ψ is large, and so the efficiency is low.

In practice, rotary converters are worked at the excitation which makes the alternating current a minimum. In this case the transverse magnetisation is small, and so the sparking at the commutator brushes is slight even in machines which spark considerably when working as dynamos.

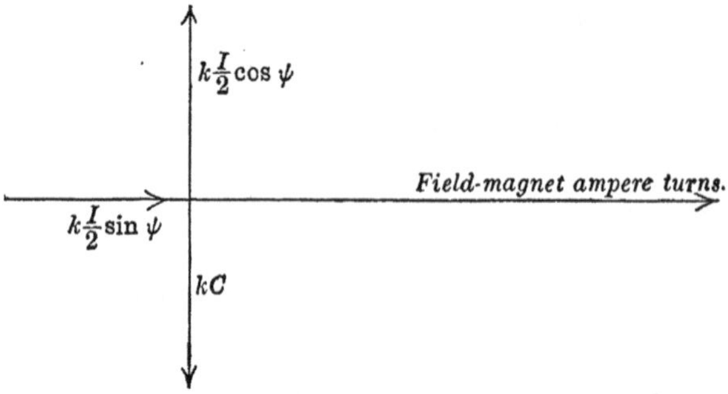

Fig. 206. Magnetising forces producing the magnetic field of a rotary converter. The resultant transverse component is almost zero.

The alternating component of the direct voltage.

In the preceding section we have considered the mean effects of the armature reaction over half of the period of revolution of the rotor. Hence any effect the period of which is this half-period or any submultiple of it will have cancelled out. We shall now show that there is an important component that has to be taken into consideration in the case of a single phase converter.

The field produced by the alternating currents is an oscillating one, and is fixed relatively to the armature. Owing to the variation of the reluctance of the path of the magnetic lines due to the magnetomotive force of the armature currents, this oscillating field does not necessarily follow the harmonic law even when the currents are sine shaped. In order, however, to simplify the problem, we shall suppose that it does. In this case the oscillating field can be resolved into two component fields rotating in opposite directions. The magnitude of each of the component fields is half that of the amplitude of the oscillating field. One of the component fields will be fixed in space, its direction making an angle ψ with the vertical, and its magnitude will be proportional to $I/2$. Hence, as we saw in the preceding section, the transverse and direct components of the field due to armature reaction are proportional to $(I/2)\cos\psi - C$ and $(I/2)\sin\psi$ respectively.

The other rotary component of the oscillating field rotates with double the angular velocity of the armature. Hence in the half-period it will have glided once round the air-gap, and therefore will introduce a component electromotive force into the direct current side. If we neglect the distortion of the field produced by the magnetising effect of the armature currents and suppose that the excitation is adjusted until the power factor is unity ($\psi = 0$), the voltage between the brushes is of the form $E + a.I \sin 2\omega t$, where a is a constant. The effective value of the voltage between the brushes is $(E^2 + a^2I^2/2)^{1/2}$ and hence it varies with the load. In single phase converters, therefore, when the armature reaction is appreciable, the ratio of the alternating to the direct voltage is not a constant even when the resistance of the armature windings is negligible.

The presence of the alternating component in the direct voltage can be shown by connecting a magnetic and a hot wire voltmeter in parallel across the commutator brushes. The difference between their readings will increase with the load. The weaker the exciting field the more marked will be this effect.

Two pole polyphase converter.

We shall now consider the case of a two pole polyphase rotary converter. Let us suppose that the machine (Fig. 207) has a ring armature with a Gramme winding (Fig. 205), and let n be the total number of turns in one phase; for instance, the number of turns between A and B. Let there be q phases and let $qn = 2m$, so that π/m is the angle between the planes of consecutive turns. Let e_1 be the maximum electromotive force developed in one turn and let E be the direct current electromotive force between the commutator brushes. Then, neglecting the resistance of the windings, we have

$$E = e_1 \sin(\pi/m) + e_1 \sin(2\pi/m) + \dots + e_1 \sin\{(m-1)\pi/m\}$$
$$= e_1 \cos(\pi/2m)/\sin(\pi/2m).$$

If v be the potential difference between the slip rings connected with A and B, and if the plane of the winding connected with A

make an angle ωt with the vertical, we have, when the resistance is negligible,

$$v = e_1 \sin \omega t + e_1 \sin (\omega t + \pi/m) + \ldots + e_1 \sin \{\omega t + (n-1)(\pi/m)\}$$
$$= e_1 \sin \{\omega t + (n-1)(\pi/2m)\} \sin (n\pi/2m)/\sin (\pi/2m)$$
$$= E \sin \{\omega t + (n-1)(\pi/2m)\} \sin (n\pi/2m)/\cos (\pi/2m).$$

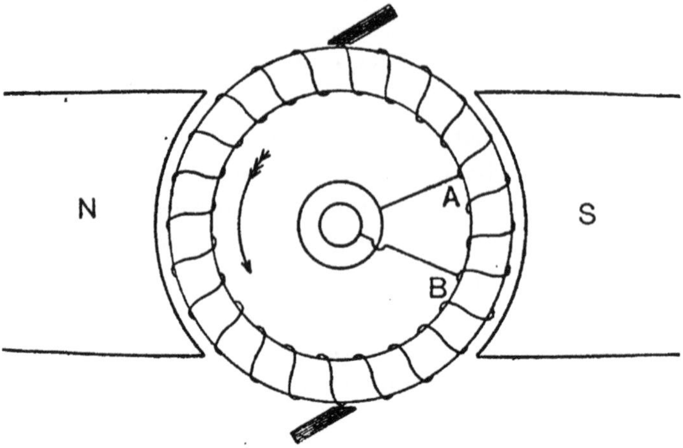

Fig. 207. Two pole rotary converter for eight phases. Slip rings and connections for one phase only are shown.

Hence, if V be the effective value of v, we have

$$V = (E/\sqrt{2}) \sin (n\pi/2m)/\cos (\pi/2m) = (E/\sqrt{2}) \sin (\pi/q)/\cos (\pi/2m).$$

In practice m is very large and so we can write unity for $\cos (\pi/2m)$. Hence, we find that

$$V = (E/\sqrt{2}) \sin (\pi/q) \quad \ldots\ldots\ldots\ldots\ldots\ldots(7).$$

In Fig. 207 the windings are connected in mesh. If we construct a regular polygon of q sides each equal to V, these sides (Fig. 208) will represent the mesh voltages in magnitude and phase. The star voltages of the system will be represented by the lines joining the centre of this polygon to the angular points. If V_s denote the effective value of the star voltage, OP_1 in Fig. 208, we have

$$V_s = V/\{2 \sin (\pi/q)\} = E/(2\sqrt{2}) \quad \ldots\ldots\ldots\ldots(8).$$

Hence the star voltage is the same whatever may be the number of phases.

In practice, the machines are generally multipolar. The general arrangement of the brushes and slip rings in this case will be understood from Fig. 209. On the direct current side, brushes of like sign are connected in parallel, and on the alternating current side, circuits of the same phase are connected in parallel between the same pair of slip rings.

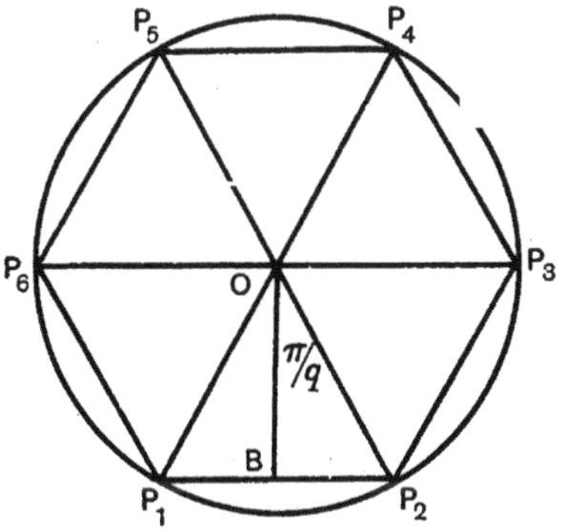

Fig. 208. Star and mesh voltages in a q phase system.

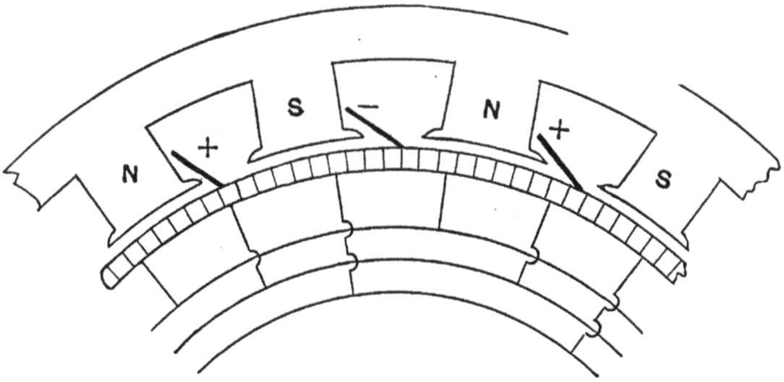

Fig. 209. Field magnets, direct current brushes, commutator and slip rings of multipolar three phase rotary converter.

Let E be the direct voltage, and let C be the direct current in the main joined to one set of brushes. The direct current output will be EC. If V be the mesh voltage of a three phase supply and A the effective value of the alternating current in each of the three mains, the electric power received by the machine is

$$\sqrt{3}\, VA \cos\psi.$$

Hence if η be the efficiency

$$EC = \eta\sqrt{3}VA\cos\psi,$$

and thus $$A = EC/(\eta\sqrt{3}V\cos\psi).$$

If, therefore, we can neglect the effect of the resistance of the windings on the ratio of E to V, we get, by (7),

$$A = \sqrt{2}C/\{\eta\sqrt{3}\sin(\pi/3)\cos\psi\} = 2\sqrt{2}C/(3\eta\cos\psi) \quad \text{...(9)}.$$

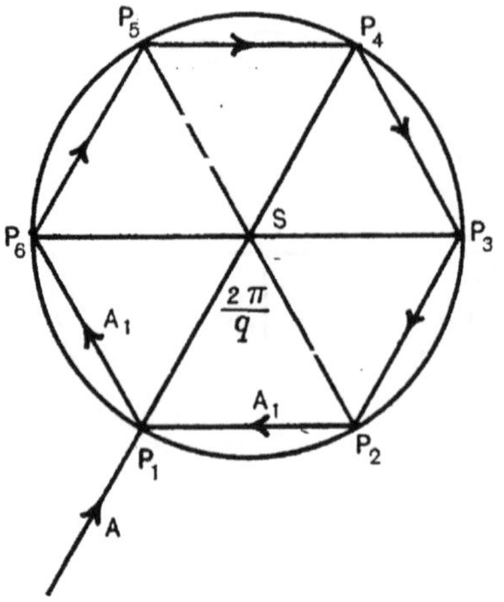

Fig. 210. Mesh currents in a q phase system.

When there are q phases (Fig. 210), we have

$$A_1 = A/\{2\sin(\pi/q)\} \quad \text{..................(10)},$$

where A is the current in each main and A_1 is the current in each winding. The power given to the rotary and expended on the q phases equals $qVA_1\cos\psi$, and hence it equals

$$qVA\cos\psi/\{2\sin(\pi/q)\}.$$

We have, therefore,

$$EC = \eta qVA\cos\psi/\{2\sin(\pi/q)\},$$

and thus by (7)

$$A = 2\sqrt{2}C/(q\eta\cos\psi) \quad \text{..................(11)},$$

and

$$I = 4C/(q\eta\cos\psi) \quad \text{.....................(12)},$$

where I denotes the maximum value of the alternating current in a main. For given values of η and $\cos\psi$, we see that qA is independent of the number of phases, and thus the weight of the copper used in the mains is the same whatever be the number of phases we adopt.

The following table is calculated by means of formulae (7), (8), (10) and (11). V_s is the star voltage and V the mesh voltage of the supply. A is the current in the main, and A_1 is the current in the winding of a mesh-connected armature.

Slip rings...	2	3	4	6	q	∞
V_s/E	0·354	0·354	0·354	0·354	$1/2\sqrt{2}$	0·354
V/E	0·707	0·612	0·500	0·354	$\sin(\pi/q)/\sqrt{2}$	0
$\eta A \cos\psi/C$	1·414	0·943	0·707	0·471	$2\sqrt{2}/q$	0
$\eta A_1 \cos\psi/C$	0·707	0·544	0·500	0·471	$\sqrt{2}/q \sin(\pi/q)$	0·450

Heating of the armature of a polyphase converter.

From the expression given above (p. 466) for the applied potential difference v between the slip rings, we deduce that the current i in the winding, when the angle of lag is ψ and the converter is bipolar, is given by

$$i = \{I/2 \sin(\pi/q)\} \sin\{\omega t - \psi + (n-1)(\pi/2m)\} \pm C/2.$$

Let us suppose that t is zero when the coil of the winding in connection with A (Fig. 207) is immediately under the top brush. The expression for the current in this coil of wire is obtained by prefixing the negative sign to $C/2$, in the above formula, for the first half of the period, and the positive sign for the second half of the period. Thus if W_1 denote the power expended in heating the coil, we get

$$W_1 = r\,[I^2/\{8\sin^2(\pi/q)\} + C^2/4 - 2\{CI/4\sqrt{2}\sin(\pi/q)\}(2\sqrt{2}/\pi) \times \cos\{(n-1)\pi/2m - \psi\}].$$

Hence, making use of the approximate equation (12), we find that

$$W_1 = r\,(C/2)^2\,[1 - 16\cos(\pi/q - \pi/2m - \psi)/\{\pi q\eta\cos\psi\sin(\pi/q)\} + 8/\{q^2\eta^2\cos^2\psi\sin^2(\pi/q)\}],$$

since $n\pi/2m = \pi/q$.

Similarly, if W_2 denote the power expended in heating the next turn of wire, we have

$$W_2 = r (C/2)^2 [1 - 16 \cos (\pi/q - 3\pi/2m - \psi)/\{\pi q \eta \cos \psi \sin (\pi/q)\} + 8/\{q^2\eta^2 \cos^2 \psi \sin^2 (\pi/q)\}].$$

If ψ be zero, we see that the coils directly connected with the slip rings get heated most. As formerly, however, when the current is lagging or leading this is not the case. For instance, when ψ equals $\pi/q - \pi/2m$, the power expended in heating a turn of the winding increases continuously as we pass from one slip ring connection to the next. It is also easy to see that for a given number of turns, the greater the number of phases, the lower will be the temperature of the coil subjected to the maximum heating.

If W denote the total heating of the armature, then

$$W = 2mr (C/2)^2 [1 - 16/\{\pi q \eta n \sin (\pi/2m)\} + 8/\{q^2\eta^2 \cos^2 \psi \sin^2 (\pi/q)\}].$$

When m is large we can write $\pi/2m$ for $\sin (\pi/2m)$, and thus

$$W = R (C/2)^2 [1 - 16/(\pi^2\eta) + 8/\{q^2\eta^2 \cos^2 \psi \sin^2 (\pi/q)\}] \quad \ldots(13),$$

where R is the resistance of the whole armature winding. Putting q equal to 2, we see that (13) reduces to (4). It has to be remembered that for a two phase converter q equals 4.

When both η and $\cos \psi$ are unity, we get from (13) the following numerical values.

Number of phases, q …………	2	3	4	6	∞
(Armature heating)/$\{R (C/2)^2\}$	1·38	0·564	0·37	0·26	0·19

The ratio of W to $R (C/2)^2$ increases very rapidly as η and $\cos \psi$ diminish. For example, when the number of phases is infinite, let us suppose that η and $\cos \psi$ are each equal to 0·9, then, W is equal to 0·43 $R (C/2)^2$, and when η and $\cos \psi$ are each equal to 0·8, W is equal to 0·95 $R (C/2)^2$. In the latter case the heating of the armature is practically the same as when the machine is acting as a direct current dynamo or motor only. In this case also for a three phase converter W equals 1·76 $R (C/2)^2$. It will be seen that the heating of the armature of a three phase converter is less than the heating of the armature of a direct current dynamo having the same output only when the efficiency and the power factor are high.

When a converter is running with the direct current circuit open, it takes a certain amount of power from the alternating current mains. This power is expended in overcoming friction and the torque due to hysteresis and eddy current losses. If we assume that this is approximately constant at all loads, we may write

$$qVA \cos\psi/\{2 \sin(\pi/q)\} = EC + qVA_0 \cos\psi_0/\{2 \sin(\pi/q)\},$$

where A_0 is the current on no load. Substituting for E its value from (7) we get

$$A \cos\psi = 2\sqrt{2}C/q + A_0 \cos\psi_0 \quad \text{............(14)},$$

and $$\eta = 2\sqrt{2}C/(2\sqrt{2}C + qA_0 \cos\psi_0).$$

This approximate equation for η shows that at small loads the efficiency will be small. An approximate value of the heating of the armature at any load can be found by calculating η by this formula and then substituting for η in equation (13).

Six phase converter.

From the table given in the preceding section it will be seen that the relative heating effect in the armature for a given output decreases as the number of phases for which the rotary is built increases. With three phase transformers (see Figs. 144 and 145) it is easy to arrange that the secondary output is six phase. We may wind the arms of the magnetic circuit with a three phase secondary winding mesh connected, represented by 1, 2, 3 in Fig. 211, and the circumference of the magnetic circuit (Fig. 144) with another three phase secondary winding 1′, 2′, 3′. A six phase supply can then be obtained from 1, 2, 3, 1′, 2′, 3′. It is more efficient therefore to wind the secondary of the transformer in this way and build the rotary for six phases. This is usually done in practice. If V be the voltage between the phases for either of the secondary windings, $V/E = 0{\cdot}612$ approx., and the voltage between consecutive slip rings is $V/\sqrt{3}$, which equals $0{\cdot}354E$.

The three secondary windings of the transformer can also be star connected (Fig. 212). In this case $V/E = 0{\cdot}707$ approx., where V is the voltage between 1 and 2′ for instance. The voltage to the neutral point in this case will be $V/2$, *i.e.* $0{\cdot}354E$.

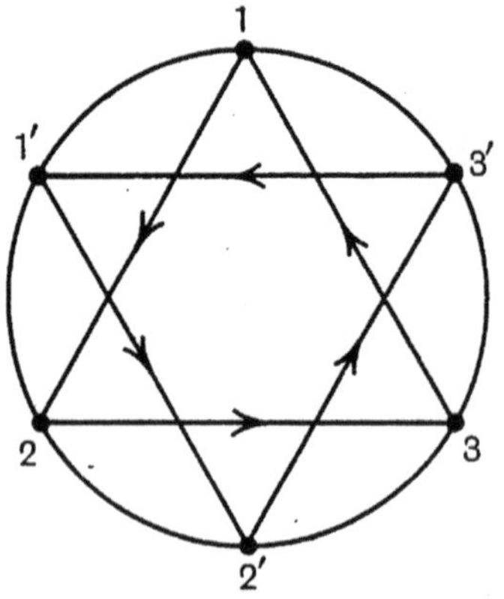

Fig. 211. 1, 2 and 3, and 1′, 2′ and 3′ are the terminals of two mesh-connected secondaries of a three phase transformer. A six phase supply is taken from 11′22′33′.

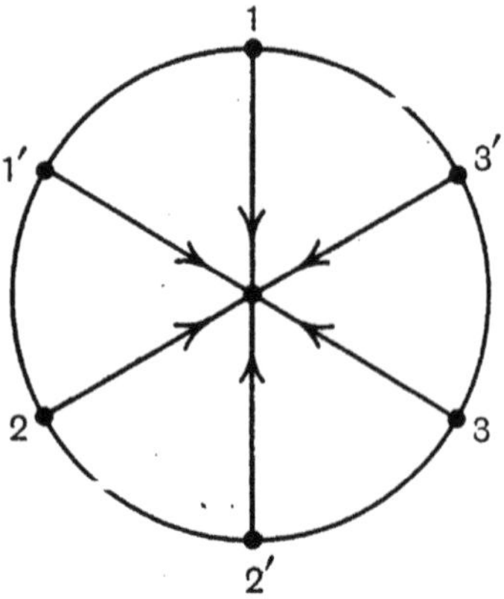

Fig. 212. 1, 2 and 3 are the terminals of one star-connected secondary and 1′, 2′ and 3′ are the terminals of the other. A six phase supply is taken from 11′22′33′.

The voltage ratio.

The voltage ratio for three phase converters in practice does not differ much from the value 0·612 we found above by making several assumptions. The poles (Fig. 207) of a bipolar converter only embrace a fraction of the armature surface and the flux under the poles is concentrated in a fairly uniform manner. Outside the air-gaps under the polar faces there are practically no lines of force, and the induced electromotive force in a turn is almost zero as soon as it leaves the extremity of a polar face. In general therefore the sine curve hypothesis is only roughly approximate. It is found by experiment that the ratio of V to E in polyphase converters varies with the ratio of the polar arc to the polar step.

The results of a test by de Marchena on a 150-kilowatt three phase rotary converter are shown in Fig. 213. The air-gap of this machine which has four poles is 0·55 cm. and the ratio of the polar arc to the polar step is 0·75. The figures obtained show that, within the limits of experimental error, the ratio of V to E is constant and equal to 0·605, whether the excitation is being increased or being diminished. This ratio is only about one per cent. less than that obtained on the sine curve hypothesis. A similar test on a 250-kilowatt four pole three phase converter gave the ratio as 0·615, which is almost in exact agreement with the theoretical number. The air-gap of this machine is 0·35 cm. and the ratio of the polar arc to the polar step 0·76.

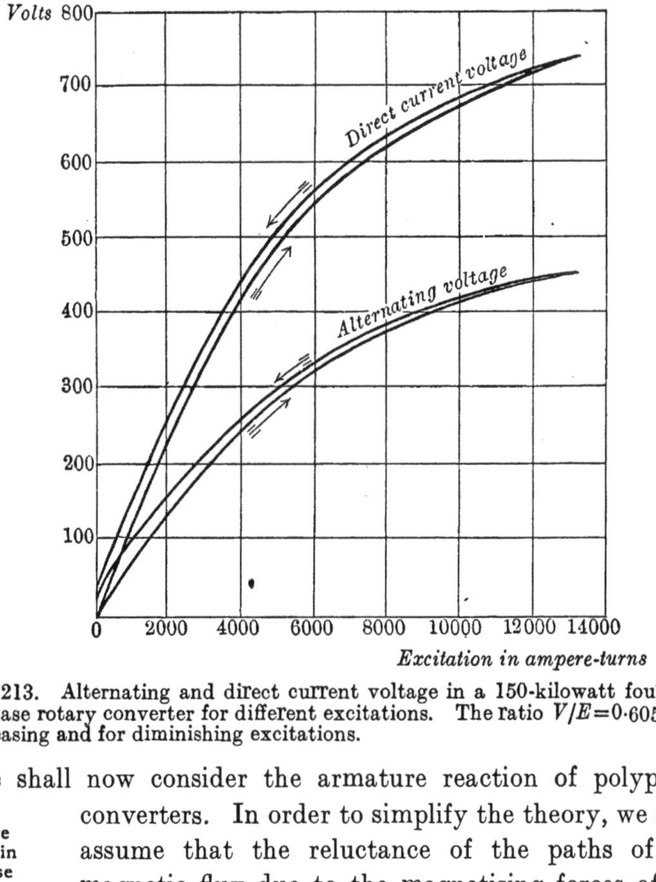

Fig. 213. Alternating and direct current voltage in a 150-kilowatt four pole three phase rotary converter for different excitations. The ratio $V/E = 0\cdot605$ both for increasing and for diminishing excitations.

Armature reaction in polyphase converters.

We shall now consider the armature reaction of polyphase converters. In order to simplify the theory, we shall assume that the reluctance of the paths of the magnetic flux due to the magnetising forces of the armature currents is constant, and that these paths are symmetrically situated relatively to one another. We shall also suppose that the number of windings is infinite, and that the currents in them follow the harmonic law. If the number of poles be $2p$, a magnetic field will be produced gliding backwards in the air-gap with an angular velocity ω/p. The field produced therefore will be fixed in space and, as in the analogous case of the bipolar single phase converter, we find that

$$\mathscr{F}_t = k\{(qI/4)\cos\psi - C\},$$

and

$$\mathscr{F} = k(qI/4)\sin\psi,$$

where $\mathscr{F}_t$ is the transverse and $\mathscr{F}$ the direct reactive magnetic force acting on a pole, and I is the maximum value of the alternating current in one of the supply mains. If the efficiency of the converter be η, we have, by (12),

$$(qI/4)\cos\psi = C/\eta,$$

and thus

$$\mathscr{F}_t = k\{(1-\eta)/\eta\}\,C.$$

Hence if the efficiency be 100 per cent. $\mathscr{F}_t$ is zero. If in addition the power factor $\cos\psi$ be unity, $\mathscr{F}$ will also be zero. A perfect rotary converter, therefore, acts like a synchronous motor in which the transverse magnetisation of the field is zero. If the current be lagging, the field is strengthened by the armature reaction $\mathscr{F}$, and if it be leading, the armature reaction weakens the field. If the efficiency of the machine were 90 per cent., the transverse magnetisation would only be one-ninth of the value it would have if there were only the direct current C in the armature. If the efficiency were 80 per cent., it would be a quarter of that which the direct current acting alone would produce.

The alternating component of the direct voltage.

In practice, instead of having an infinite number of windings symmetrically arranged, we have a finite number arranged in coils. Thus the magnitude of the field produced by the armature currents is not independent of its angular position in space, and therefore a perfect rotary field will not be produced. In this case the greater the number of phases, the higher will be the frequency, and the smaller the amplitudes of the alternating current components due to this effect.

In a q phase system, for instance, there will be q slip rings and q windings connected with them. We may suppose that the q windings are exactly similar to each other, and hence, after an interval T/q, the potential differences between the direct current brushes will be exactly the same as at the beginning of the interval. The period of the alternating component of the voltage on the direct current side, therefore, will be the qth part of the period of the applied potential difference. These pulsations may cause loss due to the eddy currents they produce in the field magnets. Thus it is advisable to have a large number of commutator bars and to make the field magnets of laminated iron stampings.

In practice the number of slip rings employed is 3, 4, 6 or 12. Single phase converters are rarely used in practical work as, on heavy loads, some of the coils usually heat excessively, and there is sparking at the brushes due to the variations in the flux in the armature caused by the armature reaction. When 6 or 12 slip rings are used, the amplitudes of the pulsations of the flux are inappreciable. It is to be noticed that a six phase converter can be operated from three phase mains (see p. 471). If the slip rings be denoted by 1, 2, 3, 4, 5 and 6, then 1, 3 and 5 will form a three phase system and so also will 2, 4 and 6. If therefore we have a three phase transformer with two distinct secondaries mesh wound, and having terminals a, b, c and a', b', c' respectively, we can run the converter by connecting a, b and c with 1, 3 and 5, and a', b' and c' with 4, 6 and 2.

Finding the armature reaction from the characteristic curve.

The general shape of the characteristic curve of a rotary converter is shown in Fig. 214. The armature is driven at constant speed and the excitation is varied, simultaneous readings being taken of the alternating voltage V between the slip rings and of the exciting direct current. In the figure PN is the value of V corresponding to the exciting ampere-turns ON.

Let V' denote the value of V when the machine is loaded, and let $P'N'$ (Fig. 214) equal V'. Then ON' is the effective value of the exciting ampere-turns and thus

$$k'I \sin \psi = NN',$$

where k' is a constant and $\cos \psi$ is the power factor. If therefore we know the characteristic curve and the values of I, ψ, V and V', we can find k'. We are thus able to predetermine the value of V, and therefore also the value of E, for any given current and power factor. The constant k' therefore can easily be determined experimentally. Its value depends on the breadth of the poles, the number of phases, the number and width of the slots, etc., and so it would be very difficult to find it by pure calculation.

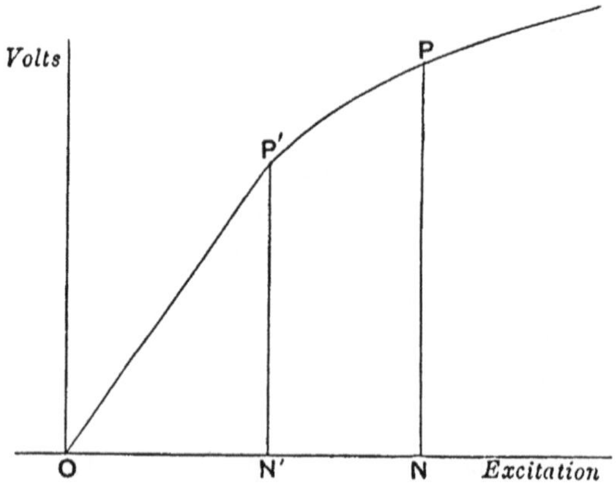

Fig. 214. The open circuit characteristic curve of a rotary converter. The armature is driven at a constant speed, and *PN* gives the voltage when the excitation is *ON*.

Compounding a rotary converter.

The field magnets of a rotary converter may have their windings connected in shunt between the direct current brushes, or they may have a compound winding. If the effective value of the potential difference applied to the slip rings is absolutely constant, that is, if the resistance and self-inductance of the mains supplying current to the rotary converter are zero, then the regulation is perfect. We can in this case annul or even reverse the excitation without sensibly altering the voltage on the direct current side.

To alter the voltage on the direct current side we must alter the voltage applied to the slip rings. This may be done by means of a variable choking coil inserted in the alternating current circuit, or by a transformer or booster which has a variable ratio of transformation. Let us suppose that the excitation is adjusted so that we have a power factor of unity on no load, then, when we alter the applied potential difference, the power factor, as a rule, will no longer be unity and the excitation will have to be altered. It is important that this be done automatically. We shall therefore consider what the resistance per turn of the shunt circuit should be, and also the number of turns of the series windings, in order that the power factor may be unity when the direct current circuit is open, and also that it should be unity when the direct current voltage is E and the current C.

Let r be the resistance per turn of the shunt winding, so that E/r gives the ampere-turns of excitation due to the current in the shunt winding when E is the direct voltage. Let also n_1 be the number of turns in the series winding. When the direct current is C the excitation of the machine is

$$n_1 C + E/r + k' \sqrt{2} A \sin \psi,$$

where A is the effective value of the alternating current, and k' is a constant. Now by hypothesis ψ is to be zero, and by (7), V equals $(E/\sqrt{2}) \sin (\pi/q)$, and thus the excitation equals

$$n_1 C + \sqrt{2} V / \{r \sin (\pi/q)\}.$$

From the characteristic curve (Fig. 214) we can find the excitation X which produces the voltage V, and hence

$$n_1 C + \sqrt{2} V / \{r \sin (\pi/q)\} = X \quad \text{..............}(a).$$

If X_0 be the excitation which produces the no-load voltage V_0, we have

$$\sqrt{2} V_0 / \{r \sin (\pi/q)\} = X_0 \quad \text{.................}(b).$$

Equation (b) determines the value of r, and from (a) we can then find n_1.

If we put an inductive coil, or, as it is generally called, a 'reactance coil,' in the circuit of the leads connecting a single phase rotary converter with the supply mains, it is found that we can both raise and lower the potential difference between the slip rings by altering the excitation of the converter. When the excitation of the converter is greater than that required for the maximum value of the power factor, the armature reaction tends to demagnetise the field, and so the phase of the armature current is in advance of the phase of the potential difference between the slip rings. The reactance coil and the armature, therefore, act like an inductive coil and a condenser in series, and partial resonance (Vol. I, p. 137) occurs. In this case the voltage V across the slip rings can be greater than the potential difference between the supply mains. Since the ratio of V to E is nearly constant, it will be seen that we can vary V, and therefore also E, through an appreciable range by varying the excitation of the machine. With polyphase converters when an inductive coil is placed in each lead connecting a main with a slip ring, a similar regulating effect on E is produced by altering the excitation.

Experimental results obtained by de Marchena are shown graphically in Figs. 215 to 218. The machines experimented on

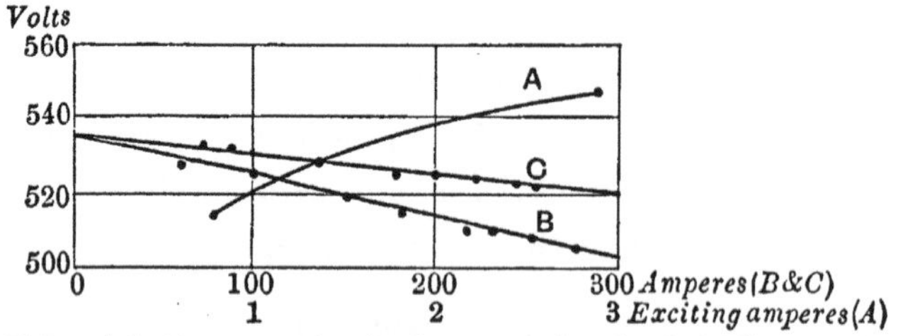

Fig. 215. *A* is the open circuit characteristic of a 150-kilowatt three phase converter. *B* shows the variation of the voltage with the load, shunt excitation only being used. *C* shows the variation of the voltage with the load with compound excitation.

had outputs of 150 and 300 kilowatts respectively. In Fig. 215, the curve A shows how the voltage on the direct current side on open circuit varies as the excitation of the 150-kilowatt converter is altered. In this test the only reactance in the circuit was that of the transformer which supplied the converter. The curve B shows the variation of E as the load is increased when a shunt winding only is employed. The curve C shows the variation of E with the load when the compound winding is used.

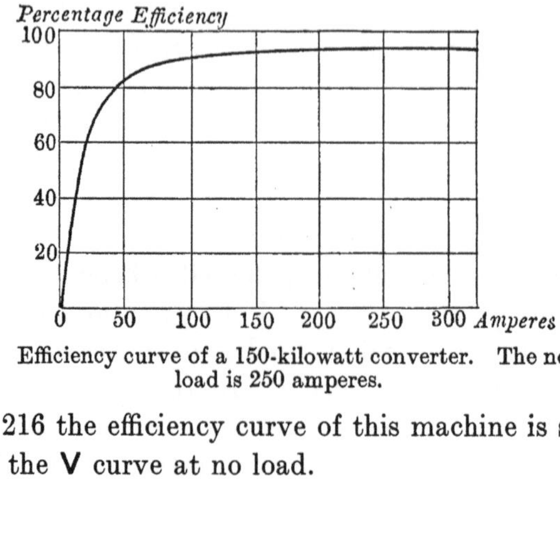

Fig. 216. Efficiency curve of a 150-kilowatt converter. The normal full load is 250 amperes.

In Fig. 216 the efficiency curve of this machine is shown, and in Fig. 217 the V curve at no load.

In a test of a 300-kilowatt machine, the results of which are shown in Fig. 218, a choking coil was put in the main circuit. The voltage drop across the terminals of this choking coil at full

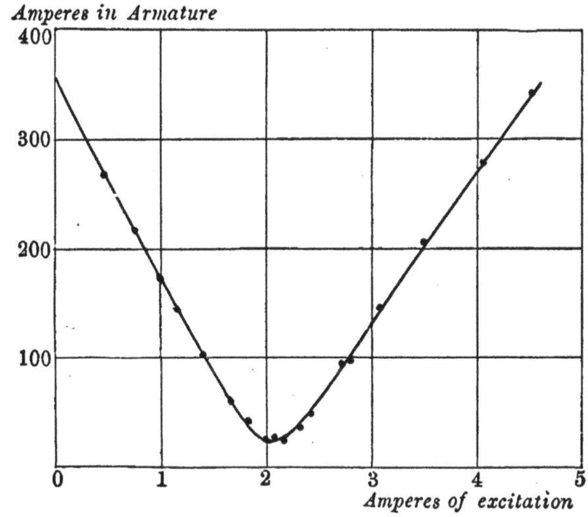

Fig. 217. V curve at no load of a 150-kilowatt converter.

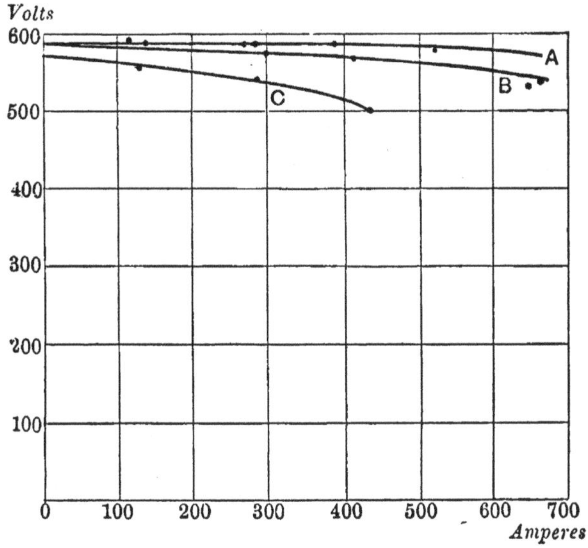

Fig. 218. Load characteristics of a 300-kilowatt rotary converter having a choking coil in the main circuit. *A* is the characteristic with compound winding. *B* is the characteristic with shunt winding alone. *C* is the characteristic with compound winding when the series windings are reversed.

load was ten per cent. of the total applied voltage. Curve *A* shows how the voltage on the direct current side varied with the load when the series winding was connected so that the ampere-turns due to it increased the total excitation. Curve *C* shows how the direct voltage varied with the load when the current passed through the series coil in the reverse direction, so that the excitation diminished as the load increased. Curve *B* shows the result of a test with the series coils short circuited.

The voltage could be varied from 512 to 584 volts at full load, and from 555 to 605 at no load, by means of the regulating inductive coils in the mains. By altering the regulating resistance in the shunt coil, when the series coil was short circuited, the voltage could be varied from 512 to 595 at full load, and from 544 to 640 at no load. The curve *A* shows that this machine can be compounded so as to give almost a straight line for the curve of the voltage on the direct current side at all loads.

Starting converters.

A converter can be started from either the direct or the alternating current side. If we start it from the direct current side, it is started in the same way as a direct current motor. Some synchronising device is employed to indicate when it has attained the proper speed and also the proper moment to close the main switch. When it is started from the alternating current side, polyphase currents are taken from the mains by means of a pressure-reducing transformer, and pass into the armature by the slip rings. A rotating magnetic field is produced, and the torque due to the eddy currents and the hysteresis in the iron is sufficient to start the armature rotating. The voltage between the brushes on the direct current side now excites the field, and the armature finally falls into step with the rotating magnetic field, the machine acting like a synchronous motor.

During the start the field magnet windings act like the secondary circuit of a transformer, and very high voltages may be generated which may spark across and break down the insulation of the field magnet windings. For this reason the field magnet windings are sometimes divided into sections which, by means of a 'break up' switch, are on open circuit during the start and are closed when the speed approaches synchronism. When the machine is running the central zero voltmeter may show that the polarity of the field magnets has been reversed. In this case the reversing switch in the shunt windings is operated and the field demagnetised. The machine now rotates as an induction motor. When the armature has slipped back by the extent of one pole the switch is closed and the machine runs normally.

Parallel running. Rotary converters as a rule work well in parallel; the rotating parts are lighter than in motor generators, and thus they respond more quickly to the regulating forces. When the loads are very variable, as in traction work, the converters are generally compound wound. In this case care must be taken to connect the series windings by equalising cables in exactly the same way as ordinary compound direct current machines are connected.

Sometimes phase swinging is set up by a sudden variation in the load. A variation of the current in the series windings alters the ampere-turns of excitation, but the flux is not instantaneously modified. Hence some machines may respond more quickly than others, and thus the voltages of the various machines may differ and current oscillations be started. To prevent this effect a 'shaped copper grid' or rectangle of thick copper is fitted round the pole piece, two or more copper bars being passed through the pole face and riveted into the sides of the grid so as to ensure good electrical contact. In some cases even a complete squirrel-cage is fitted to the field magnets. These copper gratings are called 'damper windings.' When oscillations of the magnetic and armature fields are set up, large eddy currents are induced in the damper windings which bring strong mechanical forces into play tending to restore steady running. The use of damping coils, however, slightly lowers the efficiency of converters.

Split pole converter. The direct current voltage of a rotary converter can be altered by varying the position of the brushes on the commutator. A similar effect can be produced by varying the wave shape of the magnetic flux, the brushes remaining in the same position. This principle of varying the voltage ratio is utilised in machines called 'split pole converters.' These machines (Fig. 219) have small auxiliary poles *A* placed near the main poles *N* and *S*. The ampere-turns acting on both sets of poles

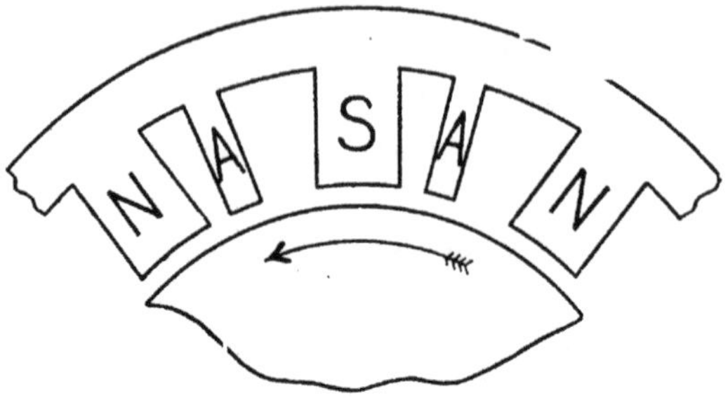

Fig. 219. Split pole converter.

can be varied and hence the distribution of the magnetic flux can be altered. For supply-circuits the frequency of which does not exceed 25, these machines are often employed. Owing to mechanical difficulties in construction and sparking troubles they are rarely used on circuits of higher frequency.

Inverted rotary converters.

When a rotary converter is used to convert direct current into alternating it is called an inverted rotary. The formulae found earlier in this chapter still apply if we write $1/\eta$ for η in them, and notice that the direct current side now acts as the motor and the alternating current side as the generator. Since the speed of a direct current motor depends on the excitation of the field magnets, it will vary with the magnitude and the power factor of the load. If the load be inductive the armature reaction will weaken the field (Fig. 21, p. 36). Hence the speed of the rotor will increase and so also will the frequency. If $I \sin \psi$ be large the speed may be dangerously high. For this reason a separate exciter is sometimes fitted on the shaft of the rotor, so that if the speed quicken the increased current in the field magnet windings may neutralise to a certain extent the armature reaction of the alternating currents. A further safety device is sometimes employed. A centrifugal tripping mechanism is mounted at the end of the shaft. When the speed exceeds a definite value centrifugal force causes a contact switch to close, and this actuates circuit breakers cutting off the supply.

Regulators. Synchronous booster.

We have seen that the regulation of the voltage on the direct current side of a rotary converter may be effected by inserting inductive coils in the mains connecting the transformer to the rotary. These coils are costly, and the efficiency of the rotary is appreciably lowered by their use. Small rotaries are sometimes supplied with transformers the voltage of which can be varied step by step by turning a handle which moves a contact piece over studs connected with different windings of the transformer. The sparking at the contacts is a drawback to this type of regulator.

Where the direct current voltage has to be varied over a wide range, the synchronous booster type of regulator is now generally employed. The synchronous booster consists simply of an alternating current generator the armature of which is mounted on the same shaft, and the windings of the armature are in series with the windings of the rotary converter, the number of poles being the same. By varying the field of the booster the voltage across the commutator can be varied. It will be seen that varying the magnetic field of the booster also alters the power factor of the rotary. The advantages of this method are ease of manipulation and a wide range of regulation, but like the use of inductive coils it increases the cost and lowers the efficiency of the machine.

Data of a 200-kilowatt rotary converter.

The following data of a 200-kilowatt three phase converter, made by the General Electric Co. of America and installed in the Brooklyn electric station, are instructive:

Number of revolutions per minute	375
Number of poles	8
Frequency of the alternating current $(375 \times 8)/(60 \times 2)$ *i.e.*	25
Alternating current in amperes at full load	1500
Applied alternating voltage	82·8
Direct voltage	125
Diameter of the armature in cms.	122
Breadth of the armature in cms.	17·8
Air-gap in centimetres	0·635
Number of slots	240
Number of conductors per slot	2

Number of commutator bars	240
Diameter of the commutator in cms.	92
Number of rows of brushes	8
Number of carbon brushes per row	9
Surface of contact of each brush in sq. cms.	8
Flux per pole in C.G.S. units	$4{\cdot}38 \times 10^6$
Flux density in the field magnets at full load	12000
Flux density in the air-gap at full load	8000
Flux density in the teeth between the slots at full load	21000
Flux density in the armature at full load	9500

(polar arc)/(polar step) = 0·637.

(alt. voltage)/(direct voltage) = V/E = 0·633.

The following table gives the losses and the maximum rise of temperature of this machine when running at full load as a converter and as a direct current dynamo:

	Converter	Dynamo
C^2R losses in watts	3500	6005
Total losses in watts	6500	9130
Temp. rise of the armature	27° C.	47° C.
Temp. rise of the commutator	36° C.	52° C.

If we assume that $\eta = 1{\cdot}00$ and $\cos\psi = 1$, and that the currents follow the harmonic law, the ratio of the C^2R losses when working as a converter to the C^2R losses when working as a direct current dynamo would be 0·564. The ratio found from the above test is 0·583. The curve of the applied potential difference was more peaky than a sine wave, and the heating losses in the converter are therefore greater than they would be if the wave were a pure sine curve. As a rule, the flatter the wave the less will be the heating of the armature windings of the converter.

Double current generators.

When the rotor of a converter is driven mechanically, we have an alternating electromotive force developed between the slip rings, and a constant potential difference between the brushes. It can therefore supply both direct and alternating current at the same time. A machine constructed on this principle is called a 'double current generator.' These machines are useful when a power station has to supply a constant load in its neighbourhood and a variable load at some distance away. It combines the economy of a direct current supply on a three wire system and the economy with which power can be transmitted to considerable distances by high voltage electric current. We shall only consider the case of polyphase double current generators, as single phase arrangements are rarely used in practice.

The theory of such machines is almost identical with that of rotary converters. The most noticeable difference is that in the double current generator, the alternating current side is acting like a generator and not like a motor. The alternating current

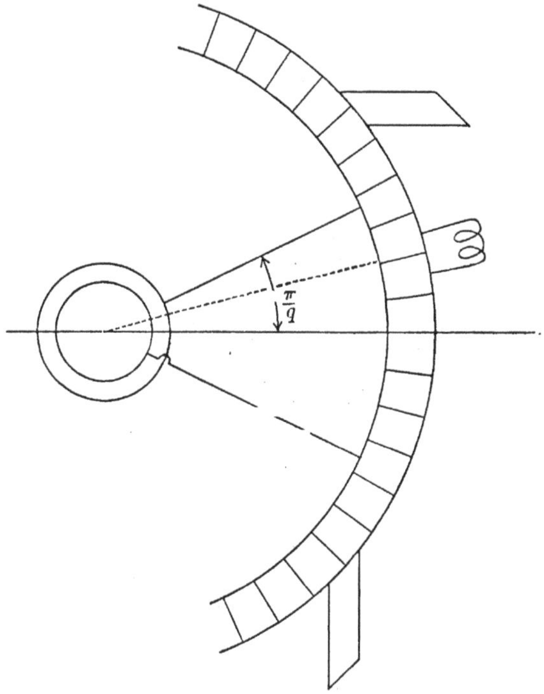

Fig. 220. Part of the commutator and two of the slip rings of a four pole three phase double current generator.

in the armature windings therefore will generally be flowing in the same direction as the direct current, and hence the heating of the armature and the transverse magnetisation of the field will be greater than for a rotary converter.

Part of the commutator, two of the slip rings, two brushes and one coil of a three phase, four pole, double current generator are shown in Fig. 220. If we make the assumption that the shape of the alternating current wave in a winding is a sine curve, and that the direct current wave is a rectangle (Fig. 221), we get, using the same notation as in the corresponding problem of the polyphase rotary (p. 469),

$$W_1 = r\,[I^2/\{8\sin^2(\pi/q)\} + C^2/4 + CI/\{2\sqrt{2}\sin(\pi/q)\}.(2\sqrt{2}/\pi).\cos(\pi/q - \pi/2m - \psi)],$$

$$W_2 = r\,[I^2/\{8\sin^2(\pi/q)\} + C^2/4 + CI/\{2\sqrt{2}\sin(\pi/q)\}.(2\sqrt{2}/\pi).\cos(\pi/q - 3\pi/2m - \psi)].$$

....

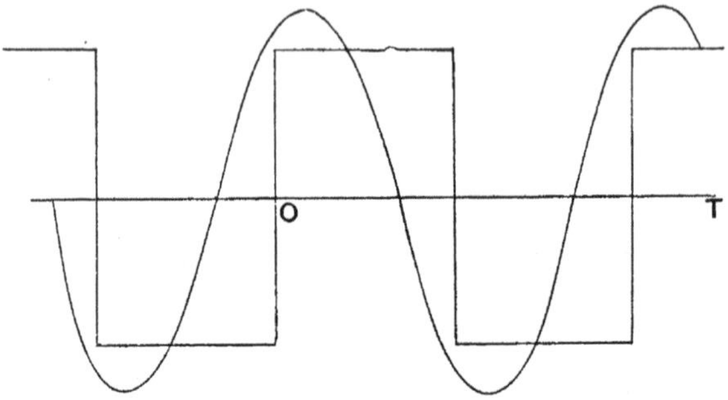

Fig. 221. Current waves in the armature of a double current generator.

Let us first consider the case when ψ is zero. In this case the smaller π/q, the greater will be the heating in the first coil. The coils in connection with the slip rings are the least heated, and those midway between them are the most heated. In the particular case, however, when ψ equals $\pi/q - \pi/2m$, the heating of the coils connected with the slip rings is a maximum. We also see that for a given number of turns, the power factor being unity, the greater the number of phases the greater will be the heating of the coils. The greater the number of phases, however, the more evenly will the heat developed be distributed over the coils.

If W_h denote the total heating of the armature, we have

$$W_h = q\{W_1 + W_2 + \dots + W_n\} = 2mr\,[C^2/4 + (CI/\pi)\cos\psi/\{n\sin(\pi/2m)\} + I^2/\{8\sin^2(\pi/q)\}].$$

In this formula I is the maximum value of the current in a main connected with a slip ring. If A be the effective value ($I/\sqrt{2}$) of this current, and if m be large, so that we can write $n\pi/2m$ or π/q for $n \sin(\pi/2m)$, we find that

$$W_h = R(C/2)^2 [1 + (4\sqrt{2}/\pi^2)\, q \cos\psi . (A/C) + \{1/\sin^2(\pi/q)\}(A/C)^2] \quad \ldots\ldots\ldots(15),$$

where R is the resistance of the whole armature winding.

We see at once from this formula that if A and C remain constant, the heating of the armature is a maximum when the power factor is unity and is a minimum when the power factor is zero.

Let us suppose that the power generated on the alternating current side is p times the power generated on the direct current side, so that $qVA_1 \cos\psi = pEC$. By (10), $2A_1 \sin(\pi/q) = A$, and by (7), $V = (E/\sqrt{2}) \sin(\pi/q)$. Hence $A/C = 2p\sqrt{2}/(q \cos\psi)$, and substituting this value in (15) we get

$$W_h = R(C/2)^2 [1 + (16/\pi^2)\, p + 8p^2/\{q^2 \sin^2(\pi/q) \cos^2\psi\}].$$

If W be the total output of the machine, we have

$$C = W/\{E(1+p)\},$$

and thus

$$W_h = \frac{RW^2}{4E^2} \cdot \frac{1 + (16/\pi^2)\, p + 8p^2/\{q^2 \sin^2(\pi/q) \cos^2\psi\}}{1 + 2p + p^2} \quad \ldots(16).$$

In the particular case of a three phase machine, we have

$$W_h = \frac{RW^2}{4E^2} \cdot \frac{1 + (16/\pi^2)\, p + 32p^2/(27 \cos^2\psi)}{1 + 2p + p^2}.$$

It is easy to see that for a given output W, at a given power factor $\cos\psi$, the heating of the armature has a minimum value when

$$p = \frac{1 - 8/\pi^2}{32/27 - 8/\pi^2 + (32/27)\tan^2\psi}$$

$$= \frac{0{\cdot}189}{0{\cdot}375 + 1{\cdot}185 \tan^2\psi}, \text{ approximately.}$$

The expression for the minimum value of the power expended in heating the armature equals

$$\frac{RW^2}{4E^2} [1 - (1 - 8/\pi^2)^2/\{1 - 16/\pi^2 + 32/(27 \cos^2\psi)\}],$$

which is approximately equal to

$$\frac{RW^2}{4E^2} [1 - 0{\cdot}0359/(0{\cdot}564 + 1{\cdot}185 \tan^2\psi)].$$

Hence this minimum value is not much smaller than when the machine is supplying W watts on the direct current side.

The transverse armature reaction in a double current generator is much greater than in a rotary converter. The formulae (5) for the transverse and direct components of the magnetomotive forces due to the currents in the armature now become

$$\left.\begin{aligned} \mathscr{F}_t &= -k\{(I/2)\cos\psi + C\} \\ \text{and} \qquad \mathscr{F} &= -k\,(I/2)\sin\psi \end{aligned}\right\} \quad \ldots\ldots\ldots\ldots\ldots(17).$$

Thus considerable distortion of the field will be produced, and the sparking at the brushes will be worse than if the machine were acting as a direct current generator having an output EC. The brushes also will require to be adjusted if the load alters, unless some compensating device be employed to neutralise partially the distorting effect of the armature currents on the field. Since the machine is now acting as a generator, the armature reaction due to a lagging current will tend to weaken the field and a leading current will strengthen it.

It is necessary to construct the generator so that it acts satisfactorily both as a direct current dynamo and as an alternator. We must therefore construct it as a high speed multipolar direct current dynamo and a low frequency slow speed alternator. The mechanical difficulties in the way of a good design are therefore considerable. The field magnets may be separately excited, otherwise a shunt or a compound winding may be used. Since the ratio of the direct to the alternating voltage cannot be altered, it is only necessary to regulate the potential difference on one side of the machine. This is generally done from the direct current side, as the alternating potential difference can easily be regulated at either end of the transmission line.

Double current generators are obviously of great use for distributing power when the direct and alternating current loads do not overlap.

The La Cour motor converter.

The La Cour motor converter (Fig. 222) consists of an ordinary induction motor I.M. rigidly coupled and also electrically connected with a direct current machine D.C.M. The three phases of the rotor are connected with a starting resistance R and also with the armature of the direct current dynamo. If the machines have the same number of poles, then at half-speed the direct current machine will be running as a synchronous motor. The rotating field due to the armature currents rotates in the opposite direction to the shaft, and hence in addition to the ordinary rotary converter action the machine acts like a dynamo. It will be seen that the induction motor exerts mechanical power to drive the shaft. It also transmits electrical energy to the armature of the D.C.M. Hence it acts partly as a motor and partly as a transformer. We also see that D.C.M. acts partly as a rotary converter and partly as a dynamo.

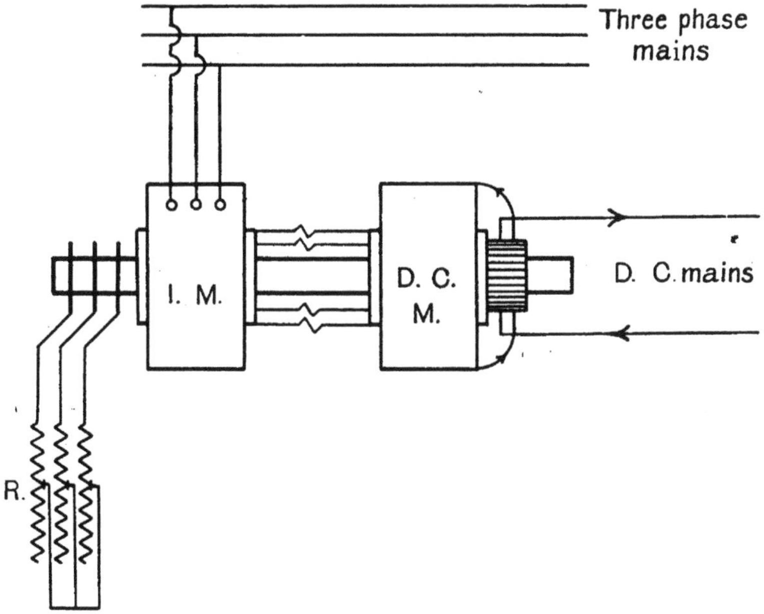

Fig. 222. La Cour motor converter.

Let us now suppose that the induction motor has $2p$ poles and that the direct current machine has $2q$ poles. If n_0 be the synchronous speed in revolutions per second of the rotor, we have $pn_0 = f$. If n be the actual speed

$$qn = f' = p(n_0 - n)$$

and thus

$$(p + q)n = pn_0 = f.$$

Hence

$$n = \frac{p}{p+q} n_0.$$

The induction machine therefore converts $p/(p + q)$ of the power supplied to the rotor into mechanical power and $q/(p + q)$ of the energy supplied into electrical power.

The motor converter is started in the same way as an induction motor. As the motor speeds up, the magnetic field of the direct current machine gradually increases in strength. The pointer of a voltmeter placed across the starting resistance fluctuates violently when synchronous speed is approached. When the starting resistances are short circuited the voltmeter reading is very small but it is steady.

The efficiency of this motor converter, although less than that of a rotary converter, is higher than that of a motor generator. It is self-starting and there is little danger of a reversal of polarity. The machine is reversible, the regulation is good and it takes up little floor space.

Three wire direct current system.

When the direct current from the rotary converter is required for lighting purposes it is usual to employ a three wire system, the difference of potential between the positive and the middle main being equal to the difference of potential between the middle and the negative main. As far as possible it is arranged that the load between the positive and the middle should be in general equal to the load between the middle and the negative. The middle main therefore has only to carry the 'out of balance' current. It is connected with the star point of the secondaries of the transformers supplying the alternating current side of the converter. The 'out of balance' direct current therefore flows through the windings of the transformer back into the armature windings. It has to be arranged so that the resultant ampere-turns of direct current acting on the limbs of the transformer equals zero; otherwise the iron would be strongly magnetised in one direction and this would injuriously affect the working of the transformer. If the 'out of balance' current does not exceed 25 per cent. of the full load in either of the outer mains, this method can be used with very satisfactory results.

When double current generators are used, the middle wire of the direct current system can be connected with the armature by means of three choking coils joined in star and with their other ends connected with alternating current brushes pressing on the slip rings.

In this case let (r, l) be the constants of each choking coil, and let I, i_1, i_2 and i_3 be the current between the positive and the middle main and the currents in the three choking coils respectively. We shall suppose that R is the resistance of the load between the positive and the middle main and that there is no load between the middle and the negative main. Let $e' + E$ and $e' - E$ be the potentials of the outer mains and let

$$e' + E \cos \omega t, \quad e' + E \cos (\omega t + 2\pi/3), \quad \text{and} \quad e' + E \cos (\omega t + 4\pi/3)$$

be the potentials of the slip rings.

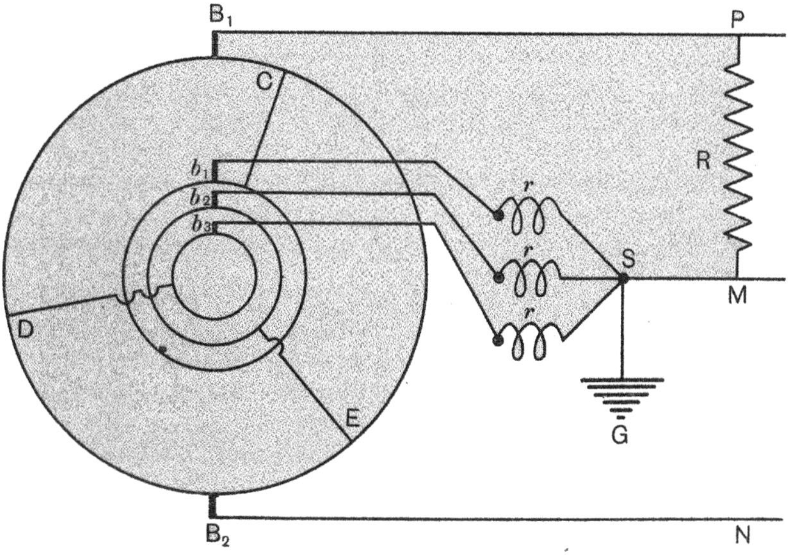

Fig. 223. Three wire direct current generator.

Then we have, assuming that the star is earthed and that the armature resistance is negligible,

$$e' + E = RI,$$

$$e' + E \cos \omega t = ri_1 + l(\partial i_1/\partial t),$$

$$e' + E \cos (\omega t + 2\pi/3) = ri_2 + l(\partial i_2/\partial t),$$

and

$$e' + E \cos (\omega t + 4\pi/3) = ri_3 + l(\partial i_3/\partial t).$$

Hence, adding the last three equations together and noticing that $i_1 + i_2 + i_3 + I = 0$, we get

$$3e' = - rI - l\,(\partial I/\partial t),$$

and hence by the first equation

$$E = (R + r/3)\,I + (l/3)\,(\partial I/\partial t),$$

and thus when the steady state is attained

$$E = (R + r/3)\,I$$

and

$$e' = -\frac{r/3}{R + r/3}\,E.$$

Thus if $r/3$ is small compared with R the regulation is good.

REFERENCES

R. M. FRIESE, 'Die Vorgänge im Gleichstromanker bei Entnahme von Wechsel- und Mehrphasenströmen,' *Elektrotechnische Zeitschrift*, Vol. 15, p. 89, 1894.

C. P. STEINMETZ, 'Der Rotierende Umformer,' *Elektrotechnische Zeitschrift*, Vol. 19, p. 138, 1898.

M. P. JANET, 'Sur les Commutatrices,' *Comptes Rendus de l'Académie des Sciences*, Vol. 127, p. 351, 1898.

DE MARCHENA, 'Emploi des Commutatrices pour la Transformation des Courants Polyphasés en Courants Continus,' *Bulletin de la Société Internationale des Électriciens*, Vol. 1, p. 205, 1901.

A. BLONDEL, 'Théorie Graphique de la Régulation des Convertisseurs Rotatifs,' *l'Éclairage Électrique*, Vol. 29, p. 206, 1901.

P. M. VERHOECKX, 'Sur la Théorie des Commutatrices,' *l'Éclairage Électrique*, Vol. 35, p. 241, 1903.

For a description of the theory and construction of permutators—a special type of converter in which the commutator is stationary and the brushes rotate at synchronous speed—see R. ROUGÉ, *La Revue Électrique*, Vol. 3, p. 33, 1905, and C. V. DRYSDALE, *Electrician*, Vol. 56, p. 305, 1905.

The British Westinghouse Electric and Manufacturing Co., Ltd., 'Rotary Converters,' Special Publication, 7390/2.

H. S. HALLO, 'The Theory and Application of Motor Converters,' *Journ. of the Inst. of El. Engin.*, Vol. 43, p. 374, 1909.

CHAPTER XVIII

The transmission of power. Direct current distribution. Single phase transmission. Graphical solution. The constants of the line. Three phase transmission. Graphical solution. Comparison of single phase with three phase. Comparison of star-connected systems. Comparison of mesh-connected systems. Maximum power transmitted by a polyphase system. Distributed capacity. Circuit in which there is no distortion. Positive and negative waves. The reflection and transmission of waves. The voltage drop on short lines. Method of getting a neutral main. References.

The transmission of power.

WE shall now consider some of the problems which arise in connection with the underground mains and overhead wires used for transmission of electrical power over considerable distances. We shall first consider the simplified problems obtained by neglecting the inductance and the capacity of the lines, and then briefly indicate how approximate solutions may be obtained in other cases. In order to simplify the problem as much as possible, let us consider first of all the efficiency of a direct current two wire system.

Direct current distribution.

Let P and P' be the terminals of the dynamo at the power station (Fig. 224), and let D and D' be the terminals of the motor at the distributing station. Let E and E_1 be the potential differences between the mains at the power and distributing stations respectively. Let also C be the current in each main and R_1 the resistance of each main. By Ohm's law we have

$$E - E_1 = 2R_1C \qquad \text{.....................(1)},$$

and therefore

$$EC = W + 2R_1C^2 \qquad \text{.....................(2)},$$

where W is the power received at the distributing station. Hence we see from (2) that, for a given power W transmitted and for given values of E and R_1, there are two possible values of C. Since from (1) E_1 equals $E - 2R_1C$, there are also two possible values of E_1. Solving the quadratic equation (2) for C, we find that

$$C = (1/4R_1)\{E \pm (E^2 - 8R_1W)^{\frac{1}{2}}\} \quad \text{...........(3)}.$$

Now the electrical efficiency η of the transmission is given by

$$\eta = E_1C/EC = 1 - 2R_1C/E \text{.................(4)},$$

and therefore, when R_1 and E are constant, the smaller the value of C the higher is the efficiency.

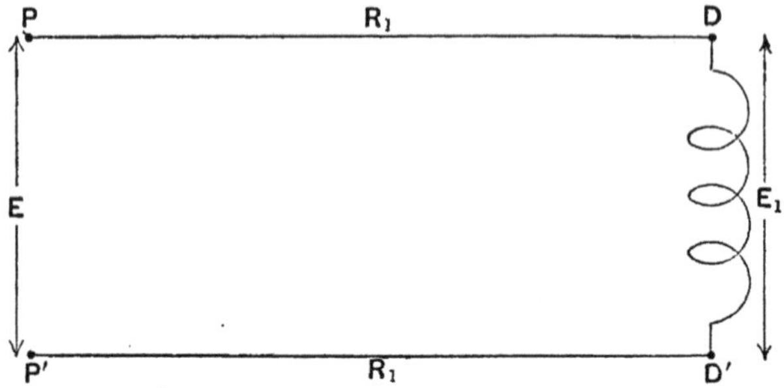

Fig. 224. Power transmission line.

From equation (3), since the quantity under the radical sign must be real, we see that the maximum power that can be transmitted is equal to $E^2/8R_1$, and in this case the current equals $E/4R_1$ and the efficiency is 50 per cent.

In general, if C_1 and C_2 are the two possible values of the current for a given power W transmitted, and if η_1 and η_2 are the corresponding efficiencies, we have, by (3),

$$C_1 + C_2 = (1/2)(E/R_1),$$

and hence by (4) $\qquad \eta_1 + \eta_2 = 1$ ....(5).

If η_1 and η_2 are not equal, one of them must be greater than 0·5. Hence we conclude that the efficiency of the electrical transmission of power over the lines in a direct current transmission plant need never be less than fifty per cent.

Single phase transmission. Let L be the self-inductance of each main and let M be the mutual inductance between the mains. Let also v and v_1 be the potentials of the terminals P and D in Fig. 224, and let v' and v_1' be the potentials of P' and D'.

Neglecting the effects of electrostatic capacity, our equations are

$$v - v_1 = R_1 i + L\frac{\partial i}{\partial t} - M\frac{\partial i}{\partial t},$$

and
$$-v' + v_1' = R_1 i + L\frac{\partial i}{\partial t} - M\frac{\partial i}{\partial t}.$$

Hence
$$e - e_1 = 2R_1 i + 2(L - M)\frac{\partial i}{\partial t},$$

where e and e_1 are the instantaneous values of the P.D. at the generating and distributing ends of the line respectively. We thus get

$$ei = e_1 i + 2R_1 i^2 + (L - M)\frac{\partial i^2}{\partial t},$$

and taking mean values we have

$$VA\cos\psi = W + 2R_1 A^2 \quad \text{..................(6)},$$

where $\cos\psi$ is the power factor of the load at the generating station.

Solving the quadratic equation (6) for A, we find that

$$A = (1/4R_1)\{V\cos\psi \pm (V^2\cos^2\psi - 8R_1 W)^{\frac{1}{2}}\} \quad \text{......(7)},$$

and if $\cos\psi_1$ be the power factor at the distributing station,

$$\eta = V_1 A\cos\psi_1/(VA\cos\psi)$$
$$= 1 - 2R_1 A/(V\cos\psi) \quad \text{....................(8)},$$

by (6), noticing that $W = V_1 A_1 \cos\psi_1$.

From (7) we see that the maximum possible value of the power transmitted is $V^2\cos^2\psi/(8R_1)$, and hence it is essential to make $\cos\psi$ as large as possible. If η_1 and η_2 be the efficiencies of the transmission for the two possible currents that can transmit a power W for a given value of $\cos\psi$, we have by (7)

$$A_1 + A_2 = (1/2)(V/R_1)\cos\psi,$$

and by (8)
$$\eta_1 + \eta_2 = 1.$$

We see therefore that, as in the case of direct current transmission, the efficiency need never be less than fifty per cent. But for small values of $\cos\psi$ the power that can be transmitted is very small.

Equation (6) may also be written in the form

$$V \cos \psi = V_1 \cos \psi_1 + 2R_1 A.$$

Graphical solution.

This suggests a graphical solution of the problem, and to make our proof rigorous we shall suppose that the electromotive forces and currents follow the harmonic law. Let OA (Fig. 225) be the vector representing V, and let OC, CB, and BA represent $2R_1 A$, $2\omega (L - M) A$, and V_1 respectively. The

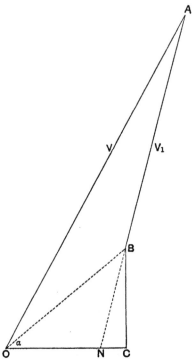

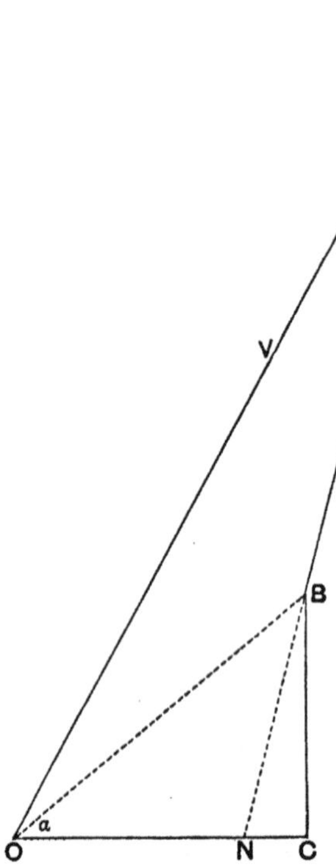

Fig. 225. V is the P.D. at the power station, and V_1 is the P.D. at the distributing station. $OC = 2R_1 A$, $CB = 2\omega (L - M) A$.

angle AOC will be ψ, the angle ANC will be ψ_1, and the angle OCB will be a right angle. We shall denote the angle BOC by α, so that tan α equals $\omega (\mathrm{L} - M)/R_1$. Then projecting OAB on OC we get

$$V \cos \psi - V_1 \cos \psi_1 = 2R_1 A.$$

Similarly $$V \sin \psi - V_1 \sin \psi_1 = 2\omega (\mathrm{L} - M) A.$$

Thus we find that

$$\tan \psi = \{2\omega (\mathrm{L} - M) A + V_1 \sin \psi_1\}/(2R_1 A + V_1 \cos \psi_1) \ldots (9).$$

If we denote the line OB in Fig. 225 by $2Z.A$, we have

$$Z = \{R_1^2 + \omega^2 (L - M)^2\}^{\frac{1}{2}},$$

and $$V^2 = V_1^2 + 4Z^2A^2 + 4V_1 ZA \cos(\psi_1 - a) \qquad \text{.........(10).}$$

If the power factor $\cos\psi_1$ of the load equals $\cos a$, then ψ_1 equals a and $\cos\psi$ is equal to $\cos\psi_1$. If ψ_1 is less than a, as, for instance, when the load at the distributing station is non-inductive, $\cos\psi$ is less than $\cos\psi_1$, and if ψ_1 is greater than a, $\cos\psi$ is greater than $\cos\psi_1$.

When $2Z.A$ (OB in Fig. 225) is small compared with V_1, we have

$$V = V_1 + 2ZA\cos(\psi_1 - a) \text{ approximately.}$$

Hence we see that for given values of V, A, and a, V_1 diminishes as ψ_1 increases from zero to a. When ψ_1 equals a, V_1 has its minimum value and V_1 increases for greater values of ψ_1.

When the constants of the line are given, equations (9) and (10) enable us to find V and ψ for given values of V_1 and ψ_1.

The constants of the line.

If we suppose that the mains are short circuited at the distributing station, we see that $2(L - M)$ is the self-inductance of the two wires in series. Hence, by the formula given on p. 89 of Vol. I, we get

$$2(L - M) = l\{0{\cdot}00148 \log_{10}(d/\rho) + 0{\cdot}000161\} \text{ henrys,}$$

where l is the distance between the power and the distributing station in statute miles, ρ the radius of each main, and d the distance between their axes. If we are calculating, however, the current produced by a high harmonic in the wave of the applied P.D., the true value of $2(L - M)$ will be a little less than that given by this formula, owing to the skin effect (see Vol. I, p. 224). In practice it is rarely necessary to take this effect into account in power transmission lines.

Let us suppose that the two mains are each No. 1 S.W.G. The diameter of each will be 0·300 inch, and the resistance per mile (R_1/l) is 0·64 ohm nearly, at 60° F. If we suppose that the distance between the axes of the wires is 18 inches we get, by the formula,

$$L - M = \{0{\cdot}00074 \log(18/0{\cdot}15) + 0{\cdot}00008\}\, l$$
$$= 0{\cdot}0016\, l.$$

When the frequency $\omega/2\pi$ is 25, we have

$$\omega(\mathrm{L}-M)=0{\cdot}25\,l, \quad \text{and} \quad R_1=0{\cdot}64\,l.$$

Hence since $\tan\alpha=\omega(\mathrm{L}-M)/R_1$ we find that α is 21·3° and the impedance Z is 0·69 l.

Three phase transmission.

Let us now consider the case of transmission by three phase currents. Let v_1, v_2, and v_3 be the potentials of the three terminals of the line at the power station, and v_1', v_2', and v_3' the potentials of the corresponding terminals at the distributing station. Using the same notation as in the last section and neglecting the electrostatic capacity, we have

$$v_1=v_1'+R_1i_1+L\frac{\partial i_1}{\partial t}+M\frac{\partial i_2}{\partial t}+M\frac{\partial i_3}{\partial t},$$

where i_1, i_2, and i_3 are the currents (Fig. 226) in the mains. We have supposed that the mains are arranged symmetrically, so that

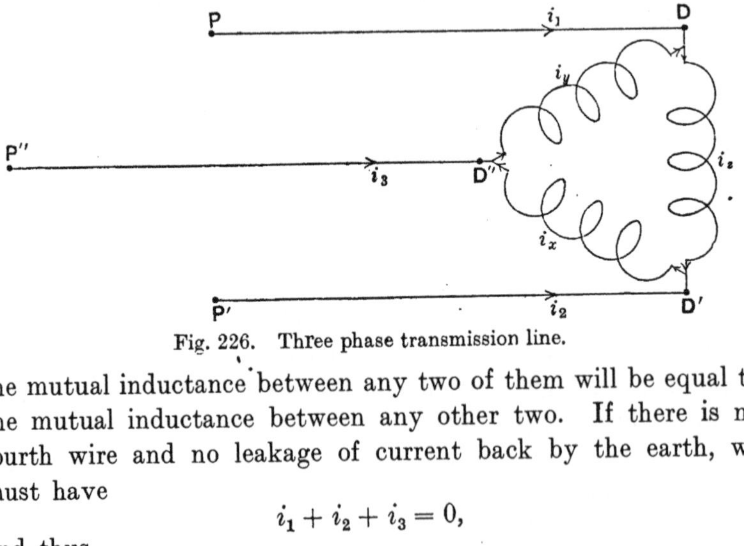

Fig. 226. Three phase transmission line.

the mutual inductance between any two of them will be equal to the mutual inductance between any other two. If there is no fourth wire and no leakage of current back by the earth, we must have

$$i_1+i_2+i_3=0,$$

and thus

$$\left.\begin{aligned}v_1&=v_1'+R_1i_1+(\mathrm{L}-M)\frac{\partial i_1}{\partial t}\\ v_2&=v_2'+R_1i_2+(L-M)\frac{\partial i_2}{\partial t}\\ \text{and}\quad v_3&=v_3'+R_1i_3+(L-M)\frac{\partial i_3}{\partial t}\end{aligned}\right\}\quad\text{...........(11).}$$

Now $v_1 i_1 + v_2 i_2 + v_3 i_3$ is the power W_g being given to the line at the power station and $v_1' i_1 + v_2' i_2 + v_3' i_3$ is the power W received at the distributing station (see Vol. I, Chap. xv). Hence, multiplying equations (11) by i_1, i_2, and i_3 respectively, adding them and taking mean values, we get

$$W_g = W + R_1 (A_1^2 + A_2^2 + A_3^2) \quad \text{...........(12)}.$$

When the load is balanced

$$A_1 = A_2 = A_3 = A\,;\ W_g = \sqrt{3} V_{1.2} A \cos\psi, \text{ and } W = \sqrt{3} V'_{1.2} A \cos\psi_1,$$

where $V_{1.2}$ and $V'_{1.2}$ are the potential differences between the mains 1 and 2 at the generating and distributing stations respectively. Hence

$$\sqrt{3} V_{1.2} A \cos\psi = W + 3R_1 A^2 \quad \text{...........(13)},$$

and

$$V_{1.2} \cos\psi = V'_{1.2} \cos\psi_1 + \sqrt{3} R_1 A \quad \text{...........(14)}.$$

Solving the equation (13) we find that

$$A = \{1/(2\sqrt{3}R_1)\}\{V_{1.2}\cos\psi \pm (V^2_{1.2}\cos^2\psi - 4R_1 W)^{\frac{1}{2}}\} \text{...(15)}.$$

The efficiency η is given by

$$\eta = W/W_g = W/(W + 3R_1 A^2) \quad \text{...........(16)},$$

or

$$\eta = 1 - \sqrt{3} R_1 A/(V_{1.2}\cos\psi) \quad \text{..............(17)}.$$

Also, if A_1 and A_2 be the two values of the current for which the power W transmitted has a given value, and if η_1 and η_2 are the corresponding efficiencies, we have

$$A_1 + A_2 = (1/\sqrt{3})(V_{1.2}/R_1)\cos\psi,$$

and thus $\eta_1 + \eta_2 = 1$. The efficiency of the transmission, therefore, need never be less than 50 per cent.

From (15), we see that the maximum value of the power transmitted is $V^2_{1.2}\cos^2\psi/(4R_1)$. We have found that for a single phase plant the maximum value of the power is $V^2\cos^2\psi/(8R_1)$. Thus by adding a third wire we have doubled the maximum amount of power that can be transmitted to the distributing station. We have assumed that the voltage between the lines and the power factor is the same in the two cases. It should be noticed, however, that the voltage between a three phase main and earth is $V_{1.2}/\sqrt{3}$, that is $0{\cdot}577\, V_{1.2}$, while the voltage between a single phase main and earth need only be $0{\cdot}5\, V$.

By subtracting the second from the first of the equations (11), we get

Graphical solution.

$$v_{1.2} = v'_{1.2} + R_1 (i_1 - i_2) + (L - M) \frac{\partial}{\partial t} (i_1 - i_2).$$

Denoting the currents in the arms of the mesh load (Fig. 225) by i_x, i_y, and i_z respectively, we have

$$i_1 = i_z - i_y \text{ and } i_2 = i_x - i_z,$$

and therefore $i_1 - i_2 = 2i_z - i_x - i_y = 3i_z$, if $i_x + i_y + i_z = 0$, that is, if the load be symmetrical and the current waves in the arms of the mesh contain no harmonics of frequency $3(2n+1)f$. Hence

$$v_{1.2} = v'_{1.2} + 3R_1 i_z + 3(L - M) \frac{\partial i_z}{\partial t}.$$

Assuming that the potential differences and the currents follow the harmonic law, the diagram (Fig. 227) will represent

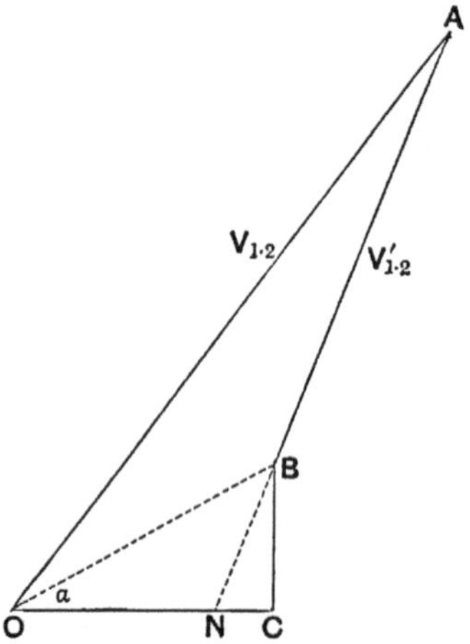

Fig. 227. $V_{1.2}$ and $V'_{1.2}$ are the potential differences between the mains 1 and 2 at the generating and distributing stations respectively. OC gives the phase of the current A' in the arm of the balanced mesh load joining 1 and 2.

the effective values of the various quantities. We have $\tan \alpha$ equal to $\omega (L - M)/R_1$ and it may be shown that

$$\tan \psi = \{3\omega (L - M) A' + V'_{1.2} \sin \psi_1\}/\{3R_1 A' + V'_{1.2} \cos \psi_1\} \ldots (18),$$

where ψ is the phase difference between $V_{1.2}$ and the current A' in the arm of the balanced mesh load at the distributing station joining the mains 1 and 2, and ψ_1 is the phase difference between

$V'_{1.2}$ and A'. Since we suppose that the load is balanced and neglect the electrostatic capacity of the mains, the current A' will be equal to the current in the phase winding of the armature of the generator which joins the terminals 2 and 1.

We also have

$$V^2_{1.2} = V'^2_{1.2} + 9Z^2A'^2 + 6V'_{1.2}ZA' \cos(\psi_1 - \alpha)$$
$$= V'^2_{1.2} + 3Z^2A^2 + 2\sqrt{3}V'_{1.2}ZA\cos(\psi_1 - \alpha) \ldots (19).$$

When $\sqrt{3}Z.A$ (OB in Fig. 227) is small compared with $V_{1.2}$ and $V'_{1.2}$, we have

$$V_{1.2} = V'_{1.2} + \sqrt{3}ZA\cos(\psi_1 - \alpha) \quad \ldots\ldots\ldots (20).$$

The single phase equation corresponding to (20) is

$$V = V_1 + 2ZA\cos(\psi_1 - \alpha).$$

Thus if we use two of the mains as a single phase system, and the voltage drop is to be the same in the two cases, the current must only equal $\sqrt{3}A/2$. In this case the power transmitted for a given voltage drop at the given power factor is $(1/2)\sqrt{3}VA\cos\psi_1$, that is, one-half the power transmitted in the three phase case.

Comparison of single phase with three phase.

Let us now compare the efficiencies of a single phase and a three phase system when the same amount of copper is used in the mains in the two cases. If R_1 be the resistance of each main in the single phase system, then $3R_1/2$ will be the resistance of each main in the three phase system. Let the power W transmitted to the distributing station, the voltage between the mains, and the power factor be the same in the two cases. Then if A_3 denote the current in each of the three phase mains, and A_1 the current in each single phase main, we have

$$W = \sqrt{3}V_1A_3\cos\psi_1 = V_1A_1\cos\psi_1,$$

and therefore $$A_1 = \sqrt{3}A_3.$$

The efficiency η_1 of the single phase transmission is given by

$$\eta_1 = W/(W + 2R_1A_1^2),$$

and the efficiency η_3 of the three phase transmission by

$$\eta_3 = W/\{W + 3(3R_1/2)A_3^2\}$$
$$= W/\{W + 1{\cdot}5R_1A_1^2\}.$$

The efficiency is therefore higher in the three phase case. Also the maximum power transmitted in the three phase case is $V^2 \cos^2 \psi/\{4(3R_1/2)\}$, that is $V^2 \cos^2 \psi/(6R_1)$, which is equal to four-thirds of the maximum power that the single phase line can transmit.

Comparison of star-connected systems.

Let us now consider the efficiency of star-connected polyphase systems, the voltage V to the centre of the star being the same in all cases. We shall suppose also that they all use the same weight of copper in the mains. Let there be q phases, and let qR be the resistance of each main, so that the resistance of all the mains in parallel is R. Let W be the power transmitted, and let A be the current in each main of the q phase system and $\cos \psi_1$ the power factor of the balanced load. Then the ratio of the power lost to the power transmitted

$$= q(qR)A^2/(qVA \cos \psi_1) = qAR/V \cos \psi_1 = WR/(V \cos \psi_1)^2.$$

The efficiency is therefore independent of the number of phases. For a two wire direct current system this ratio equals WR/V^2. Thus except when $\cos \psi_1$ is unity the polyphase systems are less economical, so far as transmission is concerned, than a direct current system.

Comparison of mesh-connected systems.

If the polyphase systems are mesh-connected, and if the voltage to earth be the same in all the systems, the problem is the same as the one discussed in the preceding section. The efficiency is therefore independent of the number of phases. If we make the hypothesis that the voltage V between consecutive mains is to be the same in all cases, then, if the same weight of copper is also used, we find that the ratio of the power lost to the power transmitted equals

$$q(qR)A^2/[qVA \cos \psi_1/\{2 \sin(\pi/q)\}] = RW\{2 \sin(\pi/q)\}^2/(V \cos \psi_1)^2,$$

when q equals 3, $= 3RW/(V \cos \psi_1)^2$,

when q equals 4, $= 2RW/(V \cos \psi_1)^2$,

and when q equals 6, $= RW/(V \cos \psi_1)^2$.

Hence, as we might have anticipated from first principles, the efficiency increases rapidly as we increase the number of phases.

To assume, however, that a constant P.D. between adjacent mains gives a proper basis for comparing the relative merits of polyphase systems is not justifiable. Let us consider, for instance, a six phase system. In this case if the voltage to the centre of the star load is V, the voltage between adjacent mains will also be V. But the voltage between opposite mains is $2V$. We should, therefore, compare this system with a single phase system the voltage of which is $2V$. Putting q equal to 2 and V equal to $2V$ in the above formula, we get $RW/(V\cos\psi_1)^2$ for the efficiency of the single phase system whose voltage is $2V$, and thus the efficiencies are the same in the two cases.

Maximum power transmitted by a polyphase system.

Let W_g be the power at the generating station, and let the distribution be q phase. If W be the power transmitted and qR the resistance of each main, we have

$$W_g = W + q.qR.A^2,$$

where A is the current in each main. Hence, if the load be balanced,

$$qV[A/\{2\sin(\pi/q)\}]\cos\psi = W + q^2RA^2,$$

where $\cos\psi$ is the power factor of the load.

Solving this equation we get

$$A = [1/\{4qR\sin(\pi/q)\}][V\cos\psi \pm \{V^2\cos^2\psi - 16RW\sin^2(\pi/q)\}^{\frac{1}{2}}].$$

The maximum possible value of the power transmitted is therefore $[V\cos\psi/\{\sin(\pi/q)\}]^2/16R$. Now it is easy to see that $V/\{\sin(\pi/q)\}$ is the maximum value of the effective voltage between any two points on the armature windings, and hence if we take this as the basis of comparison for different systems, the maximum value of the power transmitted for a given weight of copper is independent of the number of the phases.

Distributed capacity.

Let us now consider the effects of electrostatic capacity on the transmission of electrical power over a single phase line. For short distances a concentric main may be employed, but for long distances two parallel overhead wires are used. As the resistances of the mains are appreciable, we cannot imitate the electrostatic effects by a single condenser, and thus we must imagine small condensers arranged at equal distances apart all along the line (Fig. 228). Let k and $1/s$ denote the capacity and the insulation resistance between the mains per unit length, and let $r/2$ be the resistance, and $l/2$ the effective inductance of each main per unit length. Now k is a constant and we may assume also that s, the leakage conductance, is constant. When the initial disturbance that arises when the circuit is closed has died away, we may consider that r and l are also constants. A maximum value for l is obtained by making the assumption that the current is uniformly distributed over the cross section of the mains, and a minimum value by assuming that the current is entirely on the surface. In the latter case r will be infinite unless we make the further assumption that the conductivity is

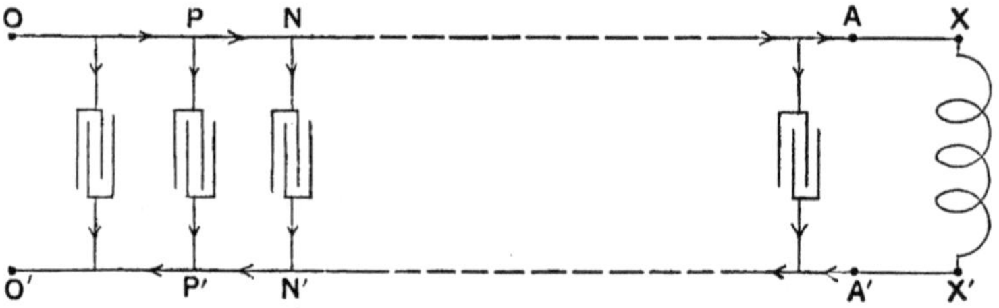

Fig. 228. Model of single phase transmission line.

infinite. In the former case r can easily be found. For bare overhead wires k can be calculated by the formulae given in Chapters IV and V of Volume I, and it can also be calculated for underground mains when the mean dielectric coefficient of the insulating material is known. In the case of concentric mains, k and s can generally be obtained from the data given in manufacturers' catalogues. If a be the length of a concentric main, the insulation resistance, $1/sa$, between the mains is given by

$$1/sa = (\sigma/2\pi a) \log_\epsilon (r_2/r_1),$$

where r_1 is the outer radius of the inner main, r_2 the inner radius of the outer conductor, and σ the mean resistivity of the insulating material in C.G.S. units. By Vol. I, p. 155, we have, in this case,

$$k'a = \lambda a/\{2 \log_\epsilon (r_2/r_1)\},$$

where k' is in electrostatic units and λ is the dielectric coefficient. We thus find that

$$k'/s = \sigma\lambda/4\pi.$$

We shall now find the differential equations which the current and electromotive force have to satisfy at points on the transmission mains of a two wire alternating current system, and, as only a very elementary discussion can be given, we shall make the assumption that we can consider r and l constant.

Let us suppose that the circuit is divided up into an infinite number of little sections, similar to $PNN'P'$ (Fig. 228), and let PN be equal to ∂x. The capacity and insulation resistance between PN and $P'N'$ are $k\partial x$ and $1/(s\partial x)$ respectively. The resistance of PN and $N'P'$ in series is $r\partial x$, and the effective inductance due to the lines linked with the currents in PN and $N'P'$ is $l\partial x$. If e be the potential difference between P and P',

$$e + (\partial e/\partial x)\,\partial x$$

will be the potential difference between N and N'. The difference between these two pressures will be equal to the sum of the E.M.F. $r\partial x . i$ required to drive a current i through a resistance $r\partial x$ and the E.M.F. $l\partial x . \partial i/\partial t$ required to overcome the inductive E.M.F. Putting this into symbols, we get

$$e - \left(e + \frac{\partial e}{\partial x}\partial x\right) = r\partial x . i + l\partial x . \frac{\partial i}{\partial t},$$

and therefore $$-\frac{\partial e}{\partial x} = ri + l\frac{\partial i}{\partial t} \quad \text{....................(21).}$$

If i be the current at P, $i + \frac{\partial i}{\partial x}\partial x$ is the current at N. The difference between these two currents will be the sum of the leakage current $se\partial x$, and the condenser current $k\partial x . \partial e/\partial t$. We therefore have

$$i - \left(i + \frac{\partial i}{\partial x}\partial x\right) = se\partial x + k\partial x . \frac{\partial e}{\partial t},$$

and hence $$-\frac{\partial i}{\partial x} = se + k\frac{\partial e}{\partial t} \quad \text{....................(22).}$$

The values of e and i at points on the line, therefore, must satisfy the differential equations (21) and (22). In addition, they must satisfy the given initial and terminal conditions.

In the general case the solution is complicated, but in the special case when l/r equals k/s, the equations can be solved easily. In this case, the equation (21) can be written in the form

Circuit in which there is no distortion.

$$-\frac{\partial e}{\partial x} = l\left(\frac{1}{\tau} + \frac{\partial}{\partial t}\right) i,$$

or

$$-\frac{\partial e}{\partial x} = \frac{1}{v}\left(\frac{1}{\tau} + \frac{\partial}{\partial t}\right) lvi \quad \text{.................}(23),$$

where $\tau = l/r = k/s$, and $v = 1/\sqrt{lk}$.

Similarly (22) becomes

$$-\frac{\partial}{\partial x}(lvi) = \frac{1}{v}\left(\frac{1}{\tau} + \frac{\partial}{\partial t}\right) e \quad \text{..............}(24).$$

From the equations given in Volume I (p. 201) it readily follows that

$$(l_{1.1} + l_{2.2} - 2l_{1.2})(k_{1.1}k_{2.2} - k^2_{1.2}) = k_{1.1} + k_{2.2} + 2k_{1.2},$$

and hence $l_0 k' = 1$, where l_0 is the inductance of the line per unit length when the resistivity of the metal forming the line is zero, and k' is the capacity of the line per unit length in electrostatic measure (Vol. I, p. 152). We saw also (Vol. I, p. 160) that k' equals $k\,(3.10^{10})^2$, and hence we find that $3.10^{10} = 1/\sqrt{l_0 k}$. In practice, neither the conductivity of the wire nor the frequency of the alternating current is infinite; the current therefore is not wholly on the surface, and so l must be greater than l_0. It follows that the maximum possible value of the quantity v in equations (23) and (24) is the velocity of light, that is, 3.10^{10} cms. per second.

Let us suppose, for instance, that we have two parallel cylindrical solid wires each of radius a, and that d is the distance between their axes. For low frequencies, we have

$$v = 3.10^{10}\left[\{4 \log_\epsilon (d + \sqrt{d^2 - 4a^2})/(2a)\}/\{4 \log_\epsilon (d/a) + 1\}\right]^{1/2}$$

(see Vol. I, pp. 88 and 165). This equation shows that v is practically equal to 3.10^{10} when d is large compared with a, and that v is very small when the wires are nearly touching. If d were equal to $10a$, we should have $v = 2{\cdot}84.10^{10}$.

By adding together the equations (23) and (24) we find that

$$\left(\frac{1}{\tau} + \frac{\partial}{\partial t}\right)(e + lvi) + v\frac{\partial}{\partial x}(e + lvi) = 0,$$

and therefore

$$\frac{\partial}{\partial t}(e + lvi)\,\epsilon^{t/\tau} + v\frac{\partial}{\partial x}(e + lvi)\,\epsilon^{t/\tau} = 0.$$

The solution of this equation is

$$(e + lvi)\,\epsilon^{t/\tau} = F_1(x - vt) \quad \text{..............(25)},$$

where $F_1(x)$ is the function of x which gives the value of $e + lvi$ at all points on the main when t is zero.

This solution may be verified at once by differentiation. Similarly by subtracting equation (23) from (24), and proceeding as before, we get

$$(e - lvi)\,\epsilon^{t/\tau} = F_2(x + vt) \quad \text{..............(26)}.$$

By adding (25) and (26) we get

$$2e = \epsilon^{-t/\tau} F_1(x - vt) + \epsilon^{-t/\tau} F_2(x + vt) \text{.........(27)}.$$

Also, by subtracting (26) from (25), we find that

$$2lvi = \epsilon^{-t/\tau} F_1(x - vt) - \epsilon^{-t/\tau} F_2(x + vt) \text{.........(28)}.$$

If we suppose that the origin is moving along the main to the right with a velocity v, then, at the time t, the value of $e + lvi$ is given by $\epsilon^{-t/\tau} F_1(x)$, where x is the distance of a point on the main from the moving origin. If, therefore, the mains are infinitely long and if their conductivity is perfect, the initial value of $e + lvi$, which is represented by $F_1(x)$, glides bodily to the right with the velocity of light. If the conductivity of the mains is not perfect, that is, if τ is not infinite, $e + lvi$ glides to the right with a velocity less than that of light, and with continually diminishing amplitude. The distribution in space, however, of $e + lvi$ is always similar to its initial value. The wave thus suffers no distortion although it may be rapidly dying away.

Equation (27) shows that the electromotive force wave, and therefore also the distribution of the electrostatic charges along the main, may be regarded as due to the motion of two waves, moving in opposite directions with velocities less than the velocity of light. For instance, suppose that initially one metre of the positive main has a charge of one microcoulomb uniformly distributed over it, and that a metre of the negative main, exactly opposite to it, has a negative charge of one microcoulomb. Let us suppose also that initially all the remaining part of the mains is at zero potential. We see by (27) that the positive charge (Fig. 229) separates into two equal charges. One half moves to the right and the other to the left with velocity v, and the shape of the wave being a rectangle initially, each of these waves is rectangular. Exactly the same phenomenon happens in the negative main and so the charges move as indicated in Fig. 229. The initial positions and values of the charges on the given assumption are indicated in I. In II, the positive and negative waves are beginning to separate. In III, they are on the point of separating, and in IV they have separated. When t equals τ the charge on either main will have diminished to $1/\epsilon$ (or approximately 37 per cent.) of its initial value.

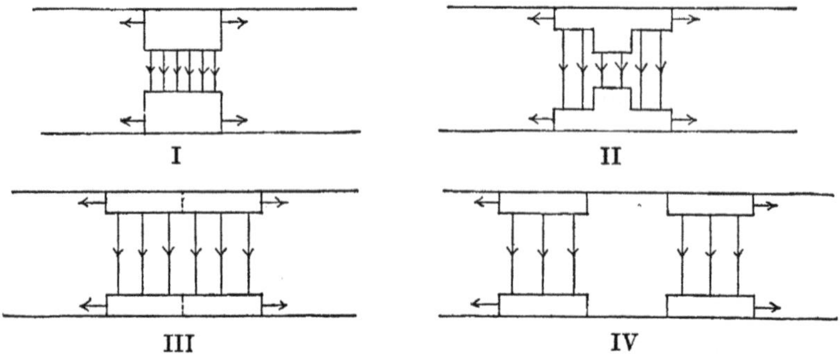

Fig. 229. Transmission of electric waves along wires. The lines represent the wires, the rectangles the magnitude and position of the charges due to the positive and negative waves.

The analogy of the electric waves in the special case we are considering with the longitudinal or torsional vibrations of an elastic bar or of a stretched string is very close. From (27) we see that

$$\frac{\partial^2}{\partial t^2}(e\epsilon^{t/\tau}) = v^2 \frac{\partial^2}{\partial x^2}(e\epsilon^{t/\tau}).$$

The equation for the longitudinal vibrations of a bar is

$$\frac{\partial^2 \xi}{\partial t^2} = v^2 \frac{\partial^2 \xi}{\partial x^2},$$

where x is the abscissa of a section of the bar in the position of equilibrium, and $x + \xi$ is its abscissa at the time t. In the equation for the vibrations of the bar v^2 is q/ρ, where q is Young's modulus and ρ is the density of the metal forming the bar.

Equations (26) and (25) show that $e - lvi$ is a negative wave, that is, one which moves to the left, and $e + lvi$ is a positive wave, that is, one which moves to the right.

Positive and negative waves.

We have supposed that the mains are of infinite length, so that there are no reflections or other interferences. If e equals lvi, there is a positive wave only. We may say, therefore, that e equal to lvi is the characteristic property of a positive wave. In this type of wave we see by (25) that

$$2e = \epsilon^{-t/\tau} F_1(x - vt), \quad \text{and} \quad i = e/lv = kve.$$

Since ke denotes the charge on the positive main per unit length, we see that this charge multiplied by the velocity v is equal to the current i. In a positive wave we have

$$(1/2)\, li^2 = (1/2)\, lk^2v^2e^2 = (1/2)\, ke^2,$$

that is, the electromagnetic energy at any instant equals the electrostatic energy. We also have $ri^2 = se^2$, which shows that the energy expended in heating the mains equals the energy expended in leakage currents.

Since the equations $e = lvi$ and $e = -lvi$ express the characteristic properties of positive and negative waves respectively, it follows that e and i have the same sign in a positive wave but opposite signs in a negative wave. In both kinds of waves the electromagnetic energy at any instant equals the electrostatic energy. When two waves travelling in opposite directions meet, the resultant e is obtained by adding the values of e for each wave together, and the resultant i is obtained by adding the values of i for each wave. If the positive and negative waves are exactly equal and similar, then, at the instant when they are superposed, e is doubled and i is zero; but, if the electrifications are opposite e is zero and i is doubled. In the first case the energy is all electrostatic, and in the second case it is all electromagnetic. It is to be noticed that the waves pass through one another without producing any interference, the attenuation of each wave during the overlapping proceeding at exactly the same rate as if the other were absent.

When the length of the mains is finite, the terminal conditions in the ideal case we are investigating can be found without difficulty. If one pair of the ends of the mains be on open circuit, the incident wave $e + lvi$

The reflection and transmission of waves.

must give rise to a reflected wave $e - lvi$, since there can be no currents at the ends. If they be connected together through a non-inductive resistance R, there will be a reflected wave $e' - lvi'$, and the current in the bridge $i + i'$ will be equal to $(e + e')/R$. Since $e = lvi$, and $e' = - lvi'$, we have, therefore,

$$e/lv - e'/lv = (e + e')/R,$$

and thus
$$e'/e = (R - lv)/(R + lv),$$

and
$$i'/i = - (R - lv)/(R + lv).$$

In proving the above equations, first given by O. Heaviside, we have assumed that the disturbance of the electrostatic and electromagnetic lines by the charges on, and the current in, the terminal bridge is negligible. When R is large this is very approximately true, but when R is small there will obviously be a considerable departure, near the end of the line, from the normal conditions. In the latter case, therefore, the solution will be only approximate.

When R is greater than lv, e' is less than e, and the reflected wave is less than the incident wave. When R equals lv, both e' and i' are zero. In this case, therefore, there is no reflected wave, the incident wave being completely absorbed, the energy of the wave, $(ke^2/2 + li^2/2)\, a' = lvi^2\, a'/v$, being converted into heat in the connecting resistance lv. We have supposed that a' is the length of the wave. When R is less than lv, e' is negative and i' is positive, and we have a recedent wave, the electrification being reversed. Finally, when R is zero, the recedent wave equals the incident wave in magnitude, the charges (Fig. 229) on the two mains simply changing places at the far end of the line and maintaining their constant velocity v round the circuit. The positive charge thus goes through the negative charge without having the slightest effect on it. They obviously cannot neutralise one another as this would have the effect of destroying the electrostatic and the electromagnetic energy of the system.

The voltage drop on short lines.

In England high tension overhead lines are sometimes used to transmit electric power between stations which are only a few miles apart. In these cases the capacity effects can, frequently, be neglected in calculating the voltage drop along the line. We shall now show how the inductance effects can be taken into account. A customary method of arranging the overhead wires in a three phase system is shown in Fig. 230.

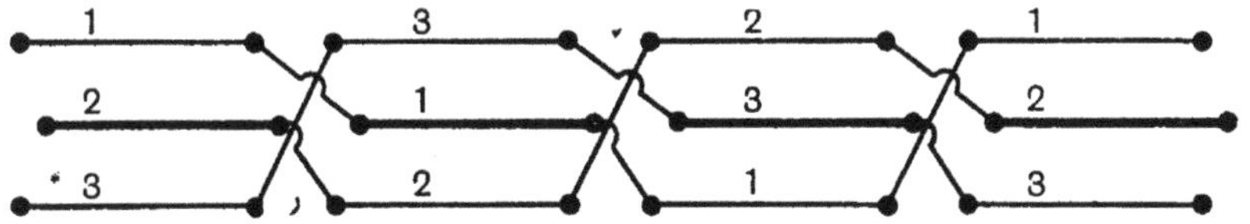

Fig. 230. Power transmission. Positive spiral.

The wires are in the same vertical plane but they are crossed at the points of suspension in such a way that the average inductance of each wire taken over three sections is the same. This equalises the voltage drop along the three mains (assuming that the total number of sections is a multiple of three). It also neutralises the inductive effect of the mains on neighbouring telephone or telegraph wires. In Fig. 230 it will be noticed that a positive 'spiral' is given to the mains and in Fig. 231 a negative spiral is given to them. The practical effects produced by the spiral given to the wires are negligibly small.

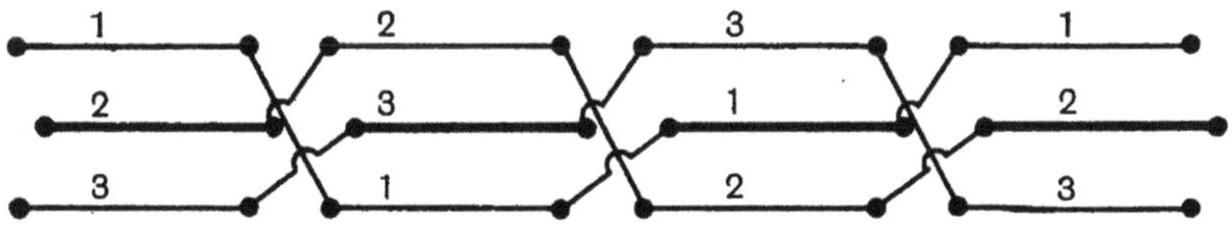

Fig. 231. Power transmission. Negative spiral.

Let us suppose that the line is arranged as in Fig. 230. If a_1 be the length of the first segment of the line and ρ its radius, then its self-inductance $L_{1.1}$ is given by (see Vol. I, p. 78)

$$L_{1.1} = 2a_1\{\log(2a_1/\rho) - 3/4\}.$$

Similarly if $L_{1.2}$ and $L_{1.3}$ be the mutual inductances between the segments 1 and 2, and the segments 1 and 3 respectively, we have

$$L_{1.2} = 2a_1\{\log(2a_1/d) - 1\} \text{ and } L_{1.3} = 2a_1\{\log(a_1/d) - 1\},$$

where d is the distance between the axes of 1 and 2, and $2d$ is the distance between the axes of 1 and 3. If v_1 and v_1' be the potentials at the ends of main 1, we have

$$v_1 - v_1' = (R_1/3)\,i_1 + nL_{1.1}\frac{\partial i_1}{\partial t} + nL_{1.2}\frac{\partial i_2}{\partial t} + nL_{1.3}\frac{\partial i_3}{\partial t}$$
$$+ (R_1/3)\,i_1 + nL_{1.1}\frac{\partial i_1}{\partial t} + nL_{1.2}\frac{\partial i_2}{\partial t} + nL_{1.2}\frac{\partial i_3}{\partial t}$$
$$+ (R_1/3)\,i_1 + nL_{1.1}\frac{\partial i_1}{\partial t} + nL_{1.3}\frac{\partial i_2}{\partial t} + nL_{1.2}\frac{\partial i_3}{\partial t},$$

where n—a multiple of 3—is the number of segments.

Hence noticing that $i_1 + i_2 + i_3 = 0$, we get

$$v_1 - v_1' = R_1 i_1 + (3nL_{1.1} - nL_{1.3} - 2nL_{1.2})\frac{\partial i_1}{\partial t}$$
$$= R_1 i_1 + L_1\frac{\partial i_1}{\partial t},$$

where
$$L_1 = n\,(3L_{1.1} - L_{1.3} - 2L_{1.2})$$
$$= 2a\,\{\log(d/\rho) + (1/3)\log 2 + (1/4)\}$$
$$= 2a\,\{\log(d/\rho) + 0{\cdot}48\}.$$

If a be in miles and L_1 be in henrys,

$$L_1 = a\{0{\cdot}00074\log(d/\rho) + 0{\cdot}000155\}.$$

If the effective value of the current in each main be A, the voltage drop along one main will be $(R_1^2 + \omega^2 L_1^2)^{1/2}A$. Hence also if $V_{1.2}$ be the voltage between the mains at the generating station and $V'_{1.2}$ be the voltage at the distributing station, then provided that $\sqrt{3}\,AZ$ is small compared with $V'_{1.2}$, we have (see Fig. 227)

$$V_{1.2} - V'_{1.2} = \sqrt{3}A\,.\,Z\cos(\psi - \alpha),\ \text{approximately},$$

where $\cos\psi$ is the power factor at the generating station and $\tan\alpha = \omega L_1/R_1$.

Method of getting a neutral main.

When we desire to obtain a four wire system from a three phase power supply system which has no fourth wire we may proceed as follows. Connect the mains with the neutral main through ring reactance coils as shown in Fig. 232. Let e_1, e_2 and e_3 be the potentials of the mains 1, 2 and 3 and let v be the potential of their common junction O. Then if i_1, i_2, i_3 and i be the currents in the wires connecting O with 1, 2 and 3, and in the neutral main respectively, we have

$$i + i_1 + i_2 + i_3 = 0.$$

We also have

$$e_1 - v = Ri_1 + 2L(\partial i_1/\partial t) - M\{\partial(i_2 + i_3)/\partial t\},$$

and two similar equations, where R is the resistance of any of the wires connecting the three power mains to the neutral, L the inductance of any coil and M the mutual inductance between any two coils on a ring core. Hence, by addition, we get

$$e_1 + e_2 + e_3 - 3v = -Ri - 2L(\partial i/\partial t) + 2M(\partial i/\partial t).$$

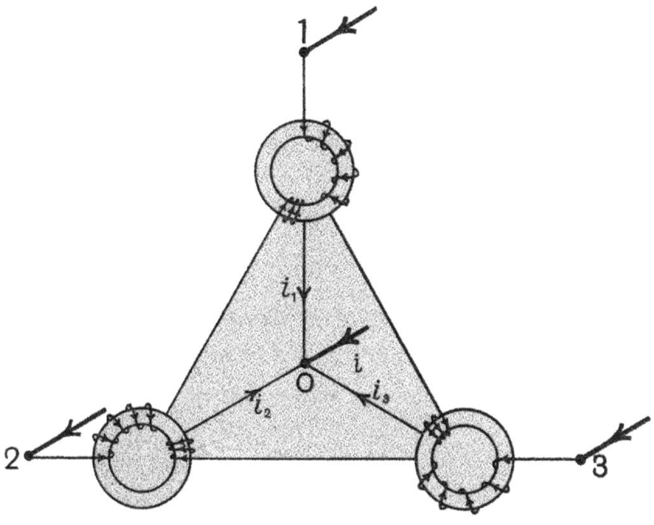

Fig. 232. Method of obtaining a fourth wire.

If L, therefore, be very nearly equal to M we may write

$$e_1 + e_2 + e_3 - 3v = -Ri.$$

When the load is balanced $e_1 + e_2 + e_3 = 0$, and thus $v = Ri/3$. Hence if R be small v will be small even when i has its full load value, and so the potential differences between the mains will vary little with the load.

REFERENCES

OLIVER HEAVISIDE, 'Effect of a Periodic Impressed Force acting at one end of a Telegraph Circuit with any Terminal Conditions. The General Solution,' *Electrical Papers*, Vol. 2, p. 245.

J. J. THOMSON, 'Electric Waves,' *Encyclopaedia Britannica* [10th Edition], Vol. 28, p. 55.

F. A. C. PERRINE and F. G. BAUM, 'The Use of Aluminium Line Wire and Some Constants for Transmission Lines,' *Trans. Am. Inst. of El. Engin.*, May, 1900.

A. BLONDEL, 'Méthode Pratique pour le Calcul des Lignes à Courants Alternatifs Présentant de la Self-Induction et de la Capacité,' *l'Éclairage Électrique*, Vol. 49, p. 121, 1906.

A. E. KENNELLY, 'The Process of Building up the Voltage and Current in Long Alternating Current Circuits,' *Proc. of the Am. Academy of Arts and Science*, Vol. 42, p. 701, 1907.

P. H. THOMAS, 'Output and Regulation in Long Distance Lines,' *Trans. Am. Inst. of El. Engin.*, June, 1909.

—— 'Calculation of High Tension Lines,' *Trans. Am. Inst. of El. Engin.*, June, 1909.

J. F. C. SNELL, *Distribution of Electrical Energy.*

A. V. ABBOTT, *Electrical Transmission of Energy.*

CHAPTER XIX

The trigonometry of imaginary quantities. Hyperbolic functions. Imaginary arguments. . Series formulae. Inverse functions. Inverse functions of complex numbers. Tables of hyperbolic sines, cosines and tangents. Trigonometrical functions. References.

The trigono- metry of imaginary quantities.

As functions of imaginary quantities are very extensively used in the practical theories of electrical power transmission, telegraphy and telephony, we shall give a summary of the main theorems with proofs of those which are not well known. The fundamental theorem on which the trigonometry of imaginary quantities is founded is the exponential theorem. This theorem is that any power or root of the Neperian base number ϵ can be computed by the series

$$\epsilon^x = 1 + x + \frac{x^2}{2!} + \frac{x^3}{3!} + \dots \qquad \dots\dots\dots\dots\dots(1).$$

By taking x equal to unity we find that $\epsilon = 2{\cdot}71828\dots$. This number is incommensurable, that is, we cannot express it as a vulgar fraction. Knowing the expansions for $\sin x$ and $\cos x$ in ascending powers of x, we easily prove from (1) that

$$\epsilon^{x\iota} = \cos x + \iota \sin x, \text{ and } \epsilon^{-x\iota} = \cos x - \iota \sin x,$$

where ι stands for $\sqrt{-1}$. Solving these equations for $\sin x$ and $\cos x$ we get Euler's formulae

$$\sin x = (\epsilon^{x\iota} - \epsilon^{-x\iota})/(2\iota) \text{ and } \cos x = (\epsilon^{x\iota} + \epsilon^{-x\iota})/2 \quad \dots(2).$$

It is to be noticed that if n be any integer

$$\epsilon^{(2n\pi+x)\iota} = \cos(2n\pi + x) + \iota \sin(2n\pi + x)$$
$$= \cos x + \iota \sin x = \epsilon^{x\iota}.$$

It follows that $\epsilon^{2\pi n\iota} = 1$. Similarly we have $\epsilon^{(2n+1/2)\pi\iota} = \iota$. We see that $\log 1 = 2\pi n\iota$ and therefore has an infinite number of imaginary values as well as the value zero, and $\log \iota$ is also a multiple valued quantity.

If (r, θ) be the polar coordinates of a point we see (Vol. I, p. 290) that

$$x + \iota y = r\epsilon^{\theta\iota} \text{ where } r = (x^2 + y^2)^{1/2} \text{ and } \tan\theta = y/x.$$

This is frequently written

$$x + \iota y = r \underline{|\theta}.$$

Similarly $x - \iota y$ is written $r \overline{|\theta}$, or better, $r \underline{|-\theta}$. If we take for example $x = 1$ and $y = 1$ we may write

$$1 + \iota = \sqrt{2}\, \underline{|45^\circ} \text{ and } 1 - \iota = \sqrt{2}\, \overline{|45^\circ} \text{ or } \sqrt{2}\, \underline{|-45^\circ}.$$

It is to be noticed that in the general case θ is a periodic quantity as we can increase or diminish the angle θ by $2n\pi$ without altering the value of its trigonometrical functions, and thus we may write

$$1 + \iota = \sqrt{2}\, \underline{|\pi/4 + 2n\pi}.$$

It is sometimes convenient to write

$$x + \iota y = x \sec\theta\, \underline{|\theta}.$$

The definitions of the hyperbolic functions are

Hyperbolic functions.

$$\sinh x = (\epsilon^x - \epsilon^{-x})/2, \qquad \cosh x = (\epsilon^x + \epsilon^{-x})/2,$$
$$\tanh x = \sinh x/\cosh x, \qquad \coth x = 1/\tanh x,$$
$$\text{sech}\, x = 1/\cosh x, \qquad \text{cosech}\, x = 1/\sinh x.$$

It readily follows that $\cosh x + \sinh x = \epsilon^x$, $\sinh(-x) = -\sinh x$, $\cosh(-x) = \cosh x$, $\tanh(-x) = -\tanh x$, etc. When speaking it is best to pronounce $\sinh x$ as sine-h-x, pronouncing it as a three-syllabled word. Similarly we pronounce $\cosh x$ as cos-h-x and the others in like manner. The following theorems readily follow from the definitions.

$$\sinh(x + y) = \sinh x \cosh y + \cosh x \sinh y,$$
$$\cosh(x + y) = \cosh x \cosh y + \sinh x \sinh y,$$
$$\tanh(x + y) = \frac{\tanh x + \tanh y}{1 + \tanh x \tanh y},$$

$$\sinh x + \sinh y = 2 \sinh \{(x+y)/2\} \cosh \{(x-y)/2\},$$
$$\sinh x - \sinh y = 2 \cosh \{(x+y)/2\} \sinh \{(x-y)/2\},$$
$$\cosh x + \cosh y = 2 \cosh \{(x+y)/2\} \cosh \{(x-y)/2\},$$
$$\cosh x - \cosh y = 2 \sinh \{(x+y)/2\} \sinh \{(x-y)/2\},$$
$$\sinh x \ \sinh y = (1/2) \{\cosh (x+y) - \cosh (x-y)\},$$
$$\sinh x \ \cosh y = (1/2) \{\sinh (x+y) + \sinh (x-y)\},$$
$$\cosh x \ \sinh y = (1/2) \{\sinh (x+y) - \sinh (x-y)\},$$
$$\cosh x \ \cosh y = (1/2) \{\cosh (x+y) + \cosh (x-y)\},$$
$$\sinh 2x = 2 \sinh x \cosh x,$$
$$\cosh 2x = 1 + 2 \sinh^2 x$$
$$= 2 \cosh^2 x - 1,$$
$$\cosh^2 x - \sinh^2 x = 1,$$
$$\tanh 2x = 2 \tanh x/(1 + \tanh^2 x),$$
$$\cosh 2x + \cos 2y = 2 + 2 (\sinh^2 x - \sin^2 y),$$
$$\cosh 2x - \cos 2y = 2 (\sinh^2 x + \sin^2 y).$$

It readily follows from Euler's theorems (2) that

Imaginary arguments.

$$\sin \iota x = (\epsilon^{-x} - \epsilon^{x})/2\iota = \iota \sinh x,$$
$$\cos \iota x = (\epsilon^{-x} + \epsilon^{x})/2 = \cosh x,$$
$$\tan \iota x = \sin \iota x / \cos \iota x = \iota \tanh x,$$
$$\cot \iota x = - \iota \coth x,$$
$$\sec \iota x = \operatorname{sech} x,$$

and
$$\operatorname{cosec} \iota x = - \iota \operatorname{cosech} x.$$

Similarly we have

$$\sinh \iota x = \iota \sin x, \quad \cosh \iota x = \cos x,$$
$$\tanh \iota x = \iota \tan x, \quad \coth \iota x = - \iota \cot x,$$
$$\operatorname{sech} \iota x = \sec x, \text{ and } \operatorname{cosech} \iota x = - \iota \operatorname{cosec} x.$$

From the ordinary trigonometrical formulae we get

$$\sin (a + \iota b) = \sin a \cos \iota b + \cos a \sin \iota b$$
$$= \sin a \cosh b + \iota \cos a \sinh b$$
$$= s_1 \epsilon^{\sigma_1 \iota},$$

where $$s_1^2 = \sin^2 a \cosh^2 b + \cos^2 a \sinh^2 b$$
$$= \sin^2 a + \sinh^2 b$$
$$= (\cosh 2b - \cos 2a)/2,$$
and $$\tan \sigma_1 = \tanh b/\tan a.$$

Similarly we get
$$\cos (a + \iota b) = c_1 \epsilon^{\gamma_1 \iota},$$
where $$c_1^2 = \cosh^2 b - \sin^2 a$$
$$= (\cosh 2b + \cos 2a)/2,$$
and $$\tan \gamma_1 = - \tan a \ \tanh b.$$

Hence also
$$\tan (a + \iota b) = (s_1/c_1)\, \epsilon^{(\sigma_1 - \gamma_1)\iota},$$
$$\cot (a + \iota b) = (c_1/s_1)\, \epsilon^{-(\sigma_1 - \gamma_1)\iota},$$
$$\sec (a + \iota b) = (1/c_1)\, \epsilon^{-\gamma_1 \iota},$$
$$\operatorname{cosec} (a + \iota b) = (1/s_1)\, \epsilon^{-\sigma_1 \iota}.$$

Similarly we have
$$\sinh (a + \iota b) = s\epsilon^{\sigma\iota}, \qquad \cosh (a + \iota b) = c\epsilon^{\gamma\iota},$$
$$\tanh (a + \iota b) = (s/c)\, \epsilon^{(\sigma-\gamma)\iota}, \qquad \coth (a + \iota b) = (c/s)\, \epsilon^{(\gamma-\sigma)\iota},$$
$$\operatorname{sech} (a + \iota b) = (1/c)\, \epsilon^{-\gamma\iota}, \qquad \operatorname{cosech} (a + \iota b) = (1/s)\, \epsilon^{-\sigma\iota},$$
where $$s^2 = \sinh^2 a + \sin^2 b = (\cosh 2a - \cos 2b)/2,$$
$$c^2 = 1 + \sinh^2 a - \sin^2 b = (\cosh 2a + \cos 2b)/2,$$
$$\tan \sigma = \tan b/\tanh a, \quad \tan \gamma = \tanh a \ \tan b,$$
and $$\tan (\sigma - \gamma) = \sin 2b/\sinh 2a.$$

Series formulae.

Circular	Hyperbolic
$\sin x = x - \frac{x^3}{3!} + \frac{x^5}{5!} - \ldots$	$\sinh x = x + \frac{x^3}{3!} + \frac{x^5}{5!} + \ldots$
$\cos x = 1 - \frac{x^2}{2!} + \frac{x^4}{4!} - \ldots$	$\cosh x = 1 + \frac{x^2}{2!} + \frac{x^4}{4!} + \ldots$
$\tan x = x + \frac{x^3}{3} + \frac{2x^5}{15} + \ldots$	$\tanh x = x - \frac{x^3}{3} + \frac{2x^5}{15} - \ldots$
$\cot x = \frac{1}{x} - \frac{x}{3} - \frac{x^3}{45} - \ldots$	$\coth x = \frac{1}{x} + \frac{x}{3} - \frac{x^3}{45} + \ldots$
$\sec x = 1 + \frac{x^2}{2} + \frac{5x^4}{24} + \ldots$	$\operatorname{sech} x = 1 - \frac{x^2}{2} + \frac{5x^4}{24} - \ldots$
$\operatorname{cosec} x = \frac{1}{x} + \frac{x}{6} + \frac{7x^3}{360} + \ldots$	$\operatorname{cosech} x = \frac{1}{x} - \frac{x}{6} + \frac{7x^3}{360} - \ldots$

In general we have

$$\tan x = \frac{2^2(2^2-1)}{2!}B_1x + \frac{2^4(2^4-1)}{4!}B_3x^3 + \dots,$$

$$\tanh x = \frac{2^2(2^2-1)}{2!}B_1x - \frac{2^4(2^4-1)}{4!}B_3x^3 + \dots,$$

$$\cot x = \frac{1}{x} - \frac{B_1}{2!}2^2x - \frac{B_3}{4!}2^4x^3 - \dots,$$

$$\coth x = \frac{1}{x} + \frac{B_1}{2!}2^2x - \frac{B_3}{4!}2^4x^3 + \dots,$$

$$\sec x = 1 + \frac{E_2}{2!}x^2 + \frac{E_4}{4!}x^4 + \dots,$$

$$\text{sech}\, x = 1 - \frac{E_2}{2!}x^2 + \frac{E_4}{4!}x^4 - \dots,$$

$$\text{cosec}\, x = \frac{1}{x} + 2(2-1)\frac{B_1}{2!}x + 2(2^3-1)\frac{B_3}{4!}x^3 + \dots,$$

$$\text{cosech}\, x = \frac{1}{x} - 2(2-1)\frac{B_1}{2!}x + 2(2^3-1)\frac{B_3}{4!}x^3 - \dots,$$

where B_1, B_3, ... are Bernoulli's numbers and E_2, E_4, ... are Euler's numbers. The values of the first five of these numbers are

$$B_1 = 1/6,\quad B_3 = 1/30,\quad B_5 = 1/42,\quad B_7 = 1/30,\quad B_9 = 5/66,$$
$$E_2 = 1,\quad E_4 = 5,\quad E_6 = 61,\quad E_8 = 1385,\quad E_{10} = 50521.$$

We see at once from the series that for small values of x we can write

$$\sinh x = x = \tanh x,\quad \cosh x = 1,$$

$$\sinh x - \sin x = x^3/3 \text{ and } \cosh x - \cos x = x^2.$$

For large values of x

$$\sinh x = \cosh x = \epsilon^x/2,$$

since ϵ^{-x} can be neglected in comparison with ϵ^x in this case. We also have $\tanh x = 1$, very approximately when x is large.

Inverse functions.

As most of the inverse hyperbolic functions have multiple values, care must be taken in practice to see that we use the value of the function suitable for the special case.

Let us first find the value of $\sinh^{-1} a$ where a is real. Putting $\sinh^{-1} a = x$ we have $a = \sinh x$, and so $\cosh x = +(1 + a^2)^{1/2}$, since $\cosh x$ must be a positive quantity. Hence

$$\epsilon^x = \sinh x + \cosh x = a + (1 + a^2)^{1/2}$$

and therefore $\sinh^{-1} a \doteq \log\{a + (1 + a^2)^{1/2}\}$.

This is true for all values of a positive or negative. For example,

$$\sinh^{-1}(12/5) = \log 5, \quad \sinh^{-1}(3/4) = \log 2,$$

and $$\sinh^{-1}(-a) = -\sinh^{-1} a.$$

Similarly when a is greater than unity

$$\cosh^{-1} a = \log\{a + (a^2 - 1)^{1/2}\}.$$

Let us now suppose that a lies in value between 1 and -1. Putting $\cosh^{-1} a = x$, we get

$$a = \cosh x = (\epsilon^x + \epsilon^{-x})/2 = (\epsilon^{x\iota \cdot \iota} + \epsilon^{-x\iota \cdot \iota})/2$$

$$= \cos(\pm x\iota) = \cos(2n\pi \pm x\iota).$$

Hence $$2n\pi \pm x\iota = \cos^{-1} a,$$

and so we may write

$$x = \cosh^{-1} a = \iota.(2n\pi \pm \cos^{-1} a),$$

where n can have any integral value, positive or negative.

When a is negative and not greater than -1, we have

$$\cosh^{-1} a = -\log\{-a + (a^2 - 1)^{1/2}\} + \iota.(2n + 1)\pi.$$

For example, we have

$$\cosh^{-1}(5/4) = \log 2, \quad \cosh^{-1}(1/2) = \iota(2n\pi \pm \pi/3),$$

and $$\cosh^{-1}(-5/4) = -\log 2 + \iota(2n + 1)\pi.$$

Putting $$a = \tanh x = (\epsilon^x - \epsilon^{-x})/(\epsilon^x + \epsilon^{-x}),$$

we get $$\epsilon^{2x} = (1 + a)/(1 - a).$$

Therefore

$$x = \tanh^{-1} a = (1/2)\log\{(1 + a)/(1 - a)\}$$

when a is not greater than unity.

Hence in this case we have

$$\tanh^{-1} a = a + a^3/3 + a^5/5 + \dots.$$

If a be greater than unity, we must have

$$\tanh^{-1} a = (1/2)\log\{(a+1)/(a-1)\} + (1/2)\log(-1)$$
$$= (1/2)\log\{(a+1)/(a-1)\} + \iota.(2n+1)\pi/2.$$

We also have

$$\tanh^{-1}(-a) = -\tanh^{-1} a.$$

Noticing that

$$\coth^{-1} a = \tanh^{-1}(1/a),$$
$$\operatorname{sech}^{-1} a = \cosh^{-1}(1/a),$$

and
$$\operatorname{cosech}^{-1} a = \sinh^{-1}(1/a),$$

we see that the values of these functions can be readily found by means of the formulae given above.

We also have

$$\sinh^{-1} x = -\iota.\sin^{-1}\iota x$$
$$= x - \frac{1}{2}.\frac{x^3}{3} + \frac{1.3}{2.4}.\frac{x^5}{5} - \dots,$$

and
$$\sinh^{-1}\iota x = \iota.\{n\pi + (-)^n \sin^{-1} x\}.$$

Inverse functions of complex numbers.

Let
$$\sinh^{-1}(a+\iota b) = x + \iota y.$$

Then
$$a + \iota b = \sinh(x + y\iota)$$
$$= \sinh x \cos y + \iota.\cosh x \sin y.$$

Hence $a = \sinh x \cos y$, and $b = \cosh x \sin y$,

and therefore
$$a^2 + b^2 + 1 = \cosh^2 x + \sin^2 y.$$

Thus
$$(\cosh x + \sin y)^2 = a^2 + (1+b)^2,$$

and
$$(\cosh x - \sin y)^2 = a^2 + (1-b)^2.$$

Noticing that neither $\cosh x + \sin y$ nor $\cosh x - \sin y$ can be negative, we get

$$2\cosh x = \{a^2 + (1+b)^2\}^{1/2} + \{a^2 + (1-b)^2\}^{1/2} = 2C \text{ (say)}$$

and

$$2\sin y = \{a^2 + (1+b)^2\}^{1/2} - \{a^2 + (1-b)^2\}^{1/2} = 2D,$$

where C is greater and D is less than unity.

Therefore

$$\sinh^{-1}(a+\iota b) = \cosh^{-1} C + \iota.\sin^{-1} D$$
$$= \log\{C + (C^2-1)^{1/2}\} + \iota.\sin^{-1} D.$$

It is to be noticed that the coefficient of ι is a multiple valued quantity, for if $\sin \alpha = D$,

$$\sin\{n\pi + (-)^n \alpha\} = D,$$

and thus

$$\sin^{-1} D = n\pi + (-)^n \alpha,$$

where n is any integer.

Similarly writing

$$\cosh^{-1}(a + \iota b) = x + \iota y,$$

we get

$$2\cosh x = \{(1+a)^2 + b^2\}^{1/2} + \{(1-a)^2 + b^2\}^{1/2} = 2A,$$

and

$$2\cos y = \{(1+a)^2 + b^2\}^{1/2} - \{(1-a)^2 + b^2\}^{1/2} = 2B,$$

and since A is greater than unity and B is less than unity, we have

$$\cosh^{-1}(a + \iota b) = \log\{A + (A^2 - 1)^{1/2}\} + \iota . \cos^{-1} B.$$

Similarly

$$\sin^{-1}(a + \iota b) = \sin^{-1} B + \iota . \log\{A + (A^2 - 1)^{1/2}\},$$

and

$$\cos^{-1}(a + \iota b) = \cos^{-1} B + \iota . \log\{A + (A^2 - 1)^{1/2}\}.$$

Let us now consider the value of the inverse hyperbolic tangent of $a + \iota b$. Writing

$$\tanh^{-1}(a + \iota b) = x + \iota y,$$

we get

$$a + \iota b = \tanh(x + \iota y) = \frac{\sinh 2x}{\cosh 2x + \cos 2y} + \iota \frac{\sin 2y}{\cosh 2x + \cos 2y}.$$

Therefore

$$a = \frac{\sinh 2x}{\cosh 2x + \cos 2y} \text{ and } b = \frac{\sin 2y}{\cosh 2x + \cos 2y}.$$

Thus

$$a^2 + b^2 + 1 = \frac{2\cosh 2x}{\cosh 2x + \cos 2y},$$

and

$$1 - a^2 - b^2 = \frac{2\cos 2y}{\cosh 2x + \cos 2y}.$$

Hence

$$\tanh 2x = 2a/(a^2 + b^2 + 1),$$

and

$$\tan 2y = 2b/(1 - a^2 - b^2).$$

Now $2a/(a^2 + b^2 + 1)$ is always less than 1, and thus,

$$\tanh^{-1}(a + \iota b) = \tfrac{1}{4}\log\frac{(a+1)^2 + b^2}{(a-1)^2 + b^2} + \iota . \tfrac{1}{2}\left(\tan^{-1}\frac{b}{1+a} + \tan^{-1}\frac{b}{1-a}\right).$$

Similarly we have

$$\begin{aligned}\coth^{-1}(a+\iota b) &= \tanh^{-1}\{1/(a+\iota b)\}\\ &= \tanh^{-1}\left\{\frac{a}{a^2+b^2}-\iota\frac{b}{a^2+b^2}\right\}\\ &= \tfrac{1}{4}\log\frac{(a+1)^2+b^2}{(a-1)^2+b^2}\\ &\quad+\iota.\tfrac{1}{2}\left(\tan^{-1}\frac{b}{1+a}+\tan^{-1}\frac{b}{1-a}\right).\end{aligned}$$

In both the equations given immediately above the coefficient of ι is a multiple valued quantity. We also have

$$\begin{aligned}\operatorname{sech}^{-1}(a+\iota b) &= \cosh^{-1}\left\{\frac{a}{a^2+b^2}-\iota\frac{b}{a^2+b^2}\right\}\\ &= \log\left[\frac{A+\{A^2-(a^2+b^2)\}^{1/2}}{(a^2+b^2)^{1/2}}\right]\\ &\quad+\iota.\cos^{-1}\frac{B}{(a^2+b^2)^{1/2}},\end{aligned}$$

and

$$\begin{aligned}\operatorname{cosech}^{-1}(a+\iota b) &= \log\left[\frac{C+\{C^2-(a^2+b^2)\}^{1/2}}{(a^2+b^2)^{1/2}}\right]\\ &\quad-\iota.\sin^{-1}\frac{D}{(a^2+b^2)^{1/2}}.\end{aligned}$$

Similarly we can prove that

$$\tan^{-1}(a+\iota b)=\tfrac{1}{2}\left\{\tan^{-1}\frac{a}{1+b}+\tan^{-1}\frac{a}{1-b}\right\}+\iota.\tfrac{1}{4}\log\frac{(1+b)^2+a^2}{(1-b)^2+a^2},$$

$$\cot^{-1}(a+\iota b)=-\tfrac{1}{2}\left\{\tan^{-1}\frac{a}{1+b}+\tan^{-1}\frac{a}{1-b}\right\}-\iota.\tfrac{1}{4}\log\frac{(1+b)^2+a^2}{(1-b)^2+a^2},$$

$$\sec^{-1}(a+\iota b)=\cos^{-1}\{B/(a^2+b^2)^{1/2}\}+\iota.\log\left[\frac{A+(A^2-a^2-b^2)^{1/2}}{(a^2+b^2)^{1/2}}\right],$$

and

$$\operatorname{cosec}^{-1}(a+\iota b)=\sin^{-1}\{B/(a^2+b^2)^{1/2}\}+\iota.\log\left[\frac{A+(A^2-a^2-b^2)^{1/2}}{(a^2+b^2)^{1/2}}\right].$$

Tables of Hyperbolic Sines, Cosines and Tangents.

x	sinh x	cosh x	tanh x	x	sinh x	cosh x	tanh x
0·00	0·0000	1·0000	0·0000	0·48	0·4987	1·1174	0·4462
0·01	0·0100	1·0001	0·0100	0·49	0·5099	1·1225	0·4542
0·02	0·0200	1·0002	0·0200	0·50	0·5211	1·1276	0·4621
0·03	0·0300	1·0005	0·0300	0·51	0·5324	1·1329	0·4700
0·04	0·0400	1·0008	0·0400	0·52	0·5438	1·1383	0·4777
0·05	0·0500	1·0013	0·0500	0·53	0·5552	1·1438	0·4854
0·06	0·0600	1·0018	0·0599	0·54	0·5666	1·1494	0·4930
0·07	0·0701	1·0025	0·0699	0·55	0·5782	1·1551	0·5005
0·08	0·0801	1·0032	0·0798	0·56	0·5897	1·1609	0·5080
0·09	0·0901	1·0041	0·0898	0·57	0·6014	1·1669	0·5154
0·10	0·1002	1·0050	0·0997	0·58	0·6131	1·1730	0·5227
0·11	0·1102	1·0061	0·1096	0·59	0·6248	1·1792	0·5299
0·12	0·1203	1·0072	0·1194	0·60	0·6367	1·1855	0·5370
0·13	0·1304	1·0085	0·1293	0·61	0·6485	1·1919	0·5441
0·14	0·1405	1·0098	0·1391	0·62	0·6605	1·1984	0·5511
0·15	0·1506	1·0113	0·1489	0·63	0·6725	1·2051	0·5581
0·16	0·1607	1·0128	0·1587	0·64	0·6846	1·2119	0·5649
0·17	0·1708	1·0145	0·1684	0·65	0·6968	1·2188	0·5717
0·18	0·1810	1·0162	0·1781	0·66	0·7090	1·2258	0·5784
0·19	0·1912	1·0181	0·1878	0·67	0·7213	1·2330	0·5850
0·20	0·2013	1·0201	0·1974	0·68	0·7336	1·2403	0·5915
0·21	0·2116	1·0221	0·2070	0·69	0·7461	1·2477	0·5980
0·22	0·2218	1·0243	0·2165	0·70	0·7586	1·2552	0·6044
0·23	0·2320	1·0266	0·2260	0·71	0·7712	1·2628	0·6107
0·24	0·2423	1·0289	0·2355	0·72	0·7838	1·2706	0·6169
0·25	0·2526	1·0314	0·2449	0·73	0·7966	1·2785	0·6231
0·26	0·2629	1·0340	0·2543	0·74	0·8094	1·2865	0·6291
0·27	0·2733	1·0367	0·2636	0·75	0·8223	1·2947	0·6352
0·28	0·2837	1·0395	0·2729	0·76	0·8353	1·3030	0·6411
0·29	0·2941	1·0424	0·2821	0·77	0·8484	1·3114	0·6469
0·30	0·3045	1·0453	0·2913	0·78	0·8615	1·3199	0·6527
0·31	0·3150	1·0484	0·3004	0·79	0·8748	1·3286	0·6584
0·32	0·3255	1·0516	0·3095	0·80	0·8881	1·3374	0·6640
0·33	0·3360	1·0550	0·3185	0·81	0·9015	1·3464	0·6696
0·34	0·3466	1·0584	0·3275	0·82	0·9150	1·3555	0·6751
0·35	0·3572	1·0619	0·3364	0·83	0·9286	1·3647	0·6805
0·36	0·3678	1·0655	0·3452	0·84	0·9423	1·3740	0·6858
0·37	0·3785	1·0692	0·3540	0·85	0·9561	1·3835	0·6911
0·38	0·3892	1·0731	0·3627	0·86	0·9700	1·3932	0·6963
0·39	0·4000	1·0770	0·3714	0·87	0·9840	1·4029	0·7014
0·40	0·4108	1·0811	0·3800	0·88	0·9981	1·4128	0·7064
0·41	0·4216	1·0852	0·3885	0·89	1·0122	1·4229	0·7114
0·42	0·4325	1·0895	0·3969	0·90	1·0265	1·4331	0·7163
0·43	0·4434	1·0939	0·4053	0·91	1·0409	1·4434	0·7211
0·44	0·4543	1·0984	0·4136	0·92	1·0554	1·4539	0·7259
0·45	0·4653	1·1030	0·4219	0·93	1·0700	1·4645	0·7306
0·46	0·4764	1·1077	0·4301	0·94	1·0847	1·4753	0·7352
0·47	0·4875	1·1125	0·4382	0·95	1·0995	1·4862	0·7398

x	sinh x	cosh x	tanh x	x	sinh x	cosh x	tanh x
0·96	1·1144	1·4973	0·7443	1·47	2·0597	2·2896	0·8996
0·97	1·1294	1·5085	0·7487	1·48	2·0827	2·3103	0·9015
0·98	1·1446	1·5199	0·7531	1·49	2·1059	2·3312	0·9033
0·99	1·1598	1·5314	0·7574	1·50	2·1293	2·3524	0·9052
1·00	1·1752	1·5431	0·7616	1·51	2·1529	2·3738	0·9069
1·01	1·1907	1·5549	0·7658	1·52	2·1768	2·3955	0·9087
1·02	1·2063	1·5669	0·7699	1·53	2·2008	2·4174	0·9104
1·03	1·2220	1·5790	0·7739	1·54	2·2251	2·4395	0·9121
1·04	1·2379	1·5913	0·7779	1·55	2·2496	2·4619	0·9138
1·05	1·2539	1·6038	0·7818	1·56	2·2743	2·4845	0·9154
1·06	1·2700	1·6164	0·7857	1·57	2·2993	2·5074	0·9170
1·07	1·2862	1·6292	0·7895	1·58	2·3245	2·5305	0·9186
1·08	1·3025	1·6421	0·7932	1·59	2·3499	2·5538	0·9202
1·09	1·3190	1·6553	0·7969	1·60	2·3756	2·5775	0·9217
1·10	1·3357	1·6685	0·8005	1·61	2·4015	2·6014	0·9232
1·11	1·3524	1·6820	0·8041	1·62	2·4276	2·6255	0·9246
1·12	1·3693	1·6956	0·8076	1·63	2·4540	2·6499	0·9261
1·13	1·3863	1·7093	0·8110	1·64	2·4806	2·6746	0·9275
1·14	1·4035	1·7233	0·8144	1·65	2·5075	2·6995	0·9289
1·15	1·4208	1·7374	0·8178	1·66	2·5346	2·7247	0·9302
1·16	1·4382	1·7517	0·8210	1·67	2·5620	2·7502	0·9316
1·17	1·4558	1·7662	0·8243	1·68	2·5896	2·7760	0·9329
1·18	1·4736	1·7808	0·8275	1·69	2·6175	2·8020	0·9342
1·19	1·4914	1·7957	0·8306	1·70	2·6456	2·8283	0·9354
1·20	1·5095	1·8107	0·8337	1·71	2·6741	2·8549	0·9367
1·21	1·5276	1·8258	0·8367	1·72	2·7027	2·8818	0·9379
1·22	1·5460	1·8412	0·8397	1·73	2·7317	2·9090	0·9391
1·23	1·5645	1·8568	0·8426	1·74	2·7609	2·9364	0·9402
1·24	1·5831	1·8725	0·8455	1·75	2·7904	2·9642	0·9414
1·25	1·6019	1·8884	0·8483	1·76	2·8202	2·9922	0·9425
1·26	1·6209	1·9045	0·8511	1·77	2·8503	3·0206	0·9436
1·27	1·6400	1·9208	0·8538	1·78	2·8806	3·0493	0·9447
1·28	1·6593	1·9373	0·8565	1·79	2·9113	3·0782	0·9458
1·29	1·6788	1·9540	0·8591	1·80	2·9422	3·1075	0·9468
1·30	1·6984	1·9709	0·8617	1·81	2·9734	3·1371	0·9478
1·31	1·7182	1·9880	0·8643	1·82	3·0049	3·1669	0·9488
1·32	1·7381	2·0053	0·8668	1·83	3·0367	3·1972	0·9498
1·33	1·7583	2·0228	0·8693	1·84	3·0689	3·2277	0·9508
1·34	1·7786	2·0404	0·8717	1·85	3·1013	3·2585	0·9518
1·35	1·7991	2·0583	0·8741	1·86	3·1340	3·2897	0·9527
1·36	1·8198	2·0764	0·8764	1·87	3·1671	3·3212	0·9536
1·37	1·8406	2·0947	0·8787	1·88	3·2005	3·3531	0·9545
1·38	1·8617	2·1132	0·8810	1·89	3·2342	3·3852	0·9554
1·39	1·8829	2·1320	0·8832	1·90	3·2682	3·4177	0·9562
1·40	1·9043	2·1509	0·8854	1·91	3·3025	3·4506	0·9571
1·41	1·9259	2·1701	0·8875	1·92	3·3372	3·4838	0·9579
1·42	1·9477	2·1894	0·8896	1·93	3·3722	3·5173	0·9587
1·43	1·9697	2·2090	0·8917	1·94	3·4075	3·5512	0·9595
1·44	1·9919	2·2288	0·8937	1·95	3·4432	3·5855	0·9603
1·45	2·0143	2·2488	0·8957	1·96	3·4792	3·6201	0·9611
1·46	2·0369	2·2691	0·8977	1·97	3·5156	3·6551	0·9619

Tables of Hyperbolic Sines, Cosines and Tangents (cont.).

x	$\sinh x$	$\cosh x$	$\tanh x$	x	$\sinh x$	$\cosh x$	$\tanh x$
1·98	3·5523	3·6904	0·9626	2·39	5·4109	5·5026	0·9834
1·99	3·5894	3·7261	0·9633	2·40	5·4662	5·5570	0·9837
2·00	3·6269	3·7622	0·9640	2·41	5·5221	5·6119	0·9840
2·01	3·6647	3·7987	0·9647	2·42	5·5785	5·6674	0·9843
2·02	3·7028	3·8355	0·9654	2·43	5·6354	5·7235	0·9846
2·03	3·7414	3·8727	0·9661	2·44	5·6929	5·7801	0·9849
2·04	3·7803	3·9103	0·9668	2·45	5·7510	5·8373	0·9852
2·05	3·8196	3·9483	0·9674	2·46	5·8097	5·8951	0·9855
2·06	3·8593	3·9867	0·9680	2·47	5·8689	5·9535	0·9858
2·07	3·8993	4·0255	0·9687	2·48	5·9288	6·0125	0·9861
2·08	3·9398	4·0647	0·9693	2·49	5·9892	6·0721	0·9864
2·09	3·9806	4·1043	0·9699	2·50	6·0502	6·1323	0·9866
2·10	4·0219	4·1443	0·9705	2·60	6·6947	6·7690	0·9890
2·11	4·0635	4·1847	0·9710	2·70	7·4063	7·4735	0·9910
2·12	4·1056	4·2256	0·9716	2·80	8·1919	8·2527	0·9926
2·13	4·1480	4·2669	0·9722	2·90	9·0596	9·1146	0·9940
2·14	4·1909	4·3086	0·9727	3·00	10·018	10·068	0·9951
2·15	4·2342	4·3507	0·9732	3·10	11·077	11·122	0·9960
2·16	4·2779	4·3932	0·9738	3·20	12·246	12·287	0·9967
2·17	4·3221	4·4362	0·9743	3·30	13·538	13·575	0·9973
2·18	4·3666	4·4797	0·9748	3·40	14·965	14·999	0·9978
2·19	4·4117	4·5236	0·9753	3·50	16·543	16·573	0·9982
2·20	4·4571	4·5679	0·9757	3·60	18·286	18·313	0·9985
2·21	4·5030	4·6127	0·9762	3·70	20·211	20·236	0·9988
2·22	4·5494	4·6580	0·9767	3·80	22·339	22·362	0·9990
2·23	4·5962	4·7037	0·9771	3·90	24·691	24·711	0·9992
2·24	4·6434	4·7499	0·9776	4·00	27·290	27·308	0·9993
2·25	4·6912	4·7966	0·9780	4·10	30·162	30·178	0·9995
2·26	4·7394	4·8437	0·9785	4·20	33·336	33·351	0·9996
2·27	4·7880	4·8914	0·9789	4·30	36·843	36·857	0·9996
2·28	4·8372	4·9395	0·9793	4·40	40·719	40·732	0·9997
2·29	4·8868	4·9881	0·9797	4·50	45·003	45·014	0·9998
2·30	4·9370	5·0372	0·9801	4·60	49·737	49·747	0·9998
2·31	4·9876	5·0868	0·9805	4·70	54·969	54·978	0·9998
2·32	5·0387	5·1370	0·9809	4·80	60·751	60·759	0·9999
2·33	5·0903	5·1876	0·9812	4·90	67·141	67·149	0·9999
2·34	5·1425	5·2388	0·9816	5·00	74·203	74·210	0·9999
2·35	5·1951	5·2905	0·9820	5·10	82·008	82·014	0·9999
2·36	5·2483	5·3427	0·9823	5·20	90·633	90·639	0·9999
2·37	5·3020	5·3954	0·9827	5·30	100·17	100·17	0·9999
2·38	5·3562	5·4487	0·9830	5·40	110·70	110·71	1·0000

For values of x greater than 5, we may write $\sinh x = \cosh x$, and $\tanh x = 1$, the maximum inaccuracy being less than 1 in 10 000.

x	sinh x	x	sinh x	x	sinh x
5·50	122·34	6·20	246·37	6·90	496·14
5·60	135·21	6·30	272·29	7·00	548·32
5·70	149·43	6·40	300·92	7·10	605·98
5·80	165·15	6·50	332·57	7·20	669·72
5·90	182·52	6·60	367·55	7·30	740·15
6·00	201·71	6·70	406·20	7·40	817·99
6·10	222·93	6·80	448·92	7·50	904·02

When Δ is not greater than 0·01 we may use the formulae

$$\sinh(x \pm \Delta) = \sinh x \pm \Delta \cosh x,$$

and

$$\cosh(x \pm \Delta) = \cosh x \pm \Delta \sinh x.$$

For example,

$$\sinh(2{\cdot}4215) = \sinh(2{\cdot}42) + 0{\cdot}0015 \cosh(2{\cdot}42)$$
$$= 5{\cdot}5870,$$

and

$$\cosh(0{\cdot}8899) = \cosh(0{\cdot}88) + 0{\cdot}0099 \sinh(0{\cdot}88)$$
$$= 1{\cdot}4227.$$

When Δ is not greater than 0·1, we have

$$\sinh(x \pm \Delta) = (1 + \Delta^2/2) \sinh x \pm (\Delta + \Delta^3/6) \cosh x,$$
$$\cosh(x \pm \Delta) = (1 + \Delta^2/2) \cosh x \pm (\Delta + \Delta^3/6) \sinh x.$$

For instance, $\sinh(4{\cdot}194) = \sinh(4{\cdot}2 - 0{\cdot}006)$

$$= \sinh(4{\cdot}2) - 0{\cdot}006 \cosh(4{\cdot}2) = 33{\cdot}14.$$

For values of x greater than 5, we may write

$$\log_{10}(2 \sinh x) = \log_{10}(2 \cosh x) = 0{\cdot}4343\, x.$$

When calculating the powers of sinh x and cosh x, the following formulae are also useful:

$$2 \sinh^2 x = \cosh 2x - 1, \qquad 2 \cosh^2 x = \cosh 2x + 1,$$
$$4 \sinh^3 x = \sinh 3x - 3 \sinh x, \qquad 4 \cosh^3 x = \cosh 3x + 3 \cosh x,$$
$$8 \sinh^4 x = \cosh 4x - 4 \cosh 2x + 3, \qquad 8 \cosh^4 x = \cosh 4x + 4 \cosh 2x + 3,$$

Trigonometrical Functions.

x	$\sin x$	$\cos x$	x	$\sin x$	$\cos x$	x	$\sin x$	$\cos x$
0·00	0·0000	1·0000	0·53	0·5055	0·8628	1·06	0·8724	0·4889
0·01	0·0100	1·0000	0·54	0·5141	0·8577	1·07	0·8772	0·4801
0·02	0·0200	0·9998	0·55	0·5227	0·8525	1·08	0·8820	0·4713
0·03	0·0300	0·9996	0·56	0·5312	0·8473	1·09	0·8866	0·4625
0·04	0·0400	0·9992	0·57	0·5396	0·8419	1·10	0·8912	0·4536
0·05	0·0500	0·9988	0·58	0·5480	0·8365	1·11	0·8957	0·4447
0·06	0·0600	0·9982	0·59	0·5564	0·8309	1·12	0·9001	0·4357
0·07	0·0699	0·9976	0·60	0·5646	0·8253	1·13	0·9044	0·4267
0·08	0·0799	0·9968	0·61	0·5729	0·8197	1·14	0·9086	0·4176
0·09	0·0899	0·9960	0·62	0·5810	0·8139	1·15	0·9128	0·4085
0·10	0·0998	0·9950	0·63	0·5891	0·8080	1·16	0·9168	0·3993
0·11	0·1098	0·9940	0·64	0·5972	0·8021	1·17	0·9208	0·3902
0·12	0·1197	0·9928	0·65	0·6052	0·7961	1·18	0·9246	0·3809
0·13	0·1296	0·9916	0·66	0·6131	0·7900	1·19	0·9284	0·3717
0·14	0·1395	0·9902	0·67	0·6210	0·7838	1·20	0·9320	0·3624
0·15	0·1494	0·9988	0·68	0·6288	0·7776	1·21	0·9356	0·3530
0·16	0·1593	0·9872	0·69	0·6365	0·7713	1·22	0·9391	0·3437
0·17	0·1692	0·9856	0·70	0·6442	0·7648	1·23	0·9425	0·3342
0·18	0·1791	0·9838	0·71	0·6518	0·7584	1·24	0·9458	0·3248
0·19	0·1889	0·9820	0·72	0·6594	0·7518	1·25	0·9490	0·3153
0·20	0·1987	0·9801	0·73	0·6669	0·7452	1·26	0·9521	0·3058
0·21	0·2085	0·9780	0·74	0·6743	0·7385	1·27	0·9551	0·2963
0·22	0·2182	0·9759	0·75	0·6816	0·7317	1·28	0·9580	0·2867
0·23	0·2280	0·9737	0·76	0·6889	0·7248	1·29	0·9608	0·2771
0·24	0·2377	0·9713	0·77	0·6961	0·7179	1·30	0·9636	0·2675
0·25	0·2474	0·9689	0·78	0·7033	0·7109	1·31	0·9662	0·2579
0·26	0·2571	0·9664	0·79	0·7104	0·7039	1·32	0·9687	0·2482
0·27	0·2667	0·9638	0·80	0·7174	0·6967	1·33	0·9712	0·2385
0·28	0·2764	0·9611	0·81	0·7243	0·6895	1·34	0·9735	0·2288
0·29	0·2860	0·9582	0·82	0·7312	0·6822	1·35	0·9757	0·2190
0·30	0·2955	0·9553	0·83	0·7379	0·6749	1·36	0·9779	0·2092
0·31	0·3051	0·9523	0·84	0·7446	0·6675	1·37	0·9799	0·1995
0·32	0·3146	0·9492	0·85	0·7513	0·6600	1·38	0·9819	0·1896
0·33	0·3240	0·9460	0·86	0·7578	0·6524	1·39	0·9837	0·1798
0·34	0·3335	0·9428	0·87	0·7643	0·6448	1·40	0·9854	0·1700
0·35	0·3429	0·9394	0·88	0·7707	0·6372	1·41	0·9871	0·1601
0·36	0·3523	0·9359	0·89	0·7771	0·6294	1·42	0·9887	0·1502
0·37	0·3616	0·9323	0·90	0·7833	0·6216	1·43	0·9901	0·1403
0·38	0·3709	0·9287	0·91	0·7895	0·6138	1·44	0·9915	0·1304
0·39	0·3802	0·9249	0·92	0·7956	0·6058	1·45	0·9927	0·1205
0·40	0·3894	0·9211	0·93	0·8016	0·5978	1·46	0·9939	0·1106
0·41	0·3986	0·9171	0·94	0·8076	0·5898	1·47	0·9949	0·1006
0·42	0·4078	0·9131	0·95	0·8134	0·5817	1·48	0·9959	0·0907
0·43	0·4169	0·9090	0·96	0·8192	0·5735	1·49	0·9967	0·0807
0·44	0·4259	0·9048	0·97	0·8249	0·5653	1·50	0·9975	0·0707
0·45	0·4350	0·9005	0·98	0·8305	0·5570	1·51	0·9982	0·0608
0·46	0·4440	0·8961	0·99	0·8360	0·5487	1·52	0·9987	0·0508
0·47	0·4529	0·8916	1·00	0·8415	0·5403	1·53	0·9992	0·0408
0·48	0·4618	0·8870	1·01	0·8468	0·5319	1·54	0·9995	0·0308
0·49	0·4706	0·8823	1·02	0·8521	0·5234	1·55	0·9998	0·0208
0·50	0·4794	0·8776	1·03	0·8573	0·5149	1·56	0·9999	0·0108
0·51	0·4882	0·8727	1·04	0·8624	0·5062	1·57	1·0000	0·0008
0·52	0·4969	0·8678	1·05	0·8674	0·4976	1·58	1·0000	−0·0092

Note that

$$\sin(1{\cdot}571 + x) = \cos x, \quad \cos(1{\cdot}571 + x) = -\sin x,$$

$$\sin(n\,.\,3{\cdot}1416 + x) = \sin x \quad \text{and} \quad \cos(n\,.\,3{\cdot}1416 + x) = \cos x.$$

We also have, when Δ is less than 0·01,

$$\sin(x \pm \Delta) = \sin x \pm \Delta \cos x,$$

and

$$\cos(x \pm \Delta) = \cos x \mp \Delta \sin x.$$

For example,

$$\cos(12{\cdot}739) = \cos(3{\cdot}314) = -\cos(0{\cdot}172)$$

$$= -\cos(0{\cdot}17) + 0{\cdot}002 \sin(0{\cdot}17) = -0{\cdot}9853;$$

$$\cos(1{\cdot}2011) = \cos(1{\cdot}20) - 0{\cdot}0011 \sin(1{\cdot}20) = 0{\cdot}3614.$$

When calculating the powers of $\sin x$ and $\cos x$ the following formulae are also useful:

$$2\sin^2 x = -\cos 2x + 1, \qquad 2\cos^2 x = \cos 2x + 1,$$

$$4\sin^3 x = -\sin 3x + 3\sin x, \qquad 4\cos^3 x = \cos 3x + 3\cos x,$$

$$8\sin^4 x = \cos 4x - 4\cos 2x + 3, \qquad 8\cos^4 x = \cos 4x + 4\cos 2x + 3,$$

....

REFERENCES

A. E. Kennelly, *Tables of Complex Hyperbolic and Circular Functions.*

—— *The Application of Hyperbolic Functions to Electrical Engineering Problems.*

J. A. Fleming, *The Propagation of Electric Currents in Telephone and Telegraph Conductors.*

E. Jahnke and F. Emde, *Funktionentafeln*, 1914.

Smithsonian Mathematical Tables, 'Hyperbolic Functions' by G. F. Becker and C. E. Van Orstrand. Five figure logarithms of $\sinh x$, $\cosh x$ and $\tanh x$ are given. Up to 0·1 the values of x proceed by steps of 0·0001, between 0·1 and 3, by steps of 0·001, and between 3 and 6, by steps of 0·01. Similar tables of the values of the hyperbolic functions are also given.

CHAPTER XX

Electric power transmission. The general equations. Line infinitely long. The distortion of the waves of P.D. and current. The attenuation factor α. The wave length factor β. The general solution. Effect of impedance of generator. Summation of vectors. Short lines. Line on open circuit. The voltage regulation. Line on short circuit. Experimental determination of constants. Direct currents. Polyphase lines. Single wire with earth return. The voltage drop. Corona loss. Numerical example. Polycyclic distribution. Dykes's system. Arnold's system. Polycyclic generator. References.

The general equations.

THE theorems given in the preceding chapter enable us to express the solution of the general equations of power transmission (21) and (22) given on p. 505 in a concise and convenient form. By differentiating these equations we easily deduce that

$$\frac{\partial^2 e}{\partial x^2} = P^2 e \quad \text{and} \quad \frac{\partial^2 i}{\partial x^2} = P^2 i \quad \text{.................(1)},$$

where

$$P^2 = (r + lD)(s + kD).$$

If we assume that e and i follow the harmonic law we have $D^2 = -\omega^2$, and thus

$$P^2 = rs - kl\omega^2 + \iota(rk + ls)\omega$$
$$= yz(\cos\theta + \iota \sin\theta),$$

where

$$y^2 = r^2 + l^2\omega^2, \quad z^2 = s^2 + k^2\omega^2,$$

$$\cos\theta = (rs - kl\omega^2)/yz \quad \text{and} \quad \sin\theta = (rk + ls)/yz \quad \text{...(2)}.$$

These equations follow since

$$y^2 z^2 = (rs - kl\omega^2)^2 + (rk + ls)^2\omega^2.$$

We also have

$$y^2 z^2 = (rs + kl\omega^2)^2 + (rk - ls)^2\omega^2.$$

By De Moivre's theorem we get

$$\pm P = y^{1/2} z^{1/2} \{\cos(\theta/2) + \iota \sin(\theta/2)\},$$

and since

$$\cos^2(\theta/2) = 1/2 + (1/2)\cos\theta, \quad \sin^2(\theta/2) = 1/2 - (1/2)\cos\theta,$$

we get by (2)

$$\pm P = \{yz/2 + (rs - kl\omega^2)/2\}^{1/2} + \iota\{yz/2 - (rs - kl\omega^2)/2\}^{1/2}$$
$$= \alpha + \iota\beta,$$

where $\alpha = \{yz/2 + (rs - kl\omega^2)/2\}^{1/2}$ ....(3),

and $\beta = \{yz/2 - (rs - kl\omega^2)/2\}^{1/2}$ ....(4).

The symbolical solutions of equations (1) are thus, using the notation of Vol. I, p. 293,

$[e] = A_1 \cosh Px + A_2 \sinh Px$ ....(5),

and $[i] = B_1 \cosh Px + B_2 \sinh Px$ ....(6),

where A_1, A_2, B_1 and B_2 are constants.

When $x = 0$, let $[e] = [e_1]$ and $[i] = [i_1]$. Hence A_1 must equal $[e_1]$ and B_1 must equal $[i_1]$.

Also since (p. 505)

$$-\partial e/\partial x = ri + l(\partial i/\partial t),$$

we get by substituting for e its value from (5), and then putting $x = 0$, that $-A_2P = (r + lD)[i_1]$, and thus

$$[e] = [e_1]\cosh Px - [i_1]\{(r + \iota l\omega)/P\}\sinh Px.$$

Similarly since $\quad -\partial i/\partial x = se + k(\partial e/\partial t),$

we get $\quad [i] = [i_1]\cosh Px - [e_1]\{P/(r + \iota l\omega)\}\sinh Px.$

If we write

$$[z_0] = \left\{\frac{r + \iota l\omega}{s + \iota k\omega}\right\}^{1/2},$$

these equations become

$[e] = [e_1]\cosh Px - [z_0][i_1]\sinh Px$ ....(7),

and $[i] = [i_1]\cosh Px - \{[e_1]/[z_0]\}\sinh Px$ ....(8).

In these equations a quantity in square brackets, $[e_1]$ for instance, stands for a vector quantity $E_1 \cos\omega t + \iota E_1 \sin\omega t$ or $E_1 \epsilon^{\omega t\iota}$, so that its real part gives the instantaneous value of the voltage at the sending end of the line. If we suppose that $[e_1]$

is given, then $[i_1]$ is determined by the terminal conditions, and hence from equations (7) and (8) we find the instantaneous values of the current and the voltage at any distance from the station in complex form.

The ratio of $[e_1]$ to $[i_1]$ is called the 'sending end impedance' of the line. We shall denote its value by $[z_1]$.

Line infinitely long.

If the line be infinitely long $[e]$ and $[i]$ must be zero when x is infinitely great, otherwise we should have an infinite amount of power expended on the line. Since in this case $\cosh Px$ and $\sinh Px$ are practically equal to one another and both are infinitely great we see by (7) and (8) that

$$[e_1]/[i_1] = [z_0].$$

Therefore $[z_0]$ is the value of the 'sending end impedance' for an infinitely long line.

Now we have

$$r + \iota l\omega = (r^2 + l^2\omega^2)^{1/2} (\cos\theta_1 + \iota \sin\theta_1)$$

$$= y\epsilon^{\theta_1 \iota},$$

and $$s + \iota k\omega = z\epsilon^{\theta_2 \iota},$$

where $$\tan\theta_1 = \omega l/r \quad \text{and} \quad \tan\theta_2 = \omega k/s.$$

Hence $$[z_0] = \left\{\frac{r + \iota l\omega}{s + \iota k\omega}\right\}^{1/2} = (y/z)^{1/2}\,\epsilon^{\{(\theta_1 - \theta_2)/2\}\iota}$$

$$= \rho_0 + \lambda_0\omega\iota \quad \text{..................(9)},$$

where

$$\rho_0 = (y/z)^{1/2}\cos\{(\theta_1 - \theta_2)/2\} \quad \text{and} \quad \lambda_0\omega = (y/z)^{1/2}\sin\{(\theta_1 - \theta_2)/2\}.$$

We see that if θ_2 is less than θ_1, λ_0 must be positive; if θ_2 is equal to θ_1, λ_0 vanishes, and if θ_2 is greater than θ_1, λ_0 must be negative. Hence when l/r is greater than k/s the line acts like an inductive coil; when $l/r = k/s$ it acts like a non-inductive resistance the value ρ_0 of which per unit length equals $(r/s)^{1/2}$, and when l/r is less than k/s it acts as if it had zero inductance but capacity k_0 per unit length, where

$$-1/(k_0\omega) = (y/z)^{1/2}\sin\{(\theta_1 - \theta_2)/2\}$$

and so $$k_0 = (z/y)^{1/2}/[\omega\sin\{(\theta_2 - \theta_1)/2\}].$$

Noticing that for an infinitely long line $[e_1] = [z_0][i_1]$ we get from (7) and (8) that

$$[e] = [e_1](\cosh Px - \sinh Px) = [e_1]\,\epsilon^{-Px}$$
$$= [e_1]\,\epsilon^{-(\alpha+\iota\beta)x},$$

and $$[i] = [i_1]\,\epsilon^{-(\alpha+\iota\beta)x}.$$

Thus the instantaneous values of the voltage and the current at a distance x along the line are given by

$$e = E_1\epsilon^{-\alpha x}\cos(\omega t - \beta x) \quad\ldots\ldots\ldots\ldots\ldots\ldots(10),$$

and $$i = E_1(z/y)^{1/2}\epsilon^{-\alpha x}\cos\{\omega t - \beta x - (\theta_1 - \theta_2)/2\}\ldots\ldots(11).$$

Equations (10) and (11) show that the amplitudes of e and i diminish according to the exponential law. For instance, the ratio of the amplitudes of e or i at distances x_2 and x_1 from the generating station is $\epsilon^{-\alpha(x_2-x_1)}$. We shall call α the 'attenuation factor' of the line.

It is easy to see from (10) and (11) that when x becomes $x + \lambda$ the phases of the potential difference and the current do not change provided that $\lambda = 2\pi/\beta$. We may therefore call $2\pi/\beta$ the 'wave length' of the disturbance and β the 'wave length factor.' The velocity of the propagation $= \lambda/T = \lambda f = \omega/\beta$.

The distortion of the waves of P. D. and current.

When the applied wave of P.D. at the origin is not sine shaped, then the shape of the wave of P.D. changes as x increases. For if the applied wave of P.D., at x equal to zero, be given by

$$e_0 = E_1\cos\omega t + E_3\cos(3\omega t - \gamma') + \ldots,$$

the values of e and i at x will be given by

$$e = E_1\epsilon^{-\alpha_1 x}\cos(\omega t - \beta_1 x) + E_3\epsilon^{-\alpha_3 x}\cos(3\omega t - \beta_3 x - \gamma') + \ldots,$$

and

$$i = E_1(z/y)^{1/2}\epsilon^{-\alpha_1 x}\cos\{\omega t - \beta_1 x - (\theta_1 - \theta_2)/2\}$$
$$+ E_3(z_3/y_3)^{1/2}\epsilon^{-\alpha_3 x}\cos\{3\omega t - \beta_3 x - (\theta_1' - \theta_2')/2\}$$
$$+ \ldots\ldots,$$

where $$z_3 = \{s^2 + 9k^2\omega^2\}^{1/2}, \quad y_3 = \{r^2 + 9l^2\omega^2\}^{1/2},$$
$$\alpha_3 = \{y_3 z_3/2 + (rs - 9kl\omega^2)/2\}^{1/2}, \quad \tan\theta_1' = 3\omega l/r, \text{ etc.}$$

These equations show that the shapes of both the voltage and current waves are, in general, different at all points of the transmission line, even when there are no reflected waves. We prove below that both β_n and ω/β_n increase as n increases. Hence the higher the order of the harmonic wave the greater is the speed with which it travels.

From (3) we see that the value of a is given by the equation

The attenuation factor a.

$$2a^2 = (r^2 + l^2\omega^2)^{1/2}(s^2 + k^2\omega^2)^{1/2} + rs - kl\omega^2 \ldots (12).$$

Assuming that l is the only variable, we get

$$\frac{\partial}{\partial l}(2a^2) = \omega^2 l \left\{\left(\frac{s^2 + k^2\omega^2}{r^2 + l^2\omega^2}\right)^{1/2} - \frac{k}{l}\right\}$$

$$= \frac{\omega^2}{(r^2 + l^2\omega^2)^{1/2}} \left\{\frac{l^2 s^2 - k^2 r^2}{l(s^2 + k^2\omega^2)^{1/2} + k(r^2 + l^2\omega^2)^{1/2}}\right\}.$$

Hence if l/r is less than k/s, a diminishes as l increases, attaining its minimum value $(rs)^{1/2}$ when $l/r = k/s$.

Similarly, if k be the only variable, a has its minimum value when $k/s = l/r$.

Again since (12) may be written

$$2a^2 = \{(ls - kr)^2\omega^2 + (rs + kl\omega^2)^2\}^{1/2} + rs - kl\omega^2,$$

we see that when k, l and ω vary, r and s remaining constant, a has its minimum value $(rs)^{1/2}$ either when $l/r = k/s$ or when ω is zero.

Finally

$$\frac{\partial(2a^2)}{\partial\omega} = \omega\left\{l\left(\frac{s^2 + k^2\omega^2}{r^2 + l^2\omega^2}\right)^{1/4} - k\left(\frac{r^2 + l^2\omega^2}{s^2 + k^2\omega^2}\right)^{1/4}\right\}^2$$

$$= \text{a positive quantity},$$

provided that $l/r - k/s$ is not zero. In the general case therefore a increases as the frequency increases. When the frequency is zero the value of a is $(rs)^{1/2}$ and when it is infinite its value is

$$(ls + kr)/\{2(kl)^{1/2}\}.$$

If $l/r = k/s$ the value of a is independent of the frequency, being always equal to $(rs)^{1/2}$.

From equation (4) for the value of β it is obvious that the greater the value of l or the greater the value of k, the greater will be the value of β. Similarly the smaller the value of r or s the smaller will be the value of β. When r and s are very small compared with $l\omega$ and $k\omega$ we have $\beta = \omega\,(kl)^{1/2}$ and this is its least possible value. Hence the maximum possible value of the velocity of propagation ω/β is $1/(kl)^{1/2}$. If we suppose that the transmission line consists of two parallel overhead cylindrical wires each of radius a and if c be the distance between their axes, we have (Vol. I, pp. 88 and 165)

The wave length factor β.

$$l = 4\log(c/a) + 1, \quad 1/k = v^2\{4\log(c/a) - 4a^2/c^2 - \dots\},$$

where v is the velocity of light. It is to be noticed that we have made the assumption that the current density is uniform over the cross section of the wires. Hence $1/(kl)^{1/2}$ is always less than v but approaches this value when a/c is very small. We see that the smaller the values of r and s, and the further the wires are apart, the greater the velocity of propagation of the electrical disturbance.

From equation (4) we get

$$\frac{\partial(2\beta^2)}{\partial\omega} = \omega\left\{l\left(\frac{s^2+k^2\omega^2}{r^2+l^2\omega^2}\right)^{1/4} + k\left(\frac{r^2+l^2\omega^2}{s^2+k^2\omega^2}\right)^{1/4}\right\}^2$$

$$= \text{a positive quantity},$$

and hence the value of β continually increases as the frequency increases from zero to infinity.

When ω is very small, it is easy to see that

$$\omega/\beta = 2\,(rs)^{1/2}/(ls+kr)$$

approximately, and when ω is very great $\omega/\beta = 1/(lk)^{1/2}$. In the special case when $l/r = k/s$, we have $\omega/\beta = 1/(lk)^{1/2}$. The velocity of propagation in this case therefore is independent of the frequency. Since $\omega/\beta = 2a/(ls+kr)$ and a increases as ω increases, we see that the velocity of propagation increases from $2\,(rs)^{1/2}/(ls+kr)$ to $1/(lk)^{1/2}$ as the frequency increases.

Let us suppose that the receiving station is at a distance a from the generating station. We shall denote the complex value $R_2 + \iota L_2\omega$ of the impedance of the

The general solution.

load by $[z_2]$, so that $[e_2] = [z_2][i_2]$, the suffix 2 denoting that the values refer to quantities at the receiving end of the line. From equations (7) and (8) we get

$$[e_2] = [e_1] \cosh Pa - [z_0][i_1] \sinh Pa \quad \text{.........(13)},$$

and
$$[i_2] = [i_1] \cosh Pa - \{[e_1]/[z_0]\} \sinh Pa \quad \text{......(14)}.$$

Writing $[z_1]$ for $[e_1]/[i_1]$, we have

$$[z_2] = \frac{[e_2]}{[i_2]} = \frac{[z_1] - [z_0] \tanh Pa}{1 - \{[z_1]/[z_0]\} \tanh Pa},$$

and so
$$[z_1] = \frac{[e_1]}{[i_1]} = [z_0] \frac{[z_2]/[z_0] + \tanh Pa}{1 + \{[z_2]/[z_0]\} \tanh Pa}$$

$$= [z_0] \tanh (Pa + \gamma) \quad \text{..........................(15)},$$

where $\tanh \gamma = [z_2]/[z_0]$.

We see from (9) that

$$[z_0] = (y/z)^{1/2} \epsilon^{\{(\theta_1 - \theta_2)/2\}\iota}.$$

If $Z_2 = (R_2^2 + L_2^2\omega^2)^{1/2}$ and $\cos \phi_2 = R_2/Z_2 =$ the power factor of the load, we have $[z_2] = Z_2 \epsilon^{\phi_2 \iota}$. Hence

$$\tanh \gamma = [z_2]/[z_0] = B_2 \epsilon^{\xi_2 \iota} \quad \text{...............(16)},$$

where $B_2 = Z_2 (z/y)^{1/2}$ and $\xi_2 = \phi_2 - (\theta_1 - \theta_2)/2$.

If we write
$$\gamma = \tanh^{-1} (B_2 \epsilon^{\xi_2 \iota}) = \delta + \iota \delta' \quad \text{...............(17)},$$

we get, by p. 522, that

$$\delta = (1/4) \log \{(1 + 2B_2 \cos \xi_2 + B_2^2)/(1 - 2B_2 \cos \xi_2 + B_2^2)\} \quad \text{.........(18)},$$

and
$$\delta' = \tfrac{1}{2} [\tan^{-1} \{2B_2 \sin \xi_2/(1 - B_2^2)\}] \quad \text{.........(19)}.$$

From the theorems given in Chapter XIX it readily follows that

$$\tanh 2\delta = 2B_2 \cos \xi_2/(1 + B_2^2), \quad \tan 2\delta' = 2B_2 \sin \xi_2/(1 - B_2^2),$$

$$\tan \xi_2 = \sin 2\delta'/\sinh 2\delta,$$

and
$$B_2^2 = (\cosh 2\delta - \cos 2\delta')/(\cosh 2\delta + \cos 2\delta').$$

We have also

$$[e_1]/[i_1] = [z_0] \tanh (Pa + \gamma)$$
$$= (y/z)^{1/2} \epsilon^{\{(\theta_1 - \theta_2)/2\}\iota} \tanh \{(\alpha a + \delta) + \iota (a\beta + \delta')\} \text{...(20)},$$

and thus by p. 518 we get finally

$$\frac{[e_1]}{[i_1]} = \left[\frac{y\{\cosh 2(a\alpha+\delta) - \cos 2(a\beta+\delta')\}}{z\{\cosh 2(a\alpha+\delta) + \cos 2(a\beta+\delta')\}}\right]^{1/2} \epsilon^{\{\xi+(\theta_1-\theta_2)/2\}\iota}$$

$$= R_1 + \iota L_1\omega \quad \text{...(21)},$$

where $$\tan\xi = \{\sin 2(a\beta+\delta')\}/\{\sinh 2(a\alpha+\delta)\} \quad \text{......(22)}.$$

It will be seen that R_1 and L_1 are the effective resistance and inductance at the generating end of the line. Hence, we have

$$R_1 = \left[\frac{y\{\cosh 2(a\alpha+\delta) - \cos 2(a\beta+\delta')\}}{z\{\cosh 2(a\alpha+\delta) + \cos 2(a\beta+\delta')\}}\right]^{1/2} \cos\{\xi+(\theta_1-\theta_2)/2\} \quad \text{.........(23)},$$

and

$$L_1\omega = \left[\frac{y\{\cosh 2(a\alpha+\delta) - \cos 2(a\beta+\delta')\}}{z\{\cosh 2(a\alpha+\delta) + \cos 2(a\beta+\delta')\}}\right]^{1/2} \sin\{\xi+(\theta_1-\theta_2)/2\} \quad \text{.........(24)}.$$

At the generating station, therefore, the angle of lag of i_1 behind e_1 is $\xi + (\theta_1 - \theta_2)/2$.

Since by (15), $[e_1] = [i_1][z_0]\tanh(Pa+\gamma)$ we easily find from (7) and (8) that

$$[e] = [e_1]\left\{\cosh Px - \frac{\sinh Px}{\tanh(Pa+\gamma)}\right\} = [e_1]\frac{\sinh\{P(a-x)+\gamma\}}{\sinh(Pa+\gamma)} \quad \text{.........(25)},$$

$$[i] = [i_1]\left\{\cosh Px - \frac{\sinh Px}{\coth(Pa+\gamma)}\right\} = [i_1]\frac{\cosh\{P(a-x)+\gamma\}}{\cosh(Pa+\gamma)} \quad \text{.........(26)},$$

and $$[z] = \frac{[e]}{[i]} = [z_0]\tanh\{P(a-x)+\gamma\} \quad \text{.........(27)}.$$

We can therefore find the effective resistance R and the effective inductance L at any place on the line at a distance x from the generating station by writing $a - x$ for a in the formulae (23) and (24), noticing that $a - x$ must also be written for a in formula (22).

If $[e_2]$ and $[i_2]$ be the values of $[e]$ and $[i]$ at the receiving station where $x = a$, and if $[e_1] = E_1\epsilon^{\omega t\iota}$, we get

$$[e_2] = [e_1]\frac{\sinh\gamma}{\sinh(Pa+\gamma)} = [e_1]\frac{\sinh(\delta+\iota\delta')}{\sinh\{a\alpha+\delta+\iota(a\beta+\delta')\}}$$

$$= E_1\left\{\frac{\cosh 2\delta - \cos 2\delta'}{\cosh 2(a\alpha+\delta) - \cos 2(a\beta+\delta')}\right\}^{1/2} \epsilon^{(\omega t+\eta-\eta')\iota},$$

where $\tan \eta = \tan \delta' / \tanh \delta$

and $\tan \eta' = \tan (a\beta + \delta')/\tanh (a\alpha + \delta)$ ....(28).

Hence the instantaneous value e_2 of the voltage at the receiving station is given by

$$e_2 = E_1 \left\{ \frac{\cosh 2\delta - \cos 2\delta'}{\cosh 2(a\alpha + \delta) - \cos 2(a\beta + \delta')} \right\}^{1/2} \cos (\omega t + \eta - \eta') \quad \text{.........(29).}$$

Similarly we get

$$i_2 = I_1 \left\{ \frac{\cosh 2\delta + \cos 2\delta'}{\cosh 2(a\alpha + \delta) + \cos 2(a\beta + \delta')} \right\}^{1/2} \times \cos \{\omega t + \zeta - \zeta' - \xi - (\theta_1 - \theta_2)/2\} \quad \text{......(30),}$$

where $\tan \zeta = \tanh \delta \, \tan \delta'$,

and $\tan \zeta' = \tanh (a\alpha + \delta) \tan (a\beta + \delta')$.

In formulae (29) and (30), α and β are given by (3) and (4), $\tan \theta_1 = l\omega/r$, $\tan \theta_2 = k\omega/s$, δ and δ' are given by (18) and (19), ξ is found from (22), and η and η' are computed from (28).

Effect of impedance of generator.

If $[e_g]$ denote the generator electromotive force and $[z_g]$ denote the generator impedance, we have

$$[e_g] = [e_1] + [i_1][z_g],$$

and thus by (27)

$$[e_g] = [i_1]\{[z_0] \tanh (Pa + \gamma) + [z_g]\},$$

and hence
$$[i_1] = \frac{[e_g]}{[z_0] \tanh (Pa + \gamma) + [z_g]} \quad \text{............(31).}$$

Since also $[e_1] = [i_1][z_0] \tanh (Pa + \gamma)$,

$[e_1]$ can be found after $[i_1]$ is calculated from (31).

These formulae give the complete solution of the problem on our assumptions, and when tables of hyperbolic functions are available the labour of evaluating them in any special case is not excessive.

We shall now consider special problems in connection with power transmission.

It readily follows from De Moivre's theorem that

Summation of vectors.

$$\left.\begin{array}{rl} & a_1\epsilon^{\alpha_1\iota} + a_2\epsilon^{\alpha_2\iota} + a_3\epsilon^{\alpha_3\iota} = B\epsilon^{\alpha\iota} \\ \text{or} & a_1 \underline{|\alpha_1} + a_2 \underline{|\alpha_2} + a_3 \underline{|\alpha_3} = B \underline{|\alpha} \end{array}\right\} \ldots\ldots\ldots\ldots\ldots(32),$$

where

$$B^2 = a_1^2 + a_2^2 + a_3^2 + 2a_2a_3 \cos(\alpha_2 - \alpha_3) + 2a_3a_1 \cos(\alpha_3 - \alpha_1) + 2a_1a_2 \cos(\alpha_1 - \alpha_2),$$

and
$$\tan \alpha = \Sigma a_1 \sin \alpha_1 / \Sigma a_1 \cos \alpha_1.$$

Short lines.

When the real and imaginary parts of $P^3a^3/6$, that is of $(a\alpha + \iota a\beta)^3/6$, can be neglected compared with unity, we may write

$$\sinh Pa = Pa = a\alpha + \iota a\beta = a\,(yz)^{1/2}\epsilon^{\iota\theta/2} \quad \ldots\ldots(33),$$

and
$$\cosh Pa = 1 + P^2a^2/2 = 1 + a^2yz\epsilon^{\iota\theta}/2 \quad \ldots\ldots\ldots(34),$$

where, by (2),

$$\theta = \tan^{-1}(\beta/\alpha) = \theta_1 + \theta_2 = \tan^{-1}(l\omega/r) + \tan^{-1}(k\omega/s).$$

In many practical cases, especially when the lines are short, these approximate formulae are sufficiently accurate. If Z_2 be the impedance and ϕ_2 be the power factor of the load, we have, by (16),

$$\coth \gamma = (1/Z_2)\,(y/z)^{1/2}\epsilon^{-\iota\{\phi_2 - (\theta_1 - \theta_2)/2\}}.$$

In practice we are generally given the voltage $[e_2]$ and the current $[i_2]$ at the receiving station and we want to calculate the voltage $[e_1]$ and the current $[i_1]$ at the generating station. By (25) we easily find that

$$\begin{aligned}[e_1] &= [e_2]\,(\sinh Pa \coth \gamma + \cosh Pa) \\ &= [e_2]\{1 + (ay/Z_2)\,\epsilon^{\iota(\theta_1 - \phi_2)} + (a^2yz/2)\,\epsilon^{\iota(\theta_1 + \theta_2)}\} \\ &= [e_2]\,B\epsilon^{\iota\psi},\end{aligned}$$

where

$$B^2 = 1 + 2\,(ay/Z_2)\cos(\theta_1 - \phi_2) + (ay/Z_2)^2 + a^2yz\cos(\theta_1 + \theta_2)$$

approximately, and

$$\tan\psi = \frac{(ay/Z_2)\sin(\theta_1 - \phi_2) + (a^2yz/2)\sin(\theta_1 + \theta_2)}{1 + (ay/Z_2)\cos(\theta_1 - \phi_2) + (a^2yz/2)\cos(\theta_1 + \theta_2)}.$$

Thus if
$$e_2 = E_2 \cos \omega t,$$

we get
$$e_1 = BE_2 \cos(\omega t + \psi) \quad \ldots\ldots\ldots\ldots\ldots\ldots(35).$$

If in the formulae for B and ψ we write x for a, then $BE_2 \cos(\omega t + \psi)$ gives the potential difference between the lines at a distance x from the distributing station.

Similarly putting $x = a$ in (26) we have

$$\begin{aligned}[i_1] &= [i_2]\{\cosh Pa + \sinh Pa \tanh \gamma\} \\ &= [i_2]\{1 + azZ_2\epsilon^{\iota(\phi_2+\theta_2)} + (a^2yz/2)\epsilon^{\iota(\theta_1+\theta_2)}\} \\ &= [i_2]B'\epsilon^{\iota\psi'},\end{aligned}$$

where

$$B'^2 = 1 + 2azZ_2 \cos(\phi_2 + \theta_2) + (azZ_2)^2 + a^2yz \cos(\theta_1 + \theta_2)$$

approximately, and

$$\tan \psi' = \frac{azZ_2 \sin(\phi_2 + \theta_2) + (a^2yz/2) \sin(\theta_1 + \theta_2)}{1 + azZ_2 \cos(\phi_2 + \theta_2) + (a^2yz/2) \cos(\theta_1 + \theta_2)}.$$

Thus if $i_2 = (E_2/Z_2) \cos(\omega t - \phi_2)$,

then $i_1 = B'(E_2/Z_2) \cos(\omega t - \phi_2 + \psi')$ ....(36).

We also have $Z_1 = (B/B') Z_2$.

For short lines we readily deduce that

$$V_1 = V_2\{1 + (ay/Z_2) \cos(\theta_1 - \phi_2)\},$$

and

$$A_1 = A_2\{1 + azZ_2 \cos(\theta_2 + \phi_2)\}.$$

In practice s is usually negligible and so we may write $z = k\omega$ and $\theta_2 = \pi/2$. Thus we have

$$V_1 = V_2\{1 + (a/Z_2)(r \cos \phi_2 + l\omega \sin \phi_2)\},$$

and

$$A_1 = A_2\{1 - ak\omega Z_2 \sin \phi_2\}.$$

For lines up to twenty miles long and frequencies not exceeding fifty the inaccuracy of these formulae is generally less than one per cent. provided that ay/Z_2 and azZ_2 are not greater than 0·1.

In this case we see that if $\theta_1 - \phi_2$ is greater than $\pi/2$, V_1 is less than V_2, and if $\theta_2 + \phi_2$ is greater than $\pi/2$, A_1 is less than A_2.

The regulation of the line is given by the formula

$$\frac{V_1 - V_2}{V_2} = \frac{ay \cos(\theta_1 - \phi_2)}{Z_2},$$

which is the same as that obtained when we neglect the capacity of the mains.

When the line is on open circuit R_2 and therefore Z_2 is infinite. Hence B_2 is infinite and thus by (17), (18) and (19) we get $\gamma = (n\pi + \pi/2)\,\iota$. Substituting this value for γ in (25), (26) and (27), we get

Line on open circuit.

$$[e] = [e_1]\frac{\cosh P(a-x)}{\cosh Pa}; \quad [i] = [i_1]\frac{\sinh P(a-x)}{\sinh Pa};$$

and
$$[z] = [z_0]\coth P(a-x) \quad \text{.................. (37).}$$

We deduce at once by the formulae given on p. 518 that if

$$e_1 = E_1 \cos \omega t,$$

$$e = E_1\left\{\frac{\cosh 2\alpha(a-x) + \cos 2\beta(a-x)}{\cosh 2\alpha a + \cos 2\beta a}\right\}^{1/2} \cos(\omega t + \gamma' - \gamma_0') \quad \text{.........(38),}$$

where
$$\tan\gamma' = \tanh\alpha(a-x)\tan\beta(a-x)$$

and
$$\tan\gamma_0' = \tanh\alpha a \tan\beta a,$$

$$R = \left\{\frac{y}{z}\cdot\frac{\cosh 2\alpha(a-x) + \cos 2\beta(a-x)}{\cosh 2\alpha(a-x) - \cos 2\beta(a-x)}\right\}^{1/2} \cos\{\gamma' - \sigma + (\theta_1 - \theta_2)/2\} \quad \text{.....(39),}$$

where
$$\tan\sigma = \tan\beta(a-x)/\tanh\alpha(a-x),$$

$$\omega L = \left\{\frac{y}{z}\cdot\frac{\cosh 2\alpha(a-x) + \cos 2\beta(a-x)}{\cosh 2\alpha(a-x) - \cos 2\beta(a-x)}\right\}^{1/2} \sin\{\gamma' - \sigma + (\theta_1 - \theta_2)/2\} \quad \text{.....(40)}$$

and finally, since

$$[i] = \frac{[e_1]}{[z_0]}\frac{\sinh P(a-x)}{\cosh Pa},$$

we get

$$i = E_1\left\{\frac{z}{y}\cdot\frac{\cosh 2\alpha(a-x) - \cos 2\beta(a-x)}{\cosh 2\alpha a + \cos 2\beta a}\right\}^{1/2}$$
$$\times \cos\{\omega t - (\theta_1 - \theta_2)/2 + \sigma - \gamma_0'\} \quad \text{.........(41).}$$

By (38) we see that if V_1 be the voltage at the generating station and V_2 be the voltage at the distributing station, we have

$$V_2 = V_1/\{(\cosh 2\alpha a + \cos 2\beta a)/2\}^{1/2} \quad \text{...........(42).}$$

Now

$$(\cosh 2\alpha a + \cos 2\beta a)/2$$
$$= 1 + a^2(\alpha^2 - \beta^2) + a^4(\alpha^4 + \beta^4)/3 + \ldots$$
$$= 1 + a^2(rs - kl\omega^2) + a^4y^2z^2/6 + a^4(rs - kl\omega^2)^2/6 + \ldots.$$

When $rs - kl\omega^2$ is positive we see that V_2 is always less than V_1. But when $rs - kl\omega^2$ is negative and a^2 is small, the number by which V_1 is divided in (42) may be less than unity, and so V_2 can be greater than unity. In practice s is generally negligibly small and hence in certain cases there can be a rise of pressure at the receiving end of the line.

The value of a that makes $\cosh 2\alpha a + \cos 2\beta a$ a minimum is given by the smallest positive root, other than zero, of the equation

$$\alpha \sinh 2\alpha a = \beta \sin 2\beta a.$$

This equation can easily be solved graphically.

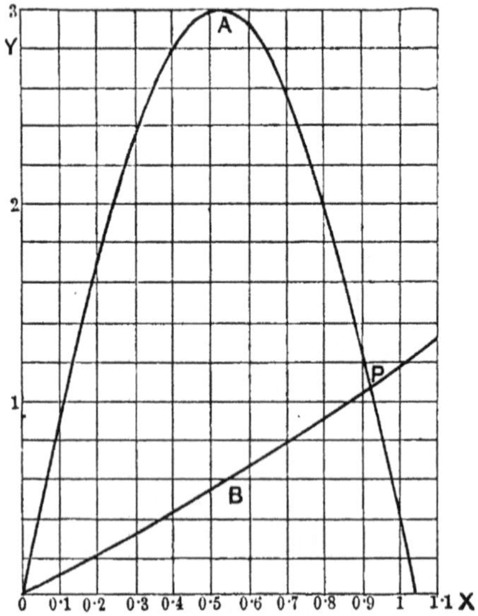

Fig. 233. The value of x (0·93) at the point where the curves OAP ($y = 3 \sin 3x$) and OBP ($y = \sinh x$) intersect makes the ordinate of the curve ABC (Fig. 234) a minimum.

Suppose, for instance, that 2α is 1, and 2β is 3. In this case the value of a which makes the voltage at the distributing station a maximum is a root of the equation $\sinh x = 3 \sin 3x$. In Fig. 233, OAP is the curve $y = 3 \sin 3x$, and OBP is the curve $y = \sinh x$. The abscissa of the point of intersection of these curves satisfies the equation $\sinh x = 3 \sin 3x$, and it therefore makes

$$\cosh x + \cos 3x$$

a minimum. In this case the value of x is nearly 0·93 and

$$\cosh x + \cos 3x$$

is then equal to 0·5 nearly. Thus for this value of a the voltage at the far end of the line would be equal to twice the voltage at the generating station, for the given values of the line constants.

In Fig. 234 the curve ABC represents the equation

$$y/3 = \cosh x + \cos 3x.$$

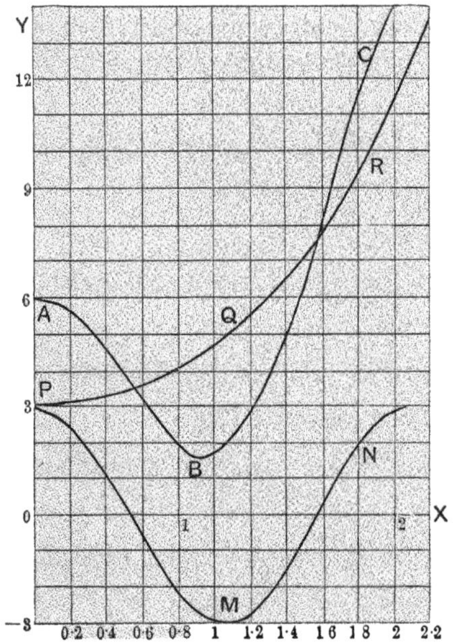

Fig. 234. The curve ABC ($y = 3\cos 3x + 3\cosh x$) is obtained by adding together the ordinates of the curves PMN ($y = 3\cos 3x$) and PQR ($y = 3\cosh x$). The length of the minimum ordinate of ABC is 1·5.

It will be seen that after attaining a minimum value, y rapidly increases.

It readily follows from (41), when the line is short and s is negligibly small compared with ωk, that

$$A = \omega K V_1 (1 - x/a) \text{ approximately,}$$

where K is the capacity between the mains and A is the effective current at a distance x from the generating end of the line. Hence the loss in heating the mains is given by

$$\int_0^a rA^2\partial x = (\omega K V_1)^2 r \int_0^a (1 - x/a)^2 \partial x$$
$$= (1/3)\,(\omega K V_1)^2 R,$$

where R is the total resistance of the two mains in series.

The voltage regulation. If $V_{2.O}$ be the value of the voltage at the distributing station on open circuit and $V_{2.L}$ be its value at full load, when the power factor is $\cos\phi_2$, the voltage at the generating station being the same in the two cases, then $(V_{2.O} - V_{2.L})/V_{2.L}$ is the voltage regulation for this power factor.

By (25) we get

$$[e_{2.L}] = [e_1]\{\sinh\gamma/\sinh(Pa+\gamma)\},$$

and

$$[e_{2.O}] = [e_1]\operatorname{sech} Pa.$$

We thus deduce that

$$[e_{2.O}]/[e_{2.L}] = 1 + \tanh Pa \coth\gamma.$$

For short lines we can write

$$\tanh Pa = Pa = a\,(yz)^{1/2}\,\epsilon^{\iota(\theta_1+\theta_2)/2}.$$

Hence in this case we have by (16)

$$[e_{2.O}]/[e_{2.L}] = 1 + (ay/Z_2)\,\epsilon^{\iota(\theta_1-\phi_2)},\ \text{approx.},$$

and thus by (32)

$$V_{2.O}/V_{2.L} = \{1 + (2ay/Z_2)\cos(\theta_1 - \phi_2) + (a^2y^2/Z_2^2)\}^{1/2}$$
$$= 1 + (ay/Z_2)\cos(\theta_1-\phi_2) + (a^2y^2/2Z_2^2)\sin^2(\theta_1-\phi_2),$$

approximately. Hence

$$(V_{2.O} - V_{2.L})/V_{2.L}$$
$$= (ay/Z_2)\cos(\theta_1-\phi_2) + (a^2y^2/2Z_2^2)\sin^2(\theta_1-\phi_2).$$

Line on short circuit. When the line is short circuited at the far end, R_2 and L_2 are zero, $\tanh\gamma$ is therefore zero and hence γ must be zero. Writing 0 for γ in (25), (26) and (27) we get

$$[e] = [e_1]\,\frac{\sinh P(a-x)}{\sinh Pa};\quad [i] = [i_1]\,\frac{\cosh P(a-x)}{\cosh Pa}$$

and

$$[z] = [z_0]\tanh P(a-x) \quad\ldots\ldots\ldots\ldots(43).$$

Hence by the formulae given on p. 518 and assuming that $[e] = E_1 \epsilon^{\omega t \iota}$, we get

$$e = E_1 \left\{\frac{\cosh 2\alpha (a - x) - \cos 2\beta (a - x)}{\cosh 2\alpha a - \cos 2\beta a}\right\}^{1/2} \cos(\omega t + \sigma - \sigma_0) \ldots (44),$$

where $\tan \sigma = \tan \beta (a - x)/\tanh \alpha (a - x)$

and $\tan \sigma_0 = \tan \beta a/\tanh \alpha a,$

$$R = \left\{\frac{y}{z} \cdot \frac{\cosh 2\alpha (a - x) - \cos 2\beta (a - x)}{\cosh 2\alpha (a - x) + \cos 2\beta (a - x)}\right\}^{1/2}$$
$$\times \cos \{(\theta_1 - \theta_2)/2 + \sigma - \gamma'\} \ldots\ldots (45),$$

$$\omega L = R \tan \{(\theta_1 - \theta_2)/2 + \sigma - \gamma'\} \ldots\ldots\ldots\ldots (46),$$

and

$$i = E_1 \left\{\frac{z}{y} \cdot \frac{\cosh 2\alpha (a - x) + \cos 2\beta (a - x)}{\cosh 2\alpha a - \cos 2\beta a}\right\}^{1/2}$$
$$\times \cos \{\omega t + \gamma' - \sigma_0 - (\theta_1 - \theta_2)/2\} \ldots\ldots (47).$$

If A_1 be the effective value of the current and V_1 be the effective value of the voltage at the generating station, when the mains are short circuited at the distributing station, we have

$$A_1 = V_1 \left\{\frac{z}{y} \cdot \frac{\cosh 2\alpha a + \cos 2\beta a}{\cosh 2\alpha a - \cos 2\beta a}\right\}^{1/2} \quad \ldots\ldots\ldots (48).$$

When $(\alpha^2 - \beta^2) a^2$, which equals $(rs - kl\omega^2) a^2$, is small compared with unity, we have

$$A_1 = \frac{V_1}{(r^2 + \omega^2 l^2)^{1/2} a} \{1 + \tfrac{1}{3} (rs - kl\omega^2) a^2\} \text{ approximately } \ldots (49).$$

In the same case it readily follows from (41) that on open circuit

$$A_1' = (s^2 + \omega^2 k^2)^{1/2} a V_1 \{1 - \tfrac{1}{3} (rs - kl\omega^2) a^2\} \text{ approximately} \ldots (50),$$

where A_1' is the charging current on open circuit. We see that when $rs - kl\omega^2$ is negative, the effect of taking leakage and capacity into account is to diminish the computed value of the short circuit current, and the effect of taking resistance and inductance into account is to increase the computed value of the charging current on open circuit.

Experimental determination of constants.

If V_1 be the effective value of the applied voltage at the generating end of the line and V_2 be the voltage at the distributing end, we get by (42) that

$$\cosh 2\alpha a + \cos 2\beta a = 2 V_1^2 / V_2^2.$$

Similarly if Z_S and Z_O be the values of the impedance of the line at the generating station when the terminals of the line at the receiving station are on short circuit and open circuit respectively, we get by (41) and (47)

$$\cosh 2aa - \cos 2\beta a = (Z_S/Z_O)(\cosh 2aa + \cos 2\beta a).$$

Hence $$\cosh 2aa = (1 + Z_S/Z_O)(V_1/V_2)^2,$$

and thus by p. 520

$$a = (1/2a)\log_\epsilon [(1 + Z_S/Z_O)(V_1/V_2)^2 + \{(1+Z_S/Z_O)^2 (V_1/V_2)^4 - 1\}^{1/2}] \quad \text{.........(51).}$$

The value of the attenuation factor a can thus be found. Again we get from (48) and (49) that

$$\cos 2\beta a = (1 - Z_S/Z_O)(V_1/V_2)^2 \quad \text{...........(52).}$$

This equation shows that there are an infinite number of values of the wave length factor β which satisfy the given conditions. If β_0 be the smallest possible value of β which satisfies equation (52), then we may write

$$\beta = \pm \beta_0 + n\pi/a \quad \text{....................(53),}$$

where n can have any positive or integral value.

It is worth noticing that if Z_S' and Z_O' be the impedances of the line measured at a distance x from the generating station first when the line is short circuited and secondly when it is on open circuit, we have

$$Z_S'Z_O' = y/z = (r^2 + \omega^2 l^2)^{1/2}/(s^2 + \omega^2 k^2)^{1/2}.$$

The value of $Z_S'Z_O'$ is therefore independent of the value of x.

In what precedes we have assumed that r, s, k and l are independent of the frequency. With telephonic frequencies of about 1000 this assumption would be a doubtful one (see Vol. I, p. 484) and with radio-telegraphic frequencies of about a million it would not be permissible without experimental investigation. With power transmission frequencies not exceeding 50 it is, however, quite permissible.

Direct currents. The corresponding equations for direct currents can be deduced at once from the equations given above by putting the frequency and therefore ω equal to zero. In this case $\tanh\gamma = R_2 (s/r)^{1/2}$; where R_2 is the resistance of the load, $P = \alpha = (rs)^{1/2}$ and hence by (25), (26) and (27)

$$E = E_1 \frac{\sinh\{(rs)^{1/2}(a-x)+\gamma\}}{\sinh\{(rs)^{1/2}a+\gamma\}}$$

$$= E_1 \frac{\sinh\{(rs)^{1/2}(a-x)\} + R_2 (s/r)^{1/2}\cosh\{(rs)^{1/2}(a-x)\}}{\sinh\{(rs)^{1/2}a\} + R_2 (s/r)^{1/2}\cosh\{(rs)^{1/2}a\}} \quad \text{.........(54).}$$

Similarly

$$R = \left(\frac{r}{s}\right)^{1/2} \frac{\tanh\{(rs)^{1/2}(a-x)\} + R_2 (s/r)^{1/2}}{1 + R_2 (s/r)^{1/2}\tanh\{(rs)^{1/2}(a-x)\}} \quad \text{...(55),}$$

and $\qquad I = E/R.$

At the generating station, where $x = 0$, we see that on open circuit

$$R_O = (r/s)^{1/2}\coth\{(rs)^{1/2}a\} \quad \text{..............(56),}$$

and when the line is short circuited

$$R_S = (r/s)^{1/2}\tanh\{(rs)^{1/2}a\} \quad \text{..............(57).}$$

Hence $\qquad (r/s)^{1/2} = (R_O R_S)^{1/2},$

and $\qquad (rs)^{1/2} = (1/a)\tanh^{-1}(R_S/R_O)^{1/2}.$

Therefore by p. 520

$$r = \frac{(R_O R_S)^{1/2}}{2a} \log_\epsilon \frac{1 + (R_S/R_O)^{1/2}}{1 - (R_S/R_O)^{1/2}} \quad \text{...........(58),}$$

and

$$s = r/(R_O R_S) \quad \text{.......................(59).}$$

Polyphase lines. The equations given above for a single phase line can easily be adapted so as to apply to a polyphase line in the special case when the load is balanced. In what precedes we have denoted the resistance of the outgoing or the return line by $r/2$. We shall now denote it by r_1. We have also denoted the conductance, the capacity, and the inductance per unit length between the lines by s, k and l respectively.

Let us now consider the equations for the electromotive forces and the currents in No. 1 main of a symmetrical polyphase system with a balanced load. Let e be the potential difference between a point at a distance x from the generating station and the earth, which since the system is balanced will be carrying no current. Let $k_{1.1}$, $k_{1.2}$, ... be the capacity coefficients and $l_{1.1}$, $l_{1.2}$, ... be the inductance coefficients per unit length between the lines. Let also s_1 be the conductance per unit length between any main and the earth. We thus find as on p. 505 that

$$e_1 - \left(e_1 + \frac{\partial e_1}{\partial x}\partial x\right) = r_1 i_1 + l_{1.1}\frac{\partial i_1}{\partial t} + l_{1.2}\frac{\partial i_2}{\partial t} + l_{1.3}\frac{\partial i_3}{\partial t},$$

and $$i_1 - \left(i_1 + \frac{\partial i_1}{\partial x}\partial x\right) = s_1 e_1 + k_{1.1}\frac{\partial e_1}{\partial t} + k_{1.2}\frac{\partial e_2}{\partial t} + k_{1.3}\frac{\partial e_3}{\partial t}.$$

Now $$i_1 + i_2 + i_3 = 0$$

and we suppose that the wires are sufficiently high above the surface of the earth to make the assumption $e_1 + e_2 + e_3 = 0$ permissible. Hence since the wires are symmetrically situated, we get

$$-\frac{\partial e_1}{\partial x} = r_1 i_1 + (l_{1.1} - l_{1.2})\frac{\partial i_1}{\partial t} \quad \text{..............(60)},$$

and $$-\frac{\partial i_1}{\partial x} = s_1 e_1 + (k_{1.1} - k_{1.2})\frac{\partial e_1}{\partial t} \quad \text{...........(61)}.$$

If d be the distance between the axes of two of the mains and ρ be the radius of each main, we have

$$l_{1.1} - l_{1.2} = 2\log(d/\rho) + 1/2 = l_1 \quad \text{...........(62)},$$

and $$k_{1.1} - k_{1.2} = 1/\{2v^2\log(d/\rho)\} = k_1 \quad \text{...........(63)},$$

where $v = 3.10^{10}$ and l_1 and k_1 are in C.G.S. units.

Comparing equations (21) and (22), p. 505 with equations (60) and (61), it will be seen that the only difference is that $r = 2r_1$, $l = 2l_1$, $k = k_1/2$ and $s = s_1/2$. All our preceding solutions therefore apply to the case of a balanced polyphase system, the values of α, β, θ_1, θ_2, yz and $rs - kl\omega^2$ remaining unaltered. To reduce (62) and (63) to henrys and farads we have to multiply them by 10^{-9} and 10^9 respectively.

Single wire with earth return.

If we have a single wire at a height h above the earth and if the current return along the surface of the earth, which we suppose to be a perfect conductor, equations (60) and (61) still apply for the potential and the current in the wire; and if we write $2h$ for d in (62) and (63) where h is the height of the wire above the earth, we get approximate values for l_1 and k_1 (see Vol. I, p. 198).

The voltage drop.

If V_1 be the effective value of the potential of No. 1 main at the generating station and V_1' be its value at the distributing station, then $V_1 - V_1'$ is called the 'voltage drop' along the main. It is to be noticed, however, that this is not the potential difference between the ends of the line, as e_1 and e_1' are not generally in phase with one another. It is easy to see that $V_1 - V_1'$ is the least possible value of the effective voltage between the ends of the line.

If V be the effective value of the potential difference between adjacent mains at the generating station of an n phase system, we have $V = 2V_1 \sin(\pi/n)$. Similarly if V' be the voltage at the distributing station we have $V' = 2V_1' \sin(\pi/n)$. It is convenient to call $V - V'$, that is $2(V_1 - V_1')\sin(\pi/n)$, the voltage drop of the system. For a three phase system its value is $(V_1 - V_1')\sqrt{3}$ and for a two phase system $(V_1 - V_1')\sqrt{2}$.

Corona loss.

We have already explained (see Vol. I, p. 170) that when the potential of an overhead wire exceeds a definite critical value E_0 a luminous discharge called the 'corona' takes place at its surface. When the corona appears the capacity between the mains as computed from the charging current increases and at the same time there is an appreciable increase in the energy lost in distribution. Peek has shown that the critical value of the potential gradient $R_{max.}$ at the surface of the wires of a single phase system at which the corona appears is given by

$$R_{max.} = p\,(30 + 9/\sqrt{pa}) \text{ approximately,}$$

where $R_{max.}$ is in kilovolts per centimetre, p is the pressure of the air in atmospheres, a the radius of either cylindrical main in centimetres and the temperature is about 25° C. The value of E_0 can thus be readily computed by the formula given on p. 171 of Volume I.

It is found by experiment that the corona loss P in watts per mile of single main is given by

$$P = 0{\cdot}024 f\,(V - V_0)^2,$$

where V is the effective pressure in kilovolts between the mains and V_0 is the value of this pressure when the corona first appears.

If a be the length of the line in miles and if $(rs - kl\omega^2)a^2$, where r, s, k and l are measured per mile instead of per centimetre, is small compared with unity, we easily deduce from (38) that

$$V/2 - V_0 = (V_1/2 - V_0) - (V_1/2)(rs - kl\omega^2)(2ax - x^2),$$

approximately, where V_1 is the voltage at the generating station. Hence if $V/2$ be everywhere greater than V_0, the power P lost in the mains owing to phenomena connected with the corona is given by

$$P = 0{\cdot}048f\int_0^a \{(V_1/2 - V_0) - (V_1/2)(rs - kl\omega^2)(2ax - x^2)\}^2 \partial x$$

$$= 0{\cdot}012fa\{(V_1 - 2V_0)^2 - (4/3)(V_1 - 2V_0)V_1(rs - kl\omega^2)a^2 + (8/15)V_1^2(rs - kl\omega^2)^2a^4\}.$$

In practice s is negligibly small and therefore P is greater than if the potential difference were constant and equal to V_1 all along the line. It is to be noticed that the corona loss only increases as the first power of the frequency, whilst the loss due to the resistance of the mains increases as the square of the frequency.

Numerical example.

Let us suppose that there are three mains, each being a cylindrical copper wire of No. 4/0 S.W.G., supported at a considerable height above the ground and symmetrically situated with regard to each other, their axes being at a distance of six feet. The radius of each wire will be 0·2 of an inch and its resistance at 16° C. will be 0·34 ohm per mile. We shall suppose that the effective value of the potential to earth at the distributing end of the line is 57700 volts, the current per main 50 amperes and the power factor 0·85. At the distributing station the pressure between the mains will be $57700\sqrt{3}$, *i.e.* 100000 volts, and the output will be $\sqrt{3} \,.\, 100000 \,.\, 50 \,.\, 0{\cdot}85$ watts, that is, 7360 kilowatts.

In practice we can make the assumption that s_1 is zero. It is convenient to express a in miles and r_1, l_1 and k_1 in ohms, henrys and farads per mile respectively. We have

$$r_1 = 0{\cdot}34 \text{ ohm},$$

and (Vol. I, p. 89)

$$l_1 = 0{\cdot}00074 \log_{10}(72/0{\cdot}2) + 0{\cdot}00008$$

$$= 0{\cdot}00197 \text{ henry},$$

$$k_1 = 0{\cdot}0388 \times 10^{-6}/\log_{10}(72/0{\cdot}2)$$
$$= 0{\cdot}0152 \times 10^{-6} \text{ farad.}$$

We also have $\omega = 2\pi f = 314, \; l_1\omega = 0{\cdot}619,$

$$y_1 = (r_1^2 + l_1^2\omega^2)^{1/2} = 0{\cdot}706, \quad z_1 = k_1\omega = 4{\cdot}77 \times 10^{-6},$$
$$y_1 z_1/2 = 1{\cdot}684 \times 10^{-6} \text{ and } k_1 l_1 \omega^2/2 = 1{\cdot}476 \times 10^{-6}.$$

When α and β are measured per mile, we get by (3) and (4)

$$\alpha = (0{\cdot}208 \times 10^{-6})^{1/2} = 0{\cdot}456 \times 10^{-3},$$
$$\beta = (3{\cdot}160 \times 10^{-6})^{1/2} = 1{\cdot}778 \times 10^{-3},$$

and $P = \alpha + \iota\beta.$

Since the power factor is 0·85 we have $\cos\phi_2 = 0{\cdot}85$ and $\phi_2 = 31^\circ\, 47'$. The impedance Z_2 per phase equals 57700/50, *i.e.* 1154, $\tan\theta_1 = 1{\cdot}82$, $\tan\theta_2 = \infty$, and thus $\theta_1 = 61^\circ\, 13'$ and $\theta_2 = 90^\circ$.

We also have (p. 536)

$$\xi_2 = \phi_2 - (\theta_1 - \theta_2)/2 = 46^\circ\, 10',$$

and by (16)

$$\tanh\gamma = Z_2 (z_1/y_1)^{1/2} \epsilon^{\xi_2\iota} = 3 \underline{|46^\circ\, 10'}$$

and $\coth\gamma = (1/3) \underline{|-46^\circ\, 10'}.$

(1). Let us first suppose that the line is 20 miles long, so that we can use the formulae (35) and (36) which are suitable for short lines. In the present case we can neglect terms containing powers of a higher than the first. Hence

$$B = 1 + (ay_1/Z_2)\cos(\theta_1 - \phi_2) = 1{\cdot}011,$$

and $\tan\psi = (ay_1/Z_2)\sin(\theta_1 - \phi_2)/B = 0{\cdot}006.$

Therefore $\psi = 20'$. Hence if $e_2 = \sqrt{2}\,.\,57700\cos\omega t$, we shall have

$$e_1 = \sqrt{2}\,.\,58300\cos(\omega t + 20').$$

Similarly by (36)

$$i_1 = \sqrt{2}\,.\,47{\cdot}1\cos(\omega t - 26^\circ\, 7').$$

The effective pressure between the mains at the generating station is thus 101000 volts, the power factor is 0·90, the power generated is 7400 kilowatts and the voltage drop is about one per cent. of the voltage at the generating station. The efficiency of the transmission line is (7360/7400)100, *i.e.* 99 per cent.

(2). Let us now suppose that the length of the line in the preceding problem is 200 miles. By (25) we have

$$[e_1] = [\sinh Pa \coth \gamma + \cosh Pa]\,[e_2]$$

and
$$Pa = 200\,(\alpha + \iota\beta) = 0{\cdot}0912 + 0{\cdot}356\iota.$$

Hence by p. 518 and noticing that 0·356 radians $= 20^\circ\,24'$,

$$\cosh Pa = [\{\cosh(0{\cdot}1824) + \cos(40^\circ\,48')\}/2]^{1/2}\,|\underline{\phi},$$

where
$$\tan\phi = \tan(20^\circ\,24')\,.\,\tanh(0{\cdot}0912).$$

Thus
$$\cosh Pa = 0{\cdot}942\,|\underline{1^\circ\,56'}.$$

Similarly
$$\sinh Pa = 0{\cdot}360\,|\underline{76^\circ\,15'}.$$

Hence substituting we get

$$\begin{aligned}[e_1] &= \{0{\cdot}360\,|\underline{76^\circ\,15'}\,.\,(1/3)\,|\underline{-46^\circ\,10'} + 0{\cdot}942\,|\underline{1^\circ\,56'}\}\,[e_2]\\ &= \{0{\cdot}120\,|\underline{30^\circ\,5'} + 0{\cdot}942\,|\underline{1^\circ\,56'}\}\,[e_2]\\ &= \{(0{\cdot}120)^2 + (0{\cdot}942)^2 + 2\,.\,0{\cdot}120\,.\,0{\cdot}942\cos 28^\circ\,9'\}^{1/2}\,|\underline{\psi}\,[e_2],\end{aligned}$$

where
$$\tan\psi = \frac{0{\cdot}120\sin 30^\circ\,27' + 0{\cdot}942\sin 1^\circ\,53'}{0{\cdot}120\cos 30^\circ\,27' + 0{\cdot}942\cos 1^\circ\,53'} = 0{\cdot}0884.$$

Therefore
$$\psi = 5^\circ\,3'.$$

Hence
$$[e_1] = 1{\cdot}051\,|\underline{5^\circ\,3'}\,[e_2].$$

Similarly
$$\begin{aligned}[i_1] &= (\sinh Pa\tanh\gamma + \cosh Pa)\,[i_2]\\ &= 1{\cdot}010\,|\underline{68^\circ\,58'}\,[i_2].\end{aligned}$$

Hence assuming that

$$e_2 = \sqrt{2}\,.\,57700\cos\omega t \quad\text{and}\quad i_2 = \sqrt{2}\,.\,50\cos(\omega t - 31^\circ\,47'),$$

we get
$$e_1 = \sqrt{2}\,.\,60640\cos(\omega t + 5^\circ\,3'),$$

$$i_1 = \sqrt{2}\,.\,50{\cdot}5\cos(\omega t + 37^\circ\,11'),$$

and
$$\cos\phi_1 = \cos(37^\circ\,11' - 5^\circ\,3') = \cos 32^\circ\,8' = 0{\cdot}847.$$

The effective voltage between the mains at the generating station is 105000 and thus the voltage drop is 5000. The power generated is $\sqrt{3}\,.\,105000\,.\,50{\cdot}5\,.\,0{\cdot}847$, *i.e.* 7780 kilowatts, and the efficiency of the transmission line is (7360/7780) 100, *i.e.* 94 per cent.

Polycyclic distribution.

In Vol. I, p. 124, it is shown that the effective value of a complex current, that is, of a current which is the resultant of a direct current C and an alternating current i, is $\sqrt{C^2 + A^2}$, where A is the effective value of i. Hence, if R be the resistance of the main in which the complex current is flowing, the heating effect will be $RC^2 + RA^2$. Each current, therefore, produces the same heating effect that it would produce if flowing singly. If C equals A, the heating effect will only be $2RC^2$, instead of $4RC^2$, which would be its value if the currents were of the same kind. This theorem has been utilised in practice for the purpose of transmitting direct and alternating currents by the same mains, and thus effecting a saving in the weight of copper required.

We may show as follows that if two currents of different frequencies are flowing in the conductor, the mean value of the power expended in heating it is practically the same as if the values of the power expended by each, acting singly, were added. Let f_1 and f_2 be the frequencies of the two currents, and, with our usual notation, let

$$i = I_1 \sin 2\pi f_1 t + I_2 \sin (2\pi f_2 t - a).$$

Taking the mean value of i^2 over a period T which we suppose to be exactly divisible by $1/f_1$ and $1/f_2$, we get

$$A^2 = A_1^2 + A_2^2 + (I_1 I_2/T)\int_0^T \{\cos(2\pi\overline{f_1 - f_2}t + a) - \cos(2\pi\overline{f_1 + f_2}t - a)\}dt$$
$$= A_1^2 + A_2^2.$$

Even if T be not a multiple of $1/f_1$ and $1/f_2$, yet by supposing it sufficiently large we may make the error in assuming that $A_1^2 + A_2^2 = A^2$ less than any assignable magnitude.

Dykes's system.

In practice, when an alternating supply is required for power purposes, the best results are obtained at very low frequencies. On the other hand, for lighting, the frequency should be greater than 30. It is sometimes desirable, therefore, to supply current at two different frequencies, and several engineers have devised systems of doing this in which the alternating currents flow along the same mains over parts of their circuits. In the system invented by F. J. Dykes a four core transformer is used (Fig. 235). By its means two distinct currents of different frequencies are sent over the same three phase lines.

It will be seen from the figure that the windings are so arranged that the fluxes due to the currents in the three phase and single phase primary windings induce electromotive forces in the secondary windings. The single phase flux, however, has no effect on the three phase E.M.F., and the flux due to the three phase currents has no effect on the single phase E.M.F. The only function of the fourth core is to provide a return path for the single phase flux. The three phase windings are connected in star, and the common junction is insulated, and so (p. 307) there are no third

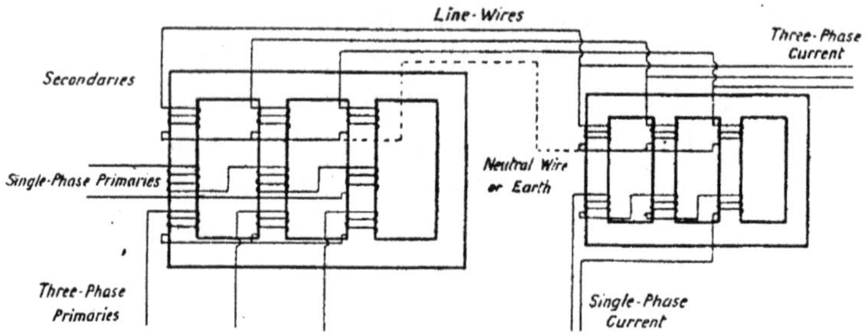

Fig. 235. Dykes's polycyclic transformer.

harmonics in the three phase currents, although owing to hysteresis there may be third harmonics in the fluxes in the four cores. The single phase primary windings consist of three equal coils connected in series, and therefore the three fluxes induced in them are in the same direction in space at the same instant and the return path is by the fourth core. The secondary windings for both the three phase and single phase currents are similar to the primary windings, but in the three phase windings the common junction is connected with a fourth wire or with the earth.

The transformer at the distributing station is exactly similar to the transformer at the power station. Let us now suppose that the terminals of a three phase machine are connected with the three phase windings of the transformer at the power station. The transformer will act like an ordinary three phase transformer, and if balanced three phase currents be taken from the secondary of the distributing transformer, the fourth wire will only have to carry a small triple frequency current. The algebraic sum of the fluxes in the first three cores at any instant will be small, and so the effects produced in the single phase windings will also be small.

Let us now consider the effect of applying an E.M.F. to the single phase primary winding. If we assume that the reluctances of the magnetic circuits linked with the three primary coils are equal and that there is no magnetic leakage, the potential differences induced between each of the three phase wires and the fourth wire will all be equal, and thus the single phase currents in the three phase wires will all be in phase and will return by the fourth wire. On the other hand the potential difference between any two of the three phase wires due to the single phase flux will be zero, as the fluxes in the two limbs of the transformer will generate equal electromotive forces in the two coils wound on them and the resultant E.M.F. therefore round the circuit of the two wires will be zero.

In actual transformers the reluctances of the three magnetic circuits linked with the three primary coils are never exactly equal, and thus, if the secondaries are on open circuit, the voltages between their terminals will be different. In one case F. J. Dykes found that the secondary voltages on open circuit were 85, 10 and 5 respectively. On connecting up, however, the secondary terminals with the transmission lines, the voltages became equal. The balance is obtained by the mutual actions of the two transformers on one another, and the required circulating currents flowing in the mains were found to be small.

In this system (Fig. 235) the three phase currents at any point can be obtained by tapping the three phase lines and using an ordinary three phase transformer. If single phase currents, or if both kinds of current are required, we can use a transformer of the type shown in the figure. The primary coils of this transformer must be four wire star connected, and the single phase current is obtained from three secondaries in series.

When a transformer is supplied in this manner with currents of different frequencies, a curious effect is produced by the periodic coincidence of the maximum magnetic fluxes due to the two types of current. This effect can be heard in the hum of the transformer, distinct 'beats' being produced. It can also be felt by the hand, as vibration is produced in the iron.

Another system of utilising three phase mains to carry single phase current of double the frequency for lighting purposes has been proposed by Arnold, Bragstad and La Cour. In Fig. 236, O' is the neutral point of the armature windings of the three phase motor. The generator G_3 is an ordinary three phase machine, but its frequency is only half that of the

Arnold's system.

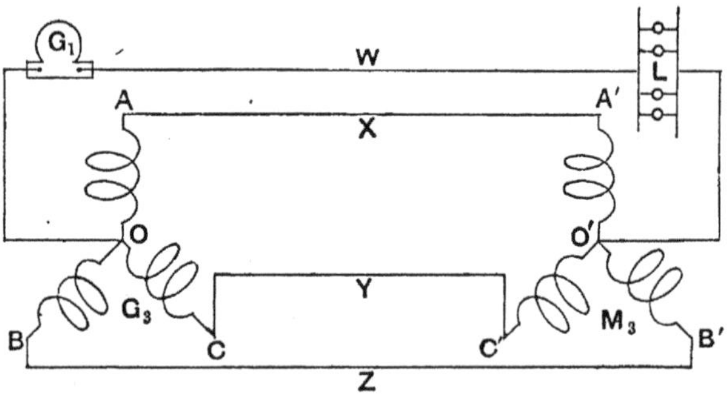

Fig. 236. Three phase polycyclic system of distribution. G_1 and G_3 are single and three phase generators. M_3 is a three phase motor, and L represents the single phase lighting load.

single phase machine G_1 which supplies the lighting load at L. The higher frequency current returns by the mains X, Y and Z, and thus economies are effected as the copper required is less than when separate mains are used.

A polycyclic system suitable for the distribution of power for motive purposes by two phase currents, and for the distribution of power for lighting by single phase currents, is shown in Fig. 237. O and O_1 are the central points of the two separate windings on the armature of the generator, and O' and O_1' are the corresponding points on the motor armature. O and O_1 are connected with the terminals of the single phase machine, and O' and O_1' with the lighting load. When two phase transformers are used to raise the pressure, the single phase machine and the lighting load are connected through transformers with the middle points of the windings on the high tension secondaries of the two phase transformers.

In order to avoid the large inductive drop in the voltage, due to the windings of the polyphase armatures or transformers which the high frequency single phase currents have to traverse, bifilar

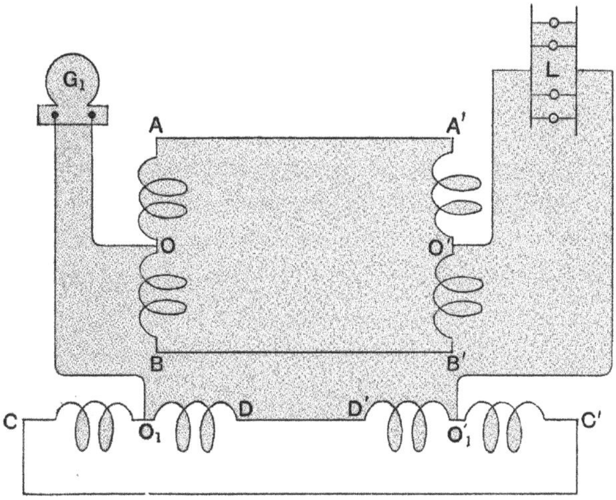

Fig. 237. Two phase polycyclic system of distribution. *AB* and *CD* are the armature windings of a two phase generator, and *A'B'* and *C'D'* are the armature windings of a two phase motor. G_1 is a single phase machine of higher frequency for the lamp load.

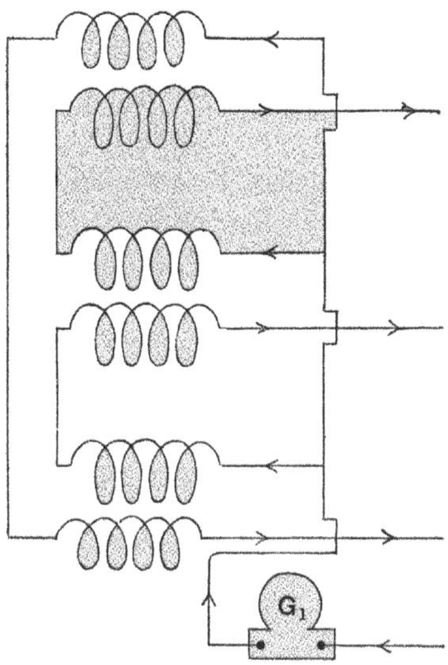

Fig. 238. Connections of the windings of a polyphase armature so as to avoid appreciable inductive drop in the single phase voltage.

windings are used, the principle of which will be understood from Fig. 238. The arrow heads show the directions of the superposed currents.

Polycyclic generator.

The single phase and three phase generators may be combined into a single machine (Fig. 239). In the polycyclic generator shown in this figure, the field poles N, S, N, ... produce ordinary three phase currents in the armature windings which, when the system is balanced, have no component along OO'. The field poles n, s, n, ... produce single phase currents of three times the frequency. These, which are represented by the double arrow heads, are superposed on the

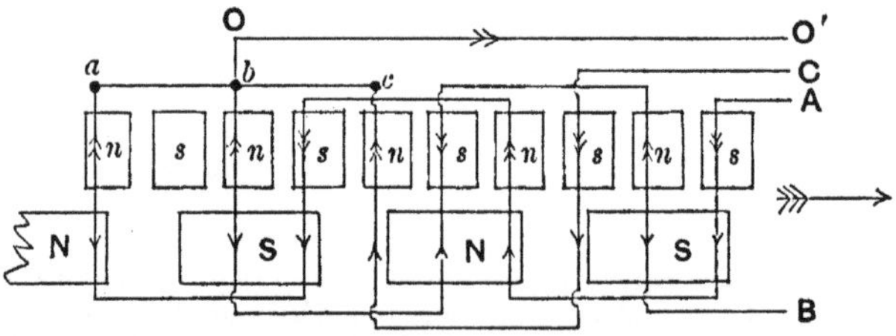

Fig. 239. Polycyclic generator with rotating field poles. N, S, ... field poles producing ordinary three phase currents in the armature windings, which, when the system is balanced, have no component along OO'. n, s, n, ... field poles producing single phase currents of three times the frequency, which are superposed on the three phase currents and flow along OO'. The double arrow heads show the directions of flow of the high frequency single phase currents. The single arrow heads apply to the three phase currents.

three phase currents represented by the single arrow heads. They are obviously in phase with one another, and will therefore flow along OO'. We have supposed that the poles rotate and that the loads are non-inductive. It will be seen that the portion of the armature windings between the single phase and three phase fields is inoperative, as it is necessary to have the magnetic fields of the two sets of poles quite distinct from one another. The winding of the machine, however, is simple.

In the method given above all the single phase current has to pass through the armature coils of the three phase machine, and so special windings have to be used which are non-inductive with respect to the single phase current. By Dykes's method the same end can be attained more cheaply, since standard generators can be used and no special apparatus is required except the comparatively inexpensive transformers.

REFERENCES

C. P. Steinmetz, *Theory and Calculation of Transient Electric Phenomena and Oscillations.*

S. Sheldon and E. Hausmann, *Electric Traction and Transmission Engineering.*

A. E. Kennelly, *The Application of Hyperbolic Functions to Electrical Engineering Problems.*

A. Blondel, 'Méthode Pratique pour le Calcul des Lignes à Courants Alternatifs Présentant de la Self-Induction et de la Capacité,' *l'Éclairage Électrique*, Vol. 49, p. 121, 1906.

J. A. Fleming, *The Propagation of Electric Currents in Telephone and Telegraph Conductors.*

L. Cohen, *The Calculation of Alternating Current Problems.*

For an instructive discussion on the Output and Regulation on Long Distance Lines and the Calculation of High Tension Lines, see the *Transactions of the Am. Inst. of El. Engin.*, June, 1909.

INDEX

CAMBRIDGE: PRINTED BY J. B. PEACE, M.A. AT THE UNIVERSITY PRESS

CAMBRIDGE PHYSICAL SERIES

The Times.—"The Cambridge Physical Series...has the merit of being written by scholars who have taken the trouble to acquaint themselves with modern needs as well as with modern theories."

Volumetric Analysis. By A. J. BERRY, M.A., Fellow of Downing College, Cambridge. Demy 8vo. pp. viii + 138. 6*s*. 6*d*. net.

"Rich in practical teaching and skilfully reflecting the lessons of its writer's practical experience in his college laboratory, the book is also thoroughgoing on its theoretical side."—*Scotsman*

Modern Electrical Theory. By N. R. CAMPBELL, Sc.D. Second edition, largely rewritten. Demy 8vo. pp. xii + 400. 9*s*. net.

"The treatment throughout is admirably clear and readable, the arrangement is logical, and the work of different investigators is carefully considered and given due weight....Possibly the most striking feature is the completeness of the book, for it would be hard to find a section of the subject which has not received due attention."—*Cambridge Review*

Sound. An Elementary Text-Book for Schools and Colleges. By J. W. CAPSTICK, M.A., D.Sc., Fellow of Trinity College, Cambridge. Crown 8vo. pp. viii + 296. With 120 figures. 4*s*. 6*d*.

"An important addition to the *Cambridge Physical Series*....The latter portion is of especial value, and gives an interesting *rationale* of the principal orchestral instruments....Dr Capstick has succeeded in making a remarkably lucid exposition of his theme, and the book will, we have little doubt, promptly take its place as a standard text-book on the subject."—*Guardian*

Mechanics. By JOHN COX, M.A., F.R.S.C. Demy 8vo. pp. xiv + 332. With four plates and 148 figures. 9*s*. net.

"Prof. Cox has his eye specially upon mechanical principles, avoiding merely mathematical difficulties so far as that is fairly possible; he 'starts from real problems, as the subject started, showing how the great investigators attacked these problems, and introducing the leading concepts only as they arise necessarily and naturally in the course of solving them'; he 'brings out incidentally the points of philosophic interest and the methods of science'; he appeals constantly to experiment for verification, 'leading up to an experimental course limited to the most important practical applications,' and eventually embodying a good deal of matter not usually found in the elementary text-books; and he adds limited sets of carefully selected examples for exercise. Students that will not learn mechanics from this work, and be fired with interest in the subject, must be hopeless."—*Educational Times*

5000
11.15

CAMBRIDGE PHYSICAL SERIES

The Study of Chemical Composition. An Account of its Method and Historical Development, with illustrative quotations. By IDA FREUND. Demy 8vo. pp. xvi+650. With 99 figures. 18*s*. net.

"The accomplished Staff Lecturer on Chemistry of Newnham College has admirably succeeded in her efforts to produce a book dealing with the historical development of theories regarding chemical composition....No part of the book is dull, and the student who starts on it will be led on to continue to the end, to his own great advantage. Many people have no time or opportunity to refer to a large mass of original papers, and the value of this book of careful selections carefully and cleverly put together will be very great....Students of physical science who wish to form a clear conception of the laws of chemical combination, of the meaning and accuracy of combining weights and combining volumes, of the doctrine of valency and of isomerism, and of the parts played by various investigators and authors in elucidating these and kindred matters cannot do better than refer to this book."—*Athenaeum*

Mechanics and Hydrostatics. An Elementary Text-book, Theoretical and Practical, for Colleges and Schools. By R. T. GLAZEBROOK, C.B., M.A., F.R.S., Director of the National Physical Laboratory and Fellow of Trinity College, Cambridge. Crown 8vo. 6*s*.

Also in separate volumes

Part I. **Dynamics.** pp. xii+256. With 99 figures. 3*s*.
Part II. **Statics.** pp. viii+182. With 139 figures. 2*s*.
Part III. **Hydrostatics.** pp. x+216. With 98 figures. 2*s*.

"A very good book, which combines the theoretical and practical treatment of mechanics very happily....The discussion of force, momentum, and motion is consistent and clear; and the experiments described are rational and inexpensive. This treatment is calculated to give much clearer ideas on dynamical concepts than a purely mathematical course could possibly do."—*Journal of Education on* Dynamics

"A clearly-printed and well-arranged text-book of hydrostatics for colleges and schools....The descriptions are clearly written, and the exercises are numerous. Moreover, the treatment is experimental; so that altogether the book is calculated to give a good grasp of the fundamental principles of hydrostatics."—*Nature on* Hydrostatics

Heat and Light. An Elementary Text-book, Theoretical and Practical, for Colleges and Schools. By R. T. GLAZEBROOK, C.B., M.A., F.R.S. Crown 8vo. 5*s*.

Also in separate volumes:

Heat. pp. xii+230. With 88 figures. 3*s*.
Light. pp. viii+210+vi. With 134 figures. 3*s*.

"The very able author of the treatise now before us is exceptionally well qualified to deal with the subject of theoretical and experimental physics, and we may at once say that he has succeeded in producing a class-book which deserves, and will doubtless receive, a full share of the patronage of our school and college authorities."—*Mechanical World*

CAMBRIDGE PHYSICAL SERIES

Electricity and Magnetism: an Elementary Text-book, Theoretical and Practical. By R. T. GLAZEBROOK, C.B., M.A., F.R.S. Crown 8vo. pp. vi + 440. 6*s*.

"As an elementary treatise on the laws of electricity and magnetism it leaves little to be desired. The explanations are clear, and the choice of experiments, intended to carry home these explanations to the mind of the student, is admirable. We have no doubt that teachers of the subject will find the volume of great use."—*Engineering*

Photo-Electricity. By A. LL. HUGHES, D.Sc., B.A., Assistant Professor of Physics in the Rice Institute, Houston, Texas. Demy 8vo. pp. viii + 144. With 40 text-figures. 6*s*. net.

"The book goes very thoroughly into all modern experimental research related to its subject....The whole question is important and may have far-reaching results, and we consider the author is to be heartily congratulated on the lucid manner in which he has dealt with it."—*Electrical Review*

The Electron Theory of Matter. By O. W. RICHARDSON, D.Sc., F.R.S., Wheatstone Professor of Physics at King's College, London. Demy 8vo. pp. viii + 612. With 58 figures. 18*s*. net.

"This book will serve as an admirably compact and yet very complete account of the subject, leading directly to many points where research, theoretical and experimental, is proceeding."—*Science Progress*

A Treatise on the Theory of Alternating Currents. By ALEXANDER RUSSELL, M.A., M.I.E.E., Principal of Faraday House. In two volumes. Demy 8vo. Vol. I. pp. xiv + 534. Second edition. 15*s*. net. Vol. II. pp. xii + 488. 12*s*. net.

"It reveals the thorough mastery which the author has of the subject, and the capable reader will find much that is of value. Recent contributions are carefully digested and generally elucidated, and, on the whole we may consider the book an up-to-date treatment of the alternating-current theory."—*Electrical Review*

Experimental Elasticity. A Manual for the Laboratory. By G. F. C. SEARLE, Sc.D., F.R.S., University Lecturer in Experimental Physics. Demy 8vo. pp. xvi + 187. 5*s*. net.

"The author...gives us in this book a collection of the experiments which he has found suitable for University students in their second and third years' preparation for the Natural Sciences Tripos....The book can be strongly recommended to all teachers and to University students. It forms an excellent monograph on elementary elasticity."—*Guardian*

Experimental Harmonic Motion. By G. F. C. SEARLE, Sc.D., F.R.S. Demy 8vo. pp. x + 92. With 32 text figures. 4*s*. 6*d*. net.

"The volume is a useful addition to the Cambridge Physical Series, and it can be well recommended to students of mathematical physics."
School World

CAMBRIDGE PHYSICAL SERIES

Air Currents and the Laws of Ventilation. Lectures on the Physics of the Ventilation of Buildings. By W. N. SHAW, Sc.D., F.R.S., Fellow of Emmanuel College, Director of the Meteorological Office. Demy 8vo. pp. xii + 94. 3*s*. net.

"The present volume by Dr Shaw is most welcome, because it draws attention forcibly to the physics of the ventilated space, and deals with the whole subject in a thoroughly scientific manner....The treatment of the subject is quite original....The whole volume is highly suggestive, and constitutes a real advance in the study of this important subject."—*Engineering*

Conduction of Electricity through Gases. By Sir J. J. THOMSON, O.M., D.Sc., LL.D., Ph.D., F.R.S., Fellow of Trinity College, Cambridge, and Cavendish Professor of Experimental Physics. Second edition enlarged and partly re-written. Demy 8vo. pp. viii + 678. 16*s*.

"It is difficult to think of a single branch of the physical sciences in which these advances are not of fundamental importance. The physicist sees the relations between electricity and matter laid bare in a manner hardly hoped for hitherto....The workers in the field of science are to-day reaping an unparalleled harvest, and we may congratulate ourselves that in this field at least we more than hold our own among the nations of the world."—*Times* (on the First Edition)

A Treatise on the Theory of Solution, including the Phenomena of Electrolysis. By WILLIAM CECIL DAMPIER WHETHAM, Sc.D., F.R.S., Fellow of Trinity College, Cambridge. Demy 8vo. pp. x + 488. 10*s*. net. [*New edition in preparation*

The Theory of Experimental Electricity. By W. C. D. WHETHAM, Sc.D., F.R.S. Second edition. Demy 8vo. pp. xii + 340. 8*s*. net.

"This book is certain to be heartily welcomed by all those who are engaged in the teaching of theoretical electricity in our University Colleges. We have no hesitation in recommending the book to all teachers and students of electricity."—*Athenaeum* (on the First Edition)

Experimental Physics. A Text-book of Mechanics, Heat, Sound and Light. By H. A. WILSON, Professor of Physics in the Rice Institute, Houston, Texas, U.S.A. Demy 8vo. pp. viii + 406. With 235 figures. 10*s*. net.

"This book is intended as a text-book for use in connection with a course of experimental lectures on mechanics, properties of matter, heat, sound and light. No previous knowledge of physics is assumed, but nevertheless the book is primarily intended for a first year college course, and the majority of the students attending such a course have studied elementary physics at school."—*From the Preface*

CAMBRIDGE UNIVERSITY PRESS
C. F. CLAY, MANAGER
London: FETTER LANE
Edinburgh: 100, PRINCES STREET
ALSO
London: H. K. LEWIS, 136, GOWER STREET, W.C

Lightning Source UK Ltd.
Milton Keynes UK
UKHW021342240119
336090UK00005B/459/P

9 780365 209874